Carbon Bonding and Structures

CARBON MATERIALS: CHEMISTRY AND PHYSICS

A comprehensive book series which encompasses the complete coverage of carbon materials and carbon-rich molecules from elemental carbon dust in the interstellar medium, to the most specialized industrial applications of the elemental carbon and derivatives. A great emphasis is placed on the most advanced and promising applications ranging from electronics to medicinal chemistry. The aim is to offer the reader a book series which not only consists of self-sufficient reference works, but one which stimulates further research and enthusiasm.

Series Editors

Dr. Prof. Franco Cataldo
Via Casilina 1626/A,
00133 Rome, Italy

Professor Paolo Milani
Department of Physics
University of Milan
Via Celoria, 26
20133, Milan, Italy

Volume 5:
Carbon Bonding and Structures
Advances in Physics and Chemistry

Volume Editor

Dr. Mihai V. Putz
Chemistry Department
West University of Timişoara
Str. Pestalozzi, No. 16
RO-300115, Timişoara
Romania

For further volumes:
http://www.springer.com/series/7825

Mihai V. Putz
Editor

Carbon Bonding and Structures

Advances in Physics and Chemistry

Editor
Mihai V. Putz
Laboratory of Computational and Structural Physical Chemistry
Chemistry Department
West University of Timişoara
Pestalozzi 16, Timişoara, RO300115
Romania
mv_putz@yahoo.com
mvputz@cbg.uvţ.ro

ISSN 1875-0745 e-ISSN 1875-0737
ISBN 978-94-007-1732-9 e-ISBN 978-94-007-1733-6
DOI 10.1007/978-94-007-1733-6
Springer Dordrecht Heidelberg London New York

Library of Congress Control Number: 2011934966

Printed on acid-free paper

Springer is part of Springer Science+Business Media (www.springer.com)

Preface

At the beginning it was Carbon; at the beginning of complex nature, complex life, and even conscience. While Hydrogen belongs to the moving Universe, Helium and Carbon are the protagonists of the universal nucleogenesis, assure the Universe's combustion, and ultimately support its evolution. As such, the Carbon was limitedly interpreted as belonging exclusively to the organic life base or to the life itself as we recognize it. Otherwise, Carbon may be part of the very-short list of the Periodic Table, i.e. {H, He, C, O, N}, that may assure for appreciable extent the inner machinery of the observed word. On the other side, Carbon has at least one special feature in each natural science (Physics, Chemistry, Biology) that makes it worthy for being in depth explored either theoretically as well as in current laboratory structural design, respectively:

- In Physics, Carbon is the preeminent resistant structure to the phenomenon of (Bose-Einstein) condensation, while being at the base of polymeric structures;
- In Chemistry, Carbon marks the unique four allotropic forms as the simple substance, diamond, graphite, and fullerene, each of these opening entire scientific chapters, plethora of nano-structures and every-day life applications;
- In Biology, Carbon assures through its tetravalent flexible bonds the backbone of polypeptides, the skeleton of amino-acids and bio-molecules themselves until the most advanced bio-responsive nano-materials.
- In Technology, Carbon, besides providing the actual challenging nano-materials and benchmark, it also opens the gates towards its relative Silicon element based composite, and hybrids.

As a consequence, the Carbon versatility seems to assure the messenger information within and in between the Natures' levels of manifestation or on its artifacts. The present volume, while approaching many parts of abovementioned fundamental research directions, brings in the International Year of Chemistry 2011 homage to the miracle of Carbon as a key element in the vast actual fields of modeling structure and bonded nanosystems with implication in all natural sciences and challenging technologies. It was possible through the exquisite contributions of eminent scientists and professors from major continents as Europe, North and South Americas, and Asia that give their best understanding of the Carbon phenomenology and advanced implication nowadays. I do thank them all for the consistent effort they encompassed in writing high-class scientific reports in providing the audience with a broad perspectives and gates to be next open in making the Carbon structure and bonding our home and reliable future!

Special thanks are due to Professor Franco Cataldo, the main coordinator of the Springer Carbon Materials Series, for kind invitation for pursuit the present editorial project, as well to the Springer Chemistry Team and to its Senior Editor Sonia Ojo for supporting all stages towards the publication of the present volume. . .on the Carbon copies!

Mihai V. Putz

Contents

Contributors

Alexandru T. Balaban Texas A&M University at Galveston, MARS, 5007 Avenue U, Galveston, TX 77551, USA
balabana@tamug.edu

Michael J. Bucknum INIFTA, Theoretical Chemistry Division, Suc. 4, C.C. 16, Universidad de La Plata, 1900 La Plata, Buenos Aires, Argentina
mjbucknum@gmail.com

Eduardo A. Castro INIFTA, Theoretical Chemistry Division, Suc. 4, C.C. 16, Universidad de La Plata, 1900 La Plata, Buenos Aires, Argentina
eacast@gmail.com

Franco Cataldo Actinium Chemical Research, Via Casilina 1626/A, 00133 Rome, Italy
franco.cataldo@fastwebnet.it

Ciprian Ciubotariu Department of Computer Sciences, University "Politehnica", P-ta Victoriei No. 2, 300006, Timişoara, Romania

Dan Ciubotariu Department of Organic Chemistry, Faculty of Pharmacy, "Victor Babes" University of Medicine and Pharmacy, P-ta Eftimie Murgu No. 2, 300041, Timişoara, Romania
dciubotariu@mail.dnttm.ro

Mircea V. Diudea Faculty of Chemistry and Chemical Engineering, "Babes-Bolyai" University, Arany Janos Str. 11, 400028 Cluj, Romania
diudea@gmail.com

Tomislav Došlić Faculty of Civil Engineering, University of Zagreb, Kačićeva 26, 10000 Zagreb, Croatia
doslic@master.grad.hr

Shinsaku Fujita Shonan Institute of Chemoinformatics and Mathematical Chemistry, Kaneko 479–7, Ooimachi, Ashigara-Kami-Gun, Kanagawa-Ken, 258–0019, Japan
shinsaku_fujita@nifty.com

Dulal C. Ghosh Department of Chemistry, University of Kalyani, Kalyani 741235, India
dcghosh1@rediffmail.com

Ante Graovac Department of Chemistry, Faculty of Science, University of Split, Nikole Tesle 12, HR-21000 Split, Croatia

NMR Center, The "Ruđer Bošković" Institute, HR-10002 Zagreb, Croatia

IMC, University of Dubrovnik, Branitelja Dubrovnika 29, HR-20000 Dubrovnik, Croatia
Ante.Graovac@irb.hr

Aleksandar Ilić Faculty of Sciences and Mathematics, University of Niš, Višegradska 33, 18000 Niš, Serbia
aleksandari@gmail.com

Nazmul Islam Department of Chemistry, University of Kalyani, Kalyani 741235, India
nazmul.islam786@gmail.com

Hagen Kleinert Institut für Theoretische Physik, Freie Universität Berlin, Arnimallee 14, D-14195 Berlin, Germany
h.k@fu-berlin.de

István László Department of Theoretical Physics, Institute of Physics, Budapest University of Technology and Economics, H-1521 Budapest, Hungary
laszlo@eik.bme.hu

Bono Lučić The Rugjer Bošković Institute, Bijenička 54, P.O.B. 180, HR-10 002 Zagreb, Croatia
lucic@irb.hr

Mihai Medeleanu Department of Organic Chemistry, University "Politehnica", P-ta Victoriei, No. 2, 300006, Timişoara, Romania
mihai.medeleanu@chim.upt.ro

Ante Miličević The Institute for Medical Research and Occupational Health, Ksaverskac. 2, P.O.B. 291, HR-10 002 Zagreb, Croatia
antem@imi.hr

Csaba L. Nagy Faculty of Chemistry and Chemical Engineering, "Babes-Bolyai" University, Arany Janos Str. 11, 400028 Cluj, Romania
nc35@chem.ubbcluj.ro

Sonja Nikolić The Rugjer Bošković Institute, Bijenička 54,
P.O.B. 180, HR-10 002 Zagreb, Croatia
sonja@irb.hr

Tudor Olariu Department of Organic Chemistry,
Faculty of Pharmacy, "Victor Babes" University of Medicine
and Pharmacy, P-ta Eftimie Murgu No. 2, 300041,
Timişoara, Romania
rolariu@umft.ro

Ottorino Ori Actinium Chemical Research,
Via Casilina 1626/A, 00133 Rome, Italy
ottorino.ori@alice.it

Sílvia Osuna Institut de Química Computacional and Departament de Química,
Universitat de Girona, Campus Montilivi, 17071 Girona, Catalonia, Spain
silvia.osuma@udg.edu

Tomaž Pisanski Department of Theoretical Computer Science, Institute
of Mathematics, Physics and Mechanics, University of Ljubljana,
Jadranska 19, SI-1000 Ljubljana, Slovenia
tomaz.pisanski@fmf.uni-lj.si

Dejan Plavšić NMR Center, The "Ruđer Bošković" Institute,
HR-10002 Zagreb, Croatia
dplavsic@irb.hr

Lionello Pogliani Dipartimento di Chimica, Università della Calabria,
via P. Bucci, 87036 Rende (CS), Italy
lionp@unical.it

Ana-Maria Putz Laboratory of Inorganic Chemistry, Institute
of Chemistry of Timişoara Romanian Academy, Ave. Mihai Viteazul,
No. 24, Timişoara, RO 300223, Romania
putzanamaria@yahoo.com

Mihai V. Putz Laboratory of Computational and Structural Physical Chemistry,
Chemistry Department, West University of Timişoara,
Pestalozzi 16, Timişoara, RO 300115, Romania
mv_putz@yahoo.com; mvputz@cbg.uvt.ro; www.mvputz.iqstorm.ro

Sandip K. Rajak Department of Chemistry, University of Kalyani,
Kalyani 741235, India
sandip1ku@gmail.com

Milan Randić National Institute of Chemistry, P.O. Box 3430,
1001 Ljubljana, Slovenia
mrandic@msn.com

Cecilia Savii Laboratory of Inorganic Chemistry, Institute of Chemistry
Timişoara of Romanian Academy, Ave. Mihai Viteazul,
No. 24, Timişoara, RO 300223, Romania
ceciliasavii@yahoo.com

Miquel Solà Institut de Química Computacional and Departament de Química,
Universitat de Girona, Campus Montilivi, 17071 Girona, Catalonia, Spain
miquel.sola@udg.edu

Marcel Swart Institut de Química Computacional and Departament de Química,
Universitat de Girona, Campus Montilivi, 17071 Girona, Catalonia, Spain

Institució Catalana de Recerca i Estudis Avançats (ICREA),
Pg. Lluís Companys 23, 08010 Barcelona, Spain
marcel.swart@udg.edu

Nenad Trinajstić The Rugjer Bošković Institute, Bijenička 54,
P.O.B. 180, HR-10 002 Zagreb, Croatia
trina@irb.hr

Vicentiu Vlaia Department of Organic Chemistry, Faculty of Pharmacy,
"Victor Babes", University of Medicine and Pharmacy,
P-ta Eftimie Murgu No. 2, 300041, Timişoara, Romania
vlaiav@gmail.com

Chapter 1
Quantum Parabolic Effects of Electronegativity and Chemical Hardness on Carbon π-Systems

Mihai V. Putz[1]

Abstract The fundamental issue of conceptually assessment of the total pi-electronic energy is here addressed towards the possibility in assuming the electronegativity and chemical hardness within a quantum parabolic energetic effect that closely resembles other pi-equivalent energy expressions within the semi-empirical computation framework as better as the carbon-based system increases its complexity. On the other side, the present analysis affirms electronegativity as the quantum observable for the states that represent full particle existence, while chemical hardness posses the second quantization degree of uncertainty in observation, although through the present study an alternative definite Hückel based resonance integral expression is advanced.

1.1 Introduction

Chemistry in general and quantum chemistry in special is nowadays affirmed as the most intriguing application of the physical and quantum mechanics principles, respectively. This because, beside offering among the first application of the quantum theory through offering the consistent picture of the chemical bonding by means of molecular orbital theory, it arrives to be developed in the powerful computational chemistry that allows the so called molecular design being performed with so many application in bio-, eco-, toxico-, and pharmaco-logy while drastically reducing the experimental costs, risks and time.

However, conceptually, modeling the chemical bonding seems to combine at best the main feature of the quantum characterization of Nature as illustrated in the flowing Figure 1.1. Basically, starting with a collection of N- electrons that evolve

[1] Laboratory of Computational and Structural Physical Chemistry, Chemistry Department, West University of Timişoara, Pestalozzi 16, Timişoara, RO 300115, Romania
e-mail: mv_putz@yahoo.com; mvputz@cbg.uvt.ro; www.mvputz.iqstorm.ro

M.V. Putz (ed.), *Carbon Bonding and Structures: Advances in Physics and Chemistry*, Carbon Materials: Chemistry and Physics 5, DOI 10.1007/978-94-007-1733-6_1,

in a given (nuclei) potential $V(\mathrm{r})$ they are qualitatively represented within the first quantization scheme by the celebrated one-electronic wave functions $\varphi(i=\overline{1,N})$ (Slater 1929) that eventually combine (viz. Hartree-Fock factorization combined with superposition principles) to produce the so called molecular orbitals $\Psi(\{\varphi(i=\overline{1,N})\})$ (Hartree 1957; Slater 1963); The quantitative realm is finally gained since the second quantization allows converting the molecular orbital many-electronic nature into the allied electronic density as prescribed by the basic principle of the Density Functional Theory (Parr 1983; Kohn et al. 1996; Parr and Yang 1989; Dreizler and Gross 1990; March 1991)

$$\rho(\mathbf{r}) = N\int \Psi^*(\mathbf{r},\mathbf{r}_2,...,\mathbf{r}_N)\Psi(\mathbf{r},\mathbf{r}_2,...,\mathbf{r}_N)d\mathbf{r}_2...d\mathbf{r}_N \tag{1.1}$$

The "quantum circle" is closed by linking the many-electronic density with the total number of electrons with the aid of the *integral* conservation law

$$\int \rho(\mathbf{r})d\mathbf{r} = N \tag{1.2}$$

this way linking the global with local quantum information. Nevertheless, when one likes to advances the *differential* connection between these two local and global quantities, the so called Fukui function resulted (Parr and Yang 1984; Yang and Parr 1985; Berkowitz 1987; Senet 1996)

$$f(\mathrm{r}) = \left(\frac{\partial\rho(\mathbf{r})}{\partial N}\right)_{V(\mathbf{r})} \tag{1.3}$$

with the main significance in characterizing the frontier "sensibility" of the studied molecular/chemical bonding system. It mainly enters in evaluation in energetic related quantities that achieve the frontier significance in the valence/chemical realm; accordingly, in the second order or *parabolic energy integral expansion*

$$\min(\Delta E) \cong -\chi\ (\Delta N) + \eta\ (\Delta N)^2 \tag{1.4}$$

in terms of the electronegativity (Sen and Jørgensen 1987)

$$\chi = -\int\left(\frac{\delta E[\rho]}{\delta\rho(\mathbf{r})}\right)_{V(\mathbf{r})} f(\mathbf{r})d\mathbf{r} \tag{1.5}$$

and chemical hardness (Sen and Mingos 1993)

$$\eta = \frac{1}{2}\iint\left(\frac{\delta^2 E[\rho]}{\delta\rho(\mathbf{r})\delta\rho(\mathbf{r}')}\right)_{V(\mathbf{r})} f(\mathbf{r})f(\mathbf{r}')d\mathbf{r}d\mathbf{r}' \tag{1.6}$$

Remarkably, expression (1.4) is so close in form with a frontier formulation of the "chemical" energy of a system, as being the energy engaged or responsible for the chemical reaction taking place or, in other terms, the energy endorsed in the systems' chemical bond that can be consumed for further reactivity, affinity, or ligation. This can be immediately become more apparent once the electronegativity and chemical hardness definitions (1.5) and (1.6) are further explicated in their differential counterparts (Parr et al. 1978; Parr and Pearson 1983):

$$\chi = -\mu = -\left(\frac{\partial E}{\partial N}\right)_{V(\mathbf{r})} \cong \frac{(E_{N_0-1} - E_{N_0}) + (E_{N_0} - E_{N_0+1})}{2} = \frac{IP + EA}{2} \cong -\frac{\varepsilon_{LUMO} + \varepsilon_{HOMO}}{2}, \quad (1.7)$$

$$\eta = -\frac{1}{2}\left(\frac{\partial \chi}{\partial N}\right)_{V(\mathbf{r})} = \frac{1}{2}\left(\frac{\partial^2 E}{\partial N^2}\right)_{V(\mathbf{r})} \cong \frac{E_{N_0+1} - 2E_{N_0} + E_{N_0-1}}{2} = \frac{IP - EA}{2} \cong \frac{\varepsilon_{LUMO} - \varepsilon_{HOMO}}{2} \quad (1.8)$$

written in terms of semi-sum and semi-difference of the ionization potentials *IP* and electron affinities *EA* for the so called "experimental" electronegativity and chemical hardness and also within the approximations of higher occupied and lower unoccupied molecular orbitals, HOMO and LUMO, respectively.

Note that, within the frontier view, the electronegativity and chemical hardness may be considered as two "orthogonal" (thus independent) chemical descriptors, see the HOMO-LUMO midlevel vs. gap of Eqs. 1.7 and 1.8, and can be therefore further used as 2D realization of the reaction coordinates to build up the chemical orthogonal space (COS) within which the chemical bond and reactivity is described.

Moreover, the present frontier picture involves, in fact, the frozen core assumption according with the Koopmans' (1934) theorem. Consequently, the present endeavor, like to explore to which extent this theorem is applicable to the chemical systems having delocalized or π- electrons available to engage in chemical reactivity; even more, we like to quest whether the energy (1.4) may be correlated and in which degree with the common semi-empirical energetic contribution to the frontier or semi-classical or chemical domain of increasingly complex molecules; from simple groups to rings, fused rings and nanostructures. In the case of relevant results, apart of offering a sort of practical energetic consequence of the Koopmans theorem, i.e. affirming the viable parabolic quantum effect of electronegativity and chemical hardness on the total energy of the system, the present study will assess the orthogonal basis set $\{\chi, \eta | \chi \perp \eta\}$ as a viable quantum set of indicators for the chemical reactivity space (Putz 2011a). Whether and in which degree it is sufficient or universal for chemical reactivity will be responded by this and subsequent communication.

1.2 Parabolic Principles of Electronegativity and Chemical Hardness

The parabolic energetic relationship (1.4) is here tested against the physical variational principle to see with which extend it is capable to unfold the popular chemical reactivity principles, while providing a consistent chemical bonding scenario. To this end one starts with setting the total energy variation

$$dE = 0 \tag{1.9}$$

as a working tool in modeling the dynamical equilibrium for natural systems. Next, one expands the left-hand side of (1.9) within the total energy functional dependency $E = E[N, V(\mathrm{r})]$, in the spirit of parabolic form (1.4), yet adding the explicit external potential influence

$$dE = -\chi dN + \eta (dN)^2 + \int \rho(\mathrm{r}) dV(\mathbf{r}) d\mathbf{r} \tag{1.10}$$

through the author's identified chemical action (see Putz 2003, 2009a and the next discussion)

$$C_A = \int \rho(\mathbf{r}) dV(\mathbf{r}) d\mathbf{r} \tag{1.11}$$

as the convolution of the density with applied potential – the two main DFT ingredients in setting the chemical frontier behavior.

Now, when Eqs. 1.9 and 1.10 are combined, one realizes that:

- either there is no action on the system ($dN = dV = 0$) so that no chemical phenomena is recorded since the physical variational principle (1.9) is fulfilled for whatever electronegativity and chemical hardness values in (1.10);
- or there is no electronic system at all ($\chi = \eta = \rho(\mathbf{r}) = 0$).

Therefore, it seems that the variational physical principle of Eq. 1.9 do not suffice to encompass the limiting cases of equilibrium, when is about the chemical frontier or valence domain; in passing, such apparently odd behavior of variational principle is nothing else than another illustration the chemical principles are not reducible to physical ones but their complement (Putz 2011b). This is also the present case when the double variational procedure on the total energy is needed, i.e. through applying the additional differentiation on physical energy expansion (1.10), within the so called "chemical variational mode" (and denoted as $\delta[]$) where the total differentiation will be taken only over the scalar-global (extensive χ, η, dN) and local (intensive $\rho(\mathrm{r}), V(\mathrm{r})$) but not over the vectorial (physical – as the coordinate itself r) quantities, and yields (Putz 2011c)

$$\delta[dE] = -\delta[\chi dN] + \delta[\eta (dN)^2] + \int \delta[\rho(\mathbf{r}) dV(\mathbf{r})] d\mathbf{r} \tag{1.12}$$

Now, the *chemical* variational principle applied to Eq. 1.12 takes the form

$$\delta[dE] \geq 0 \tag{1.13}$$

when certain amount of charge transfer and the system's potential fluctuations (departing from equilibrium) are involved in producing chemical reactivity and/or binding

$$dN = |dN| = ct. \neq 0,\ dV(\mathbf{r}) \neq 0 \tag{1.14}$$

Now, Eq. 1.12 with condition (1.13) releases the quantitative basis of the individual reactivity principles:

- for electronegativity contribution we have the general inequality:

$$-\delta[\chi dN] \geq 0 \Leftrightarrow -|dN|\delta[\chi] \geq 0 \Rightarrow \delta[\chi] \leq 0 \tag{1.15}$$

  containing both the equality and minimum electronegativity fluctuations around chemical equilibrium (Mortier et al. 1985; Tachibana 1987; Tachibana and Parr 1992);
- for chemical hardness contribution one yields the twofold principles resumed into the inequality

$$\delta[\eta(dN)^2] \geq 0 \Leftrightarrow (dN)^2\delta[\eta] \geq 0 \Rightarrow \delta[\eta] \geq 0 \tag{1.16}$$

  as corresponding with the hard-and-soft-acids-and-basis (HSAB) and maximum hardness (MH) principles (Pearson 1990; Chattaraj and Schleyer 1994; Chattaraj and Maiti 2003; Putz et al. 2004; Chattaraj et al. 1991, 1995; Putz 2008a);
- for chemical action contribution there remains the sufficient exact equality

$$\delta\int\rho(\mathbf{r})dV(\mathbf{r})d\mathbf{r} = 0 \tag{1.17}$$

The last expression leaves with the successive equivalent forms

$$0 = \delta\int\rho(\mathbf{r})\{V(\mathbf{r}) - V(\mathbf{r}_0)\}d\mathbf{r} = \delta\left\{\int\rho(\mathbf{r})V(\mathbf{r})d\mathbf{r}\right\} - \delta\left\{V(\mathbf{r}_0)\int\rho(\mathbf{r})d\mathbf{r}\right\} \tag{1.18}$$

with $V(\mathrm{r}_0)$ being the constant potential at equilibrium. However, through employing the basic DFT relationship for electronic density (1.2) and Eq. 1.14 produces the so called chemical action principle that represents the chemical specialization for the physical variational principle of Eq. 1.9.

The hierarchy of the electronegativity, chemical hardness and chemical action principles was recently advanced as describing the paradigmatic stages of bonding through the sequence (Putz 2011a, c):

$$\delta\chi = 0 \rightarrow \delta C_A = 0 \rightarrow \Delta\chi{<}0 \rightarrow \delta\eta = 0 \rightarrow \Delta\eta{>}0 \tag{1.19}$$

Table 1.1 Synopsis of the basic principles of reactivity towards chemical equilibrium with environment in terms of electronegativity, chemical action, and chemical hardness (Putz 2008b, 2011a)

Chemical Principle	Principle of Bonding
$\delta\chi = 0$	*Electronegativity equality*: "Electronegativity of all constituent atoms in a bond or molecule have the same value" (Sanderson 1988)
$\delta C_A = 0$	*Chemical action minimum variation*: Global minimum of bonding is attained by optimizing the convolution of the applied potential with the response density (Putz 2003, 2008a, 2009a, 2011a)
$\Delta\chi < 0$	*Minimum (residual) electronegativity*: "the constancy of the chemical potential is perturbed by the electrons of bonds bringing about a finite difference in regional chemical potential even after chemical equilibrium is attained globally" (Tachibana et al. 1999)
$\delta\eta = 0$	*Hard-and-soft acids and bases*: "hard likes hard and soft likes soft" (Pearson 1973, 1990, 1997)
$\Delta\eta > 0$	*Maximum (residual) hardness*: "molecules arranges themselves as to be as hard as possible" (Pearson 1985, 1997)

as corresponding to the *encountering* (or the electronegativity equality) *stage*, followed by *chemical action minimum variation* (i.e. the global minimum of bonding interaction), then by *the charge fluctuation stage* (due to minimum or residual electronegativity), ending up with *the polarizability stage* (or HSAB) and with the *final steric* (due to maximum or residual hardness) *stage*. Nevertheless, from Eq. 1.19 one observes the close laying chemical action with electronegativity influence in chemical reactivity and bonding principles.

Having conceptually advocating on electronegativity and chemical hardness different influences on various levels of quantum reactivity of atoms and molecules, the global scenario of reactivity may be advanced implying five stages of chemical bonding hierarchy by referring to the principles resumed in Table 1.1:

(i) The *encountering stage*, associated with the charge flow from the more electronegativity regions to the lower electronegativity regions in a molecular formation, is thus dominated by the difference in electronegativity between reactants and consumed when the electronegativity equalization principle is fulfilled among all constituents of the products: it is the *covalent* binding step (Mortier et al. 1985; Sanderson 1988);

(ii) The *global optimization stage*, associates with the variational principle of the total energy of ground/valence state in bonding that can be resumed by the corresponding chemical action principle (Putz 2003, 2009a, 2011a; Putz and Chiriac 2008) that adjust the applied potential and the response electronic density to be convoluted/coupled in optimum/unique way, i.e. establishing the global minima on the potential surface of the system.

(iii) The *charge fluctuation stage*, relies on the fact that partial fractional instead of integer charges are associated with atoms-in-molecules; therefore, even if the chemical equilibrium is attained globally the electrons involved in bonds acts as foreign objects between pairs of regions, at whatever level of molecular partitioning procedure, grounded by the quantum fluctuations in special and by quantum nature of the electron in general; it produces the degree of *ionicity* occurred in bonds (Tachibana and Parr 1992; Tachibana et al. 1999);

(iv) The *polarizability stage*, in which the induced ionicity character of bonds is partially compensated by the chemical forces through the hardness equalization between the pair regions in molecule; at this point the *HSAB principle* (Pearson 1973, 1990, 1997; Chattaraj and Schleyer 1994; Chattaraj and Maiti 2003; Putz et al. 2004) is involved as a second order effect in charge transfer – see the step (i) above;

(v) The *steric stage*, where the second order of quantum fluctuations provides a further amount of finite difference, this time in attained global hardness, that is transposed in relaxation effects among the nuclear and electronic distributions so that the remaining unsaturated chemical forces to be dispersed by stabilization of the molecular structure; this is covered by the *maximum hardness principle* and the fully stabilization of the molecular system in a given environment (Pearson 1985, 1997; Chattaraj et al. 1991, 1995; Putz 2008a).

Having this way proved the efficiency of the parabolic energy expression in terms of electronegativity and chemical hardness indices, with the regulatory effects in chemical reactivity principles, the next step consists in discussing their observability character in order to can be employed as viable quanto-computational tools linking the density with many-electronic information and with the energetic parabolic behavior.

1.3 On Quantum Character of Electronegativity and Chemical Hardness

As Fig. 1.1 suggests the second quantization stays as the key step in recovering the observable quantities in chemical domain, since assuring the passage from orbital to density description of open systems. As such, one would next proceed with expressing the electronegativity and chemical hardness within the framework of second quantization as well, through relaying on the general *parabolic* Hamiltonian (Surján 1989),

$$\hat{H} = \sum_{pq} h_{pq} \hat{a}_p^+ \hat{a}_q + \frac{1}{2} \sum_{pqts} g_{pq,ts} \hat{a}_p^+ \hat{a}_t^+ \hat{a}_q \hat{a}_s \tag{1.20}$$

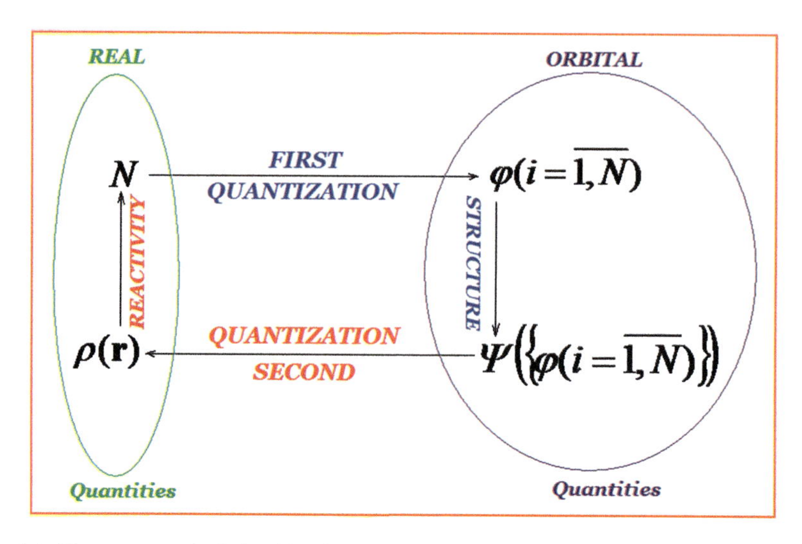

Fig. 1.1 The conceptual relationships between the real and orbital quantities as being linked by the first (particle to wave) and second (field to particle) quantization thus respectively characterizing the structure and reactivity of the many-electronic chemical systems

built within the fermionic Fock space with the help of creation and annihilation particle operators

$$\hat{a}^{+} = |1\rangle\langle 0|, \tag{1.21}$$

$$\hat{a} = |0\rangle\langle 1| \tag{1.22}$$

as can be easily recognized through the respective actions:

$$\hat{a}^{+}|0\rangle = |1\rangle\langle 0 \mid 0\rangle = |1\rangle, \tag{1.23}$$

$$\hat{a}|1\rangle = |0\rangle\langle 1 \mid 1\rangle = |0\rangle \tag{1.24}$$

for the vacuum $|0\rangle$ and uni-particle $|1\rangle$ sectors. These sectors resemble, however, the entire particle projection space:

$$\hat{1} = |0\rangle\langle 0| + |1\rangle\langle 1| = \hat{a}\hat{a}^{+} + \hat{a}^{+}\hat{a} = \left\{\hat{a}, \hat{a}^{+}\right\} \tag{1.25}$$

while fulfilling the dot product rules

$$\langle 0 \mid 1\rangle = \langle 1 \mid 0\rangle = 0 \tag{1.26}$$

$$\langle 0 \mid 0\rangle = \langle 1 \mid 1\rangle = 1 \tag{1.27}$$

Now, the passage from the orbital to density picture may be immediately illustrated with the aid of the second quantization presented rules by employing the inner-normalization of the Eq. 1.1 under the form

$$\begin{aligned} 1 &= \langle \Psi_0 \mid \Psi_0 \rangle = \langle \Psi_0 | \hat{1} | \Psi_0 \rangle \\ &= \langle \Psi_0 | (\hat{a}\hat{a}^+ + \hat{a}^+\hat{a}) | \Psi_0 \rangle = \langle \Psi_0 | \hat{a}\hat{a}^+ | \Psi_0 \rangle + \langle \Psi_0 | \hat{a}^+\hat{a} | \Psi_0 \rangle \\ &= |\langle 0 \mid \Psi_0 \rangle|^2 + |\langle 1 \mid \Psi_0 \rangle|^2 = (1 - \rho_0) + \rho_0, \; \rho_0 \in [0, 1], \end{aligned} \tag{1.28}$$

for unperturbed frontier molecular state $|\Psi_0\rangle$ with associated eigen-energy E_0 for a given valence system, reciprocally related by the conventional eigen-equation

$$\hat{H}|\Psi_0\rangle = E_0|\Psi_0\rangle \tag{1.29}$$

Observe that here the molecular orbital state is placed on the frontier domain such that further reactivity will be accounted by means of expressing the ionization and affinity actions, namely as (Putz 2009b, 2011d)

$$\begin{aligned} |\Psi_\lambda^I\rangle &= (1 + \lambda \hat{a}\hat{a}^+)|\Psi_0\rangle = |\Psi_0\rangle + \lambda|0\rangle\langle 1 \mid 1\rangle\langle 0 \mid \Psi_0\rangle \\ &= |\Psi_0\rangle + \lambda\sqrt{1 - \rho_0}|0\rangle, \end{aligned} \tag{1.30}$$

$$\begin{aligned} |\Psi_\lambda^A\rangle &= (1 + \lambda \hat{a}^+\hat{a})|\Psi_0\rangle = |\Psi_0\rangle + \lambda|1\rangle\langle 0 \mid 0\rangle\langle 1 \mid \Psi_0\rangle \\ &= |\Psi_0\rangle + \lambda\sqrt{\rho_0}|1\rangle \end{aligned} \tag{1.31}$$

obtaining therefore the perturbed frontier states through the perturbation factor λ. In these conditions, the frontier indices of electronegativity and chemical hardness are formed from the perturbed energy

$$\langle E_{\lambda \in \Re}^{I \leftrightarrow A} \rangle := \frac{\langle \Psi_\lambda^I | \hat{H} | \Psi_\lambda^A \rangle}{\langle \Psi_\lambda^I \mid \Psi_\lambda^A \rangle} \tag{1.32}$$

and electronic density

$$\rho_{\lambda \in \Re}^{I \leftrightarrow A} := \frac{\langle \Psi_\lambda^I | \hat{a}^+\hat{a} | \Psi_\lambda^A \rangle}{\langle \Psi_\lambda^I \mid \Psi_\lambda^A \rangle} \tag{1.33}$$

to be respectively:

$$\chi_\lambda = -\frac{\partial \langle E_\lambda \rangle}{\partial \rho_\lambda} = -\frac{\partial \langle E_\lambda \rangle}{\partial \lambda} \frac{\partial \lambda}{\partial \rho_\lambda} \tag{1.34}$$

and

$$\eta_\lambda = \frac{1}{2} \frac{\partial^2 \langle E_\lambda \rangle}{\partial \rho_\lambda^2} = \frac{1}{2} \left\{ \left[\frac{\partial}{\partial \lambda} \left(\frac{\partial \langle E_\lambda \rangle}{\partial \lambda} \right) \right] \frac{\partial \lambda}{\partial \rho_\lambda} + \frac{\partial \langle E_\lambda \rangle}{\partial \lambda} \left[\frac{\partial}{\partial \lambda} \left(\frac{\partial \lambda}{\partial \rho_\lambda} \right) \right] \right\} \frac{\partial \lambda}{\partial \rho_\lambda} \tag{1.35}$$

Explicit dependency on the perturbation factor is thus necessary on both perturbed density and energy, in order the involved derivatives $\partial\lambda/\partial\rho_\lambda$ and $\partial\langle E_\lambda\rangle/\partial\lambda$ be appropriately formulated and combined in working Eqs. 1.34 and 1.35.

For computing the density (1.33) with ionization and affinity states (1.30) and (1.31) one uses the above second quantization rules to firstly yield the expressions

$$\langle\Psi_0|\hat{a}\hat{a}^+\hat{a}^+\hat{a}|\Psi_0\rangle = \langle\Psi_0 \mid 0\rangle\langle 1 \mid 1\rangle\langle 0 \mid 1\rangle\langle 0 \mid 0\rangle\langle 1 \mid \Psi_0\rangle = 0, \tag{1.36}$$

$$\langle\Psi_0|\hat{a}^+\hat{a}\hat{a}^+\hat{a}|\Psi_0\rangle = \langle\Psi_0 \mid 1\rangle\langle 0 \mid 0\rangle\langle 1 \mid 1\rangle\langle 0 \mid 0\rangle\langle 1 \mid \Psi_0\rangle = \rho_0, \tag{1.37}$$

$$\langle\Psi_0|\hat{a}\hat{a}^+\hat{a}^+\hat{a}\hat{a}^+\hat{a}|\Psi_0\rangle = \langle\Psi_0 \mid 0\rangle\langle 1 \mid 1\rangle\langle 0 \mid 1\rangle\langle 0 \mid 0\rangle\langle 1 \mid 1\rangle\langle 0 \mid 0\rangle\langle 1 \mid \Psi_0\rangle = 0 \tag{1.38}$$

to be then organized in the frontier density (Putz 2009b; 2011d)

$$\begin{aligned}
\rho_{\lambda\in\Re}^{I\leftrightarrow A} &= \frac{\langle\Psi_\lambda^I|\hat{a}^+\hat{a}|\Psi_\lambda^A\rangle}{\langle\Psi_\lambda^I \mid \Psi_\lambda^A\rangle_{0<<\rho_0\le 1}}\\
&= \frac{\langle\Psi_0|(1+\lambda\hat{a}\hat{a}^+)\hat{a}^+\hat{a}(1+\lambda\hat{a}^+\hat{a})|\Psi_0\rangle}{\langle\Psi_0|(1+\lambda\hat{a}\hat{a}^+)(1+\lambda\hat{a}^+\hat{a})|\Psi_0\rangle_{0<<\rho_0\le 1}}\\
&= \frac{\langle\Psi_0|\left(\hat{a}^+\hat{a}+\lambda\hat{a}^+\hat{a}\hat{a}^+\hat{a}+\lambda\hat{a}\hat{a}^+\hat{a}^+\hat{a}+\lambda^2\hat{a}\hat{a}^+\hat{a}^+\hat{a}\hat{a}^+\hat{a}\right)|\Psi_0\rangle}{\langle\Psi_0|\left(1+\lambda\hat{a}^+\hat{a}+\lambda\hat{a}\hat{a}^++\lambda^2\hat{a}\hat{a}^+\hat{a}^+\hat{a}\right)|\Psi_0\rangle_{0<<\rho_0\le 1}}\\
&= \rho_0\frac{1+\lambda}{1+\lambda\rho_0}
\end{aligned} \tag{1.39}$$

Now, the density related expressions in electronegativity and chemical hardness cast as

$$\frac{\partial\lambda}{\partial\rho_\lambda} = \frac{(1+\lambda\rho_0)^2}{\rho_0(1-\rho_0)}, \tag{1.40}$$

$$\frac{\partial}{\partial\lambda}\left(\frac{\partial\lambda}{\partial\rho_\lambda}\right) = 2\frac{1+\lambda\rho_0}{1-\rho_0} \tag{1.41}$$

When passing to the energy related quantities, similarly, one employs the ionization and affinity states (1.30) and (1.31) into the perturbed energy (1.32) that, at its turn, needs the pre-evaluation of the expressions calling the eigen-equation (1.29)

$$\langle\Psi_0|\hat{H}\hat{a}^+\hat{a}|\Psi_0\rangle = \langle\Psi_0|\hat{H}|1\rangle\langle 0 \mid 0\rangle\langle 1 \mid \Psi_0\rangle = E_0\langle\Psi_0 \mid 1\rangle\langle 1 \mid \Psi_0\rangle = E_0\rho_0, \tag{1.42}$$

$$\begin{aligned}\langle\Psi_0|\hat{a}\hat{a}^+\hat{H}|\Psi_0\rangle &= \langle\Psi_0 \mid 0\rangle\langle 1 \mid 1\rangle\langle 0|\hat{H}|\Psi_0\rangle = \langle\Psi_0 \mid 0\rangle\langle 0 \mid \Psi_0\rangle E_0 \\ &= E_0(1-\rho_0),\end{aligned} \tag{1.43a}$$

as well as of the term based on the usual second quantization form of the Eq. 1.20.

$$\langle\Psi_0|\hat{a}\hat{a}^+\hat{H}\hat{a}^+\hat{a}|\Psi_0\rangle = \langle\Psi_0 \mid 0\rangle\langle 1 \mid 1\rangle\langle 0|\hat{H}|1\rangle\langle 0 \mid 0\rangle\langle 1 \mid \Psi_0\rangle = 0 \tag{1.43b}$$

in the virtue of the immediate cancelation

$$\langle 0|\hat{H}|1\rangle \sim \langle 0|\hat{a}_p^+...|1\rangle = \langle 0 \mid 1\rangle\langle 0|...|1\rangle = 0. \tag{1.44}$$

Now, the energy (1.32) may be evaluated successively with the result (Putz 2009b, 2011d)

$$\begin{aligned}\left\langle E_{\lambda\in\Re}^{I\leftrightarrow A}\right\rangle &= \frac{\langle\Psi_\lambda^I|\hat{H}|\Psi_\lambda^A\rangle}{\left\langle\Psi_\lambda^I \mid \Psi_\lambda^A\right\rangle_{0<<\rho_0\leq 1}} \\ &= \frac{\langle\Psi_0|(1+\lambda\hat{a}\hat{a}^+)\hat{H}(1+\lambda\hat{a}^+\hat{a})|\Psi_0\rangle}{\langle\Psi_0|(1+\lambda\hat{a}\hat{a}^+)(1+\lambda\hat{a}^+\hat{a})|\Psi_0\rangle_{0<<\rho_0\leq 1}} \\ &= \frac{\langle\Psi_0|\hat{H}|\Psi_0\rangle + \lambda\langle\Psi_0|\hat{H}\hat{a}^+\hat{a}|\Psi_0\rangle + \lambda\langle\Psi_0|\hat{a}\hat{a}^+\hat{H}|\Psi_0\rangle + \lambda^2\langle\Psi_0|\hat{a}\hat{a}^+\hat{H}\hat{a}^+\hat{a}|\Psi_0\rangle}{1+\lambda\rho_0} \\ &= E_0\frac{1+\lambda}{1+\lambda\rho_0}\end{aligned} \tag{1.45}$$

With the help of (1.45) the first and second derivative with respect the perturbation factor look like

$$\frac{\partial\langle E_\lambda\rangle}{\partial\lambda} = E_0\frac{1-\rho_0}{(1+\lambda\rho_0)^2}, \tag{1.46}$$

$$\frac{\partial}{\partial\lambda}\left(\frac{\partial\langle E_\lambda\rangle}{\partial\lambda}\right) = -2E_0\rho_0\frac{1-\rho_0}{(1+\lambda\rho_0)^3} \tag{1.47}$$

Combining the expressions (1.40) and (1.46) in (1.34) will leave the frontier orbital electronegativity with the form and of its density limits (Putz 2009b, 2011a, d)

$$\chi_\lambda = -\frac{E_0}{\rho_0} = -\mu_0 = \begin{cases}\infty & , \ \rho_0 \to 0\,(E_0<0) \\ -E_0 = -\langle\psi_0|H|\psi_0\rangle & , \ \rho_0 \to 1\end{cases} \tag{1.48}$$

while for the chemical hardness the combination of Eqs. 1.40, 1.41, 1.46, and 1.47 in 1.35 produces the frontier orbital chemical hardness with its density limits (Putz 2010a, 2011a, d)

$$\eta_\lambda = 0 \cdot E_0 \frac{1 + \lambda\rho_0}{\rho_0(1 - \rho_0)} = \begin{cases} 0, & \rho_0 \in (0, 1) \\ 0 \cdot \infty = ?, & \rho_0 \to 0 \\ 0 \cdot \infty = ?, & \rho_0 \to 1 \end{cases} \tag{1.49}$$

The results (1.48) and (1.49) enlighten on the following quantum perspectives of electronegativity and chemical hardness, in most general cases:

- electronegativity is behaving like the associated orbital (eigen) energy for density approaching the integer quantum particle realization, otherwise being manifestly field;
- chemical hardness does not manifest as a quantum index (or it has the zero value) for densities that are not integer representation of fermionic existence; in other words it has no quantum observable character for electronic states unless they are fully equivalent with (integer) particle manifestations; on the other side, for such integer density states, i.e. the second and the third branches of the limit (1.49), the chemical hardness has not definite (universally observable) quantity; if the result in such cases is infinite it act like a field (like the electronegativity, i.e. like a super-potential since the electronegativity is seen as the minus of the chemical potential); if it is zero then in all cases the chemical hardness is not observable and the parabolic form itself of the frontier energy is superfluous; for the non-zero results chemical hardness preserves the parabolic form of the frontier energy (1.4) with the meaning registering its curvature, i.e. how fast it changes from donor to acceptor character, in accordance with the high of the HOMO-LUMO gap of Eq. 1.8.

Overall, beside the fact the quantum observable character of the chemical hardness remain an open issue, being neither informed nor definitely confirmed by the present analysis, there was this way nevertheless argue on the parabolic sufficiency on the quantum expansion in the frontier energy; in other terms, the second quantization firmly prescribes the manifestation of the electronegativity as an energy for the quantum states that characterizes full existence of the particles, being this virtually accompanied by the generally not definite second order contribution coming from the chemical hardness observable indeterminacy. Even shorter, for precisely defined particle-quantum states their energy may be represented as a superposition of an observable and a not observable (hidden variable) chemical contribution; such unique manifestation may be regarded as a special or complementary uncertainty principle for chemical behavior, in the spirit of above enounced physics-to-chemistry non-reductionism.

However, there is clear that if the second order or parabolic manifestation of chemical phenomena inscribes quantity with non-observable character, the cubic or even higher order of quantum energy manifestation may be conceptually discarded.

There remains however to establish the specific behavior of the assumed parabolic quantum effect on chemical reactivity in an quantitative manner – an issue that will be in next addressed, after surveying the most common semi-empirical orbital pictures, as another level of treating the frontier effects of orbitals, this time on an inter-orbitalic basis.

1.4 Survey on Semiempirical Quantum Energies (Putz 2010b)

1.4.1 General Mono-Electronic Molecular Orbitals' Equations

For better understanding how the semi-empirical methods were introduced, worth reconsidering the molecular eigen-equation (1.29) under the so called independent-electron problem

$$\hat{H}_i^{eff}\psi_i = E_i\psi_i \tag{1.50}$$

with the aid of effective electron Hamiltonian partitioning

$$\hat{H} = \sum_i \hat{H}_i^{eff} \tag{1.51}$$

and the correspondent *molecular* monoelectronic wave-functions (orbitals) fulfilling the conservation rule of probability

$$\int \psi_i^2(\mathbf{r})d\mathbf{r} = 1 \tag{1.52}$$

Note that while in Fig. 1.1 the so called many-electronic molecular orbitals are about, in Eq. 1.50 the uni- (or *i*-) electronic molecular orbitals are assumed, each of them moving with its Hamiltonian that is effective since it contains also the influence of the all other electrons in the system upon it. Even more, these molecular mono-orbitals are further decomposed in mathematical atomic orbitals, see ϕ's in Eq. 1.53 that are essentially not so different from the individual φ's wave-functions of the *N*-electrons considered in Fig. 1.1. The actual atomic orbitals are mathematical rather physical objects (for which reason they are often called as basis set) and reflect the atomic participation in the molecular or bonding system rather that the total or valence number of electrons in the system. As such, viewed as the linear combination over the *atomic orbitals*, the resulted MO-LCAO wave-function

$$\psi_i = \sum_v C_{vi}\phi_v \tag{1.53}$$

replaced in Eq. 1.50 and followed by integration over the electronic space allows for matrix version of Eq. 1.50

$$(H^{eff})(C) = (S)(C)(E) \tag{1.54}$$

having the diagonal energy-matrix elements as the eigen-solution

$$(E)_{ij} = E_{ij} = E_i\delta_{ij} = \begin{cases} E_i ... i = j \\ 0 \; ... i \neq j \end{cases} \tag{1.55}$$

to be found in terms of the expansion coefficients matrix (C), the matrix of the Hamiltonian elements

$$H_{\mu\nu} = \int \phi_\mu H^{eff} \phi_\nu d\tau \tag{1.56}$$

and the matrix of the (atomic) overlapping integrals

$$S_{\mu\nu} = \int \phi_\mu \phi_\nu d\tau \tag{1.57}$$

where all indices in Eqs. (1.55)–(1.57) refers to matrix elements since the additional reference to the "i" electron was skipped for avoiding possible confusion.

Yet, the solution of the matrix Eq. (1.54) may be unfolded through the Löwdin orthogonalization procedure (Löwdin 1950, 1993), involving the diagonalization of the overlap matrix by means of a given unitary matrix (U), $(U)^+(U) = (1)$, by the resumed procedure

$$(s) = (U)^+(S)(U) \tag{1.58}$$

$$\left(s^{-1/2}\right)_{ii} = \left[(s)_{ii}\right]^{-1/2} \tag{1.59}$$

$$\left(S^{-1/2}\right) = (U)\left(s^{-1/2}\right)(U)^+ \tag{1.60}$$

$$\left(\left(S^{1/2}\right)(C)\right)^+ \left(\left(S^{-1/2}\right)\left(H^{eff}\right)\left(S^{-1/2}\right)\right)\left(\left(S^{1/2}\right)(C)\right) = (E) \tag{1.61}$$

However, the solution given by Eq. 1.61 is based on the form of effective independent-electron Hamiltonians that can be quite empirically constructed – as in Extended Hückel Theory (Hoffmann 1963); such "arbitrariness" can be nevertheless avoided by the so called *self-consistent field* (SCF) in which the one-electron effective Hamiltonian is considered such that to depend by the solution of Eq. 1.53 itself, *i.e.*, by the matrix of coefficients (C); the resulted "Hamiltonian" is called Fock operator, while the associated eigen-problem is consecrated as the Hartree-Fock equation

$$F\psi_i = E_i\psi_i \tag{1.62}$$

In matrix representation Eq. 1.62 looks like

$$(F((C)))(C) = (S)(C)(E) \tag{1.63}$$

that may be iteratively solved through diagonalization procedure starting from an input (C) matrix or – more physically appealing – from a starting electronic distribution quantified by the density matrix

$$P_{\mu\nu} = \sum_i^{occ} C_{\mu i} C_{i\nu} \tag{1.64}$$

with major influence on the Fock matrix elements

$$F_{\mu\nu} = H_{\mu\nu} + \sum_{\lambda\sigma} P_{\lambda\sigma}\left[(\mu\nu|\lambda\sigma) - \frac{1}{2}(\mu\lambda|\nu\sigma)\right] \tag{1.65}$$

Note that now the one-electron Hamiltonian effective matrix components $H_{\mu V}$ differ from those of Eq. 1.56 in what they truly represent this time the kinetic energy plus the interaction of a single electron with the core electrons around all the present nuclei. The other integrals appearing in Eq. 1.65 are generally called as the two-electrons-multi-centers integrals and write as

$$(\mu\nu|\lambda\sigma) = \int \phi_\mu^A(\mathbf{r}_1)\phi_\nu^B(\mathbf{r}_1)\frac{1}{r_{12}}\phi_\lambda^C(\mathbf{r}_2)\phi_\sigma^D(\mathbf{r}_2)d\mathbf{r}_1 d\mathbf{r}_2 \tag{1.66}$$

From definition (1.66) there is immediate to recognize the special integral $J = (\mu\mu|\nu\nu)$ as the Coulomb integral describing repulsion between two electrons with probabilities ϕ_μ^2 and ϕ_ν^2.

Moreover, the Hartree-Fock Eq. 1.63 with implementations given by Eqs. 1.64 and 1.65 are known as Roothaan equations (Roothaan 1951) and constitute the basics for closed-shell (or restricted Hartree-Fock, RHF) molecular orbitals calculations. Their extension to the spin effects provides the equations for the open-shell (or unrestricted Hartree-Fock, UHF) known also as the Pople-Nesbet Unrestricted equations (Pople and Nesbet 1954).

1.4.2 Semiempirical Approximations

The second level of approximation in molecular orbital computations regards the various ways the Fock matrix elements of Eq. 1.65 are considered, namely the approximations of the integrals (1.66) and of the effective one-electron Hamiltonian matrix elements $H_{\mu\nu}$.

The main route for such endeavor is undertaken through neglecting at different degrees certain *differential* overlapping terms (integrals) – as an offset ansatz – although with limited physical justification – while the adjustment with experiment is done (post-factum) by fitting parameters – from where the semiempirical name of such approximation. Practically, by emphasizing the (nuclear) centers in the electronic overlapping integral (1.57)

$$S_{\mu\nu} = \int \phi_{\mu}^{A}(\mathbf{r}_1)\phi_{\nu}^{B}(\mathbf{r}_1)d\mathbf{r}_1 \tag{1.67}$$

the differential overlap approximation may be considered by two situations.

- By *neglecting the differential overlap* (NDO) through the mono-atomic orbitalic constraint

  $$\phi_{\mu}\phi_{\nu} = \phi_{\mu}\phi_{\mu}\delta_{\mu\nu} \tag{1.68}$$

  leaving with the simplified integrals

  $$S_{\mu\nu} = \delta_{\mu\nu}\int \phi_{\mu}^{A}(\mathbf{r}_1)\phi_{\mu}^{A}(\mathbf{r}_1)d\mathbf{r}_1 = \delta_{\mu\nu} \tag{1.69}$$

  $$\begin{aligned}(\mu\nu|\lambda\sigma) &= \delta_{\mu\nu}\delta_{\lambda\sigma}\int \phi_{\mu}^{A}(\mathbf{r}_1)\phi_{\mu}^{A}(\mathbf{r}_1)\tfrac{1}{r_{12}}\phi_{\lambda}^{B}(\mathbf{r}_2)\phi_{\lambda}^{B}(\mathbf{r}_2)d\mathbf{r}_1 d\mathbf{r}_2 \\ &= \delta_{\mu\nu}\delta_{\lambda\sigma}(\mu_A\mu_A|\lambda_B\lambda_B) \equiv \delta_{\mu\nu}\delta_{\lambda\sigma}\gamma^{AB}\end{aligned} \tag{1.70}$$

  thus reducing the number of bielectronic integrals, while the tri- and tetra-centric integrals are all neglected;
- By *neglecting the diatomic differential overlap* (NDDO) of the bi-atomic orbitals

  $$\phi_{\mu}^{A}\phi_{\nu}^{B} = \phi_{\mu}^{A}\phi_{\nu}^{A}\delta_{AB} \tag{1.71}$$

  that implies the actual simplifications

  $$S_{\mu\nu} = \delta_{AB}\int \phi_{\mu}^{A}(\mathbf{r}_1)\phi_{\nu}^{A}(\mathbf{r}_1)d\mathbf{r}_1 = \delta_{AB}\delta_{\mu\nu} \tag{1.72}$$

  $$(\mu\nu|\lambda\sigma) = \delta_{AB}\delta_{CD}\int \phi_{\mu}^{A}(\mathbf{r}_1)\phi_{\nu}^{A}(\mathbf{r}_1)\frac{1}{\mathbf{r}_{12}}\phi_{\lambda}^{C}(\mathbf{r}_2)\phi_{\sigma}^{C}(\mathbf{r}_2)d\mathbf{r}_1 d\mathbf{r}_2 \tag{1.73}$$

  when overlaps (or contractions) of atomic orbitals on different atoms are neglected. For both groups of approximations specific methods are outlined below.

1.4.2.1 NDO Methods

The basic NDO approximation was developed by Pople and is known as the Complete Neglect of Differential Overlap CNDO semiempirical method (Pople et al. 1965;

Pople and Segal 1965, 1966; Pople and Beveridge 1970). It employs the approximation (1.68)–(1.70) such that the molecular rotational invariance is respected through the requirement the integral (1.70) depends only on the atoms A or B where the involved orbitals reside – and not by the orbitals themselves. That is the integral γ^{AB} in (1.69) is seen as the average electrostatic repulsion between an electron in any orbital of A and an electron in any orbital of B

$$V_{AB} = Z_B \gamma^{AB} \tag{1.74}$$

In these conditions, the working Fock matrix elements of Eq. 1.65 become within RHF scheme

$$F_{\mu\mu}^{CNDO} = H_{\mu\mu}^{CNDO} + \left(P_{AA} - \frac{1}{2} P_{\mu\mu} \right) \gamma^{AA} + \sum_{B \neq A} P_{BB} \gamma^{AB} \tag{1.75}$$

$$F_{\mu\nu}^{CNDO} = H_{\mu\nu}^{CNDO} - \frac{1}{2} P_{\mu\nu} \gamma^{AB} \tag{1.76}$$

from Eqs. 1.75 and 1.76 there appears that the core Hamiltonian has as well the diagonal and off-diagonal components; the diagonal one represents the energy of an electron in an atomic orbital of an atom (say A) written in terms of ionization potential and electron affinity of that atom (Oleari et al. 1966)

$$U_{\mu\mu}^{CNDO} = -\frac{1}{2}(I_\mu + A_\mu) - \left(Z_A - \frac{1}{2} \right) \gamma^{AA} \tag{1.77}$$

added to the attraction energy respecting the other (B) atoms to produce the one-center-one-electron integrals

$$H_{\mu\mu}^{CNDO} = U_{\mu\mu}^{CNDO} - \sum_{B \neq A} V_{AB} \tag{1.78}$$

overall expressing the energy an electron in the atomic orbital φ_μ would have if all other *valence electrons* were removed to infinity. The non-diagonal terms (*the resonance integrals*) are parameterized in respecting the overlap integral and accounts (through β_{AB} parameter averaged over the atoms involved) on the diatomic bonding involved in overlapping

$$H_{\mu\nu}^{CNDO} = \beta_{AB}^{CNDO} S_{\mu\nu} \tag{1.79}$$

The switch to the UHF may be eventually done through implementing the spin equivalence

$$P^T \equiv P^{\uparrow+\downarrow} = \frac{1}{2} P^{\uparrow} = \frac{1}{2} P^{\downarrow} \tag{1.80}$$

although the spin effects are not at all considered since no exchange integral involved. This is in fact the weak point of the CNDO scheme and it is to be slightly improved by the next semiempirical methods.

The exchange effect due to the electronic spin accounted within the Intermediate Neglect of Differential Overlap (INDO) method (Slater 1960) through considering in Eqs. 1.75 and 1.77 the exchange one-center integrals $\gamma^{AA} \equiv K = (\mu\nu|\mu\nu)$ is evaluated as

$$(sp_x|sp_x)^{INDO} = \frac{1}{3}G^1,\ (p_x p_y|p_x p_y)^{INDO} = \frac{3}{25}F^2, \ldots \tag{1.81}$$

in terms of the Slater-Condon parameters G^1, F^2, ... usually used to describe atomic spectra.

The INDO method may be further modified in parameterization of the spin effects as developed by Dewar's group and lead with the Modified Intermediate Neglect of Differential Overlap (MINDO) method (Pople and Beveridge 1970; Baird and Dewar 1969; Dewar and Hasselbach 1970; Dewar and Lo 1972; Bingham et al. 1975a, b, c, d; Dewar et al. 1975; Murrell and Harget 1971) whose basic equations look like

$$F_{\mu\nu}^{\uparrow(MINDO)} = \begin{cases} H_{\mu\nu}^{MINDO} - (\mu\mu|\nu\nu)P_{\mu\nu}^{\uparrow} & \ldots\mu|_A,\ \nu|_{B\neq A} \\ \left(2P_{\mu\nu}^{\uparrow+\downarrow} - P_{\mu\nu}^{\uparrow}\right)(\mu\nu|\mu\nu) - P_{\mu\nu}^{\uparrow}(\mu\mu|\nu\nu) & \ldots\mu|_A \neq \nu|_A \end{cases} \tag{1.82}$$

$$F_{\mu\mu}^{\uparrow(MINDO)} = H_{\mu\mu}^{MINDO} + \sum_{\nu|A}\left[(\mu\mu|\nu\nu)P_{\nu\nu}^{\uparrow+\downarrow} - (\mu\nu|\mu\nu)P_{\nu\nu}^{\uparrow}\right] + \sum_{B}\gamma_{MINDO}^{AB}\sum_{A}^{B}P_{\mu\mu}^{\uparrow+\downarrow} \tag{1.83}$$

Apart of specific counting of spin effects, other particularity of MINDO respecting the CNDO/INDO is that all the non-zero two-center Coulomb integrals are set equal and parameterized by the appropriate one center two electrons integrals A_A and A_B within the Ohno-Klopman expression (Ohno 1964; Klopman 1964)

$$\gamma_{MINDO}^{AB} = (s_A s_A|s_B s_B) = (s_A s_A|p_B p_B) = (p_A p_A|p_B p_B) = \frac{1}{\sqrt{r_{AB}^2 + \frac{1}{4}\left(\frac{1}{A_A} + \frac{1}{A_B}\right)^2}} \tag{1.84}$$

The one-center-one-electron integral $H_{\mu\mu}$ is preserved from the CNDO/INDO scheme of computation, while the resonance integral (1.79) is modified as follows

$$H_{\mu\nu}^{MINDO} = \left(I_\mu + I_\nu\right)\beta_{AB}^{MINDO}S_{\mu\nu} \tag{1.85}$$

with the parameter β_{AB}^{MINDO} being now dependent on the atoms-in-pair rather than the average of atomic pair involved. As in INDO, the exchange terms, i.e., the one-center-two-electron integrals, are computed employing the atomic spectra and the G^k, F^k, Slater-Condon parameters, see Eq. 1.81 (Pople et al. 1967). Finally, worth mentioning that the MINDO (also with its MINDO/3 version) improves upon the CNDO and INDO the molecular geometries, heats of formation, being particularly suited for dealing with molecules containing heteroatoms.

1.4.2.2 NDDO Methods

This second group of neglecting differential overlaps semiempirical methods includes along the interaction quantified by the overlap of two orbitals centered on the same atom also the overlap of two orbitals belonging to different atoms. It is manly based on the Modified Neglect of Diatomic Overlap (MNDO) approximation of the Fock matrix, while introducing further types of integrals in the UHF framework (Dewar and Thiel 1977; Dewar and McKee 1977; Dewar and Rzepa 1978; Davis et al. 1981; Dewar and Storch 1985; Thiel 1988; Clark 1985)

$$F_{\mu v}^{\uparrow(MNDO)} = \begin{cases} H_{\mu v}^{MNDO} - \sum_{\lambda|A}\sum_{\sigma|B}(\mu\lambda|v\sigma)P_{\lambda\sigma}^{\uparrow} & \ldots\mu|_A, v|_{B\neq A} \\ H_{\mu v}^{MNDO} + P_{\mu v}^{\uparrow}[3(\mu v|\mu v) - (\mu\mu|vv)] + \sum_{B}\sum_{\lambda|B}\sum_{\sigma|B}(\mu v|\lambda\sigma)P_{\lambda\sigma}^{\uparrow+\downarrow} & \ldots\mu|_A \neq v|_A \end{cases} \tag{1.86}$$

$$F_{\mu\mu}^{\uparrow(MNDO)} = H_{\mu\mu}^{MNDO} + \sum_{v|A}\left[(\mu\mu|vv)P_{vv}^{\uparrow+\downarrow} - (\mu v|\mu v)P_{vv}^{\uparrow}\right] + \sum_{B}\sum_{\lambda|B}\sum_{\sigma|B}(\mu\mu|\lambda\sigma)P_{\lambda\sigma}^{\uparrow+\downarrow} \tag{1.87}$$

Note that similar expressions can be immediately written within RHF once simply replacing

$$P^{\uparrow(\downarrow)} = -\frac{1}{2}P^{\uparrow+\downarrow} \tag{1.88}$$

in above Fock (1.86) and (1.87) expressions.

Now, regarding the (Coulombic) two-center-two-electron integrals of type (1.73) appearing in Eqs. 1.86 and 1.87 there were indentified 22 unique forms for each pair of non-hydrogen atoms, i.e., the rotational invariant 21 integrals $(ss|ss)$, $(ss|p_\sigma p_\sigma)$, $(ss|p_\pi p_\pi)$, ..., $(p_\sigma p_\sigma|p_\sigma p_\sigma)$, $(p_\pi p_\pi|p_\pi p_\pi)$, ..., $(sp_\sigma|sp_\sigma)$, $(sp_\pi|sp_\pi)$,..., $(p_\pi p_\sigma|sp_\pi)$, $(p_\pi p_\sigma|p_\pi p_\sigma)$, and the 22nd one that is written as a combination of two of previously ones, namely $(p_\pi p_\pi'|p_\pi p_\pi') = 0.5[(p_\pi p_\pi|p_\pi p_\pi) - (p_\pi p_\pi|p_\pi' p_\pi')]$, with the typical integral approximation relaying on the Eq. (1.84) structure, however slightly modified as

$$(ss|ss)^{MNDO} = \frac{1}{\sqrt{(r_{AB} + c_A + c_B)^2 + \frac{1}{4}\left(\frac{1}{A_A} + \frac{1}{A_B}\right)^2}} \tag{1.89}$$

where additional parameters c_A and c_B represent the distances of the multipole charges from their respective nuclei. The MNDO one-center one-electron integral has the same form as in NDO methods, *i.e.*, given by Eq. 1.78 with the average potential of Eq. 1.74 acting on concerned center; still, the resonance integral is modified as

$$H_{\mu\nu}^{MNDO} = \frac{\beta_{\mu}^{MNDO} + \beta_{\nu}^{MNDO}}{2} S_{\mu\nu} \tag{1.90}$$

containing the atomic adjustable parameters β_{μ}^{MNDO} and β_{ν}^{MNDO} for the orbitals ϕ_{μ} and ϕ_{ν} of the atoms A and B, respectively. The exchange (one-center-two-electron) integrals are mostly obtained from data on isolated atoms (Oleari et al. 1966). Basically, MNDO improves MINDO through the additional integrals considered the molecular properties such as the heats of formations, geometries, dipole moments, HOMO and LUMO energies, etc., while problems still remaining with four-member rings (too stable), hypervalent compounds (too unstable) in general, and predicting out-of-plane nitro group in nitrobenzene and too short bond length (~0.17 Å) in peroxide – for specific molecules.

The MNDO approximation is further improved by aid of the Austin Model 1 (AM1) method (Dewar et al. 1985; Dewar and Dieter 1986; Stewart 1990) that refines the inter-electronic repulsion integrals

$$(s_A s_A|s_B s_B)^{AM1} = \frac{1}{\sqrt{r_{AB}^2 + \frac{1}{4}\left(\frac{1}{AM_A} + \frac{1}{AM_B}\right)^2}} \tag{1.91}$$

while correcting the one-center-two-electron atomic integrals of Eq. 1.84 by the specific (AM) monopole-monopole interaction parameters. In the same line, the nuclei-electronic charges interaction adds an energetic correction within the α_{AB} parameterized form

$$\Delta E_{AB} = \sum_{A,B} \left\{ Z_A Z_B (s_A s_A|s_B s_B) \left[1 + \left(1 + \frac{1}{r_{AB}} \right) e^{-\alpha_{AB} r_{AB}} \right] - Z_A Q_B (s_A s_A|s_B s_B) \right\} \tag{1.92}$$

The AM1 scheme, while furnishing better results than MNDO for some classes of molecules (e.g., for phosphorous compounds), still provides inaccurate modeling of phosphorous-oxygen bonds, too positive energy of nitro compounds, while the peroxide bond is still too short. In many case the reparameterization of AM1 under the Stewart's PM3 model (Stewart 1989a, b) is helpful since it is based on a wider plethora of experimental data fitting with molecular properties. The best use of PM3 method lays in the organic chemistry applications.

Going to systematically implement the transition metal orbitals in semiempirical methods the INDO method is augmented by Zerner's group either with non-spectroscopic and spectroscopic (viz. fitting with UV spectra) parameterization (Del Bene and Jaffé 1968a, b, c), known as ZINDO/1 and ZINDO/S methods, respectively (Ridley and Zerner 1976; Bacon and Zerner 1979; Stavrev et al. 1995; Stavrev and Zerner 1995; Cory et al. 1997; Anderson et al. 1986, 1991). The working equations are formally the same as those for INDO except for the energy of an atomic electron of Eq. 1.77 that now uses only the ionization potential instead of electronegativity of the concerned electron. Moreover, for ZINDO/S the core Hamiltonian elements $H_{\mu\mu}$ is corrected

$$\Delta H_{\mu\mu}^{ZINDO} = \sum_{B} (Z_B - Q_B)\gamma_{(\mu\mu|ss)}^{AB(ZINDO)} \tag{1.93}$$

by the f_r parameterized integrals

$$\gamma_{(\mu\mu|ss)}^{AB(ZINDO)} = \frac{f_r}{\dfrac{2f_r}{\gamma_{\mu\mu}^{A} + \gamma_{ss}^{B}} + r_{AB}}, f_r = 1.2 \tag{1.94}$$

in terms of the one-center-two-electron Coulomb integrals $\gamma_{\mu\mu}^{A}, \gamma_{ss}^{B}$. Equation 1.94 conserves nevertheless the molecular rotational invariance through making the difference between the *s*- and *d*- Slater orbitals exponents. The same type of integrals correct also the nuclei-electronic interaction energy by quantity

$$\Delta E_{AB} = \sum_{A,B} \left\{ \frac{Z_A Z_B}{r_{AB}} - Z_A Q_B \gamma_{(\mu\mu|ss)}^{AB} \right\} \tag{1.95}$$

Since based on fitting with spectroscopic transitions the ZINDO methods are recommended in conjunction with single point calculation and not with geometry optimization that should be consider by other off-set algorithms.

Beyond of either NDO or NDDO methods the self-consistent computation of molecular orbitals can be made by the so called ab initio approach directly relaying on the HF equation or on its density functional extension, as will be in next sketched.

1.5 From Hückel- to Parabolic- to π-Energy Formulations

The Hückel method is simple and has been in use for decades (Hückel 1931a, b). It is based on the σ-π separation approximation while accounting for the pi-electrons only, i.e. the atomic orbitals involved refer to those $2p_z$ for Carbon atoms as well to the $2p_z$ and $3p_z$ orbitals for the second and third period elements as (N,O, F) and (S, Cl) respectively; further discussion on the d-orbitals involvement may be also undertaken, yet the method essence reside in non explicitly counting on the electronic repulsion with an effective, not-defined, mono-electronic Hamiltonian, as the most simple semi-empirical approximation. In these conditions, for the mono-electronic Hamiltonian matrix elements two basic assumptions are advanced, namely:

- In the case of hydrocarbures (C containing only π-systems) one has:

$$H_{\mu\nu} = \int \phi_\mu H^{eff} \phi_\nu d\tau = \begin{cases} \alpha \ \ldots \ \phi_\mu = \phi_\nu \\ 0 \ \ldots \ \phi_\mu, \phi_\nu \in non - bonded \ \ atoms \\ \beta \ \ldots \ \phi_\mu, \phi_\nu \in bonded \ \ atoms \end{cases} \quad (1.96a)$$

  where all Coulombic integrals are considered equal among them and equal with the quantity α representing the energy of the electron on atomic orbital $(2p_z)_C$; non-diagonal elements are neglected, i.e. for the non-bonding atoms; and the exchange or resonance integral is set equal with the non-definite β integral for neighboring bonding atoms.
- In the case heteroatoms (X) are present in the system one has to consider the Coulombic parameter h_X correlating with the electronegativity difference between the heteroatom X and carbon, along the resonance parameter k_{CX} that may include correlation with the binding energy; the form of matrix elements of monoelectronic effective Hamiltonian looks therefore as

$$H_{XC} = \int \phi_X H^{eff} \phi_C d\tau = \begin{cases} \alpha + h_X \beta \ \ldots \ \phi_X = \phi_X \\ 0 \qquad\quad \ldots \ \phi_X, \phi_C \in non - bonded \ \ atoms \\ k_{CX}\beta \qquad \ldots \ \phi_X, \phi_C \in bonded \ \ atoms \end{cases} \quad (1.96b)$$

As a consequence of these approximations, the total pi-energy with Hückel approach may be written in the virtue of the Eq. 1.51 summative as

$$E_\pi = \sum_C \alpha + \sum_{C,X} (h_X + k_{CX})\beta = \alpha N_\pi + \left(\sum_X h_X + \sum_{\substack{C-X, \\ C=X}} k_{CX} \right) \beta \quad (1.97)$$

Remarkably, the comparison of the Hückel energy (1.97) with the form (1.4) allows in advancing the electronegativity and chemical hardness related parabolic form

$$E_{para(bolic)} \cong -\chi \, N_\pi + \eta \, N_\pi^2 \quad (1.98)$$

when identified the reactive frontier electrons with the pi-electrons in the system, $\Delta N = N_\pi$. Even more, the present discussion permits the identification of the

Coulombic and resonance integrals in terms or electronegativity and chemical hardness

$$\alpha = -\chi, \tag{1.99}$$

$$\beta = \frac{\eta}{\sum_{X} h_X + \sum_{\substack{C-X, \\ C=X}} k_{CX}} N_{\pi}^2 \tag{1.100}$$

However, beside the possibility of assessing the Hückel integrals, the parabolic energy (1.98) may be useful in testing the constructed pi-energy abstracted from the total energy according with the recipe

$$\begin{aligned} E_{pi}(molecule) &\cong E_{Total}(molecule) - E_{Bind}(molecule) - E_{Heat}(molecule) \\ &= \sum_{atoms}^{molecule} E_{Total}(atom) - \sum_{atoms}^{molecule} E_{Heat}(atom) \end{aligned} \tag{1.101}$$

since:

- The total energy is relative to a sum of atomic energies for semi-empirical computations

$$E_{Total}(molecule) = \sum_{atoms}^{molecule} E_{Total}(atom) \tag{1.102}$$

- Binding energy is the energy of the molecular atoms separated by infinity minus the energy of the stable molecule at its equilibrium bond length

$$E_{Bind}(molecule) = E_{\infty}(atoms) - E_{equilibrium}(molecule) \tag{1.103}$$

- The heat of formation is calculated by subtracting atomic heats of formation from the binding energy:

$$E_{Heat}(molecule) = E_{Bind}(molecule) - \sum_{atoms}^{molecule} E_{Heat}(atom) \tag{1.104}$$

Through its form the energy (1.101) may have the frontier meaning, thus appropriately assessing the pi-formed system, while the remaining challenge is to test whether it can be well represented by the parabolic chemical reactivity descriptors related energy (1.98). To this end, four carbon based systems are analyzed due to their increased structure complexity, namely the butadiene, benzene, naphthalene, and fullerene that have been characterized by Hückel and most common semiempirical methods through the data in Tables 1.2–1.5, while

Table 1.2 The Butadiene π-system, with $\Delta N = N_\pi = 4$, frontier energetic quantities, ionization potential (*IP*), electron affinity (*EA*), electronegativity (χ), and chemical hardness (η) of Eqs. 1.7 and 1.8 – in electron volts (eV), and the resulted parabolic energy of Eq. 1.98, alongside with the π-related energy based on the Hückel simplified (with Coulomb integrals set to zero, $\alpha = 0$) expression of (1.97) for the experimental/Hückel method and on the related energy form of Eq. 1.101 and the other semi-empirical methods (*CNDO, INDO, MINDO, MNDO, AM1, PM3, ZINDO*) as described in the previous section – expressed in kilocalories per mol (kcal/mol); their ratio in the last column reflects the value of the actual departure of the electronegativity and chemical hardness parabolic effect from the pi-bonding energy, while for the first (*Exp/Hückel*) line it expresses the resonance contribution (and a sort of β factor integral) in (**1.97**) for the π-bond in this system; the eV to kcal/mol conversion follows the rule 1 eV ≅ 23.069 kcal/mol

Quantity→ Method↓	*IP* [eV]	*EA* [eV]	χ [eV]	η [eV]	|*E-(parabolic*| [kcal/mol]	*E-pi* [kcal/mol]	|*E-pi/E-para*|
Exp/Hückel	9.468[(a)]	−0.263[a]	4.6025	4.8655	473.2375	−990.214	2.09243
E-Hückel	12.50681	9.107174	10.80699	1.699818	683.521	−8982.58	13.14163
CNDO	13.32281	−3.35577	4.983522	8.33929	1079.173	−16,204	15.0152
INDO	12.75908	−3.90238	4.428352	8.330732	1128.823	−15684.9	13.8949
MINDO3	9.101508	−1.12356	3.988974	5.112535	575.4419	−12782.5	22.2134
MNDO	9.138431	−0.38813	4.37515	4.763282	475.3518	−12791.7	26.9101
MNDO/d	9.138306	−0.38791	4.375199	4.763108	475.3152	−12791.7	26.9121
AM1	9.333654	−0.44841	4.442624	4.891031	492.7019	−12751.2	25.8801
PM3	9.468026	−0.26348	4.602275	4.865752	473.3047	−12100.1	25.5651
ZINDO-1	9.156922	−7.45074	0.853094	8.303829	1453.768	−14210.2	9.77473
ZINDO-S	8.623273	−0.48256	4.070358	4.552916	464.6534	−10639.4	22.8974

[a] Calculated as the negative of the HOMO and LUMO energies (University Illinois 2011)

Table 1.3 The same quantities as those reported in Table 1.2, here for the Benzene π-system, with $\Delta N = N_\pi = 6$

Quantity→ Method↓	*IP* [eV]	*EA* [eV]	χ [eV]	η [eV]	|*E-(parabolic*| [kcal/mol]	*E-pi* [kcal/mol]	|*E-pi/E-para*|
Exp/Hückel	9.24384[a]	−1.60817[b]	3.817835	5.426005	1724.663	−5021.68[c]	2.91169
E-Hückel	12.81724	8.229032	10.52314	2.294105	503.941	−12,231	24.27077
CNDO	13.8859	−4.06892	4.908487	8.97741	3048.394	−23007.2	7.54733
INDO	13.48267	−4.58566	4.448502	9.034166	3135.63	−22231.9	7.09009
MINDO3	9.179751	−1.24984	3.964955	5.214796	1616.597	−18289.3	11.3135
MNDO	9.39118	−0.36809	4.511543	4.879637	1401.77	−18341.8	13.0848
MNDO/d	9.391201	−0.36807	4.511564	4.879638	1401.767	−18341.8	13.0848
AM1	9.653243	−0.55504	4.549103	5.10414	1489.794	−18315.5	12.294
PM3	9.751339	−0.3962	4.677572	5.073768	1459.4	−17222.6	11.8012
ZINDO-1	9.865785	−8.12621	0.869786	8.996	3615.126	−20453.7	5.6578
ZINDO-S	8.995844	−0.86318	4.066331	4.929514	1484.104	−15255.9	10.2795

[a] From National Institute of Standard and Technology (NIST 2011a)

[b] From interpolation data presented in Fig. 1.3

[c] From the Hückel total π-energy: 2 ×(2+2)=8β[a.u.] … × 627.71…~5021.68 kcal/mol, see Cotton (1971a)

Table 1.4 The same quantities as those reported in Table 1.2, here for the Naphthalene π-system, with $\Delta N = N_{\pi} = 10$

Quantity→ Method↓	*IP* [eV]	*EA* [eV]	χ [eV]	η [eV]	|*E-(para) bolic*| [kcal/mol]	*E-pi* [kcal/mol]	|*E-pi/ E-para*|
Exp/ Hückel	8.12[a]	−0.2[b]	3.96	4.16	3884.82	−8589.58[b]	2.21106
E-Hückel	12.18617	9.287281	10.73673	1.449445	804.993	−19567.4	24.30753
CNDO	11.57309	−2.27415	4.64947	6.923617	6913.459	−37529.4	5.42846
INDO	10.99398	−2.82895	4.082517	6.911462	7030.23	−36238.6	5.15468
MINDO3	8.21238	−0.47596	3.868211	4.34417	4118.425	−29913.4	7.26332
MNDO	8.574443	0.331878	4.453161	4.121283	3726.394	−30023.4	8.05696
MNDO/d	8.574308	0.331923	4.453116	4.121193	3726.3	−30023.4	8.05716
AM1	8.711272	0.265637	4.488455	4.222818	3835.367	−30004.1	7.82302
PM3	8.83573	0.407184	4.621457	4.214273	3794.829	−28,103	7.4056
ZINDO-1	7.512728	−6.39221	0.560258	6.952471	7890.081	−33558.4	4.25324
ZINDO-S	7.868645	−0.04134	3.913653	3.954993	3659.046	−24987.1	6.82887

[a] From National Institute of Standard and Technology (NIST 2011b)
[b] From the Hückel total π-energy: $2 \times (2.303 + 1.618 + 1.303 + 1.000 + 0.618) = 13.684\beta$[a.u.] ... × 627.71...~8589.58 kcal/mol, see Cotton (1971b)

Table 1.5 The same quantities as those reported in Table 1.2, here for the Fullerene π-system, with $\Delta N = N_{\pi} = 60$

Quantity→ Method↓	*IP* [eV]	*EA* [eV]	χ [eV]	η [eV]	|*E-(para) bolic*| [kcal/mol]	*E-pi* [kcal/mol]	|*E-pi/ E-para*|
Exp/ Hückel	7.58[a]	2.7[b]	5.14	2.44	94204.57	−58478.5[c]	0.62076
E-Hückel	11.43288	9.864988	10.64894	0.783948	17813.2	−96998.9	5.44534
CNDO	8.86603	−0.3482	4.258917	4.607113	185411.8	−206,145	1.11182
INDO	8.072314	−1.17385	3.449232	4.623082	187195.6	−198,445	1.06009
MINDO3	7.162502	0.530575	3.846539	3.315964	132368.6	−166,316	1.25646
MNDO	9.130902	2.562977	5.84694	3.283963	128270.9	−167,601	1.30662
MNDO/d	9.130722	2.563319	5.847021	3.283702	128,260	−167,601	1.30673
AM1	9.642135	2.948629	6.295382	3.346753	130257.6	−168,141	1.29083
PM3	9.482445	2.885731	6.184088	3.298357	128,402	−154,715	1.20493
ZINDO-1	−2.57843	−12.4464	−7.5124	4.933972	215277.5	−238,864	1.10956
ZINDO-S	1.89132	−3.96099	−1.03483	2.926153	122938.5	−126,998	1.03302

[a] From De Vries et al. (1992)
[b] From Yang et al. (1987)
[c] From Hückel total π-energy: 93.161602 β[a.u.] ... × 627.71...~ 58478.5 kcal/mol, see Haddon et al. (1986); Haymet (1986); Fowler and Woolrich (1986); Byers-Brown (1987)

the bivariate correlation between the obtained parabolic- and pi- energies are in Figs. 1.2, 1.4, 1.5, and 1.6 represented. From these results one may note the systematic increasing of the $E_{para(bolic)}$ vs. E_{pi} correlation up to its almost parallel

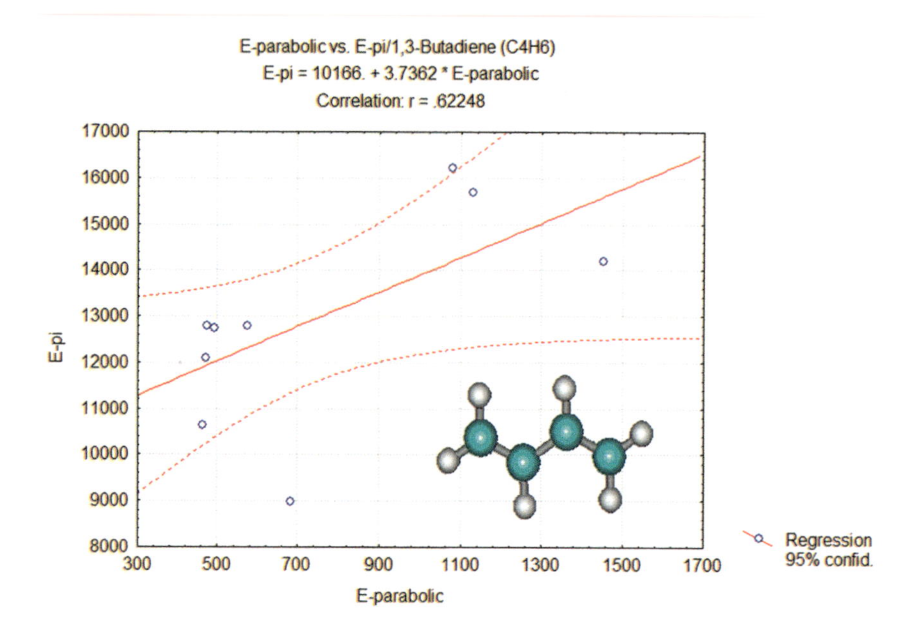

Fig. 1.2 The bivariate correlation of the parabolic- with π-energies as reported in Table 1.2 for Butadiene system

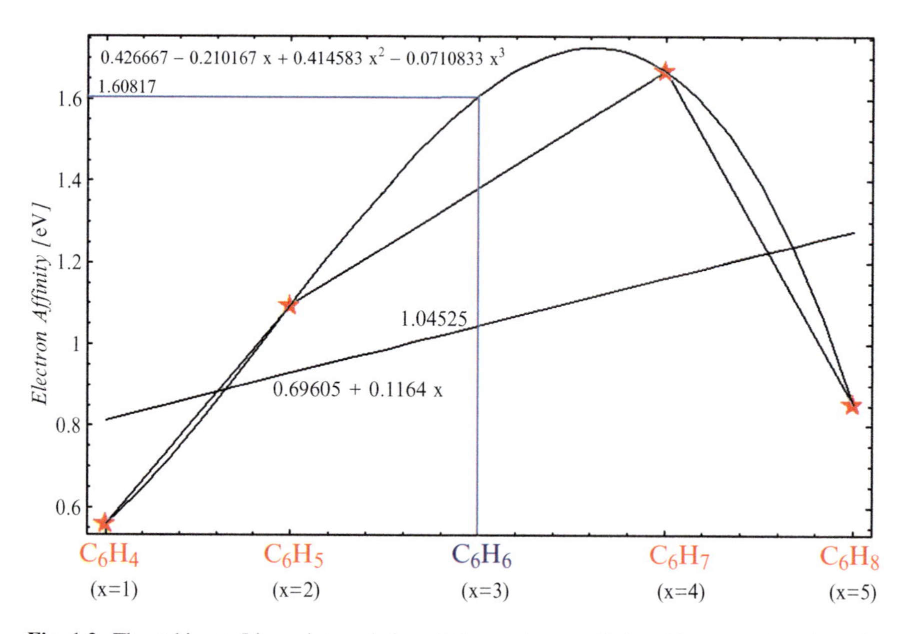

Fig. 1.3 The cubic vs. Linear interpolation of the electronic affinity of Benzene based on the data on four adiacent points for o-benzyne (C_6H_4, 0.560 eV), phenyl (C_6H_5, 1.096 eV), methylchylopentadienyl (C_6H_7, 1.67 eV), and for $(CH_2)_2C\text{-}C(CH_2)_2$ (C_6H_8, 0.855 eV), as reported in Lide (2004)

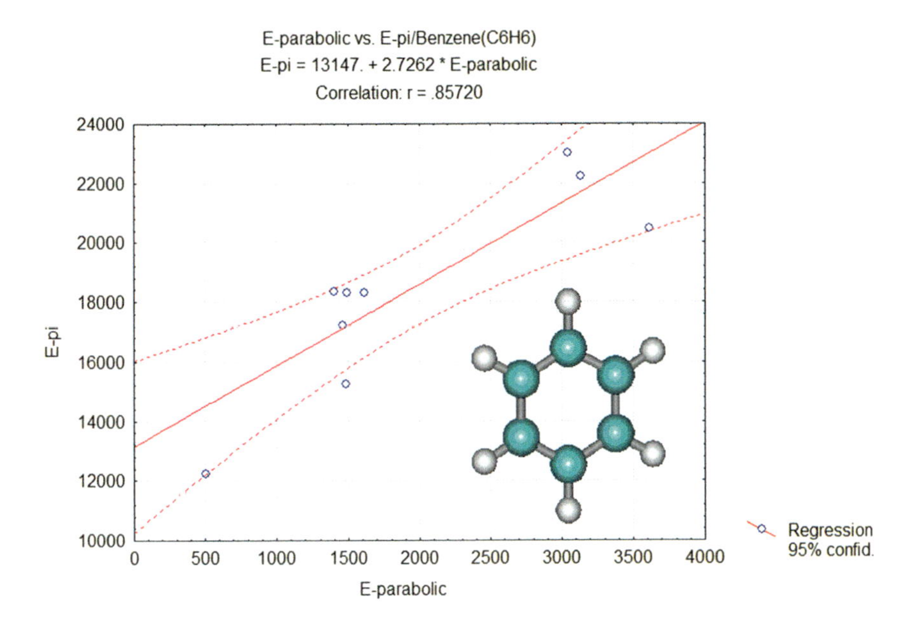

Fig. 1.4 The bivariate correlation of the parabolic- with π- energies as reported in Table 1.3 for Benzene system

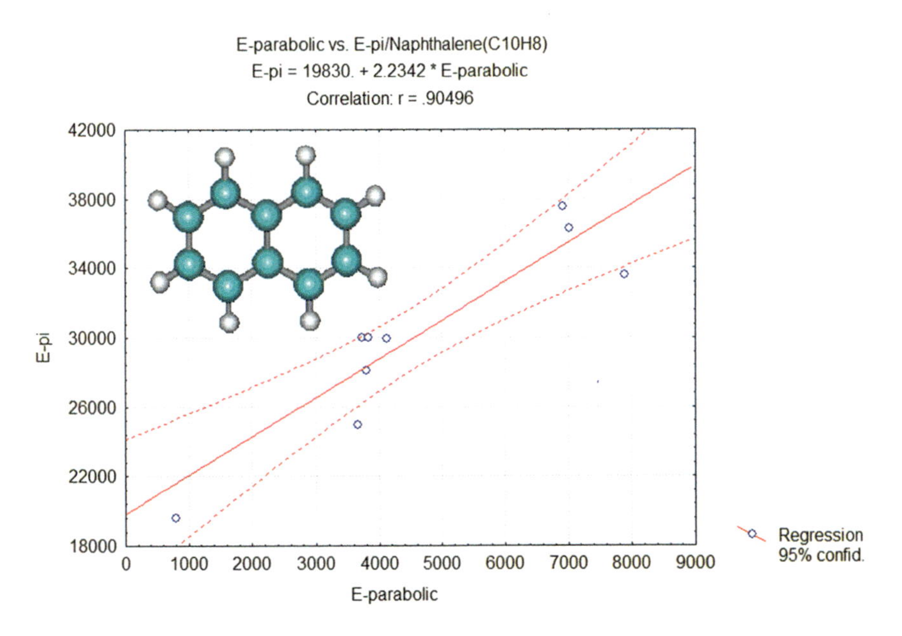

Fig. 1.5 The bivariate correlation of the parabolic- with π- energies as reported in Table 1.4 for Naphthalene system

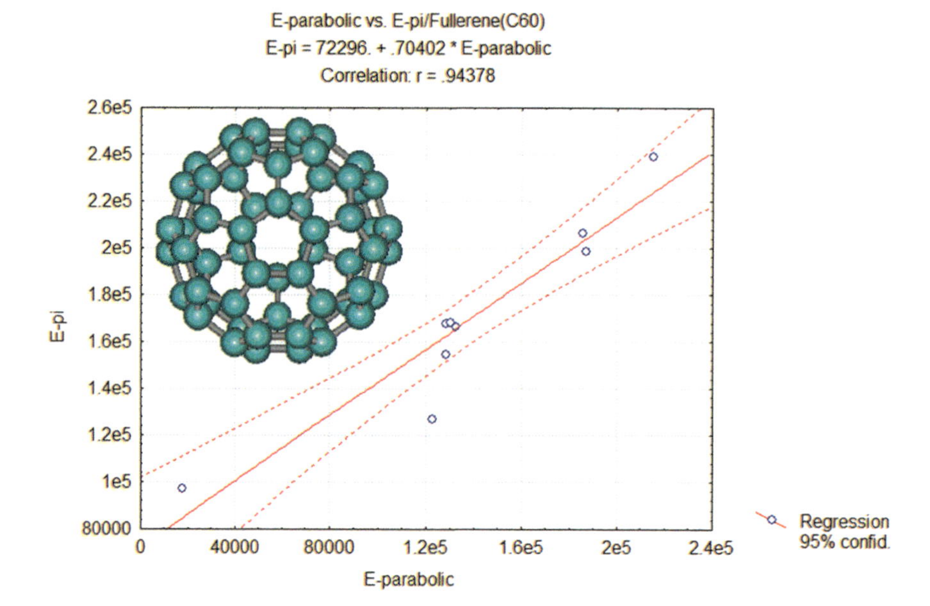

Fig. 1.6 The bivariate correlation of the parabolic- with π- energies as reported in Table 1.5 for Fullerene system

behavior as going from simple pi-systems with few frontier electrons until complex nanomolecules such as fullerene. On the other side the actual study may give an impetus in characterizing nanostructures by electronegativity and chemical hardness reactivity indicators, parabolically combined and almost fitting with the total energy of pi-electrons.

However, one should mention also the open issues remained, such as:

- The correct identification of the energy (1.101) with the pi-energy as a suitable generalization of the Hückel one (1.97);
- The physical meaning of the parabolic energy (1.98) since, through its correlation with the so called pi-energy (1.101) widely includes exchange-correlation effects, especially with its chemical hardness dependence though the explicit resonance relationship (1.100);
- The type and the complexity of the carbon system, hydrogenated or not, the quantum parabolic effect of reactivity indices in Eq. 1.98 overcomes other inner structural quantum influences to produce best correlation with atoms-in-molecules frontier energy (1.101).

Overall, for the moment we remain with the fact electronegativity and chemical hardness Zmay be worth combined to produce an energy that is better and better representing the pi-electronic systems with the increasing complexity of the system on focus; the way in which this depend on the carbon containing system, alone or in combination with heteroatoms or for the nanosystems systems without carbon, along the above enounces open issues, remain for the future research and forthcoming communications.

1.6 Conclusion

There is already long and reach scientific history the primer chemical concepts such as electronegativity help in properly modeling the structure, reactivity, and bonding of many-electronic systems in the range of tens of electron volts or hundred of kcal/mod domains, i.e. within the chemical realm or nature manifestation. However, although highly celebrated it becomes soon clear it cannot be alone standing in comprehensively characterizing the chemical reactivity space, in a generalized sense or reactivity and bonding. As such, the companion of chemical hardness was advanced, as a sort of super-potential for the energy equilibrium, since electronegativity was customarily associated with the minus of the chemical potential of the envisaged molecular system. Together, electronegativity and chemical hardness help in building both the so called chemical orthogonal space and reactivity (COSR) and in providing the consistent algorithm and hierarchy of chemical reactivity principles due to the identified double variation principle of energy density functional – that was affirmed as the non-reductive chemical variational specificity to the expected physical simple energy variation. Latter, the quantum observable character comes into discussion, and that issue was only partly solved by means of the second quantization formalism, in the case of electronegativity, while it remains undecided for the chemical hardness; fortunately, the chemical hardness observational character was possible to be settle down when the Hückel formalism was approached, in which frame it was identified with the resonance integral corrected with the number of frontier- or pi-electrons and by additional electronegativity and exchange-and correlation factors, see Eq. 1.100. Nevertheless, when parabolically combined with electronegativity the chemical hardness provides a quantum energy that accounts for the frontier electronic effects as better as the concerned system has more complex carbon structure – a behavior systematically revealed by the semi-empirical computations on paradigmatic butadiene, benzene, naphthalene and fullerene systems. Although there remain several open issues, among which checking for similar behavior for non-carbon nanosystems, as well as in depth exploring the physical meaning of the actually proposed parabolic form of the pi-related energy only on the base of electronegativity and chemical hardness, there seems that these reactivity indices have still great potential in modeling the next era of nanosystems in simple and powerful manner, providing their quantum observational character will be directly or implicitly clarified. Overall, the "beauty" of electronegativity and chemical hardness versatility makes us optimistic in this respect; or, in Van Gogh wise words, while assuming they are like "stars" on the "chemical conceptual sky", "*for my part I know nothing with any certainty. . . but the sight of* [such, n.a.] *stars makes me dream*"!

Acknowledgements Author thanks Prof. Mircea Diudea from Babes-Bolyai University of Cluj-Napoca for courtesy in providing the Hyper file for the Fullerene structure and to Romanian Ministry of Education and Research for supporting the present work through the CNCS-UEFISCDI (former CNCSIS-UEFISCSU) project <Quantification of The Chemical Bond Within Orthogonal Spaces of Reactivity. Applications on Molecules of Bio-, Eco- and Pharmaco- Logical Interest>, Code PN II-RU-TE-2009-1 grant no. TE-16/2010-2011.

References

Anderson WP, Edwards WD, Zerner MC (1986) Inorg Chem 25:2728–2732
Anderson WP, Cundari TR, Zerner MC (1991) Int J Quantum Chem 39:31–45
Bacon AD, Zerner MC (1979) Theor Chim Acta 53:21–54
Baird NC, Dewar MJS (1969) J Chem Phys 50:1262–1275
Berkowitz M (1987) J Am Chem Soc 109:4823–4825
Bingham RC, Dewar MJS, Lo DH (1975a) J Am Chem Soc 97:1285–1293
Bingham RC, Dewar MJS, Lo DH (1975b) J Am Chem Soc 97:1294–1301
Bingham RC, Dewar MJS, Lo DH (1975c) J Am Chem Soc 97:1302–1306
Bingham RC, Dewar MJS, Lo DH (1975d) J Am Chem Soc 97:1307–1311
Byers-Brown W (1987) Chem Phys Lett 136:128–133
Chattaraj PK, Maiti B (2003) J Am Chem Soc 125:2705–2710
Chattaraj PK, Schleyer PvR (1994) J Am Chem Soc 116:1067–1071
Chattaraj PK, Lee H, Parr RG (1991) J Am Chem Soc 113:1854–1855
Chattaraj PK, Liu GH, Parr RG (1995) Chem Phys Lett 237:171–176
Clark TA (1985) Handbook of computational chemistry. Wiley, New York
Cory MG, Stavrev KK, Zerner MC (1997) Int J Quantum Chem 63:781–795
Cotton FA (1971a) Chemical application of group theory. Wiley, New York, p 137
Cotton FA (1971b) Chemical application of group theory. Wiley, New York, p 163
Davis LP, Guidry RM, Williams JR, Dewar MJS, Rzepa HS (1981) J Comput Chem 2:433–445
De Vries J, Steger H, Kamke B, Menzel C, Weisser B, Kamke W, Hertel IV (1992) Chem Phys Lett 188:159–162
Del Bene J, Jaffé HH (1968a) J Chem Phys 48:1807–1814
Del Bene J, Jaffé HH (1968b) J Chem Phys 48:4050–4056
Del Bene J, Jaffé HH (1968c) J Chem Phys 49:1221–1229
Dewar MJS, Dieter KM (1986) J Am Chem Soc 108:8075–8086
Dewar MJS, Hasselbach E (1970) J Am Chem Soc 92:590–598
Dewar MJS, Lo DH (1972) J Am Chem Soc 94:5296–5303
Dewar MJS, McKee ML (1977) J Am Chem Soc 99:5231–5241
Dewar MJS, Rzepa HS (1978) J Am Chem Soc 100:58–67
Dewar MJS, Storch DM (1985) J Am Chem Soc 107:3898–3902
Dewar MJS, Thiel W (1977) J Am Chem Soc 99:4899–4907
Dewar MJS, Lo DH, Ramsden CA (1975) J Am Chem Soc 97:1311–1318
Dewar MJS, Zoebisch EG, Healy EF, Stewart JJP (1985) J Am Chem Soc 107:3902–3909
Dreizler RM, Gross EKU (1990) Density functional theory. Springer Verlag, Heidelberg
Fowler PW, Woolrich J (1986) Chem Phys Lett 127:78–83
Haddon RC, Brus LE, Raghavachari K (1986) Chem Phys Lett 125:459–464
Hartree DR (1957) The calculation of atomic structures. Wiley, New York
Haymet AD (1986) J Am Chem Soc 108:319–321
Hoffmann R (1963) J Chem Phys 39:1397–1412
Hückel E (1931a) Z Physik 71:204–286
Hückel E (1931b) Z Physik 72:310–337

Klopman G (1964) J Am Chem Soc 86:4550–4557
Kohn W, Becke AD, Parr RG (1996) J Phys Chem 100:12974–12980
Koopmans T (1934) Physica 1:104–113
Lide DR (2004) CRC handbook of chemistry and physics, 85th edn. CRC Press, Boca Raton, Section 10–147
Löwdin PO (1950) J Chem Phys 18:365–376
Löwdin PO (1993) Int J Quantum Chem 48:225–232
March NH (1991) Electron density theory of many-electron systems. Academic, New York, 1991
Mortier WJ, Genechten Kv, Gasteiger J (1985) J Am Chem Soc 107:829–835
Murrell JN, Harget AJ (1971) Semi-empirical self-consistent-field molecular orbital theory of molecules. Wiley Interscience, New York
NIST (2011a) http://webbook.nist.gov/cgi/cbook.cgi?ID=C71432&Mask=20#Ion-Energetics
NIST (2011b) http://webbook.nist.gov/cgi/cbook.cgi?ID=C91203&Mask=20#Ion-Energetics
Ohno K (1964) Theor Chim Acta 2:219–227
Oleari L, DiSipio L, De Michelis G (1966) Mol Phys 10:97–109
Parr RG (1983) Annu Rev Phys Chem 34:631–656
Parr RG, Pearson RG (1983) J Am Chem Soc 105:7512–7516
Parr RG, Yang W (1984) J Am Chem Soc 106:4049–4050
Parr RG, Yang W (1989) Density functional theory of atoms and molecules. Oxford University Press, New York
Parr RG, Donnelly RA, Levy M, Palke WE (1978) J Chem Phys 68:3801–3808
Pearson RG (1973) Hard and soft acids and bases. Dowden, Hutchinson & Ross, Stroudsburg
Pearson RG (1985) J Am Chem Soc 107:6801–6806
Pearson RG (1990) Coord Chem Rev 100:403–425
Pearson RG (1997) Chemical hardness. Wiley-VCH, Weinheim
Pople JA, Beveridge DV (1970) Approximate molecular orbital theory. McGraw-Hill, New York
Pople JA, Nesbet RK (1954) J Chem Phys 22:571–572
Pople JA, Segal GA (1965) J Chem Phys 43:S136–S151
Pople JA, Segal GA (1966) J Chem Phys 44:3289–3297
Pople JA, Santry DP, Segal GA (1965) J Chem Phys 43:S129–S135
Pople JA, Beveridge DL, Dobosh PA (1967) J Chem Phys 47:2026–2034
Putz MV (2003) Contributions within density functional theory with applications to chemical reactivity theory and electronegativity. Dissertation.com, Parkland
Putz MV (2008a) MATCH Commun Math Comput Chem 60:845–868
Putz MV (2008b) Absolute and chemical electronegativity and hardness. Nova Science, New York
Putz MV (2009a) J Mol Struct (THEOCHEM) 900:64–70
Putz MV (2009b) Int J Quantum Chem 109:733–738
Putz MV (2009c) Int J Chem Model 1:141–147
Putz MV (2010a) Studia Univ Babes Bolyai Chemia 55:47–50
Putz MV (2010b) Int J Mol Sci 11:1269–1310
Putz MV (2011a) Curr Phys Chem 1:111–139
Putz MV (2011b) Int J Chem Model 3:15–22
Putz MV (2011c) MATCH Commun Math Comput Chem 66:35–63
Putz MV (2011d) In: Putz MV (ed) Quantum frontiers of atoms and molecules. NOVA Science, New York, pp 251–270
Putz MV, Chiriac A (2008) In: Putz MV (ed) Advances in quantum chemical bonding structures. Transworld Research Network, Kerala, pp 1–43
Putz MV, Russo N, Sicilia E (2004) J Comput Chem 25:994–1003
Ridley JE, Zerner MC (1976) Theor Chim Acta 42:223–236
Roothaan CCJ (1951) Rev Mod Phys 23:69–89
Sanderson RT (1988) J Chem Educ 65:112–119
Sen KD, Jørgensen CK (eds) (1987) Electronegativity, vol 66, Structure and Bonding. Springer Verlag, Berlin

Sen KD, Mingos DMP (eds) (1993) Chemical hardness, vol 80, Structure and bonding. Springer Verlag, Berlin
Senet P (1996) J Chem Phys 105:6471–6490
Slater JC (1929) Phys Rev 34:1293–1322
Slater JC (1960) Quantum theory of atomic structure. McGraw-Hill, New York
Slater JC (1963) Theory of molecules and solids, vol 1, Electronic structure of molecules. McGraw-Hill, New York
Stavrev KK, Zerner MC (1995) J Chem Phys 102:34–39
Stavrev KK, Zerner MC, Meyer TJ (1995) J Am Chem Soc 117:8684–8685
Stewart JJP (1989a) J Comput Chem 10:209–220
Stewart JJP (1989b) J Comput Chem 10:221–264
Stewart JJP (1990) J Comp Aided Mol Des 4:1–103
Surján PR (1989) Second quantized approach to quantum chemistry. Springer, Berlin
Tachibana A (1987) Int J Quantum Chem 21:181–190
Tachibana A, Parr RG (1992) Int J Quantum Chem 41:527–555
Tachibana A, Nakamura K, Sakata K, Morisaki T (1999) Int J Quantum Chem 74:669–679
Thiel W (1988) Tetrahedron 44:7393–7408
University of Illinois (2011) http://butane.chem.illinois.edu/jsmoore/Experimental/piMOs/PiMOAnalysis.html
Yang W, Parr RG (1985) Proc Natl Acad Sci USA 82:6723–6726
Yang SH, Pettiette CL, Conceicao J, Cheshnovsky O, Smalley RE (1987) Chem Phys Lett 139:233–238

Chapter 2
Stiff Polymers at Ultralow Temperatures

Hagen Kleinert[1]

Abstract Polymers can be cooled down to ultralow temperatures by embedding them in a gas whose temperature is lowered by laser and evaporation cooling. In this regime their fluctuations show quantum behavior. This is calculated here for the second and fourth moments of the end-to-end distribution in the large stiffness expansion. The result should soon be measurable.

2.1 Introduction

Present-day laser techniques make it possible to build optical traps, and lattices of traps, in which one can host a variety of atoms or molecules and study their behavior at very low temperatures. Gases of bosons, fermions, and their simple bound states have been investigated in this way with interesting insights into the quantum physics of many-body systems – see Pitaevskii and Stringari (2002) and the foregoing notes and references. In this note we would like to propose to use these traps for the study of the quantum behavior of stiff polymers. The low temperatures can be reached by buffer-gas cooling with He which permits reaching temperatures of the order of mK. This should in be possible, for example, with carbohydrates or polyacetylene. We shall assume the traps to be much wider than the length of the polymer so that we may ignore the distortions coming from the trap potential. The end-to-end distribution $P_L(\mathbf{R})$ of a polymer of length L contains information on various experimentally observable properties, in particular moments

$$\langle R^m \rangle = S_D \int_0^\infty dR R^{D-1} R^m P_L \tag{2.1}$$

[1] Institut für Theoretische Physik, Freie Universität Berlin, Arnimallee 14,
D-14195 Berlin, Germany
e-mail: h.k@fu-berlin.de

M.V. Putz (ed.), *Carbon Bonding and Structures: Advances in Physics and Chemistry*, Carbon Materials: Chemistry and Physics 5, DOI 10.1007/978-94-007-1733-6_2,

where $S_D = 2\pi^{D/2}/\,\Gamma(D/2)$is the surface of a unit sphere in D dimensions. The classical temperature behavior of these moments is well known (Yamakawa 1997; Kleinert 2006). Here we shall calculate the modifications caused by quantum fluctuations.

Let us briefly recall the calculation of the classical end-to-end distribution in the Kratky-Porod chain with N links of length a in D dimensions (Yamakawa 1997; Kleinert 2006). Its bending energy is

$$E^N_{bend} = \frac{\kappa a}{2} \sum_{n=1}^{N-1} (\nabla \mathbf{u}_n)^2 \tag{2.2}$$

where κ the stiffness, $\mathbf{u}_n$ are unit vectors on a sphere in D dimensions specifying the directions of the polymer links, and $\nabla \mathbf{u}_n \equiv (\mathbf{u}_{n+1} - \mathbf{u}_n)/a$ is the difference between neighboring $\mathbf{u}_n$'s. The initial and final link directions have a distribution

$$P(\mathbf{u_2}, \mathbf{u_1}|L) \equiv (\mathbf{u_b} L|u_\mathbf{b} 0) = \int D^d \mathbf{u} e^{-\beta E^N_{bend}} \tag{2.3}$$

where $D^D\mathbf{u}$ is the product of integrals over the unit spheres of $\mathbf{u}_n$ ($n = 2,\ldots,$ $N-1$), and $\beta \equiv 1/k_B T$ (T = temperature, k_B = Boltzmann constant). The normalization is irrelevant and will be fixed at the end.

2.2 Classical Treatment of Stiff Polymer

If L denotes the length of the polymer, the bending energy reads $E^L_{bend} = \kappa \int_0^L ds(\partial_S \mathbf{u})^2/2$. Then the probability (2.3) coincides with the Euclidean path of integral of a particle on the surface of a unit sphere. The end-to-end distance in space is $\mathbf{R} = \int_0^L ds\mathbf{u}(s)$, and its distribution is given by the path integral

$$P_L(\mathbf{R}) \propto \int D^D \mathbf{u} \delta^{(D)}(\mathbf{R} - L\mathbf{u_o}) e^{-\bar{\kappa} \int_0^L ds \mathbf{u}'^2(s)/2} \tag{2.4}$$

where $\bar{\kappa} \equiv \beta\kappa$ and $\mathbf{u}_0 \equiv L^{-1} \int_0^L ds\mathbf{u}(s)$. Introducing the dimensionless vectors $\mathbf{q}^T$ transverse to $\mathbf{R}$, we parametrize $\mathbf{u}$ as $(\mathbf{q}, \sqrt{1-\mathbf{q}^2})$ and see that the δ – function enforces

$$\int_0^L ds\mathbf{q}(s) = 0, R = L - \int_0^L ds\left[\mathbf{q}^2(s)/2 + \ldots\right] \tag{2.5}$$

At large stiffness, the distribution can be calculated from the one-loop approximation to the path integral which leads to the Fourier integral (Kleinert 2006;

Wilhelm and Frey 1996; Dhar and Chaudhuri 2002; Stepanow and Schuetz 2002; Samuel and Sinha 2002)

$$P_{L,\beta}(\mathbf{R}) \underset{small\ \beta}{\propto} \int_{-i\infty}^{i\infty} \frac{dk^2}{2\pi i} e^{\beta\kappa k^2(L-R)} F_{L,0}\left(k^2L^2\right) \tag{2.6}$$

where $F_{L,0}(k^2L^2)$is the partition function

$$F_{L,0}\left(k^2L^2\right) \equiv \int_{NBC} D'^{D-1}\mathbf{q}^T e^{-(\beta\kappa a/2)\sum_{n=1}^{N}[(\nabla q)_n^{T2}+k^2q_n^{T2}]}$$

$$\propto \left[\frac{\prod_{n=1}^{\infty}|K_n|^2}{\prod_{n=1}^{\infty}\left(|K_n|^2+k^2\right)}\right]^{\frac{D-1}{2}} = \left(\frac{N\sinh\tilde{k}a}{\sinh\tilde{k}L}\right)^{\frac{D-1}{2}} \tag{2.7}$$

with $\tilde{k}$ defined by $\sinh\tilde{k}a = ka$ – see Sect. 2.3.2 in Kleinert (2006). The symbol NBC indicates that the open ends of the path integral may be accounted for by Neumann boundary conditions – see Sects. 2.12, 3.4, and 15.9.4 in Kleinert (2006).

For a classical polymer, we may use the model in the continuum limit where $a \to 0$. Then $\mathbf{u}_n$ is replaced by the tangent vector $\mathbf{u}(s) = \partial_s\mathbf{x}(s)$ of the space curve $\mathbf{x}(s)$ of the polymer, where s is the distance of the link from one of the endpoints measured along the polymer. In this limit, $\tilde{k}L$ coincides with $kL = \bar{k}$, and the right-hand side of (2.7) can be expanded as in a power series of $\bar{k}$:

$$F_{L,0}\left(\bar{k}^2\right) = 1 - \frac{D-1}{2^2\cdot 3}\bar{k}^2 + \frac{(D-1)(5D-1)}{2^5\cdot 3^2\cdot 5}\bar{k}^4 + \ldots \tag{2.8}$$

Inserting this into (2.6) and setting $r \equiv R/L$, we may calculate the unnormalized moments $\langle r^m\rangle = \int dr r^{D-1+m} P_{L,\beta}$ from the integrals

$$\langle r^m\rangle = \int dz(1+z)^{D-1+m} f\left(\hat{k}^2\right)\delta(z) \tag{2.9}$$

where $\hat{k}^2L^2$ is the differential operator $-(L/\beta\kappa)\partial_z = [-2l/(D-1)]\partial_z$, and $l \equiv (D-1)L/\beta\kappa$ is the *flexibility* of the polymer. From this we find

$$\langle r^0\rangle = \tilde{N}\left[1 - \frac{D-1}{6}l + \frac{(5D-1)(D-2)}{360}l^2 + \ldots\right],$$

$$\langle r^2\rangle = \tilde{N}\left[1 - \frac{D+1}{6}l + \frac{(5D-1)D(D+1)}{360(D-1)}l^2 + \ldots\right],$$

$$\langle r^4\rangle = \tilde{N}\left[1 - \frac{D+3}{6}l + \frac{(5D-1)(D+2)(D+3)}{360(D-1)}l^2 + \ldots\right] \tag{2.10}$$

where $\tilde{N}$ is some constant. Dividing these by $\langle r^0 \rangle$, we arrive at the normalized moments – see Sect. 15.9.4 in Kleinert (2006)

$$\langle r^2 \rangle = 1 - \frac{1}{3}l + \frac{13D-9}{180(D-1)}l^2 + \dots \tag{2.11}$$

$$\langle r^4 \rangle = 1 - \frac{2}{3}l + \frac{23D-11}{90(D-1)}l^2 + \dots \tag{2.12}$$

2.3 Quantum Fluctuations

Quantum effects are now taken into account by adding for each mass point of the polymer at $\mathbf{x}_n$ a kinetic action

$$A_{kin} \equiv \frac{M}{2}\int_0^{\hbar\beta} dt[\dot{\mathbf{x}}_n(t)]^2 \tag{2.13}$$

where M is the mass. Since $\mathbf{u}_n(t) = \nabla \mathbf{x}_n(t)$ the Euclidean action with time $\tau = it$ reads

$$A = \frac{\kappa a}{2}\int_0^{\hbar\beta} d\tau \sum_{n=1}^{N}\left[g^{-2}\left(\partial_\tau \nabla^{-1}\mathbf{u}_n\right)^2 + \left(\nabla \mathbf{u}_n\right)^2\right] \tag{2.14}$$

where $g \equiv \sqrt{\kappa a/M}$, and $F_{L,0}\left(\bar{k}^2\right)$ is replaced by $F_{L,\beta}\left(\bar{k}^2\right) = e^{-(D-1)\Gamma_{L,\beta}\left(\bar{k}^2\right)}$ with

$$\Gamma_{L,\beta}\left(\bar{k}^2\right) = \frac{1}{2}\mathrm{Tr}\log\left[-g^{-2}\partial_\tau^2\left(\nabla\overline{\nabla}\right)^{-1} - \nabla\overline{\nabla} + k^2\right] \tag{2.15}$$

The eigenvalues of $i\partial_\tau$ are the Matsubara frequencies $\omega_m = 2\pi m/\hbar\beta$, $(m = 0, \pm 1, \pm 2, \dots)$ leading to the finite-temperature generalization of (2.7):

$$F_{L,0}(\bar{k})^2 = \left[\frac{\prod_{m,n}\left(g^{-2}\omega_m^2 + |K_n|^4\right)}{\prod_{m,n}\left(g^{-2}\omega_m^2 + |K_n|^4 + k^2|K_n|^2\right)}\right]^{(D-1)/2} \tag{2.16}$$

Performing the product over the m's, we arrive at

$$F_{L,\beta}(k) = \prod_{n=1}^{\infty}\left[\frac{\sinh K_n^2 \hbar g\beta/2}{\sinh\sqrt{K_n^4 + k^2|K_n|^2}\hbar g\beta/2}\right]^{D-1} \tag{2.17}$$

2.4 Stiffness Expansion

In the product (2.17) we perform an expansion in powers of $\bar{k} \equiv kL$, and find

$$F_{L,\beta}(\bar{k}) = \exp\left[(D-1)\left(f_1\bar{k}^2 + f_2\bar{k}^4 + \dots\right)\right] \tag{2.18}$$

where

$$f_1(b) = -\frac{b}{4\pi^2}\sum_{n=1}^{\infty}\coth\frac{n^2 b}{2} \tag{2.19}$$

$$f_2(b) = \frac{b^2}{32\pi^4}\sum_{n=1}^{\infty}\left[\frac{2}{bn^2}\coth\frac{n^2 b}{2} + \left(\coth^2\frac{n^2 b}{2} - 1\right)\right] \tag{2.20}$$

The parameter b is the reduced inverse temperature $b \equiv \pi^2\hbar g/k_B TL^2$.

As a cross check of the above results we go to the high-temperature limit where $\coth(n^2b/2) \to 2/n^2b$ and thus $f_1(b) \to -1/12$, $f_2(b) \to 1/360$. Inserting these into (2.18), we recover (2.8).

Quantum behavior sets in if b becomes larger than unity. To estimate when this happens we measure the lengths a, L in Å, the mass M in units of proton mass, the temperature T in mK, and the constant g in units of Å^2/s, we find that $b \approx 7.380 \times 10^6\sqrt{\kappa a/A}/TL^2$, where A is the atomic number M. In these natural units, κ, a, T, are of order unity, experimentalists should be able to observe the quantum behavior for not too long chains.

At very low temperatures where quantum effects become most visible we find the asymptotic behavior

$$f_2(b) \to \frac{b}{16\pi^4}\sum_{n=1}^{\infty}\frac{1}{n^2} = \frac{b}{96\pi^2} \tag{2.21}$$

In this regime, the sum in $f_1(b)$ diverges linearly. It is made finite by remembering that we are dealing with the continuum limit of a discrete polymer with $N = L/a$ links. Hence we must carry the sum only to $n = N$, and obtain

$$f_1(b) \to -\frac{b}{4\pi^2}\sum_{n=1}^{N} 1 = -\frac{b}{4\pi^2}N \tag{2.22}$$

Setting $r - 1 \equiv z$, we replace $\left(\bar{k}^2\right)^n$ in Eq. 2.18 by $[-2l/(D-1)]^n\partial_z^n\delta(z)$, and insert the resulting expansion into the integral $\int dz(1+z)^{D-1+m}$ to find the unnormalized moments of r^0, r^2, r^4 at zero temperature. From their ratios we obtain the normalized moments:

$$\langle r^2 \rangle = 1 + 4lf_1 + l^2\left(4f_1^2 + 8\frac{2D-1}{D-1}f_2\right) + \dots \tag{2.23}$$

$$\langle r^4 \rangle = 1 + 8lf_1 + l^2\left(24f_1^2 + 16\frac{2D+1}{D-1}f_2\right) + \dots \tag{2.24}$$

From these we find

$$\langle 1 - r \rangle = -2lf_1 - 8l^2 f_2 + \dots \tag{2.25}$$

$$\left\langle (1-r)^2 \right\rangle = l^2\left(4f_1^2 + \frac{8}{D-1}f_2\right) + \dots \tag{2.26}$$

and the cumulant

$$\left\langle (1-r)^2 \right\rangle_c = l^2\frac{8}{D-1}f_2 - 32l^3 f_1 f_2 - 64l^4 f_2^2 + \dots \tag{2.27}$$

Hence we find in the zero-temperature limits

$$\langle r^2 \rangle \approx 1 - \frac{bl}{\pi^2}N, \quad \langle r^4 \rangle = 1 - 2\frac{bl}{\pi^2}N \tag{2.28}$$

where $bl \equiv (D-1)\hbar c/\kappa$. For large c, the polymer at zero temperature may appear considerably shorter than expected from the linear extrapolation of the high-temperature behavior to zero temperature.

The quantum effect can be studied most easily by measuring for a polymer of high stiffness κ the peak value of 1 − r which behaves like

$$\langle 1 - r \rangle \approx -2lf_1(b) = -2(D-1)\frac{\hbar c}{\kappa}\frac{1}{b}f_1(b) \tag{2.29}$$

One may plot the function $C(b) \equiv 6\kappa/(D-1)\hbar c\langle 1 - r \rangle$, for which our result implies the behavior shown in Fig. 2.1 for various link numbers N.

We challenge experimentalists to detect this behavior.

2.5 Further Quantum Effect

Further quantum effect can be observed if the links of the polymer contain a spin $S = 1/2, 1, 3/2, 2, \dots$ along the link direction. This can be taken into account by adding the kinetic action (2.30) a Berry phase. For each link $[\mathbf{u}_n(\tau)$, it corresponds to the interaction of the particle on the surface of a unit sphere in $\mathbf{u}$ space with a

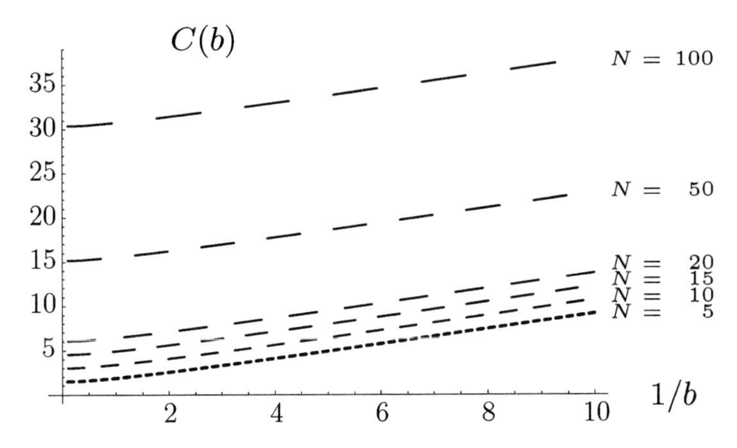

Fig. 2.1 Temperature behavior of $C(b) \equiv 6\kappa/(D-1)\hbar c\langle 1-r\rangle$ for various link numbers N. The classical limit of these curves are their straight-line asymptotes starting out at the origin with slope $(6/\pi^2)\sum_{n=1}^{N} n^{-2}$

magnetic monopole of quantized charge q lying at the center of the sphere – see Sect. 8.13 in Kleinert (2006):

$$A_0 = \hbar S \sum_{n=1}^{N-1} \int_0^{\hbar\beta} d\tau \frac{\mathbf{n} \times \mathbf{u}_n(\tau)}{1 - \mathbf{n} \times \mathbf{u}_n(\tau)} \cdot \dot{\mathbf{u}}_n(\tau) \tag{2.30}$$

The irrelevant Dirac string is chosen to export the magnetic flux of strength S along the $\mathbf{n}$ direction to infinity. This action creates a radial magnetic field $B = -S\mathbf{u}_n$of the surface of the sphere. If we assume $\mathbf{R}$ to run along the positive z direction, the small transverse fluctuation $\mathbf{q}^T$in (2.7) will take place near the north pole of the sphere and receive a an additional magnetic interaction $\hbar S \sum_{u=1}^{N-1} \int_0^{\hbar\beta} d\tau\, \mathbf{q}_n^T \times \dot{\mathbf{q}}_n^T/2a$. This will change each factor in the product (2.17) to a product of two square roots – see Eq. 2.679 in Kleinert (2006)

$$\left[\frac{\sinh K_n K_n^+(0)\hbar c\beta/2}{\sinh K_n K_n^+(k)\hbar c\beta/2}\right]^{\frac{1}{2}} \left[\frac{\sinh K_n K_n^-(0)\hbar c\beta/2}{\sinh K_n K_n^-(k)\hbar c\beta/2}\right]^{\frac{1}{2}} \tag{2.31}$$

where

$$K_n^{\pm}(k) \equiv \sqrt{K_n^2 + k^2 + k_S^2} \pm k_S, \quad k_S \equiv \frac{\hbar c S}{2\kappa a} \tag{2.32}$$

the stretched polymer. For arbitrary temperatures, this changes Eqs. 2.19 and 2.20 to

$$f_1(b) = -\frac{b}{8\pi^2}\sum_{n=1}^{\infty}\frac{n}{n_S}\left(\coth\frac{nn_S^+ b}{2} + \coth\frac{nn_S^- b}{2}\right) \tag{2.33}$$

$$f_2(b) = \frac{b^2}{64\pi^4} \sum_{n=1}^{\infty} \left[\begin{array}{l} \frac{2n}{bn_S^3} \left(\coth \frac{nn_S^+ b}{2} + \coth \frac{nn_S^- b}{2} \right) \\ + \frac{n^2}{n_S^2} \left(\coth^2 \frac{nn_S^+ b}{2} + \coth^2 \frac{nn_S^- b}{2} - 2 \right) \end{array} \right] \tag{2.34}$$

where $n_S^{\pm} = n_S \pm \kappa_S, n_S \equiv \sqrt{n^2 + \kappa_S^2}$, and $\kappa_S \equiv \hbar cSL/2\pi\kappa a = k_S L/\pi$. At high temperatures, these become

$$f_1^S(b) \to -\frac{1}{4\pi^2} \sum_{n=1}^{\infty} \frac{1}{n_S} \left(\frac{1}{n_S^-} + \frac{1}{n_S^+} \right) = -\frac{1}{12}, \tag{2.35}$$

$$f_2^S(b) \to \frac{1}{16\pi^4} \sum_{n=1}^{\infty} \left[\frac{1}{n_S^3} \left(\frac{1}{n_S^-} + \frac{1}{n_S^+} \right) + \frac{1}{n_S^2} \left(\frac{1}{(n_S^-)^2} + \frac{1}{(n_S^+)^2} \right) \right] = \frac{1}{360} \tag{2.36}$$

The classical limit is independent of κ_S, as could have been anticipated. At low temperatures, we obtain for small κ_S to lowest order

$$f_1^S(b) \to -\frac{b}{4\pi^2} \sum_{n=1}^{\infty} \frac{n}{n_S} = -\frac{b}{4\pi^2} \left(N - \frac{\pi^2 \kappa_S^2}{12} \right) \tag{2.37}$$

$$f_2^S(b) \to -\frac{b}{16\pi^4} \sum_{n=1}^{\infty} \frac{n}{n_S^3} = \frac{b}{96\pi^2} \left(1 - \frac{\pi^2 \kappa_S^2}{10} \right) \tag{2.38}$$

Thus $f_1(b)$depends only very weakly on κ_S so that the curves in Fig. 2.1 are practically unchanged by an extra spin S along the links. The spin dependence becomes visible only in measurements of $f_2(b)$ which can be extracted from suitable combinations of the moments $\langle 1 - r \rangle$ and $\langle (1 - r^2) \rangle_c$obtained by solving Eqs. 2.25 and 2.27.

2.6 Summary

Our discussion has shown that at ultralow temperatures quantum fluctuations cause observable effects in polymers. We have calculated these effects for the lowest moments $\langle r^2 \rangle$ and $\langle r^4 \rangle$ of the end-to-end distribution for ordinary polymers as well as for polymers in which each link carries a spin S. In the latter case the polymers are flexible one-dimensional quantum Heisenberg ferromagnets. With the presently available tramps and cooling techniques, experimentalists should be able to detect these effects.

Acknowledgement The author is grateful to I. Bloch, W. Ketterle, G. Meijer, F. Noguiera, and T. Pfau for valuable comments.

References

Dhar A, Chaudhuri D (2002) Phys Rev Lett 89:065502

Kleinert H (2006) Path integrals in quantum mechanics, statistics, polymer physics, and financial markets, 4th edn. World Scientific, Singapore, pp. 1–15470 (www.physik.fu-berlin.de/~kleinert/b5)

Pitaevskii L, Stringari S (2002) Bose-Einstein condensation. Oxford University Press, Oxford

Samuel J, Sinha S (2002) Phys Rev E 66:050801(R)

Stepanow S, Schuetz M (2002) Europhys Lett 60:546–551

Wilhelm J, Frey E (1996) Phys Rev Lett 77:2581–2584

Yamakawa H (1997) Helical wormlike chains in polymer solution. Springer Verlag, Berlin

Chapter 3
On Topological Modeling of 5|7 Structural Defects Drifting in Graphene

Ottorino Ori[1], **Franco Cataldo**[1], **and Ante Graovac**[2,3,4]

Abstract Rearrangements of the graphene layers under iterated Stone-Wales rotations are studied here from a pure topological point of view. The analysis provides indications about the relative chemical stability of different isomeric distributions of pentagon–heptagon pairs, the 5|7 structural defects, diffusing in the graphene layer. Our computations are performed in the dual topological space, the Wiener index being taken as the topological potential of the system that rules the migration of the defects in the graphenic lattice.

3.1 Introduction

This study investigates some topological mechanisms that govern the *diffusion* (as well the *annihilation*) of pentagon–heptagon pairs, the so-called 5|7 defects, in the graphene lattice. *Only isomeric* modifications of the lattice are considered in our simulations that only adopts Stone-Wales (SW) rotations to generate and propagate the 5|7 pairs in the graphenic layers. In particular, the rotation $SW_{6/6}$ transforms four adjacent hexagons in two 5|7 pairs (Fig. 3.1a) whereas the transformation of $SW_{6/7}$ type is able to split these two pairs rotating (Fig. 3.1b) the bond between the heptagon and the nearby hexagon with the overall structural result of swapping one of the 5|7 pair with a nearby 6|6 couple. Evidently, iterated sequences of $SW_{6/7}$ rotations are capable to diffuse the 5|7 pairs in the lattice, describing various wave-like

[1] Actinium Chemical Research, Via Casilina 1626/A, 00133 Rome, Italy
e-mail: ottorino.ori@alice.it; franco.cataldo@fastwebnet.it

[2] Department of Chemistry, Faculty of Science, University of Split, Nikole Tesle 12, HR-21000 Split, Croatia

[3] NMR Center, The "Ruđer Bošković" Institute, HR-10002 Zagreb, Croatia
e-mail: Ante.Graovac@irb.hr

[4] IMC, University of Dubrovnik, Branitelja Dubrovnika 29, HR-20000 Dubrovnik, Croatia

M.V. Putz (ed.), *Carbon Bonding and Structures: Advances in Physics and Chemistry*, Carbon Materials: Chemistry and Physics 5, DOI 10.1007/978-94-007-1733-6_3,

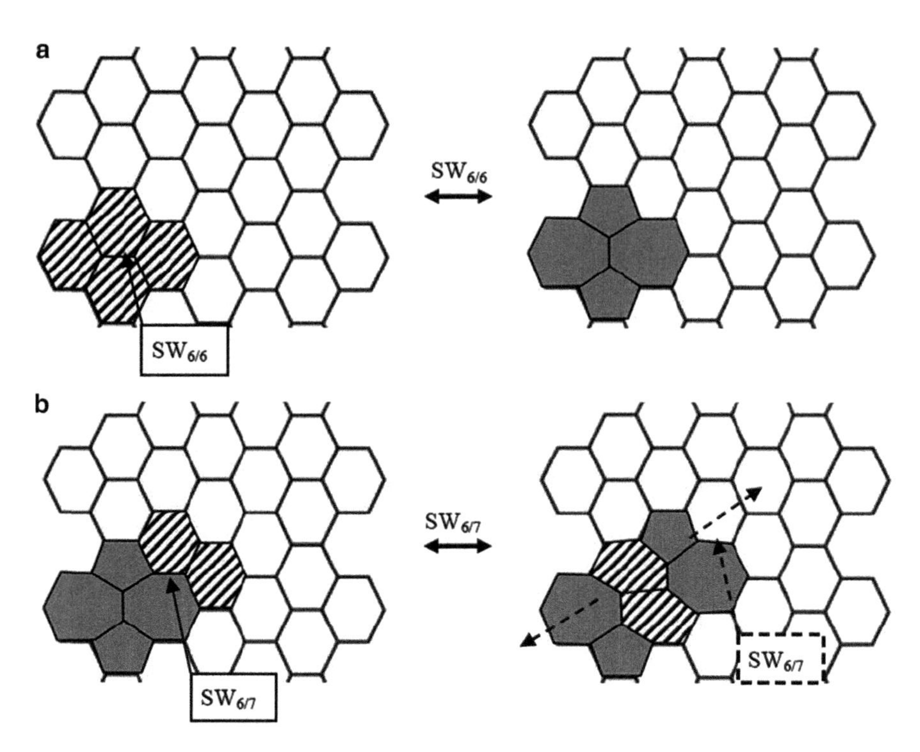

Fig. 3.1 The action of the basic Stone-Wales rotations in both the direct and dual graphene lattice representations is shown. (**a**) $SW_{6/6}$ flips the arrowed central bond of the four shaded hexagons originating two 5|7 pairs in gray. (**b**) $SW_{6/7}$ splits the two pairs along dashed direction swapping one of them with two nearby hexagons (*shaded*). The dashed $SW_{6/7}$ operator represents the next available rotation

topological effects. All the Stone-Wales (SW) structural rearrangements are *isomeric and reversible* and one may denote with $SW_{7/7}$ the inverse rotation of $SW_{6/6}$; $SW_{7/7}$ *annihilates* the 5|7 defects transforming the two 5|7 pairs in four hexagons (Fig. 3.1a).

Experimentally, high-resolution transmission electron microscopy (TEM) studies made on single-walled carbon nanotubes (Hashimoto et al. 2004) evidence *in situ* formation of 5|7 pairs in single graphene layers. Pentagon–heptagon pairs are described to be stable graphene defects in the theoretical works (Nordlund et al. 1996; Krasheninnikov et al. 2001) that explain how energetic electrons and ions are able to generate 5|7 defects in carbon nanotubes and graphite layers as a result of knock-on atom displacements. All these recent investigations evidence the structural stability of large portions of the graphene plane modified by the presence and the propagation of 5|7 defects. The scope of this work is then to search for stable configurations of the graphenic layers modified by the presence of 5|7 dislocations *only* by imposing the minimization of a suitable topological invariant of the lattice. Our results demonstrate that topological-based techniques constitute a new effective tool for modeling diffusion processes in extended systems like the graphenic structures, as reported in the next paragraphs.

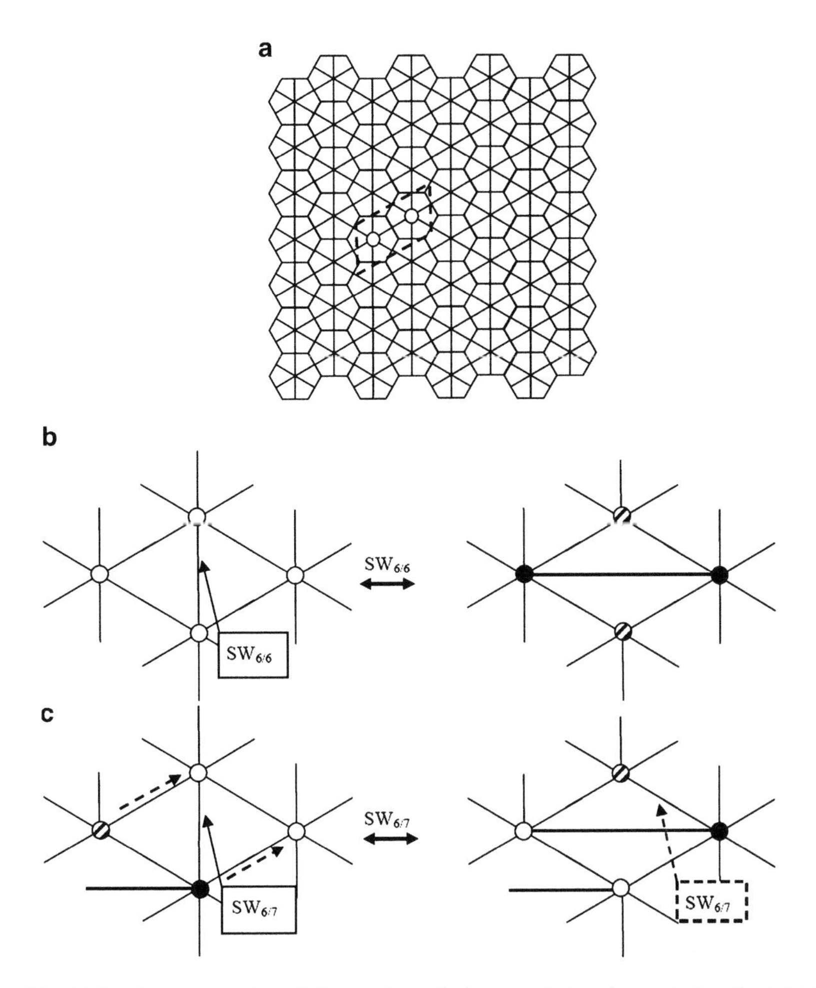

Fig. 3.2 (**a**) Dual representation of the graphene lattice armchair orientated, the direct lattice being also represented. The dashed region individuate the lattice supercell along the diagonal direction. (**b–c**) Dual representation of the SW mechanisms (**a**, **b**) given Fig. 3.1. Hexagons, pentagons, heptagons are represented by *white, shaded, black circles* respectively. The dashed $SW_{6/7}$ operator represents the next available rotation

Being the current study based on a pure graph-theoretical approach, two are its basic elements introduced here e.g. the peculiar type of the chemical graph G used to represent the chemical structures under study (graphene and graphenic layers) and the choice of the topological graph-invariant W capable to model the structural modifications induced by the presence of pentagon–heptagon defects in those structures. About G, we decided to operate in the *dual topological representation* G of the graphene layer (Fig. 3.2) allowing a very simple representations the operators $SW_{6/6}$ and $SW_{6/7}$ extensively used in the current research. G derives from the 3-connected direct graphene lattice just transforming the hexagonal faces in the corresponding 6-connected vertices (6-nodes). The number of bonds is preserved by this correspondence, since two dual nodes of *G* have an edge in common only if the corresponding hexagons share a chemical bond in the direct graphene layer. Indicating with *N* the nodes of *G*, the Wiener index *W(N)* provides a quite effective topological graph-invariant to model the topological diffusion of the 5|7 pair in the dual graphene layer *G*. In the following, the main computational characteristics of our *W*-based model are introduced by describing how the $SW_{6/6}$ rotation topologically modifies the graphene dual plane.

3.2 Topological Potential

The adopted theoretical model assigns to the Wiener index *W(N)* of the dual graph *G* (Todeschini and Consonni 2000) the role of the *topological potential* of the system subject to a minimum principle. As a function of *the graph chemical distances* Eq. 3.1, *W* privileges those transformations that increase the overall topological compactness of the system and it has been recently used in simulating the growing steps of fullerene-like nanostructures on the graphene dual plane (Cataldo et al. 2010). Indicating with d_{ij} the minimum distance between vertices V_i, V_j in the chemical graph *G, W* is defined as:

$$W(N) = {}^1\!/_2\,\Sigma_{ij} d_{ij} \tag{3.1}$$

where the summations over *i* and *j* are extended from 1 to *N*. Evidencing the individual contribution w_i originated by the site V_i:

$$w_i = {}^1\!/_2\,\Sigma_j d_{ij} \tag{3.2}$$

Denoting with *M* the diameter of the graph *G*, $M = max\{d_{ij}\}$, the structural-meaningful formula is produced, with *k* ranging from 1 to *M*:

$$w_i = {}^1\!/_2\,\Sigma_k k b_{ik} \tag{3.3}$$

where the integer b_{ik} gives the number of *k*-neighbors of V_i , coefficients b_{ik} obey to the closure condition $1 + \Sigma_k b_{ik} = N$. Relation (3.3) allows us to rewrite Eq. 3.1 as:

$$W(N) = \Sigma_i w_i \tag{3.4}$$

Molecular graphs made of equivalent atoms (e.g. the graphene lattice or the C_{60} stable fullerene molecule) all w_i have the same value $\underline{w}$, becoming *W*:

$$W(N) = N\underline{w} \tag{3.5}$$

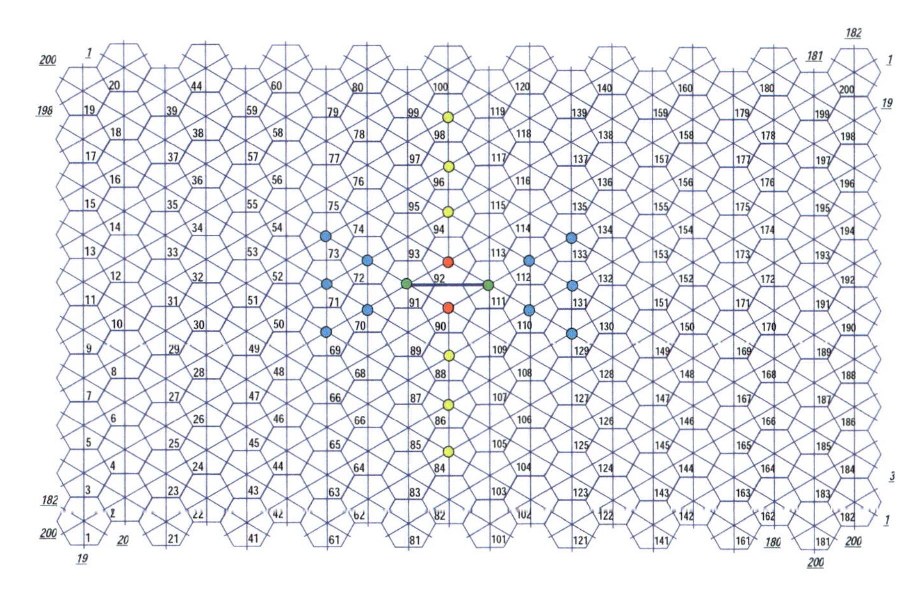

Fig. 3.3 The $N = 200$ dual closed graphene graph modified by the presence of a Stone-Wales defect with two pentagons (*red*) and two heptagons (*green*); this defect favors the side regions of the heptagons (*pale-blue vertices*) whereas the *yellow* hexagons correspond to unstable vertices. Underlined nodes labels illustrate how closed ends conditions are applied

More generally, each graph G presents at least one site with the minimum $\underline{w}$ value, being the corresponding vertex called the *minimal vertex of the graph* $\underline{V}$. For the graphene closed graph in Fig. 3.3 all the $N = 200$ sites play the same role of minimal vertex having $\underline{w} = 582.5$.

Following the method previously adopted to simulate the stability of graphene layers with nanocones (Cataldo et al. 2010) and the relative stability of the isomers of the C_{66} fullerene (Vukicevic et al. 2011), we apply our *topological minimum principle* to simulate the defective configurations of the graphene layer by stating that *chemically stable structures minimize the Wiener index of the corresponding chemical graphs G*. Equation 3.1 reaches its minimum values in correspondence of the most compact lattice configurations with the shortest possible distances d_{ij}, suggesting that *the most-compact structures will present the highest chemical stability*. Moreover, our topological approximation implies *that minimal vertices of G are likely to be the most stable of the structure*. The lattice in fact tends to keep unchanged the coordination shells around these vertices, preserving in such a way their low contributions to the overall Wiener index according to Eq. 3.4.

As we remarked in previous works (Cataldo et al. 2010, 2011; Ori and Cataldo 2011) the distance-based index $W(N)$ shows, on D-dimensional structures, a polynomial-like behavior $W(N) \approx N^s$ (with $s = 2 + 1/D$) whose validity has been recently extended to $D > 1$ fractals (Ori et al. 2011). In case of $D = 2$ graphenic planes, the Wiener index formula is:

$$W(N) = a_5N^{5/2} + a_4N^2 + .. + a_1N^{1/2} + a_0 \quad (3.6)$$

In Fig. 3.2 the graphene lattice is represented under an up-down *armchair* orientation; it has N starred nodes and the closed form for its Wiener index is (Cataldo et al. 2011):

$$W(N) = \delta(7N^{5/2} - 2N^{3/2})/24 \qquad \text{where } \delta = 2^{-1/2} \quad (3.7)$$

Formula (3.7) considers periodic boundary conditions on the dual graphenic lattices, see Ori and Cataldo (2011).

The influence of the topological potential W on the propagation of 5|7 pairs in the graphene dual lattice are reported in the next paragraphs. In the following, the topological effects induced in the plane by the creation of the 5|7 double-pair (the so-called SW defect) are described. We consider then the dual graphene periodic lattice with $N = 200$ vertices and $B = 600$ bonds represented in Fig. 3.3. In this initial status, all the N lattice vertices are equivalent with $\underline{w}_0 = 582.5$, $M = 10$, the common string of coordination numbers being $\{b_{ik}\} = \{6\ 12\ 18\ 24\ 29\ 28\ 26\ 24\ 22\ 10\}$. From Eqs. 3.5 and 3.7 the starting value of the topological potential W_0 for the regular graphene dual layer $W_0 = 116{,}500$ is derived. Once the $SW_{6/6}$ rotation is applied the presence of the SW defect causes an immediate reduction of the plane symmetry, as one may see monitoring the different distribution of the w_i values detailed in Table 3.1. In Fig. 3.3 the green circle are the two heptagonal vertices newly generated by $SW_{6/6}$; these nodes are minimal vertices of the lattice contributing to the topological potential given by Eq. 3.4 with $\underline{w} = 560$, producing a net stabilization effect of about 4% if compared to the previously computed $\underline{w}_0$. The overall value of the Wiener index for the whole lattice is also reduced to $W = 116{,}015$, pointing out that the creation of the 5|7 double-pair is, from the topological point of view, a favored modifications of the graphene plane. The two new heptagons are connected by the rotated bold-blue edge in Fig. 3.3, resulting from the $SW_{6/6}$ reversible flip. On the other side, the red circles indicate the two associated pentagons corresponding to the two less stable nodes of the graphenic lattice having $w_{max} = 584.5$. All the N nodes are sorted by the action of the topological potential in 13 distinct sets of topologically equivalent vertices, ranked in Table 3.1 according to their w_i values ranging from $\underline{w}$ (related to the most stable vertices) to the less favored w_{max}. The yellow circles evidence in Fig. 3.3 the hexagonal nodes that, after the pentagons, individuate the *less stable regions* of the lattices, corresponding to the 10,11,12 sets of Table 3.1; this group of six nodes individuates a line of unstable vertices perpendicular to the heptagon-heptagon new bond of the SW defect. Pale-blue circles mark stable nodes belonging to the 2, 3 sets of Table 3.1.

Our topological approach has been able, so far, of predicting the stability of the pentagon-heptagon double pairs arising from the $SW_{6/6}$ isomeric transformations of the graphene dual plane. In what follows we will illustrate moreover some topological mechanisms, still based on the applications of the SW operators, allowing the *isomeric diffusion* of the 5|7 defect in graphenic lattices.

Table 3.1 Topological characterization of the 200 nodes of the reference dual graphene lattice when a SW defect is present. Nodes are sorted by their w_i values. Dual graph is made of 200 vertices with $W = 116{,}015$; $M = 10$; total bonds in the graph $B = 600$; summing b_{i1} entries gives B value twice; sum of w_i values is W. The most stable vertices are the two heptagons V91, V111; on the contrary the two adjacent pentagons V90, V92 are the less stable ones

Set	Vertex	$\{b_{im}\}$ $M = 10$	w_i
1	V91 V111	7 13 19 25 30 29 27 25 23 1	560
2	V70 V72 V110 V112	6 13 19 25 30 29 27 25 23 2	564.5
3	V69 V71 V73 V129 V131 V133	6 12 19 25 30 29 27 25 23 3	568.5
4	V48 V50 V52 V54 V128 V130 V132 V134	6 12 18 25 30 29 27 25 23 4	572
5	V26 V28 V30 V32 V34 V36 V146 V148 V150 V152 V154 V156	6 12 18 24 29 29 27 25 23 6	577.5
6	V25 V27 V29 V31 V33 V35 V37 V165 V167 V169 V171 V173 V175 V177	6 12 18 24 29 28 27 25 23 7	579.5
7	V4 V6 V8 V10 V12 V14 V16 V18 V164 V166 V168 V170 V172 V174 V176 V178	6 12 18 24 29 28 26 25 23 8	581
8	V3 V5 V7 V9 V11 V13 V15 V17 V19 V183 V185 V187 V189 V191 V193 V195 V197 V199	6 12 18 24 29 28 26 24 23 9	582
9	V1 and the remaining vertices	6 12 18 24 29 28 26 24 22 10	582.5
10	V84 V98	6 12 18 23 30 28 26 24 22 10	583
11	V86 V96	6 12 17 24 30 28 26 24 22 10	583.5
12	V88 V94	6 11 18 24 30 28 26 24 22 10	584
13	V90 V92	5 12 18 24 30 28 26 24 22 10	584.5

3.3 Radial Diffusion of the 5|7 Pairs

Figure 3.4 represents, on the left, the structure of the graphene lattice after the sequence of SW rearrangements evidenced by the letters **a** to **c**. On the right, the behavior of the topological potential W is also reported, the initial point of the curve corresponding to the $SW_{6/6}$ rotation of Fig. 3.4. This sequence of topological transformations is favored by the decreasing of the topological potential W that reaches a final value of $W = 115{,}510$, lower than the starting level $W = 116{,}015$ previously computed. To better understand how the curve in Fig. 3.4 is computed, one has to consider that, for example, the plotted value $W = 115{,}870$ in **a** gives the Wiener index value for the lattice modified by the initial $SW_{6/6}$ rotation *and* the rotation of the **a** bond in Fig. 3.4. The total effect on the lattice of the $SW_{6/6}$ rotation plus the three rotations **a, b, c** stabilizes the topological potential of the system at the final level of $W = 115{,}510$, corresponding to the configuration in Fig. 3.4 that is therefore more stable than the initial one of Fig. 3.3 generated by the simple Stone-Wales rotation.

Moreover, this final configuration features the increased distances between the four circled vertices that represent the pentagonal (red) and the heptagonal (green) faces of the original SW dislocation. This result allow us to define the general characteristics of the structural, isomeric modifications of the graphene lattice – called here *radial dislocations*- produced during the *radial diffusion* of the 5|7 pairs: at each step R_i the two pentagons and the two heptagons interchange their positions

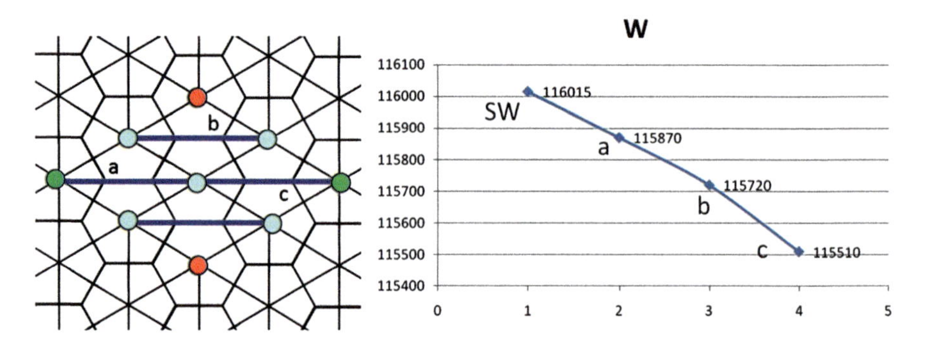

Fig. 3.4 From the starting configuration with the SW defect of Fig. 3.3, the radial diffusion of the pentagons (*red circles*) and the heptagons (*green circles*) is represented on the left driven by the decreasing values of the topological potential W (*right*). The Wiener index presents for the $N = 200$ dual closed graphene graph, a constant decreasing favoring the pentagons-heptagons separations. The radial dislocation on the left is called R_1 being equal to 1 the number of hexagons (*blue*) aligned with the heptagons

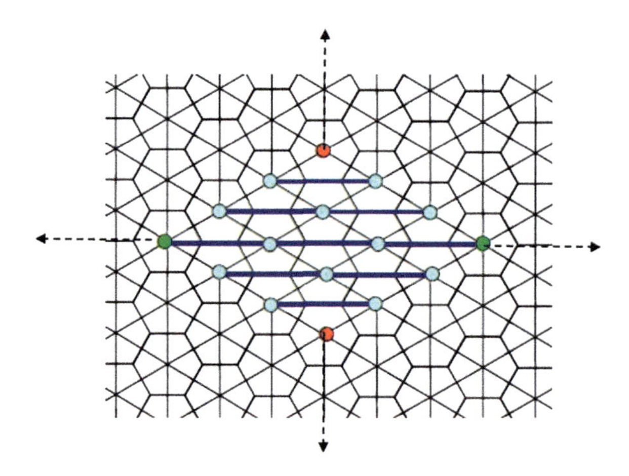

Fig. 3.5 The radial dislocation R_2 is represented here in the graphenic dual plane. Pentagons (*red*) and heptagons (*green*) *diffuse in the arrowed directions* to minimize the overall topological potential of the structure $W = 114{,}905$. This radial configuration is called R_2 being in this case the number of hexagons in line with the two heptagons equal to 2

with other local hexagons forming a diamond-shaped region that grows along two perpendicular directions.

This endless *radial diffusion* not only separates the two nodes of a given 5|7 pair (e.g. it splits the pentagon from the heptagon) but it is also capable to *orient,* perpendicularly to the original direction, *all bonds shared by the hexagonal faces* inside the diamond region, as show by the bolded blue lines in Figs. 3.4 and 3.5. *The bond orientation is a peculiar effect* involved by the 5|7 pairs propagation and it will affect also the diagonal diffusion mechanisms presented in the next paragraph.

Conventionally, the number of the hexagons aligned with the two heptagons defines the order R_i of the radial dislocation; Figs. 3.4 and 3.5 represent then R_1 and R_2 iteration respectively. It is important noticing the main prediction coming from our topological simulations, e.g. *the larger is the radial dislocation the bigger is its stabilization effect on W*. We have then that, for each pair $j < i$, the overall topological potential of the system varies respecting the stabilizing condition $W(R_j) < W(R_i)$. The reported cases confirm this trend for the sizes $i = 1, j = 2$, with $W(R_1) = 115{,}510$ lower than $W(R_2) = 114{,}905$. We end this paragraph evidencing how the dual space provides the natural arena for generating and simulating all sorts of SW transformations of the graphene plane, avoiding the graphical difficulties that one usually encounters in modifying, in the direct lattice, many rings of carbon atoms.

3.4 Diagonal Diffusion of the 5|7 Pairs

The some-how arbitrary definition of *diagonal* direction refers to the *armchair* orientation adopted in Fig. 3.2. The diagonal direction assumes in the graphene plane a peculiar importance recently investigated and confirmed by several authors. Ab-initio extended calculations based on density-functional theoretical methods in Jeong et al. (2008) assign to large *diagonal* dislocations dipoles a particular energetic stability. Analogous anisotropic properties are reported in the extended study of Samsonidze et al. (2002) where the authors state that the *diagonal* distribution of *multiple SW defects* is energetically favored by the presence of the cylindrical curvature associated to small tensile strain, making the extended dislocation of the 5|7 pairs thermodynamically favorable for a sufficiently large separation distance, being the starting SW dipole split by the insertion of a certain number η of hexagon-hexagon pairs 6|6.

Topologically, the extension of the graphene region interested by the diagonal diffusion of the 5|7 pairs may infinitely grow applying an arbitrary number of $SW_{6/7}$ rotations. The simplest propagation *diagonal* mechanisms available for the 5|7 pairs has been called in Ori and Cataldo (2011) the SW wave, with the following *diagonal* diffusion mechanism: given a 5|7 pair (originated for example by a $SW_{6/6}$ transformation) the operator $SW_{6/7}$ rotates the *vertical* bond that, in the graphene dual lattice, connects the 7-node of the pair to an adjacent 6-node causing the *diagonal* swap of the 5|7 pair with a 6|6 pair with the creation of a new horizontal hexagon-hexagon bond (Fig. 3.2 gives more details). This iterated propagation mechanism originates in the graphene plane a diagonal (or vertical as well) *extended dislocation dipole*. Comparing it to the previously described *radial dipole*, two structural peculiarities characterize this case: (i) the SW wave propagates along a linear region of the graphene layer; (ii) the 7- and the 5-node of the pair remains connected during the whole diffusion process. In both cases however the 6-nodes involved by the diffusion process present the rotated bonds already evidenced in the above Figs. 3.2b, 3.2c, 3.4, 3.5.

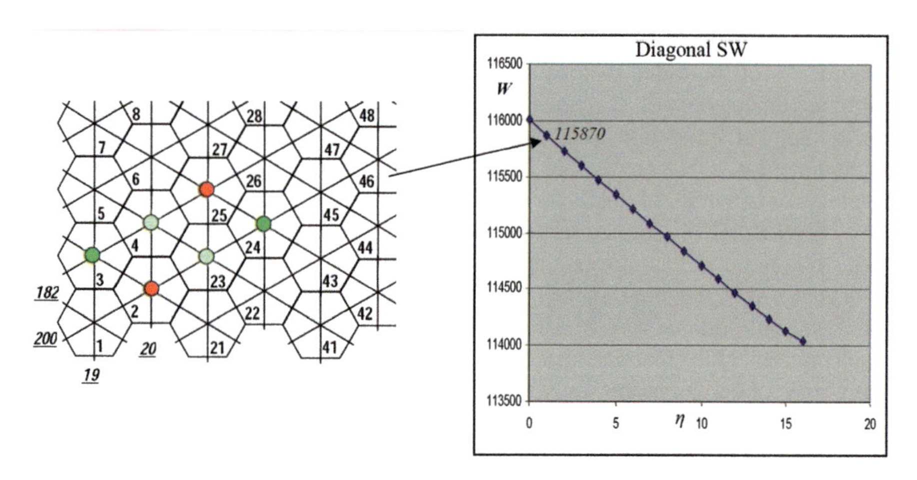

Fig. 3.6 *Left*: diagonal SW wave for the $N = 200$ dual closed graphene at the diffusion step $\eta = 1$ of the pentagon-heptagon pair (*red-green circles*). The swapping hexagons are circled in pale blue. *Right*: Diagonal dislocations freely flow in the lattice minimizing, with an almost linear dependence from η, the topological potential W of the structure. At $\eta = 1, W = 115{,}870$

The behavior of the diagonal SW wave has been simulated by computing the variations of the topological potential W on the $N = 200$ dual lattice previously considered. Let's thus restart from the configuration given in Fig. 3.3 whose topological potential values $W = 116{,}015$ corresponds to the point marked as SW in the Fig. 3.4 curve. At this moment then a 5|7 double pair has been generated in the plane. The first diffusion step of the diagonal SW wave (Fig. 3.6) is induced by the $SW_{6/7}$ rotation (Fig. 3.2c) moving one of the 5|7 pair with a further reduction of $W = 115{,}870$. This stabilizing effect on W is confirmed at each propagation steps, allowing an iterated reduction in the topological potential of the system according to an almost linear dependence from the number of the propagation steps η, see the curve in Fig. 3.6 where the $\eta = 0$ case corresponds to the SW defect $W = 116{,}015$. These results clearly evidence the tendency of the graphene lattice to allow an endless topological diffusion the 5|7 pair *along the diagonal* direction, generating extended diagonal dislocation dipoles. It is worth noticing that this basal diffusion mechanism predicted by our topological model for the 5|7 pair just relies on the connectivity properties of the pentagon-heptagon pairs embedded in the graphene hexagonal mesh, and it may therefore be related to the topological properties of the graphenic lattices only.

The influence of the $SW_{6/7}$ operator in modifying the graphene structure has been further investigated by considering, on the closed dual lattice G, the possible *annihilation* (obtained by the $SW_{7/7}$ operator mentioned in the introduction) of the two 5|7 pairs appearing in Fig. 3.6. In this case the diagonal propagation of the 5|7 dislocation on the periodic lattice reaches a very stable configuration where, after the *collision and the annihilation* of the two 5|7 pairs, all vertices modified during the diffusion process are represented by 6-nodes with rotated bonds, see the sequence in Fig. 3.7. The peculiar stability of this configuration is confirmed by

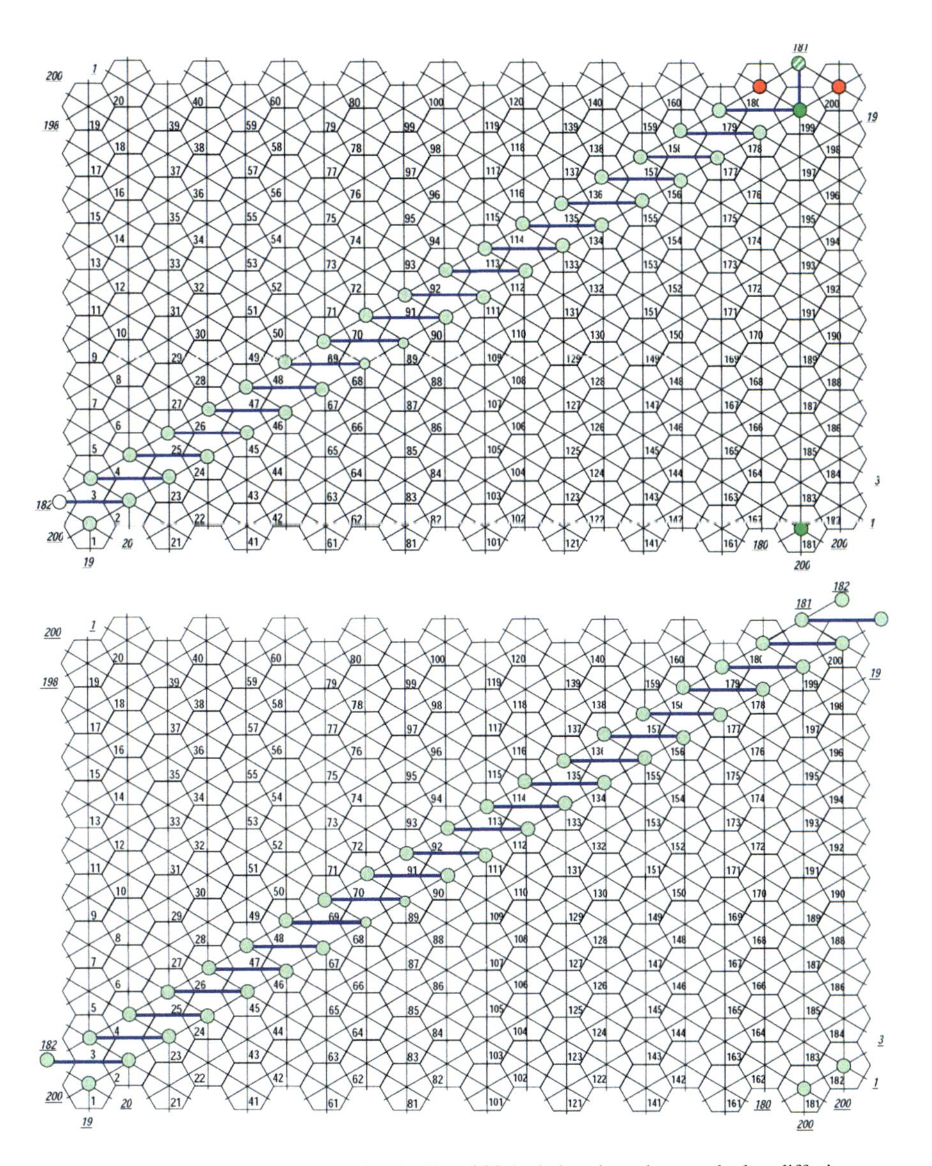

Fig. 3.7 *Top*: the diagonal SW wave for the $N = 200$ dual closed graphene at the last diffusion step *before the annihilation of the 5/7 pairs*; it has $W = 113{,}925$. *Bottom*: Diagonal dislocations at the annihilation point $W = 114{,}000$ when only rotated bonds (*the bold-blue ones*) populate the region modified by the diffusion process. Underlined vertices evidence the periodic imposed conditions

the value of its topological potential $W = 114{,}000$ that, compared to the starting value of the graphene ideal dual lattice $W_0 = 116{,}500$, produces a 2% gain in the topological stability of the system. The action of the $SW_{6/6}$, $SW_{6/7}$ and $SW_{7/7}$ flips may also produce in the plane parallel *multi-diagonal* 5|7 dislocations as shown the example in Fig. 3.8; the same illustration evidences the status of the layer after

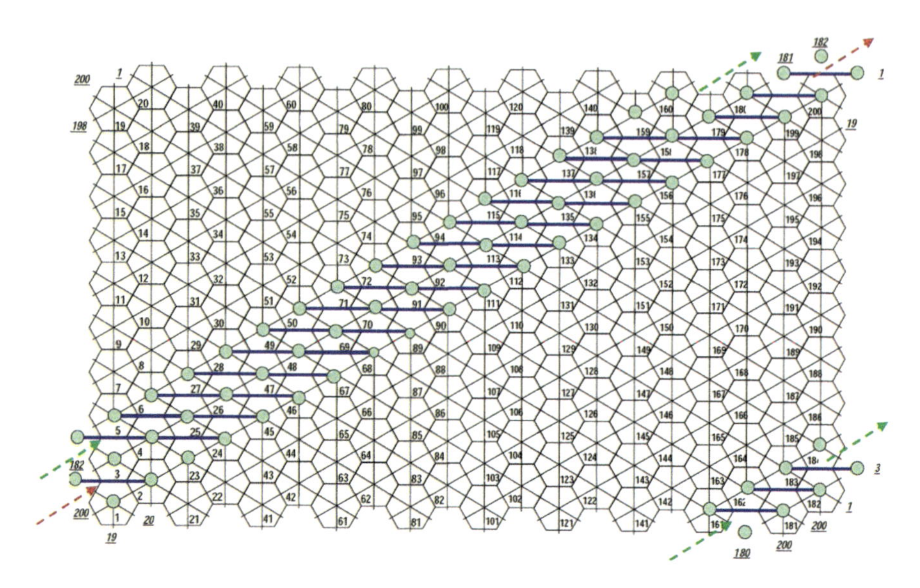

Fig. 3.8 The diagonal double SW wave generate the dislocation parallel to the green arrows; after the *annihilation of the 5/7 pair* this lattice configuration has $W = 112{,}500$. Underlined vertices and the colored arrows evidence the periodic imposed conditions

the annihilations of all the 5|7 pairs; this iterated effect may produce new stable graphenic configurations with more and more extended regions tiled by 6-nodes connected by rotated bonds (in bold-blue in Fig. 3.8). The imposed periodic conditions force the dipolar dislocation to move on the circular path signaled with the green arrows in Fig. 3.8, also represented in Fig. 3.7 where the 5|7 pairs diffuse along the circumference corresponding to the main diagonal of the periodic lattice along a path signaled by the purple arrows in Fig. 3.8. The curve in Fig. 3.9 summarize this original effect, by showing the uniform decreasing of the topological potential W when the multi-diagonal diffusion and annihilation of 5|7 pairs affects larger and larger portions of the graphenic layer as in Figs. 3.7 and 3.8. The first portion of the curve describes the behavior of W when the two 5|7 pairs migrate, as in Fig. 3.7, along the lattice main diagonal (the path marked with purple arrows in Fig. 3.8) until their collision/annihilation in correspondence to the local peak D_1; the second part of the curve follows the generation /migration of other two 5|7 defects along the lattice diagonal closed circuit marked with green arrows in Fig. 3.8, until also these two pairs collide/annihilate generating the second local peak D_2. It is quite interesting evidencing how this effect may continue until the coverage of the whole plane with orientated bonds is completed, producing, at each migration step, lattice configurations with a lower value of the topological potential W.

The comparison with literature assigns a certain grade of relevance to our theoretical results. Future studies of the proposed isomeric transformations will be undertaken to better understand the chemical implications of the proposed diffusion mechanisms for the 5|7 dislocations.

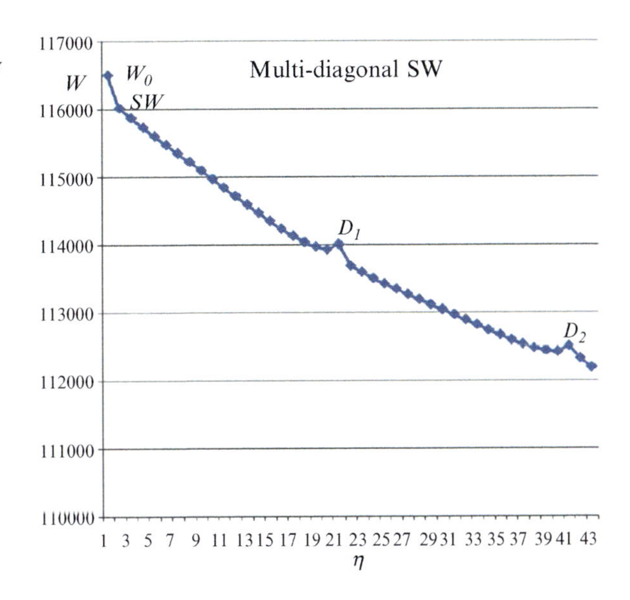

Fig. 3.9 The dependence of the topological potential W from the number of the rotated bonds is reported during the multi-diagonal diffusion of the 5|7 defects in the closed lattice G. Ideal graphene starting ideal graphene lattice has $W_0 = 116{,}500$, whereas configurations D_1 (Fig. 3.7) and D_2 (Fig. 3.8) have $W = 112{,}500$ and $W = 112{,}500$ respectively

3.5 Conclusions

In the graphene plane, the approximated topological potential W favors both the radial and the diagonal diffusion processes of the 5|7 dislocations whose chemical relevance may be matter for future experimental and theoretical studies. This work shows that the topological modeling is able to simulate complex mechanism on large chemical structures, as defective graphenic layers, confirming its ability to promptly sieve interesting chemical structures for subsequent ab-initio studies. The topological methods proposed here are applicable to the theoretical characterization of various defective lattices like fullerenes, schwarzites, etc.

References

Cataldo F, Ori O, Iglesias-Groth S (2010) Mol Simulat 36(5):341–353
Cataldo F, Ori O, Graovac A (2011) Int J Chem Mod 3(1–2):45–64
Hashimoto A, Suenaga K, Gloter A, Urita K, Iijima S (2004) Nature 430:870–873
Jeong BW, Ihm J, Lee G-D (2008) Phys Rev B 78:165403(5)
Krasheninnikov AV, Nordlund K, Sirviö M, Salonen E, Keinonen J (2001) Phys Rev B 63:245405(4)
Nordlund K, Keinonen J, Mattila T (1996) Phys Rev Lett 77:699–702
Ori O, Cataldo F (2011) Chem Phys Lett. in press
Ori O, Cataldo F, Vukicevic D, Graovac A (2011) Iran J Math Chem, in press
Samsonidze GeG, Samsonidze GG, Yakobson BI (2002) Comp Mat Sci 23:62–72
Todeschini R, Consonni V (2000) Handbook of molecular descriptors. Wiley-VCH, Weinheim
Vukicevic D, Cataldo F, Ori O, Graovac A (2011) Chem Phys Lett 501:442–445

Chapter 4
The Chemical Reactivity of Fullerenes and Endohedral Fullerenes: A Theoretical Perspective

Sílvia Osuna[1], Marcel Swart[1,2], and Miquel Solà[1]

Abstract We report here a review of our recent efforts at understanding and predicting reactivity and regioselectivity of endohedral (metallo)fullerenes and their parent free fullerenes. The effect of encapsulation of trimetallic nitrides or noble gas atoms/dimers was shown to have a profound impact on the stability and reactivity of fullerene compounds. It is not just the encapsulation, but in particular the nature of the species that is encapsulated that determines how (un)reactive the fullerene becomes. These findings have important consequences for future applications of (metallo)fullerenes in biomedicine and (nano)technology.

4.1 Introduction

Soon after the discovery of the buckminsterfullerene (C_{60}), it was hypothesized that the diameter of this molecule was large enough to encapsulate a variety of atoms inside the carbon cage (Kroto et al. 1985). Indeed, the first stable metallofullerene based on the "bucky ball" (La@C_{60}) was detected in the same year as the initial discovery (Heath et al. 1985), and some years later the formation of higher fullerenes encapsulating a lanthanum atom, such as La@C_{70}, La@C_{74}, and

---

[1] Institut de Química Computacional and Departament de Química,
Universitat de Girona, Campus Montilivi, 17071 Girona, Catalonia, Spain
e-mail: silvia.osuna@udg.edu; miquel.sola@udg.edu

[2] Institució Catalana de Recerca i Estudis Avançats (ICREA), Pg. Lluís Companys 23,
08010 Barcelona, Spain
e-mail: marcel.swart@udg.edu

M.V. Putz (ed.), *Carbon Bonding and Structures: Advances in Physics and Chemistry*,
Carbon Materials: Chemistry and Physics 5, DOI 10.1007/978-94-007-1733-6_4,

La@C_{82} was reported in macroscopic quantities (Chai et al. 1991). It was in the same study, where the symbol @ was introduced to denote that the former atom or molecule is encaged inside the fullerene molecule. These species containing atoms or molecules inside the fullerene cages are called endohedral fullerenes (EFs). The most studied compounds are the so-called endohedral metallofullerenes (EMFs) due to their high abundancy (Akasaka and Nagase 2002). In 1991, the first structural characterization by X-ray diffraction of an EMF was reported (Chai et al. 1991).

EMFs can be classified in several classes (Chaur et al. 2009): (i) classical EFs of the type M@C_{2n} and M_2@C_{2n} (*M = metal, noble gas, small molecule and* $60 \leq 2n \leq 88$); (ii) trimetallic nitride EMFs ($M_3N@C_{2n}$, M = *metal and* $68 \leq 2n \leq 96$); (iii) metallic carbide EMFs ($M_2C_2@C_{2n}$, $M_3C_2@C_{2n}$, $M_4C_2@C_{2n}$, *and* $M_3CH@C_{2n}$ *with* M = *metal and* $68 \leq 2n \leq 92$); (iv) metallic oxide EMFs ($M_4O_2@C_{2n}$ *and* $M_4O_3@C_{2n}$); and (v) metallic sulfide ($M_2S@C_{2n}$). Of course, as the number of EFs families and the elements of each family increase rapidly this current classification will be modified in the future.

The high interest awakened by these EMFs is in part motivated by the promising potential applications of these compounds in a wide range of research topics such as magnetism, superconductivity, and nonlinear optical (NLO) properties (Whitehouse and Buckingham 1993; Hu et al. 2008). For instance, the opto-electronic properties of EMFs might be tailored without changing the outer carbon cage by changing the metal cluster encapsulated inside the cage. The encapsulation of the metals inside the carbon fullerene cage makes these compounds ideal for medical applications, for example as hosts of radioactive atoms for use in nuclear medicines (Diener et al. 2007; Shultz et al. 2010) or as effective magnetic resonance imaging (MRI) contrast agents (Dunsch and Yang 2007; Laus et al. 2007). In addition, EMFs used as electron acceptor in electron donor–acceptor dyads can lead to promising photovoltaic materials to be used in solar energy conversion/storage systems (Guldi et al. 2010). The use of some EFs with long spin lifetime for quantum computing or spintronic devices has also been suggested (Pietzak et al. 2002; Harneit 2002). Finally, they have also been used for monitoring chemical reactions of the fullerene cages via changes in the electron paramagnetic resonance (EPR) signals (Pietzak et al. 2002; Wang et al. 2001).

The stability and reactivity of the fullerene cages is highly modified after the encapsulation due to the interaction between the inner species and the carbon cage. For instance, the C_{80} cage of I_h symmetry is the most unstable among the seven structures of C_{80} that satisfy the isolated pentagon rule (IPR) (Kroto 1987; Schmalz et al. 1988). However, this is the preferred cage for encapsulating two La atoms or a Sc_3N unit. Indeed, Sc_3N@I_h-C_{80} is the third most abundant fullerene after C_{60} and C_{70} (Valencia et al. 2009). Similar situations are found for other fullerene cages such as the D_{3h}-C_{78} and C_{2v}-C_{82}. On the other hand, while the IPR is strictly followed by all pure-carbon fullerenes isolated to date (Lu et al. 2008), an increasing number of EMFs have been synthesized that present non-IPR cages. Therefore, the IPR appears to be more a suggestion than a rule for these species (Kobayashi et al. 1997) and, in general, for any

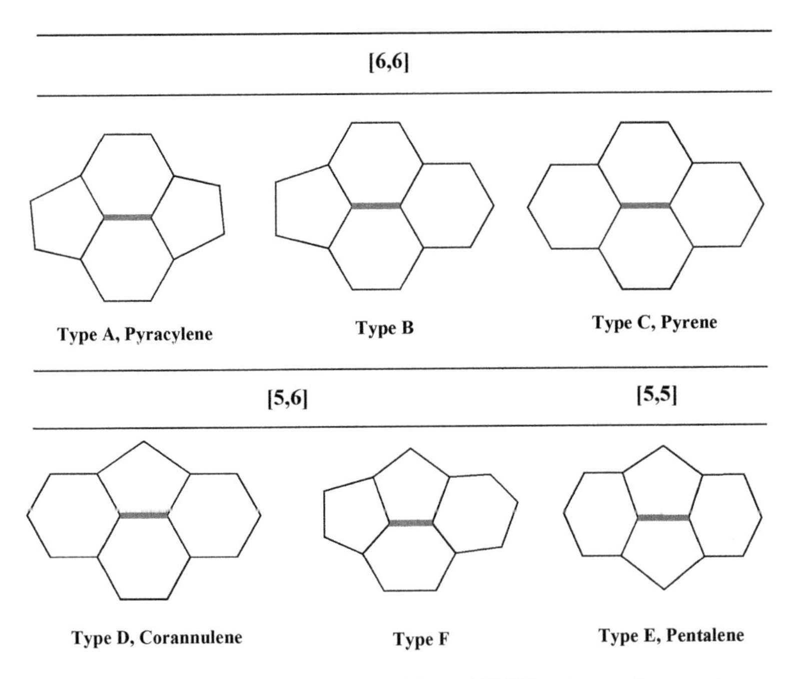

Fig. 4.1 Representation of the different [6,6], [5,6], and [5,5] bond types that may be present in any fullerene structure

charged fullerene (Aihara 2001). In non-IPR carbon cages, there are six possible different C–C bond types A to F (represented in Fig. 4.1), where types E and F are not present in IPR structures.

4.1.1 Trimetallic Nitride Template (TNT) Fullerenes

A rule for the stabilization of fullerene cages in metallic nitride EMFs has been recently established based on the fact that there is a formal charge transfer of six electrons from the nitride to the carbon cage, i.e. $M_3N^{6+}@C_{2n}^{6-}$ (Popov and Dunsch 2007, 2009). According to the maximum hardness principle (Pearson 1997, 1999; Parr and Chattaraj 1991; Torrent-Sucarrat et al. 2001), Campanera and coworkers concluded that the most stable metallic nitride EMFs are those encapsulated in fullerene cages that present large (LUMO-4)-(LUMO-3) gaps (Campanera et al. 2005; Valencia et al. 2007).

The metal cluster encapsulated inside the fullerene cages has an enormous influence on the reactivity of these compounds (Guha and Nakamoto 2005; Martín 2006), which is, in general, reduced in TNT EMFs as compared to free fullerenes

(Campanera et al. 2005, 2006; Osuna et al. 2008). Many recent experimental and theoretical studies show that not only the chemical reactivity but also the regioselectivity of TNT EMFs is strongly affected by the encapsulated cluster, metal species, carbon cage size, and symmetry (Akasaka and Nagase 2002; Thilgen and Diederich 2006; Martín 2006; Cardona et al. 2005a; Iiduka et al. 2005).

The production of endohedral compounds such as $Sc_3N@C_{80}$, $Sc_3N@C_{78}$ or $Sc_3N@C_{68}$ in high yields using the so-called trimetallic nitride template (TNT) process, led to the organic functionalization of these fascinating molecules (Stevenson et al. 2000). In 2002, the Diels-Alder reaction was successfully produced for the first time on the I_h isomer of $Sc_3N@C_{80}$ (Lee et al. 2002). The crystallographic characterization of the first Diels-Alder adduct performed on an endohedral metallofullerene indicated that a symmetric adduct was obtained after reaction with 6,7-dimethoxyisochroman-3-one (Lee et al. 2002). The addition was shown to occur on the corannulene-type [5,6] bonds. Campanera et al. performed theoretical calculations (Campanera et al. 2006) that correctly describe the reactive exohedral sites of $Sc_3N@C_{80}$ for the Diels-Alder reaction. Based on these studies, the most reactive sites were those with high Mayer Bond Order (MBO) (Mayer 1983) and high pyramidalization angles (Haddon 2001; Haddon and Chow 1998). The geometry of the Diels-Alder adduct was similar to that found for the reaction on C_{60}, where the usual reactive bonds are the pyracylene-type [6,6] bonds. It should be mentioned here that the I_h isomer of $Sc_3N@C_{80}$ does not possess the reactive pyracylene-type bonds. In 2005, the 1,3-dipolar cycloaddition of N-ethylazomethine ylide and $Sc_3N@C_{80}$ was reported (Cardona et al. 2005b) where the same addition pattern, i.e. over the [5,6] bonds, was produced. In contrast, the 1,3-dipolar cycloaddition involving N-tritylazomethine ylide and $Sc_3N@C_{80}$ led to two different monoadducts. The thermodynamic control product was shown by both NMR and X-ray crystallography to correspond to the [5,6] addition, however the [6,6] adduct was the kinetic control addition product (Cai et al. 2006). Interestingly, the 1,3-dipolar cycloaddition to the parent $Y_3N@C_{80}$ lead to the [6,6] addition (Cardona et al. 2005b). The same regioselectivity was observed in the cyclopropenation reaction of diethyl bromomalonate and $Y_3N@C_{80}$ (Cardona et al. 2005b). The X-ray structure indicated that the C–C bond attacked is open rather than closed and that one of the yttrium atoms is positioned near the site of the cleaved bond. It should be noted here that the carbene addition to $Y@C_{82}$ was shown to yield an open fulleroid [6,6] regioisomer whose attacked bond was situated close to one of the yttrium atoms (Lu et al. 2009). Moreover, the 1,3-dipolar cycloaddition performed on the $M_2@I_h$-C_{80} (M = La, Ce) yielded two regioisomeric adducts corresponding to the [6,6] and [5,6] addition (Yamada et al. 2009). Interestingly, the M_2 cluster is facing the attacked bond in the case of the [5,6] product. The isomerization from the [6,6] to the [5,6] regioisomer was observed in the case of Y_3N@(N-Ethylpyrrolidino-C_{80}) (Cardona et al. 2006; Echegoyen et al. 2006). This was also the case for the N-ethylazomethine ylide addition to $Er_3N@C_{80}$ (Cardona et al. 2006). Theoretical calculations at the BP86/TZ2P level for the Y_3N@(N-Ethylpyrrolidino-C_{80}) indicated that the isomerization process takes place through a pirouette-kind of mechanism instead of involving the retro-cycloaddition reaction from the [6,6] adduct (Rodríguez-Fortea et al. 2006). The 1,3-dipolar cycloaddition was also produced to the encapsulated mixed-metal clusters $Sc_2YN@C_{80}$ and $ScY_2N@C_{80}$ (Chen et al. 2007a). As mentioned above, the reaction on $Sc_3N@C_{80}$ and $Y_3N@C_{80}$ was shown to give [5,6] and [6,6] cycloaddition adducts, respectively. Interestingly, the major adduct obtained in the case of the metal mixed Sc_2YN endohedral compound corresponded to the [5,6] addition as it happens with its parent $Sc_3N@C_{80}$. A change on the regioselectivity is produced with the encapsulation of two or more yttrium atoms (i.e. ScY_2N and Y_3N) inside the cage. The [6,6] product is the minor adduct in the case of $ScY_2N@C_{80}$, whereas in $Y_3N@C_{80}$ only the [6,6] regio-isomer is obtained. These findings suggest that the nature of the metal cluster encapsulated inside dictates the exohedral functionalization of the endohedral metallofullerenes.

The Diels-Alder reaction on the D_{5h} isomer of $Sc_3N@C_{80}$ and $Lu_3N@C_{80}$, and 1,3-dipolar cycloadditions to the D_{5h} isomer of $Sc_3N@C_{80}$ indicated a higher reactivity of the D_{5h} isomer as compared to the I_h isomer (Cai et al. 2006). The latter increase on the reactivity was explained in terms of the HOMO-LUMO energy gaps for both isomers (Cai et al. 2006). The LUMO orbitals of the D_{5h} isomers of $Sc_3N@C_{80}$ and $Lu_3N@C_{80}$ are comparable to those of the I_h isomer, whereas a destabilization of the HOMOs for the D_{5h} isomer is produced. The D_{5h} isomer was shown to be 21.1 kcal·mol^{-1} less stable at PBE/TZ2P than the corresponding I_h isomer (Popov and Dunsch 2007). Interestingly, the D_{5h} isomer presents the reactive pyracylene-type [6,6] bonds (the most reactive bond in the case of C_{60}). The 1,3-dipolar cycloaddition reaction performed on $Sc_3N@D_{5h}$-C_{80} yielded two possible monoadducts (Cai et al. 2006). The NMR spectrum for the thermodynamically stable cycloadduct was thought to be resulting from reaction at the pyracylene-type [6,6] bond. The other regioisomer obtained corresponding to the addition to an asymmetric bond, presumably of [6,6] type, was partially converted after heat treatment to other unidentified monoadducts.

The Diels-Alder reaction with ortho-quinodimethane and the gadolinium based metallofullerene $Gd_3N@C_{80}$ was achieved in 2005 (Stevenson et al. 2005). The latter compound is of significant interest because of its potential applications as MRI contrast agent (Agnoli et al. 1987). It was shown that two o-quinodimethane molecules were attached to the $Gd_3N@C_{80}$ surface (i.e. the formation of a bisadduct). The yield of the reaction was modest as only 0.5–1 mg of bisadduct was obtained from 5 mg of $Gd_3N@C_{80}$ (10–20%). For comparison, the yield of the 1,3-dipolar reaction on $Sc_3N@C_{80}$ is 30–40% (Cai et al. 2005). A combined theoretical and experimental investigation of the change on the regioselectivity of the 1,3-dipolar cycloaddition and a series of gadolinium and scandium mixed endohedral metallofullerenes ($Sc_xGd_{(3-x)}N@C_{80}$) was performed in 2007 (Chen et al. 2007b). The regioselectivity of the reaction was changed upon introduction of gadolinium atoms. The [5,6] product was the major adduct in $Sc_3N@C_{80}$, $Sc_2GdN@C_{80}$, and $ScGd_2N@C_{80}$, however the [6,6] adduct was also obtained in $Sc_2GdN@C_{80}$ and $ScGd_2N@C_{80}$. The [6,6] regioisomer was the major cycloaddition product in the case of $Gd_3N@C_{80}$. Interestingly, the thermal treatment of the final products led to the partial or total isomerization of the [6,6] adducts formed in the case of $Sc_2GdN@C_{80}$ and $ScGd_2N@C_{80}$ to the [5,6] regioisomers, respectively. Experimental and theoretical findings showed that the difference in stability between [6,6] and [5,6] products in the case of $ScGd_2N@C_{80}$ is very small (*the [5,6] adduct is at PBE/DNP just 2.3* kcal·mol^{-1} *more stable than the [6,6] one*). This energy difference ranges from 11.7 to −0.4 kcal·mol^{-1} along the series $Sc_3N@C_{80} > Sc_2GdN@C_{80} > ScGd_2N@C_{80} > Gd_3N@C_{80}$. The reactivity of the gadolinium based endohedral compounds $Gd_3N@C_{80}$, $Gd_3N@C_{84}$, and $Gd_3N@C_{88}$ was investigated to study the effect of the cage size on the exohedral reactivity (Chaur et al. 2008). They observed that among all considered compounds $Gd_3N@C_{80}$ was the most reactive cage through reaction with bromomalonate. The flattened shape of C_{84} and C_{88} cages makes them less pyramidalized and thus less reactive upon the Bingel reaction.

The synthesis and characterization of the first N-tritylpyrrolidino derivative of $Sc_3N@C_{78}$ utilizing the Prato reaction was successfully produced in 2007 (Cai et al. 2007). On the basis of NMR spectra and DFT calculations, Cai and coworkers concluded that the two monoadducts obtained corresponded to the addition to two different type B [6,6] bonds (called c-f and b-d). The X-ray diffraction of one of the obtained compounds (c-f addition) confirmed that the 1,3-dipole was attached to a [6,6] bond (Cai et al. 2007). It is interesting to remark here that the cyclopropanation reaction of $Sc_3N@C_{78}$ and diethyl bromomalonate yielded one monoadduct and one dominant symmetric bisadduct, which corresponded to the same type B [6,6] addition (Cai et al. 2008). The photochemical addition reaction of adamantylidene to $La@C_{78}$ was produced on both a [6,6] and a [5,6] bond (Cao et al. 2008).

Tremendous efforts have been devoted to the functionalization of EMFs and to get a better understanding of the reactivity of EMFs. Nevertheless, the development of regioselective reactions for EMFs is still in its infancy. The main reason is the difficulty to produce and isolate sufficient quantities to investigate their reactivity. Although this has been improved over the years (Tellgmann et al. 1996) and now there are companies offering EMFs at affordable prices (see e.g. http://sesres.com), still the low EMF yields limit the investigations on the EMFs reactivity to mainly the most abundant EMFs. Theoretical studies are important to predict or to give support to possible addition sites. In this chapter, studies on the exohedral reactivity of the free cages D_{3h}-C_{78}, C_2-C_{78} and the corresponding TNT EMFs derivatives $X_3N@D_{3h}$-C_{78} (X = Sc, Y) and $X_3N@C_2$-C_{78} (X = Sc, Y) will be discussed. First, the theoretical exploration of the exohedral reactivity of the free cages upon the Diels-Alder reaction with 1,3-cis-butadiene will be explored. Afterwards, the effect of the encapsulation of different TNT units on the regioselectivity and reactivity of the fullerene cage will be presented. Then, the Diels-Alder cycloaddition reaction will be investigated on the preferred isomer of C_{78} for encapsulating large TNT units such as Y_3N. Finally, the effect of the encapsulation of noble gas atoms or dimers on the reactivity of C_{60} will be briefly explored.

4.2 Chemical Reactivity of D_{3h}-C_{78} and Its Endohedral Derivative $Sc_3N@C_{78}$

Although the exohedral reactivity of free fullerenes is quite well-understood, how TNT endohedral metallofullerenes react is still unclear as different factors counteract. An increase of the reactivity might be expected taking into account that the insertion of the TNT unit leads to a higher pyramidalization of some carbon atoms. The more pyramidalized the C–C bond being attacked, the closer it is to the sp^3 bonding situation of the final adduct and the lower the deformation energy of the cage. On the other hand, the electronic transfer produced from the TNT unit to the fullerene reduces the electron affinity of the cage which implies that a reduction of the reactivity might be produced. Moreover, the LUMO orbitals of the endohedral compound are destabilized because of the charge transfer of six electrons from the metal cluster to the fullerene cage, thus disfavoring the interaction with the HOMO of the diene. In contrast to C_{80}, the rotation of the TNT unit encapsulated inside the C_{78} cage is highly impeded (Campanera et al. 2002), and therefore the study of how the reactivity of the different bonds is affected by the metal insertion can be directly investigated. The Diels-Alder [4 + 2] reaction has been studied with the BP86/TZP//BP86/DZP method over the 13 non-equivalent bonds of the D_{3h}-C_{78} and $Sc_3N@D_{3h}$-C_{78} compounds (see Fig. 4.2) (Osuna et al. 2008).

In Fig. 4.2, all non-equivalents bonds are marked in the fullerene compound, as well as the activation barriers obtained for every addition site. D_{3h}-C_{78} has seven non-equivalent [6,6] type bonds that can be classified in three subtypes: (i) pyracylenic of type A (*bonds called **1** and **7**, see Fig.* 4.2), (ii) type B (*bonds **3**, **4**, **5**, and **6***) and (iii) pyrenic of type C (*bond **2***). Furthermore, there are six [5,6]

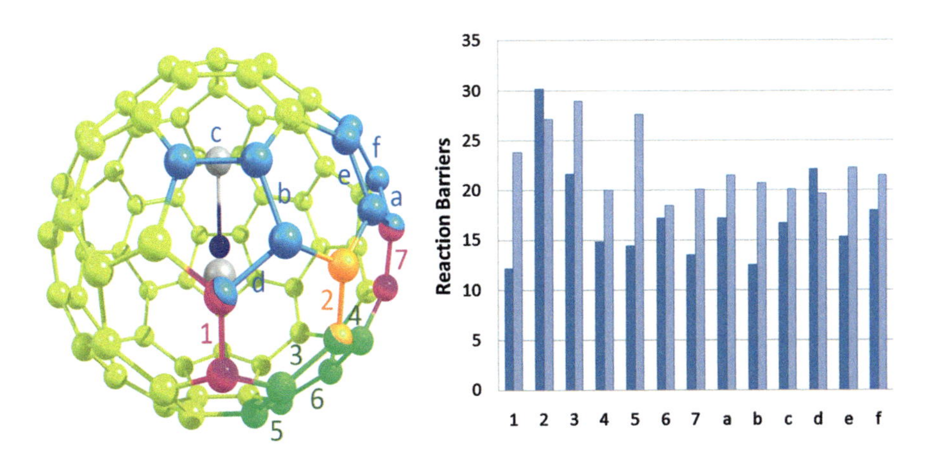

Fig. 4.2 Representation of all non-equivalents bonds of the $Sc_3N@C_{78}$ endohedral fullerene, and the activation barriers obtained for the Diels-Alder reaction on the free C_{78} (*represented in dark blue*) and the $Sc_3N@C_{78}$ (*represented in light blue*) compounds. All energies are represented in kcal·mol^{-1}

type bonds of type D (*corannulene, **a**–**f***). We will refer to each different bond according to this nomenclature, where for example number **1** is used to denote the pyracylenic or type A bond situated in the position indicated in Fig. 4.2.

The Diels-Alder reaction on the free D_{3h}-C_{78} cage is basically favored over a [5,6] bond called **b** and two type A [6,6] bonds (bonds **1** and **7**) (Osuna et al. 2008). The corannulene (type D) [5,6] bond **b** presents a reaction energy of -23.9 kcal·mol^{-1} and an activation barrier of 12.5 kcal·mol^{-1}. Pyracylenic (type A) bonds called **1** and **7** do also present favorable reaction and activation barriers (for bond **1**: $\Delta E_R = -16.0$ kcal·mol^{-1}, $\Delta E^{\ddagger} = 12.2$ kcal·mol^{-1}, and for bond **7**: $\Delta E_R = -18.8$ kcal·mol^{-1}, $\Delta E^{\ddagger} = 13.5$ kcal·mol^{-1}). It is important to remark here that pyracylene bonds correspond to the most favorable addition sites for C_{60}, which is indeed also the case for the free D_{3h}-C_{78} cage.

The encapsulation of the scandium based metal cluster inside the D_{3h}-C_{78} cage (i.e. $Sc_3N@D_{3h}$-C_{78}) involves a change in the regioselectivity of the reaction. The most reactive bonds are two type B [6,6] bonds called **6** and **4**, and one [5,6] type D bond called **c**. The reaction energies obtained are -12.7, -9.7 and -10.4 kcal·mol^{-1} for the addition over **6**, **4** and **c**, respectively. Furthermore, a huge destabilization is produced after the TNT encapsulation, as most of the considered bonds become less reactive by approximately 12–20 kcal·mol^{-1}. A high destabilization is observed for those bonds situated close to the scandium atoms (especially for **1** and **b**). The only case where the cycloaddition reaction is enhanced after the encapsulation is over bond **6** which is stabilized by ca. 17 kcal·mol^{-1}. Interestingly, the lowest activation barrier is found for bond **6** (18.5 kcal·mol^{-1}), which also presents the most exothermic reaction energy. It should be noted that the most favorable addition site in $Sc_3N@D_{3h}$-C_{78} presents an activation barrier which is 6.3 kcal·mol^{-1} higher in energy than the lowest found for the free cage (18.5 kcal·mol^{-1} for bond **6** in $Sc_3N@D_{3h}$-C_{78} as compared to 12.2 kcal·mol^{-1} for bond **1** in D_{3h}-C_{78}). However, the reaction becomes less regioselective as bonds **4**, **7**, **b**, **c**, and **d** present activation barriers within the range of 19.7–20.7 kcal·mol^{-1}. Interestingly, the 1,3-dipolar cycloaddition reaction on $Sc_3N@D_{3h}$-C_{78} yielded two cycloaddition products corresponding to the addition to the [6,6] bonds called **6** and **4** (Cai et al. 2007). Although our calculations indicate that the reaction might also be favorable over bond **c** (**7** and **b** could also be formed even though they present less exothermic reaction energies), they are in good agreement with the experimental findings.

The reactivities found for the free cage and its endohedral derivative can be described in terms of C–C bond distances, pyramidalization angles, and shapes of the lowest-lying unoccupied molecular orbitals. The most favorable addition sites present short C–C bond distances, that is indeed the case for bonds **1**, **7**, and **b** for D_{3h}-C_{78} (see Table 4.1). Similarly, bond **2** presents the longest C–C bond distance and gives a significantly endothermic reaction energy. However, there are some bonds that present similar bond distances and their reaction energies are exothermic (for example bond **c**). The most reactive bond in the case of $Sc_3N@D_{3h}$-C_{78} does also present the shortest C–C bond distance, but for instance bond **7** has the same C–C distance and its reaction energy is approximately 5 kcal·mol^{-1} less exothermic.

Table 4.1 Bond distances R_{CC} (Å) and pyramidalization angles θ_p (degrees)[a] for the bond-types in free and endohedral fullerene[b]

			C_{78}		$Sc_3N@C_{78}$	
Product	Bond-type		R_{CC}	θ_p[a]	R_{CC}	θ_p[a]
1	*A*	*[6,6]*	**1.369**	10.46	1.440	**13.80**
2	*C*	*[6,6]*	1.465	8.58	1.466	8.33
3	*B*	*[6,6]*	1.432	9.62	1.450	9.26
4	*B*	*[6,6]*	1.415	9.60	**1.426**	9.44
5	*B*	*[6,6]*	1.418	9.53	1.432	8.97
6	*B*	*[6,6]*	1.420	9.44	**1.400**	9.99
7	*A*	*[6,6]*	**1.388**	**11.64**	**1.400**	**11.21**
a	*D*	*[5,6]*	1.438	**11.64**	1.437	**11.21**
b	*D*	*[5,6]*	**1.410**	10.49	1.446	9.73
c	*D*	*[5,6]*	1.465	10.32	**1.423**	9.27
d	*D*	*[5,6]*	1.446	10.56	1.452	**12.00**
e	*D*	*[5,6]*	1.438	10.38	1.449	10.92
f	*D*	*[5,6]*	1.442	**11.13**	1.432	10.88

[a] Pyramidalization angles averaged over both atoms that constitute the bond under consideration
[b] In **boldface** the bonds that are predicted to be most reactive

In the case of the endohedral compound, the longest bond (**2**) does not possess the least favorable reaction energy. Hence, there is not an overall correlation between C–C bond distances and reaction energies, apart from the fact that the most reactive bonds do exhibit short C–C bond distances.

As it happens with bond distances, the prediction of the fullerene reactivity in terms of the pyramidalization angles is not straightforward. The most reactive sites exhibit from moderately to high values, however a large pyramidalization angle does not always correspond to an enhanced reactivity of the bond (see Table 4.1). The encapsulation of the Sc_3N moiety inside the cage leads to an increase of the pyramidalization angles, especially for those bonds situated close to the scandium influence. For instance, bond **1** in $Sc_3N@D_{3h}$-C_{78} presents the highest pyramidalization angle (13.8°), but the cycloaddition reaction over it is endothermic by 4 kcal·mol^{-1}. Therefore, the use of pyramidalization angles to predict fullerene reactivity does not always lead to the correct answer.

Finally, the cycloaddition reaction between 1,3-cis-butadiene and the fullerene compounds might also be understood in terms of the molecular orbitals of both reacting species. The most prominent interaction occurs between the HOMO of the diene and the LUMO of the fullerene, therefore those bonds presenting suitable shaped orbitals to interact with the HOMO of the diene might be the most favorable addition sites. In C_{78}, bonds **1**, **7** and **b** present suitable orbitals to interact, and are indeed the most reactive sites of the fullerene compound (see Fig. 4.3). However, several bonds present similar suitable antibonding orbitals to react with diene **1**, **2**, **3**, **4**, **6**, **7**, **c**, and **e** in the case of $Sc_3N@C_{78}$ (see Fig. 4.4). Among all bonds with suitable orbitals to interact only **6**, **4**, **7** and **c** present favorable reaction and activation energies. Moreover bond **d** does not possess suitable shaped orbitals and its reaction and activation barriers are substantially favorable. Hence, the

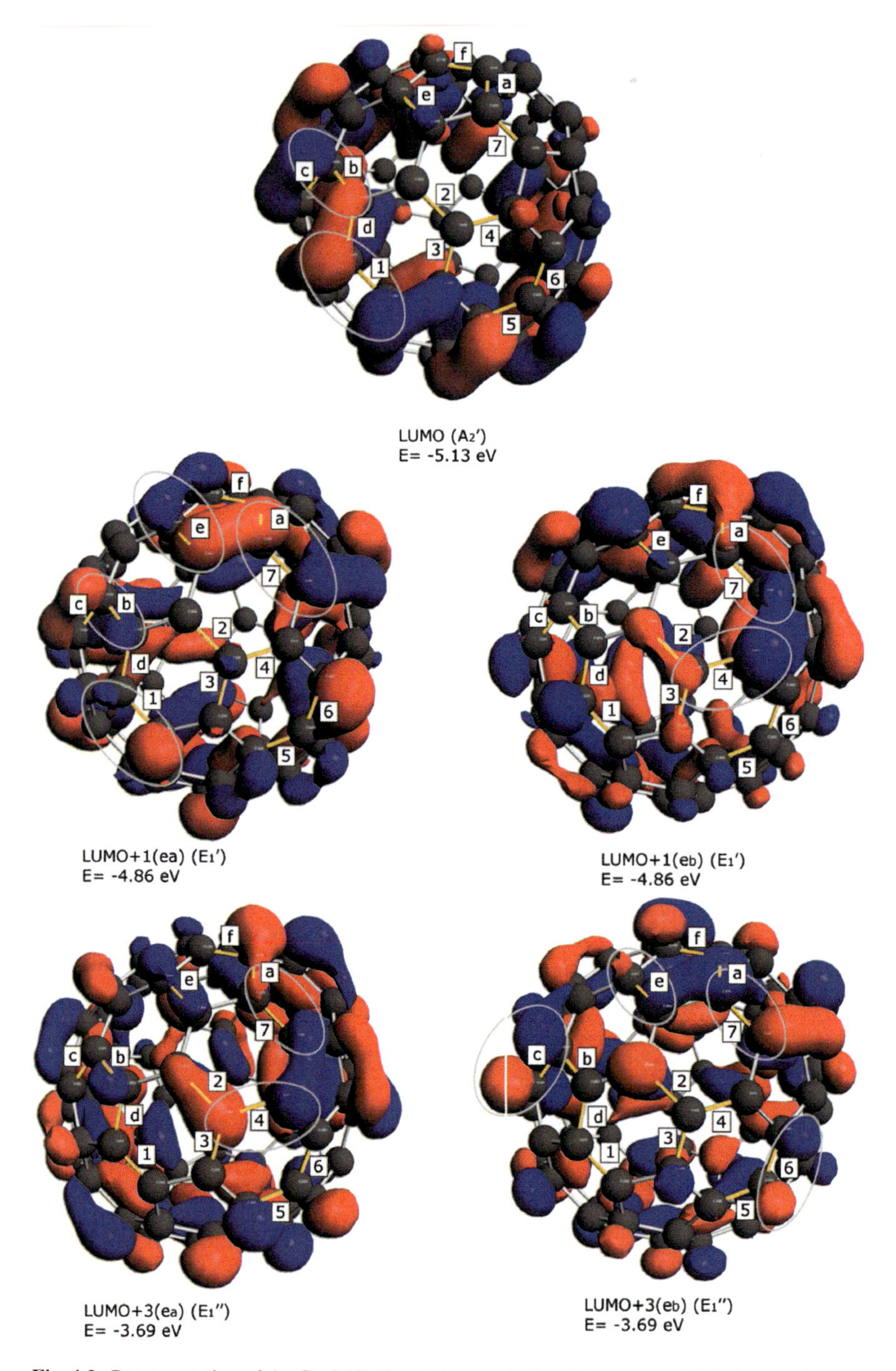

Fig. 4.3 Representation of the C_{78} LUMO and degenerate LUMO + 1 and LUMO + 3 molecular orbitals (isosurface value 0.02 a.u) where all non-equivalent [6,6] and [5,6] bonds have been marked. Those bonds with favorable orbitals to interact with the HOMO of the diene are marked with ellipses

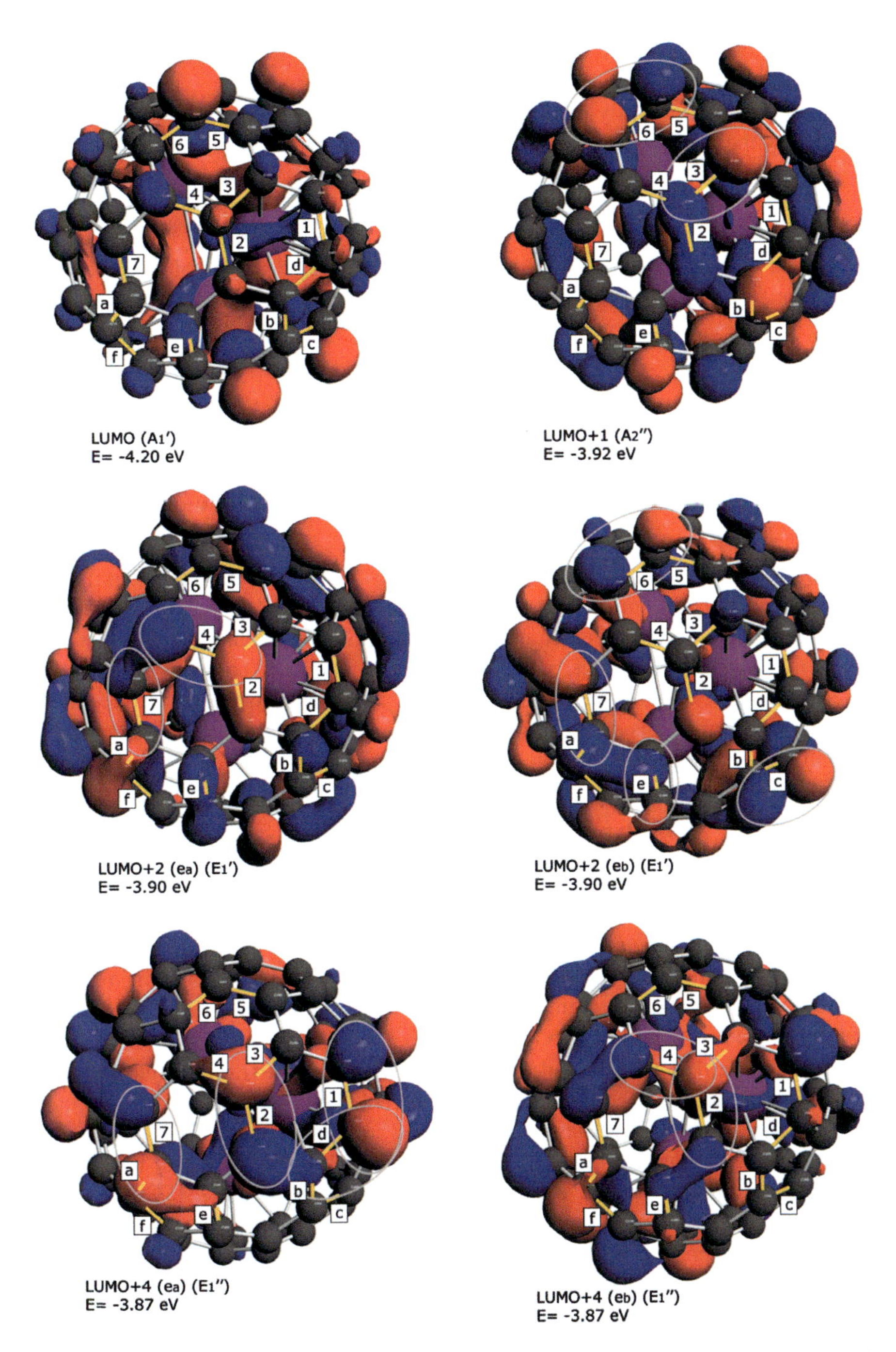

Fig. 4.4 Representation of the $Sc_3N@C_{78}$ LUMO, LUMO + 1, and degenerate LUMO + 2 and LUMO + 4 molecular orbitals (isosurface value 0.02 a.u.) where all non-equivalent [6,6] and [5,6] bonds have been marked. Those bonds with favorable orbitals to interact with the HOMO of the diene are marked with ellipses

predictions of reactivity for fullerene compounds using the LUMO orbitals are too imprecise, as one finds many bonds suitable to interact.

Although the previously mentioned descriptors do not give accurate results for describing the exohedral reactivity of the cages, the combination of all three descriptors (*C–C bond distances, pyramidalization angles and molecular orbital analysis*) gives quite successful results. Only bonds **1**, **7** and **b** in the case of C_{78}, and bonds **4**, **6**, **7**, and **c** in $Sc_3N@C_{78}$ fulfill the three criteria. They exhibit short C–C bond distances, relatively high pyramidalization angles and suitable orbitals to interact with diene. And in fact, our thermodynamic and kinetic study indeed shows these bonds to be most reactive.

4.3 The Diels-Alder Reaction on Endohedral $Y_3N@C_{78}$: The Importance of the Fullerene Strain Energy

In some experimental studies, it was observed that the exohedral reactivity of the TNT endohedral metallofullerenes is highly affected by the nature of the encapsulated cluster (Cardona et al. 2005a). Our initial study involving the Diels-Alder reaction on the endohedral scandium based fullerene compound has been extended to directly compare how the reactivity is affected by encapsulating either scandium or yttrium inside the cage (Osuna et al. 2009a). In the first part of this section, the preferred addition sites for the $Y_3N@D_{3h}$-C_{78} molecule will be thoroughly described as well as compared to the previously reported $Sc_3N@D_{3h}$-C_{78} and D_{3h}-C_{78} (Osuna et al. 2008). Finally, an insight into the exohedral reactivity of the most favorable isomer for the encapsulation of the large Y_3N unit is presented.

The large yttrium based TNT cluster is forced to adopt a pyramidal configuration inside the D_{3h}-C_{78} cage, and two clearly differentiated areas are present (see Fig. 4.5): the so-called *up* region, which is more influenced by the nitrogen atom, and the *down* part which has the yttrium atoms in close contact. In every region, 13 non equivalent bonds might be considered to take into account all possible addition sites: two type A [6,6] bonds (**1** and **7**), four type B [6,6] bonds (**3**, **4**, **5**, **6**), one type C [6,6] bond (**2**), and 6 type D [5,6] bonds (**a-f**). Our study of the Diels-Alder reaction at the ZORA-BP86/TZP//ZORA-BP86/DZP level on both faces of the fullerene indicates that both areas are equally reactive with energy differences of at most 1.6 kcal·mol^{-1}. The most stable regioisomer for the Diels-Alder cycloaddition reaction over the endohedral compound $Y_3N@D_{3h}$–C_{78} is shown to be favored over the [5,6] bond **d** that exhibits the longest bond distance in the initial fullerene ($\Delta E_R = -15.0$ kcal·mol^{-1}, $\Delta E^{\ddagger} = 17.1$ kcal·mol^{-1}). As far as we know, this is the first case of a cycloaddition reaction where the most stable addition is obtained over one of the longest C–C bonds in the cage. This observation is of significance as those bonds with the shortest bond distances are usually related with the most reactive positions. Therefore, bond distances cannot be considered a predictor of fullerene reactivity anymore, and as a consequence those studies

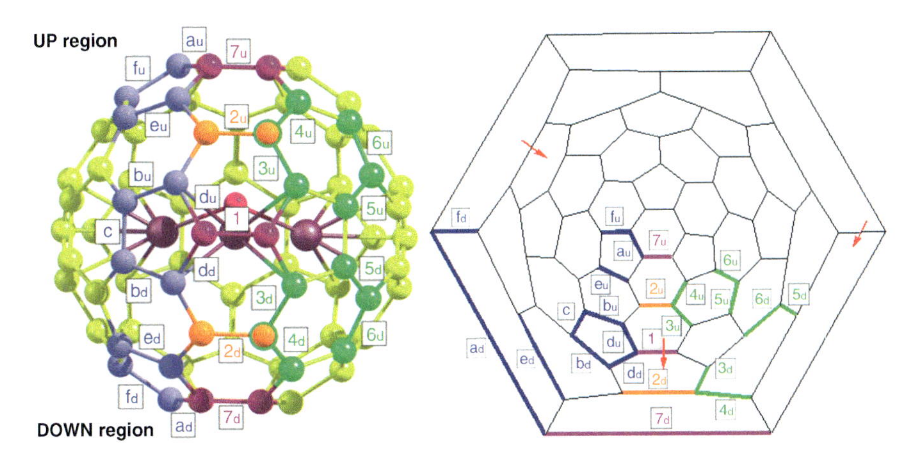

Fig. 4.5 Representation of all non-equivalents bonds of the $Y_3N@D_{3h}$-C_{78}. The Schlegel diagram of the fullerene (2D representation) is also depicted where the non-equivalent bonds are marked. The Y_3N cluster presents a pyramidal configuration and therefore two clearly differentiated areas exist. The up region is more affected by the nitrogen atom, whereas the *down* area is more influenced by the yttrium atoms

where only short bonds were investigated might not give the correct picture of the reactivity of endohedral fullerene compounds.

The second most favorable regioisomer corresponds to the addition over a type B [6,6] bond called **6** ($\Delta E_R = -11.0$ kcal·mol^{-1}, $\Delta E^{\ddagger} = 18.3$ kcal·mol^{-1}). Finally, although the reaction energy for the cycloaddition reaction to the [5,6] bond called **e** is hardly exothermic (−4.1 kcal·mol^{-1}), it does present a low activation barrier (17.2 kcal·mol^{-1}). Moreover, there is a difference of 4.1 kcal·mol^{-1} between the activation barrier of bond **e** situated in the *down* and *up* areas. The enhanced reactivity of bond **e** situated in the down region is basically attributed to the presence of suitable shaped orbitals to interact with diene at lower energy. Moreover, the cycloaddition reaction over bond $\mathbf{e_u}$ (i.e. situated in the up region) is disfavored as it breaks an attractive interaction between the N atom and this $\mathbf{e_u}$ bond.

By comparing the same Diels-Alder reaction over the related compounds D_{3h}-C_{78}, $Sc_3N@D_{3h}$-C_{78}, and $Y_3N@D_{3h}$-C_{78} different reactivity patterns are observed (see Fig. 4.6). For the free cage, the reaction is favored over the [5,6] bond called **b**. The second and third most stable regiosiomers correspond to the addition to the pyracylenic [6,6] bonds called **7** and **1**, respectively. Once the scandium based TNT cluster is encapsulated inside, the addition is basically preferred to the type B [6,6] bond called **6**. The other favorable interactions are over the type B [6,6] bond **4** and the type D [5,6] **c**. It should be emphasized here, that the most reactive bonds in $Sc_3N@D_{3h}$-C_{78} exhibit short C–C bond distances, relatively high pyramidalization angles and are situated far away from the scandium influence. In contrast to $Sc_3N@D_{3h}$-C_{78}, the reaction in the case of $Y_3N@D_{3h}$-C_{78} is basically favored over bond **d** having one of the yttrium atoms in close contact. This preference for reacting with a bond situated close to the yttrium atoms is due to two different factors. First, the D_{3h} cage is extremely deformed, especially in the

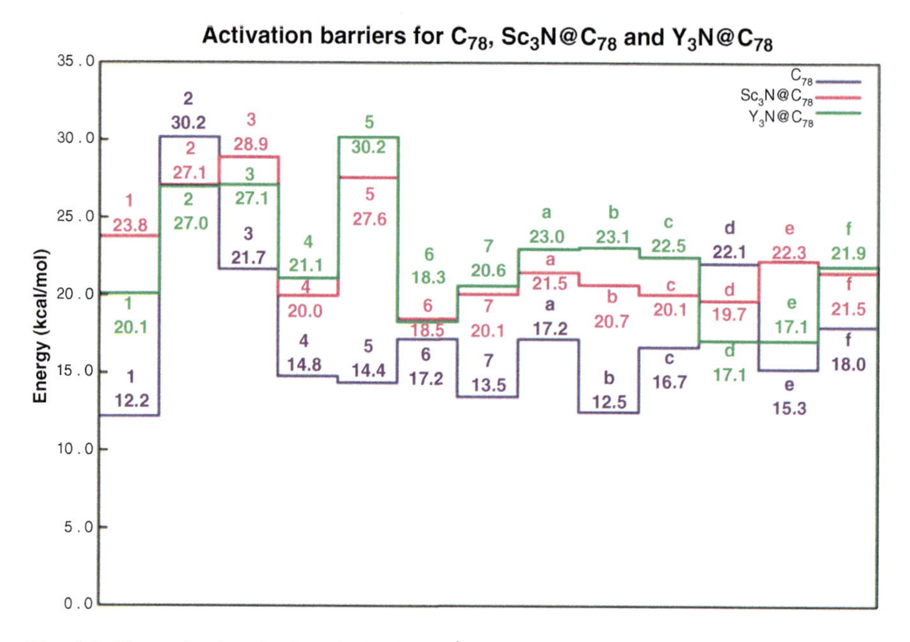

Fig. 4.6 The activation barriers in kcal·mol^{-1} obtained for D_{3h}-C_{78} (*represented in lilac*), $Sc_3N@D_{3h}$-C_{78} (*in pink*) and $Y_3N@D_{3h}$-C_{78} (*in green*)

pyracylenic areas situated close to the yttrium atoms which contain the most reactive bonds, thus the attack reduces the strain energy of the cage. Second, in the final adduct the Y_3N cluster gets additional space to adopt a more planar configuration. The C–C bond of the attacked bond **d** is practically broken and an open fulleroid is obtained. The addition to bond **d** is preferred as the diene has to be deformed to a lesser extent to react (*in the case of bonds **1** and **3** situated close to the yttrium atoms, the deformation of the diene is approximately 22 kcal·mol^{-1}, whereas only 14 in the case of **d***).

As observed in the previous section, the encapsulation of Sc_3N inside the D_{3h} cage produces a decrease of the exohedral reactivity. It is basically governed by the electronic charge transfer from the TNT to the fullerene that leads to LUMOs higher in energy. Most of the considered bonds in the case of $Y_3N@D_{3h}$-C_{78} slightly decrease their reactivity, which is consistent with the relatively larger HOMO-LUMO gap found for $Y_3N@D_{3h}$-C_{78} (1.26 and 1.22 eV for the yttrium and scandium based metallofullerenes, respectively) and the higher electron transfer produced in the case of yttrium.

4.4 The Diels-Alder Reaction on the C_2: 22010 Cage

The most favorable C_{78} cage to encapsulate the large Y_3N cluster is the non-IPR C_2: **22010** isomer where the TNT moiety can adopt a planar configuration (Popov and Dunsch 2007). The difference in energy between $Y_3N@D_{3h}$-C_{78} and $Y_3N@C_2$-C_{78}

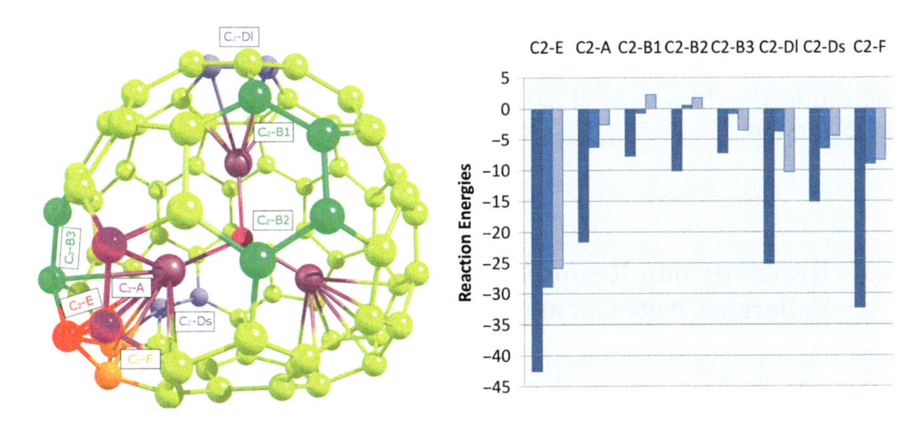

Fig. 4.7 Representation of the selected bonds of the $Y_3N@C_2$-C_{78} compound. The reaction energies obtained for the different cases studied: C_2-C_{78} (*represented in dark blue*), $Sc_3N@C_2$-C_{78} (*in blue*), and $Y_3N@C_2$-C_{78} (*in light blue*) are expressed in kcal·mol^{-1}. Different colors are used to indicate the different bond types studied: *pink*, [6,6] type A; *green*, [6,6] type B; *blue*, [5,6] type D; *red*, [5,5] type E; *orange*, [5,6] type F

is 20.2 kcal·mol^{-1} at ZORA-BP86/TZP//ZORA-BP86/DZP. The latter is similar to the difference of 21.1 kcal·mol^{-1} between the two synthesized and exohedrally functionalized D_{5h} and I_h cages of the C_{80} fullerene, which are both experimentally attainable (Popov and Dunsch 2007). Among all non-equivalent bonds of the C_2: **22010** cage, eight bonds were selected on the basis of the reactivity trends observed in the D_{3h} cage: one type E [5,5] bond only present in the non-IPR cages (called **C_2-E**), one type F [5,6] bond (**C_2-F**), two type B [6,6] bonds with short bond distances and situated far away from the metals (**C_2-B1**, **C_2-B2**), another type B [6,6] bond situated near one of the yttrium atoms (**C_2-B3**), one type D [5,6] bond with large C–C bond distances and positioned close to the yttrium metal (**C_2-Dl**), another type D [5,6] bond with short bond distance and situated far away from the yttrium influence (**C_2-Ds**), and finally one pyracylene [6,6] bond called **C_2-A** close to the yttrium atom (see Fig. 4.7).

Interestingly, the Diels-Alder reaction on $Y_3N@C_2$-C_{78} is favored over the [5,5] bond called **C_2-E** which has one of the yttrium atoms directly coordinated towards it. As far as we know, the reactivity of these [5,5] bonds was never assessed before. Although Campanera and coworkers predicted a low reactivity of these non-IPR bonds on the basis of the Mayer Bond Order analysis (Campanera et al. 2006), our theoretical findings indicate that the reaction is substantially exothermic (−25.9 kcal·mol^{-1}) and highly stereoselective. The reaction over the rest of the considered bonds is from 15.6 to 28.1 kcal·mol^{-1} less favorable. This observed tendency to react with those bonds situated close to the metal atoms might either be influenced by the presence of the yttrium atoms or be dictated by the C_2 cage. Hence, the Diels-Alder reaction was also assessed in the case of the free C_2 and the scandium based endohedral derivative. Interestingly, the reaction is found to be favored over the same [5,5] bond called **C_2-E** in both C_2-C_{78} and $Sc_3N@C_2$-C_{78} compounds (the reaction energies obtained are −42.6 and −28.9 kcal·mol^{-1}, respectively).

Therefore, our theoretical calculations indicate that the exohedral functionalization of synthesized $Tm_3N@C_{78}$ (Krause et al. 2005), $Dy_3N@C_{78}$ (Popov et al. 2007) and $Gd_3N@C_{78}$ (Beavers et al. 2009) might be stereoselectively produced over the [5,5] bonds.

4.5 Reactivity and Regioselectivity of Noble Gas Endohedral Fullerenes $Ng@C_{60}$ and $Ng_2@C_{60}$ (Ng = He-Xe)

Krapp and Frenking performed a theoretical study on the noble gas dimers endohedral fullerenes $Ng_2@C_{60}$ (Ng = He-Xe) (Krapp and Frenking 2007). Interestingly, they observed that an electron transfer of 1–2 electrons is produced in the case of the larger noble gas homologues, in particular for the Xe_2 dimer. Free noble gas dimers are rarely observed, however a genuine chemical bond is formed once the Xe_2 unit is trapped inside the fullerene moiety. In addition to that, the encapsulation of Ar_2, Kr_2 and Xe_2 was found to affect the C–C bond distances of the C_{60} compound as well as the pyramidalization angles. Therefore, a change on the exohedral reactivity might be observed. In this section, the Diels-Alder reaction is discussed either for the single noble gas endohedral compounds $Ng@C_{60}$ (Ng = He-Xe) and the noble gas dimers endohedral fullerenes $Ng_2@C_{60}$ at the ZORA-BP86/TZP level of theory (Osuna et al. 2009b).

First, the Diels-Alder reaction between 1,3-cis-butadiene and C_{60} has been studied as reference. The reaction is favored over the pyracylene [6,6] bond that presents a reaction energy of -20.7 kcal·mol^{-1} and an activation barrier of 12.7 kcal·mol^{-1}. The [5,6] bonds are substantially less reactive as the reaction and activation energies obtained are 15.4 and 8.3 kcal·mol^{-1} less favorable. The noble gas encapsulation hardly affects the exohedral reactivity of the cage, i.e. differences of less than 0.4 kcal·mol^{-1} were observed for both the reaction energies and barriers.

More interesting results were obtained for the case of the noble gas dimer encapsulation. Krapp and Frenking studied the cage isomerism of the noble gas endohedral derivatives and observed that the most stable structure was the D_{3d} isomer for He-Kr, and the D_{5d} isomer for Xe (Krapp and Frenking 2007). However, the energy differences between the different isomers were found to be very low. Therefore, we decided to study the Diels-Alder reaction on the D_{5d} isomer for all noble gases for many reasons. First, the comparison of the different bonds can only be done considering the same isomer for all cases studied. Second, the most interesting compound to study is the xenon-based endohedral fullerene because of the electron transfer produced. Finally, the energies for the encapsulation of the He-Kr atoms inside the D_{5d} isomer differed by less than 2 kcal·mol^{-1} from the D_{3d} equivalents. Note that for the D_{5d} isomer there are six non-equivalent type D [5,6] bonds (called **a**, **b**, **c**, **d**, **e**, and **f**) and three type A [6,6] bonds (called **1**, **2**, and **3**) (see Fig. 4.8).

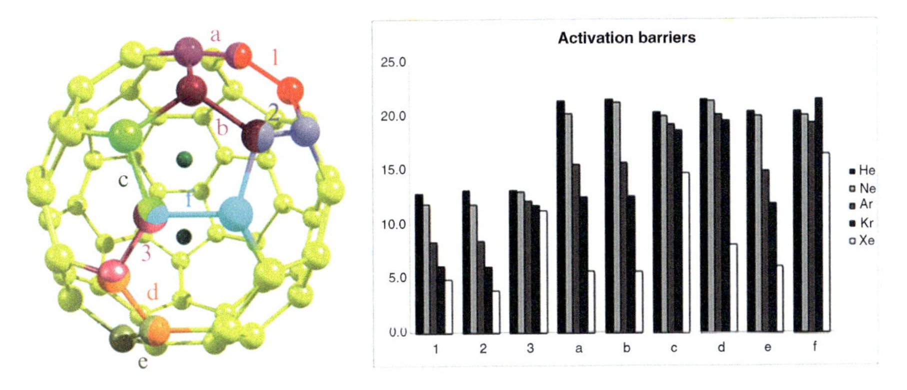

Fig. 4.8 Representation of all non-equivalent bonds of the $Ng_2@C_{60}$ compound. The activation energies (in kcal·mol^{-1}) corresponding to the Diels-Alder cycloaddition reaction between 1,3-butadiene and all non-equivalent bonds for all considered noble gas endohedral compounds. $Ng_2@C_{60}$ has been represented on the *right*. A grey scale has been used to represent the different noble gases endohedral compounds: *black color* is used to represent the helium-based fullerene, *light grey* for neon, *medium grey* for argon, *dark grey* for krypton, and *white* for xenon

The Diels-Alder reaction produced on the lighter noble gas dimer compounds (i.e. $He_2@C_{60}$ and $Ne_2@C_{60}$) presents reaction and activation barriers that are close to the ones obtained for free C_{60}. I.e., the reaction energies for the most reactive bond **1** are compared to the free fullerene 0.2 and 2.4 kcal·mol^{-1} more favorable for the helium and neon dimer compounds, respectively. Likewise, the activation barrier for the addition to bond **1** is 12.8 and 11.9 kcal·mol^{-1} for the $He_2@C_{60}$ and $Ne_2@C_{60}$ cases, respectively. The other [6,6] bonds present similar reaction and activation energies, whereas [5,6] bonds are much more less reactive. It is important to remark that the addition of 1,3-butadiene produces a rotation of the noble gas dimer which is reoriented during the course of the reaction from the initial position to face the attacked bond.

Once Ar_2 and Kr_2 are inserted inside C_{60}, the reaction becomes substantially more exothermic (−32.2 and −39.9 kcal·mol^{-1} for bonds **1** and **2**, respectively), and the activation barriers are largely reduced (to ca. 8 and 6 kcal·mol^{-1} for the Ar_2 and Kr_2 compounds, respectively). The addition to the [6,6] bond **3** is less favored, as the noble gas moiety is not totally reoriented to face the attacked bond. Of course, the larger the noble gas atom, the more impeded the rotation of the noble gas dimer inside the cage. Hence, for the larger noble gas endohedral compounds the addition is favored over those bonds situated close to the C_5 axis where the dimer is initially contained. This lack of rotation leads to substantially less favored reaction and activation barriers.

The preferred addition site for the Xe-based compound corresponds to [6,6] bond **1** (−44.9 kcal·mol^{-1}), however the [5,6] bonds **a**, **b** and **e** do also present favorable reaction energies (−44.6,−44.5, and−45.5 kcal·mol^{-1}, respectively). On the other hand, the lowest activation energy is found for the [6,6] bond **2** (3.8 kcal·mol^{-1}), nonetheless bonds **1**, **a**, **b**, and **e** also present low energy barriers (4.9, 5.7, 5.6, 6.1 kcal·mol^{-1}, respectively). Therefore, the reaction is no longer regioselective as five (!) regioisomers might be formed during the reaction between 1,3-butadiene and $Xe_2@C_{60}$.

The enhanced reactivity along the series $He_2@C_{60}$ < $Ne_2@C_{60}$ < $Ar_2@C_{60}$ < $Kr_2@C_{60}$ < $Xe_2@C_{60}$ can be attributed to several factors. First, the HOMO-LUMO gap is reduced from 1.63 eV for $He_2@C_{60}$ to 0.75 eV for $Xe_2@C_{60}$ (for the free cage it is 1.66 eV), which is basically produced by a slight stabilization of the LUMO and a major destabilization of the HOMO. The latter is a complex situation as the HOMO for the lighter noble gas compounds (a_{1u} orbital, for He-Ar) is different to that of xenon and krypton fullerenes (a_{2u} orbital that primarily presents antibonding σ^* orbitals in the noble gas dimer unit). The destabilization of the a_{2u} orbital increases from He to Xe because of the reduction of the Ng-Ng distance along the series. Second, the deformation energy of the cage also plays an important role. The encapsulation of He_2 and Ne_2 inside C_{60} hardly affects the cage as the calculated deformation energies are 0 and less than 1 kcal·mol^{-1}, respectively. However, the insertion of the larger Ar_2, Kr_2, and Xe_2 leads to a deformation energy of 11.2, 22.5 and 34.1 kcal·mol^{-1}, respectively. The high deformation energy found, especially for the xenon-based compound, leads to a highly strained cage where all [5,6] and [6,6] bonds situated close to the initial position of the Xe_2 dimer are equally reactive. The reaction is then extremely exothermic and unselective as the strain of the cage is partially released after reaction. Finally, the Ng–Ng bond distance elongation does also contribute to the enhanced reactivity for the heavier noble gas compounds. After reaction, the Ng-Ng distance is increased by 0.028, 0.043, 0.040, 0.035, and 0.054 Å along the He_2–$Xe_2@C_{60}$ series which corresponds to an stabilization of −0.2, −1.0, −4.1, −5.3, and −10.4 kcal·mol^{-1}. This decompression represents an important contribution to the exothermicity of the reaction for those bonds where the Ng dimer is reoriented facing the attacked bond.

4.6 Conclusions

The effect of the encapsulation of trimetallic nitride (TNT) complexes or noble gas dimers on the exohedral reactivity of fullerene cages is profound. Not only does the encapsulation affect the reactivity, it also changes the regioselectivity patterns. For the TNT complexes, a reduction in the reactivity is observed corresponding to an increase of the barriers by some 6 kcal·mol^{-1}, and a decrease of the reaction energy by some 12–20 kcal·mol^{-1}. The preferred addition sites for the free C_{78} fullerene are totally different from those for $Sc_3N@C_{78}$, which are again radically different for $Y_3N@C_{78}$. Both the free and $Sc_3N@C_{78}$ fullerenes prefer to react over C–C bonds with short distances, which in the case of $Sc_3N@C_{78}$ are located far away from (the influence of) the scandium atoms. In contrast, the $Y_3N@C_{78}$ fullerene preferably reacts over long C–C bonds, close to the yttrium atoms. This latter is in part attributed to the deformation of the cage.

The deformation of the cage also plays a role for the encapsulation of noble gas dimers in C_{60}, but there it leads to drastically more reactive compounds. I.e. the larger the noble-gas atoms, the smaller the reaction barrier and the more exothermic are the products. Similar to $Y_3N@C_{78}$ this results primarily from a strained fullerene, which is (partially) released upon reaction. Also the decompression of the noble gas dimer contributes, as is the major destabilization of the HOMO orbital. For the xenon-dimer fullerene, which is characterized by a charge transfer of one to two electrons to the fullerene, many reactive bonds are found and there is almost no regioselectivity anymore.

Acknowledgments The following organizations are thanked for financial support: the Ministerio de Ciencia e Innovación (MICINN, project numbers CTQ2008-03077/BQU and CTQ2008-06532/BQU), and the DIUE of the Generalitat de Catalunya (project numbers 2009SGR637 and 2009SGR528). Excellent service by the Centre de Supercomputació de Catalunya (CESCA) is gratefully acknowledged. The authors also are grateful to the computer resources, technical expertise, and assistance provided by the Barcelona Supercomputing Center – Centro Nacional de Supercomputación (BSC-CNS, MareNostrum). Support for the research of M. Solà was received through the ICREA Academia 2009 prize for excellence in research funded by the DIUE of the Generalitat de Catalunya.

Appendix: Computational Details

All Density Functional Theory (DFT) calculations were performed with the Amsterdam Density Functional (ADF) program (Baerends et al. 2009; te Velde et al. 2001) and the related QUILD (QUantum-regions Interconnected by Local Descriptions) (Swart and Bickelhaupt 2008). The molecular orbitals (MOs) were expanded in an uncontracted set of Slater type orbitals (STOs) of triple-ζ (TZP) quality containing diffuse functions and one set of polarization functions. Core electrons (1s for 2nd period, 1s2s2p for 3rd-4th period) were not treated explicitly during the geometry optimizations (frozen core approximation) (te Velde et al. 2001), as it was shown to have a negligible effect on the obtained geometries (Swart and Snijders 2003). An auxiliary set of s, p, d, f, and g STOs was used to fit the molecular density and to represent the Coulomb and exchange potentials accurately for each SCF cycle. Energies and gradients were calculated using the local density approximation (Slater exchange and VWN correlation) (Vosko et al. 1980) with non-local corrections for exchange (Becke 1988) and correlation (Perdew 1986) included self-consistently (i.e. the BP86 functional). For the studies with heavier elements, relativistic corrections were included self-consistently using the Zeroth Order Regular Approach (ZORA) (van Lenthe et al. 1993; te Velde et al. 2001).

The actual geometry optimizations and transition state searches were performed with the QUILD program (Swart and Bickelhaupt 2008). QUILD constructs all input files for ADF, runs ADF, and collects all data; ADF is used only for the generation of the energy and gradients. Furthermore, the QUILD program uses improved geometry optimization techniques, such as adapted delocalized coordinates (Swart and Bickelhaupt 2006) and specially constructed model Hessians with the appropriate number of eigenvalues (Swart and Bickelhaupt 2006, 2008). The latter is of particular use for TS searches. All TSs have been characterized by computing the analytical vibrational frequencies, to have one (and only one) imaginary frequency corresponding to the approach of the reacting molecules.

References

Agnoli AL, Jungmann D, Lochner B (1987) Neurosurg Rev 10:25–29

Aihara J-i (2001) Chem Phys Lett 343:465–469

Akasaka T, Nagase S (2002) Endofullerenes: a new family of carbon clusters. Kluwer Academic, Dordrecht

Baerends EJ, Autschbach J, Bashford D, Berger JA, Bérces A, Bickelhaupt FM, Bo C, de Boeij PL, Boerrigter PM, Cavallo L, Chong DP, Deng L, Dickson RM, Ellis DE, van Faassen M, Fan L, Fischer TH, Fonseca Guerra C, Giammona A, Ghysels A, van Gisbergen SJA, Götz AW, Groeneveld JA, Gritsenko OV, Grüning M, Harris FE, van den Hoek P, Jacob CR, Jacobsen H, Jensen L, Kadantsev ES, van Kessel G, Klooster R, Kootstra F, Krykunov MV, van Lenthe E, Louwen JN, McCormack DA, Michalak A, Mitoraj M, Neugebauer J, Nicu VP, Noodleman L, Osinga VP, Patchkovskii S, Philipsen PHT, Post D, Pye CC, Ravenek W, Rodríguez JI, Romaniello P, Ros P, Schipper PRT, Schreckenbach G, Seth M, Snijders JG, Solà M, Swart M, Swerhone D, te Velde G, Vernooijs P, Versluis L, Visscher L, Visser O, Wang F, Wesolowski T.A, van Wezenbeek EM, Wiesenekker G, Wolff SK, Woo TK, Yakovlev AL, Ziegler T (2009) ADF 2009.01. SCM, Amsterdam

Beavers CM, Chaur MN, Olmstead MM, Echegoyen L, Balch AL (2009) J Am Chem Soc 131:11519–11524

Becke AD (1988) Phys Rev A 38:3098–3100

Cai T, Ge ZX, Iezzi EB, Glass TE, Harich K, Gibson HW, Dorn HC (2005) Chem Commun 3594–3596

Cai T, Xu L, Anderson MR, Ge Z, Zuo T, Wang X, Olmstead MM, Balch AL, Gibson HW, Dorn HC (2006) J Am Chem Soc 128:8581–8589

Cai T, Xu L, Gibson HW, Dorn HC, Chancellor CJ, Olmstead MM, Balch AL (2007) J Am Chem Soc 129:10795–10800

Cai T, Xu L, Shu C, Champion HA, Reid JE, Anklin C, Anderson MR, Gibson HW, Dorn HC (2008) J Am Chem Soc 130:2136–2137

Campanera JM, Bo C, Olmstead MM, Balch AL, Poblet JM (2002) J Phys Chem A 106:12356–12364

Campanera JM, Bo C, Poblet JM (2005) Angew Chem Int Ed 44:7230–7233

Campanera JM, Bo C, Poblet JM (2006) J Org Chem 71:46–54

Cao B, Nikawa H, Nakahodo T, Tsuchiya T, Maeda Y, Akasaka T, Sawa H, Slanina Z, Mizorogi N, Nagase S (2008) J Am Chem Soc 130:983–989

Cardona CM, Kitaygorodskiy A, Echegoyen L (2005a) J Am Chem Soc 127:10448–10453

Cardona CM, Kitaygorodskiy A, Ortiz A, Herranz MA, Echegoyen L (2005b) J Org Chem 70:5092–5097

Cardona CM, Elliott B, Echegoyen L (2006) J Am Chem Soc 128:6480–6485

Chai Y, Guo T, Jin C, Haufler RE, Chibante LPF, Fure J, Wang L, Alford JM, Smalley RE (1991) J Phys Chem 95:7564–7568

Chaur MN, Melin F, Athans AJ, Elliott B, Walker BC, Holloway K, Echegoyen L (2008) Chem Commun 2665

Chaur MN, Melin F, Ortiz AL, Echegoyen L (2009) Angew Chem Int Ed 48:7514–7538

Chen N, Fan LZ, Tan K, Wu YQ, Shu CY, Lu X, Wang C-R (2007a) J Phys Chem C 111:11823–11828

Chen N, Zhang E-Y, Tan K, Wang C-R, Lu X (2007b) Org Lett 9:2011–2013

Diener MD, Alford JM, Kennel SJ, Mirzadeh S (2007) J Am Chem Soc 129:5131–5138
Dunsch L, Yang S (2007) Small 3:1298–1320
Echegoyen L, Chancellor CJ, Cardona CM, Elliott B, Rivera J, Olmstead MM, Balch AL (2006) Chem Commun 2653–2655
Guha S, Nakamoto K (2005) Coord Chem Rev 249:1111–1132
Guldi DM, Feng L, Radhakrishnan SG, Nikawa H, Yamada M, Mizorogi N, Tsuchiya T, Akasaka T, Nagase S, Herranz MA, Martín N (2010) J Am Chem Soc 1332:9078–9086
Haddon RC (2001) J Phys Chem A 105:4164–4165
Haddon RC, Chow SY (1998) J Am Chem Soc 120:10494–10496
Harneit W (2002) Phys Rev A 65:032322
Heath JR, O'Brien SC, Zhang Q, Liu Y, Curl RF, Kroto HW, Tittel FK, Smalley RE (1985) J Am Chem Soc 107:7779–7780
Hu H, Cheng W-D, Huang S-H, Xie Z, Zhang H (2008) J Theor Comput Chem 7:737–749
Iiduka Y, Ikenaga O, Sakuraba A, Wakahara T, Tsuchiya T, Maeda Y, Nakahodo T, Akasaka T, Kako M, Mizorogi N, Nagase S (2005) J Am Chem Soc 127:9956–9957
Kobayashi K, Nagase S, Yoshida M, Osawa E (1997) J Am Chem Soc 119:12693–12694
Krapp A, Frenking G (2007) Chem Eur J 13:8256–8270
Krause M, Wong J, Dunsch L (2005) Chem Eur J 11:706–711
Kroto HW (1987) Nature 329:529–531
Kroto HW, Heath JR, O'Brien SC, Curl RF, Smalley RE (1985) Nature 318:162–163
Laus S, Sitharaman B, Tóth É, Bolskar RD, Helm L, Wilson LJ, Merbach AE (2007) J Phys Chem C 111:5633–5639
Lee HM, Olmstead MM, Iezzi E, Duchamp JC, Dorn HC, Balch AL (2002) J Am Chem Soc 124:3494–3495
Lu X, Nikawa H, Nakahodo T, Tsuchiya T, Ishitsuka MO, Maeda Y, Akasaka T, Toki M, Sawa H, Slanina Z, Mizorogi N, Nagase S (2008) J Am Chem Soc 130:9129–9136
Lu X, Nikawa H, Feng L, Tsuchiya T, Maeda Y, Akasaka T, Mizorogi N, Slanina Z, Nagase S (2009) J Am Chem Soc 131:12066–12067
Martín N (2006) Chem Commun 2093–2104
Mayer I (1983) Chem Phys Lett 97:270–274
Osuna S, Swart M, Campanera JM, Poblet JM, Solà M (2008) J Am Chem Soc 130:6206–6214
Osuna S, Swart M, Solà M (2009a) J Am Chem Soc 131:129–139
Osuna S, Swart M, Solà M (2009b) Chem Eur J 15:13111–13123
Parr RG, Chattaraj PK (1991) J Am Chem Soc 113:1854–1855
Pearson RG (1997) Chemical Hardness: applications from molecules to solids. Wiley-VCH, Oxford
Pearson RG (1999) J Chem Educ 76:267–275
Perdew JP (1986) Phys Rev B 33:8822–8824, Erratum: ibid. **34**, 7406–7406 (1986)
Pietzak B, Weidinger K-P, Dinse A, Hirsch A (2002) In: Akasaka T, Nagase S (eds) Endofullerenes: a new family of carbon clusters. Kluwer Academic, Amsterdam, pp 13–66
Popov AA, Dunsch L (2007) J Am Chem Soc 129:11835–11849
Popov AA, Dunsch L (2009) Chem Eur J 15:9707–9729
Popov AA, Krause M, Yang S, Wong J, Dunsch L (2007) J Phys Chem B 111:3363–3369
Rodríguez-Fortea A, Campanera JM, Cardona CM, Echegoyen L, Poblet JM (2006) Angew Chem Int Ed 45:8176–8180
Schmalz TG, Seitz WA, Klein DJ, Hite GE (1988) J Am Chem Soc 110:1113–1127
Shultz MD, Duchamp JC, Wilson JD, Shu C-Y, Ge J, Zhang J, Gibson HW, Fillmore HL, Hirsch JI, Dorn HC, Fatouros PP (2010) J Am Chem Soc 132:4980–4981
Stevenson S, Fowler PW, Heine T, Duchamp JC, Rice G, Glass T, Harich K, Hajdu E, Bible R, Dorn HC (2000) Nature 408:427–428
Stevenson S, Stephen RR, Amos TM, Cadorette VR, Reid JE, Phillips JP (2005) J Am Chem Soc 127:12776–12777
Swart M, Bickelhaupt FM (2006) Int J Quantum Chem 106:2536–2544

Swart M, Bickelhaupt FM (2008) J Comput Chem 29:724–734
Swart M, Snijders JG (2003) Theor Chem Acc 110:34–41, Erratum, (2004) Theor Chem Acc 111:56
te Velde G, Bickelhaupt FM, Baerends EJ, Fonseca Guerra C, van Gisbergen SJA, Snijders JG, Ziegler T (2001) J Comput Chem 22:931–967
Tellgmann R, Krawez N, Lin S-H, Hertel IV, Campbell EEB (1996) Nature 382:407–408
Thilgen C, Diederich F (2006) Chem Rev 106:5049–5135
Torrent-Sucarrat M, Luis JM, Duran M, Solà M (2001) J Am Chem Soc 123:7951–7952
Valencia R, Rodríguez-Fortea A, Poblet JM (2007) Chem Commun 4161–4163
Valencia R, Rodríguez-Fortea A, Clotet A, de Graaf C, Chaur MN, Echegoyen L, Poblet JM (2009) Chem Eur J 15:10997–11009
van Lenthe E, Baerends EJ, Snijders JG (1993) J Chem Phys 99:4597–4610
Vosko SH, Wilk L, Nusair M (1980) Can J Phys 58:1200–1211
Wang G-W, Saunders M, Cross RJ (2001) J Am Chem Soc 123:256–259
Whitehouse DB, Buckingham AD (1993) Chem Phys Lett 207:332–338
Yamada M, Okamura M, Sato S, Someya CI, Mizorogi N, Tsuchiya T, Akasaka T, Kato T, Nagase S (2009) Chem Eur J 15:10533–10542

Chapter 5
High Pressure Synthesis of the Carbon Allotrope Hexagonite with Carbon Nanotubes in a Diamond Anvil Cell

Michael J. Bucknum[1] **and Eduardo A. Castro**[1]

Abstract In a previous report, the approximate crystalline structure and electronic structure of a novel, hypothetical hexagonal carbon allotrope has been disclosed. Employing the approximate extended Hückel method, this C structure was determined to be a semi-conducting structure. In contrast, a state-of-the-art density functional theory (DFT) optimization reveals the hexagonal structure to be metallic in band profile. It is built upon a bicyclo[2.2.2]-2,5,7-octatriene (barrelene) generating fragment molecule, and is a Catalan network, with the Wells point symbol $(6^6)_2(6^3)_3$ and the corresponding Schläfli symbol (6, 3.4). As the network is entirely composed of hexagons and, in addition, possesses hexagonal symmetry, lying in space group P6/mmm (space group #191), it has been given the name hexagonite. The present report describes a density functional theory (DFT) optimization of the lattice parameters of the parent hexagonite structure, with the result giving the optimized lattice parameters of $\mathbf{a} = 0.477$ nm and $\mathbf{c} = 0.412$ nm. A calculation is then reported of a simple diffraction pattern of hexagonite from these optimized lattice parameters, with Bragg spacings enumerated for the lattice out to fourth order. Results of a synchrotron diffraction study of carbon nanotubes which underwent cold compression in a diamond anvil cell (DAC) to 100 GPa, in which the carbon nanotubes have evidently collapsed into a hitherto unknown hexagonal C polymorph, are then compared to the calculated diffraction pattern for the DFT optimized hexagonite structure. It is seen that a close fit is obtained to the experimental data, with a standard deviation over the five matched reflections being given by $\sigma_x = 0.003107$ nm/reflection.

[1] INIFTA, Theoretical Chemistry Division, Suc. 4, C.C. 16,
Universidad de La Plata, 1900 La Plata, Buenos Aires, Argentina
e-mail: mjbucknum@gmail.com; eacast@gmail.com

M.V. Putz (ed.), *Carbon Bonding and Structures: Advances in Physics and Chemistry*,
Carbon Materials: Chemistry and Physics 5, DOI 10.1007/978-94-007-1733-6_5,

5.1 Introduction

As a potential allotropic structure of C, the crystalline and electronic structure of the so-called, 3-dimensional (3D) hexagonite lattice[1] and some of its expanded 3D derivatives, were first reported by Karfunkel and Dressler (1992). The description of the parent structure of hexagonite in their report (Karfunkel and Dressler 1992), was substantially refined and clarified later on by Bucknum et al. in a paper published in 2001, where an identification of the space group symmetry (P6/mmm, space group #191), and a complete set of crystallographic coordinates for the hexagonite unit cell were given (Bucknum and Castro 2006).

Such a 3-dimensional (3D) hexagonite structure can be expanded into an indefinitely large number of derivative 3D structures, by the insertion of 1,4-dimethylene-2,5-cyclohexadieneoid organic spacers into the parent hexagonite structure (Karfunkel and Dressler 1992; Bucknum and Castro 2006). Expanded hexagonites include 3D crystalline materials with arbitrarily large pores directed along the crystallographic **c**-axis, they occur in infinite families possessing orthorhombic (Pmmm), trigonal (P3m1) and hexagonal (P6/mmm) space group symmetries. It was also reported in this paper (Bucknum and Castro 2006), that hexagonite could be realized from the elaboration of a bicyclo[2.2.2]-2,5,7-octatriene (barrelene) generating fragment molecule (Cotton 1990; Zimmerman and Paufler 1960; Wilcox, Jr. et al. 1960) in 3D, as is shown in Fig. 5.1.

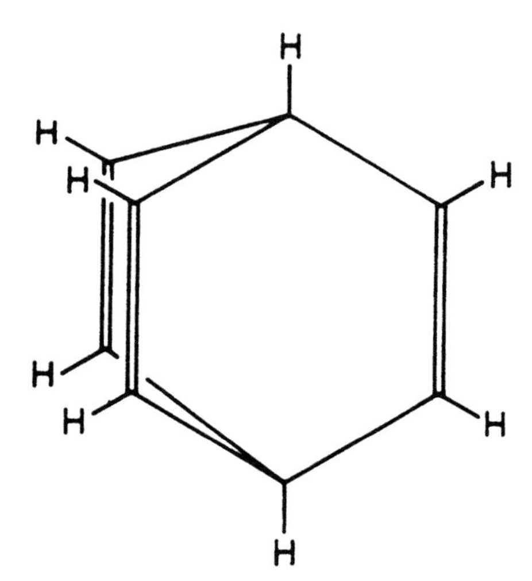

Fig. 5.1 Structure of bicyclo [2.2.2]-2,5,7-octatriene

[1] The C structure described in this communication, and elsewhere, with the name hexagonite is not to be confused with the inorganic mineral structure of the same name. The authors felt it appropriate to name the C structure, described herein, as hexagonite because of the special circumstance of its hexagonal symmetry space group (P6/mmm, #191), combined with its further 6-ness, as distinguished by its topological polygonality, given by n = 6, in which all smallest circuits in the network are hexagons.

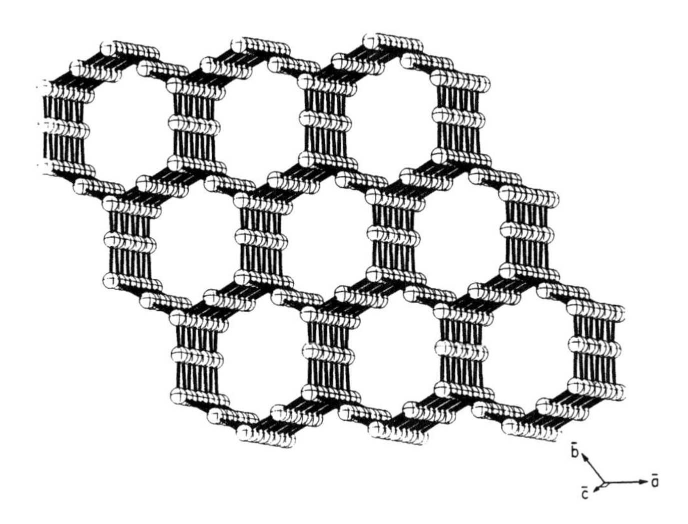

Fig. 5.2 Extended drawing of the hexagonite lattice, viewed approximately normal to the **ab**-plane of the lattice

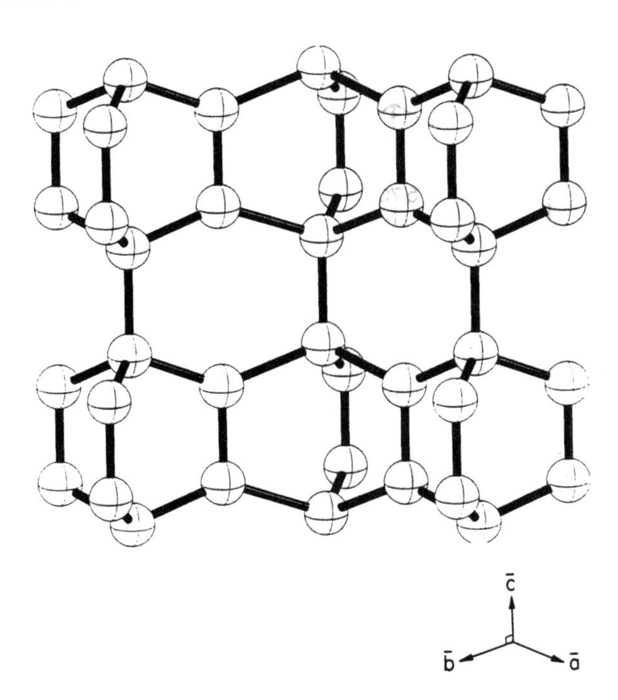

Fig. 5.3 View of the hexagonite lattice from the perspective of the crystallographic **ab**-plane

Thus the full elaboration of the 3D hexagonite network, from the barrelene generating fragment, can be seen in Fig. 5.2 from a perspective normal to the **ab**-plane of the lattice.

Yet another perspective of this hexagonite lattice is shown in Fig. 5.3, where there is a view of it parallel to the **ab**-plane (Bucknum and Castro 2006).

Table 5.1 Fractional hexagonal crystallographic coordinates of hexagonite from original report

Atom#	x/a	y/b	z/c	a	c
1	2/3	1/3	0.1935	4.89 Å	3.88 Å
2	2/3	1/3	0.8065	4.89 Å	3.88 Å
3	1/3	2/3	0.1935	4.89 Å	3.88 Å
4	1/3	2/3	0.8065	4.89 Å	3.88 Å
5	1/2	1/2	0.3265	4.89 Å	3.88 Å
6	1/2	1/2	0.6735	4.89 Å	3.88 Å
7	1/12	2/3	0.3265	4.89 Å	3.88 Å
8	1/12	2/3	0.6735	4.89 Å	3.88 Å
9	5/12	5/6	0.3265	4.89 Å	3.88 Å
10	5/12	5/6	0.6735	4.89 Å	3.88 Å

One can see in these perspective views of the hexagonite lattice, given in Figs. 5.2 and 5.3, the omnipresence of 6-ness in the structure. The organic tunnels apparent in Fig. 5.2, are indeed hexagonal macrocyclic tunnels which are further built upon component hexagons. Thus in Fig. 5.3, which is in the crystallographic **ab**-plane, we see illustrated the hexagon nature of these rings, that are components of the larger rings directed along the **c**-axis and apparent in the view of Fig. 5.2.

In the 2006 report by Bucknum et al. on hexagonite's structure (Bucknum and Castro 2006), the C-C single bonds were assumed to be 0.1500 nm, and the C=C double bonds were assumed to be 0.1350 nm, and all bond angles were assumed to be tetrahedral at 109.5°, except the trigonal C-C-C angles, which bisect the crystallographic **c**-axis, that were constrained to be 141°. This resulted in a crudely defined unit cell, with the lattice parameters given by **a** = **b** = 0.4890 nm and **c** = 0.3880 nm, and the set of fractional hexagonal coordinates, as listed in Table 5.1 below.

5.2 Chemical Topology of Hexagonite

Some of the topological character of the hexagonite lattice has been described previously (Bucknum and Castro 2006). An introduction to the topological characterization of crystalline networks is given by us elsewhere (Bucknum and Castro 2005). From a perspective entirely normal to the **ab**-plane, as shown in Fig. 5.2, the lattice reveals itself in this aspect to be reminiscent of the familiar hexagonal tiling of the plane, represented by the Schläfli symbol (6, 3) (or the Wells point symbol 6^3) called the honeycomb tessellation. Remarkably, a view perpendicular to the **c**-axis, inclined by 30° from the **a**-axis of the unit cell, reveals yet a second perspective from which a perfect honeycomb tessellation emerges from the pattern of bonds within the hexagonite lattice. There are thus two views of this hexagonite pattern that reveal its high hexagonal symmetry, in space group P6/mmm, as manifested in two independent honeycomb motifs that are patterned in directions entirely perpendicular to each other from the perspective of the unit cell.

As hexagonite is a 3-,4-connected network, it contains an admixture of 3-connected and 4-connected vertices in the unit cell. The overall connectivity of the lattice (Bucknum et al. 2005), a weighted average of the 3- and 4-connected points taken from the stoichiometry of the network, is given by $p = 3^{2/5}$. While the other key topological parameter in this analysis, called the polygonality (Bucknum et al. 2005), is indeed simply $n = 6$, as inspection of Figs. 5.2 and 5.3 will reveal. One can thus represent the topology of hexagonite by the Wells point symbol $(6^6)_2(6^3)_3$ and this, then, has the corresponding Schläfli symbol (n, p) = (6, 3.4) (Bucknum et al. 2005). It is a Catalan C-network, that can be expanded infinitely by insertion of 1,4-dimethylene-2,5-cyclohexadieneoid organic spacers between the barrelene moieties that make up the parent hexagonite lattice. This has been described already by Karfunkel et al. in their 1992 paper (Karfunkel and Dressler 1992; Bucknum and Castro 2006).

It is interesting here to see that hexagonite, and the expanded hexagonites, are represented by the collective Schläfli symbol given by $(n, p) = (6, 3^{x/x+y})$, where "x" represents the number of 4-connected points in the unit of pattern, which will always be 4, and "x + y" represents the sum of the numbers of 3- and 4-connected points in the unit of pattern, which will increase in increments, as the 1,4-dimethylene-2,5-cyclohexadieneoid organic spacers are added to the unit of pattern in the expanded hexagonites. Hexagonite, and its expanded derivatives, therefore represent a related family of Catalan 3D C-based networks that provide an interesting contrast to the Archimedean family of C-based fullerenes (Bucknum and Castro 2009). In contrast to the Catalan hexagonites, the fullerenes collectively have the Schläfli index $(n, p) = (5^{x/x+y}, 3)$, where "x" is the number of hexagons in the polyhedron, and "x + y" is the sum of the numbers of pentagons and hexagons in the polyhedron (Bucknum and Castro 2009).

A Schläfli relation exists for the polyhedra, shown as Eq. 5.1 below, that is entirely rigorous for the innumerable fullerene-like structures which collectively possess the Schläfli index cited above. In Eq. 5.1, the parameter E is the number of edges in the fullerene-like polyhedron (or polyhedron), "n" is the weighted average polygon size over the polygons in the polyhedron (for fullerene-like structures it will always be an admixture of pentagons and hexagons), and "p" is the weighted average connectivity over the vertices in the polyhedron (for fullerene-like structures, this will always be 3). The number of edges E is related to the number of vertices, V, and the number of faces, F, by the Euler identity (Bucknum et al. 2005; Bucknum and Castro 2009), given as $V - E + F = 2$.

$$\frac{1}{n} - \frac{1}{2} + \frac{1}{p} = \frac{1}{E} \tag{5.1}$$

In Sect. 5.3 that follows, we report on the electronic structural characteristics of the C-based hexagonite structure from the point of view of the extended Hückel molecular orbital method (EHMO), which is an approximate solid state electronic structure algorithm based upon the tight binding methodology (Hoffmann 1963;

Hoffmann and Lipscomb 1962; Whangbo and Hoffmann 1978; Whangbo et al. 1979). Next, in Sect. 5.3, we report on the details of a density functional theory[2] (DFT) geometry optimization of the parent hexagonite structure, and provide a calculation of a simple diffraction pattern of hexagonite (Warren 1990), out to fourth order in Bragg spacings of the crystalline structure. We then compare this theoretically calculated data to experimental data, reported in 2004, for a hexagonal polymorph of carbon synthesized by cold compression of C nanotubes to 100 GPa pressure in a diamond anvil cell (DAC) by Wang et al. (2004).

5.3 Electronic Structure of Hexagonite by Extended Hückel Method

Using the approximate electronic structure algorithm called the extended Hückel method (EHMO) (Hoffmann and Lipscomb 1962; Hoffmann 1963; Whangbo and Hoffmann 1978; Whangbo et al. 1979), a calculation of the approximate band structure and density of states (DOS) of hexagonite, as an allotrope of C, were carried out (Bucknum and Castro 2006). Figure 5.4 shows a representation of the approximate band structure of the C-based hexagonite. Figure 5.5 shows the corresponding density of states (DOS), this is derived from the band structure shown in Fig. 5.4. This approximate electronic structure algorithm (EHMO) reports the hexagonite lattice to be semi-conducting, but a separate density functional theory (DFT) method, described in Sect. 5.4, calculates the extended C-based structure to be metallic in band profile, though these results will not be pursued further here.[3]

Thus Figs. 5.4 and 5.5 indicate that the hexagonite structure should be a C-based semi-conductor (Bucknum and Castro 2006), in this approximation, where three unoccupied π^* bands are relatively low-lying, and separated from the σ^* manifold

[2] CASTEP (Cambridge Serial Total Energy Package) is a plane wave pseudopotential code, based upon density functional theory (DFT), that was used to optimize the hexagonite structure in the present report. Therefore, for the present implementation of CASTEP, used to optimize the structural parameters of hexagonite, the local density approximation (LDA) was used, ultrasoft pseudopotentials were employed, the basis set had an energy cutoff of 400 eV and k-point sampling was done with a $10 \times 10 \times 4$ mesh. The ultrasoft pseudopotentials used in the calculation are due to Vanderbilt (D. Vanderbilt, "Soft Self-Consistent Pseudopotentials in a Generalized Eigenvalue Formalism," *Phys. Rev. B*, **1990**, *41* (Rapid Communications), 7892–7895.) The Brillouin zone was sampled at a density of 0.004 nm^{-1}.

[3] The CASTEP-DFT method calculates the band structure of hexagonite to be metallic, in contrast the approximate EHMO method calculates the hexagonite structure to have a semi-conducting band profile. It is believed that the EHMO calculations of semi-conducting hexagonite are closer to a true reflection of the electronic structure of the lattice than that provided by the DFT results, based upon the fact that hexagonite can be viewed as a layering of delocalized π bonding (sp^2) sandwiched between insulating layers of C σ bonding (sp^3), and thus it cannot realistically be represented as a 3D metallic structure.

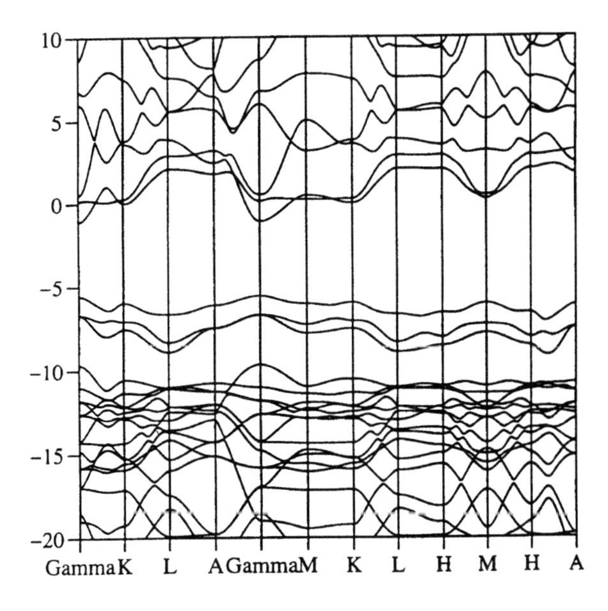

Fig. 5.4 Electronic band structure of the hexagonite crystal structure

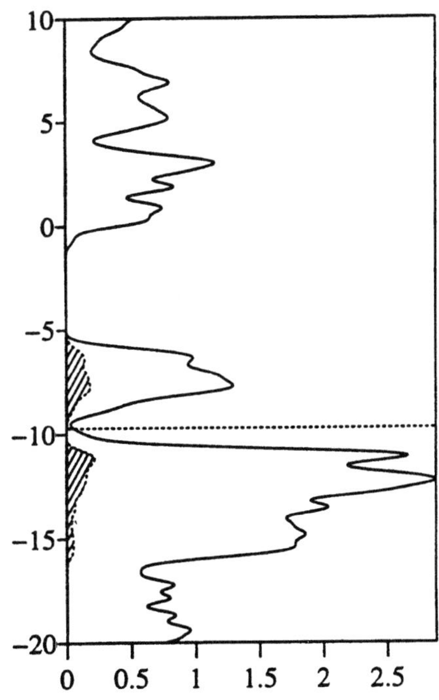

Fig. 5.5 Density of states (DOS) of the hexagonite crystal structure

by several eV of energy (Bucknum and Castro 2006). One could therefore envision the doping of the pores of hexagonite with univalent alkali metals, divalent alkaline earth metals or trivalent rare earth metals to form conducting or indeed, with the corresponding stoichiometry, insulating metal-organic composite structures, in which the metal atoms readily ionize in the parent C-lattice, and donate their valence shell electrons to the lower-lying, unoccupied π^* bands of the host hexagonite lattice (Bucknum and Castro 2006). The electronic features of such doped hexagonites, and their potential applications in high technology remain an unexplored vista of both theory and, potentially some day, experiment (Kaner and MacDiarmid 1988; Burroughes et al. 1988; Epstein and Miller 1979).

5.4 High Pressure Synthesis of Hexagonite from Compression of Nanotubes in a Diamond Anvil Cell

The C-based hexagonite structure, as described previously in Sect. 5.1, has been optimized with the DFT algorithm named CASTEP (Segall et al. 2002), the details of the optimization calculations are reported herein. The CASTEP code has been adequately described elsewhere (Segall et al. 2002). The DFT-CASTEP results indicate a slightly different optimized geometry for hexagonite than was assumed to pertain in the initial report by Bucknum et al. of the structure (Bucknum and Castro 2006). The C-C single bonds in the lattice, are found to be fairly closely related to each other with a uniform length of about 0.1521 nm. The C=C double bonds in hexagonite are 0.1326 nm in length. The C=C–C trigonal angles are about 115°, while the C-C-C trigonal angles are 130°, and the tetrahedral angles within the cage of the barrelene substructures are 103° while outside this cage the tetrahedral angles are 115°.

Finally, and most importantly, the lattice parameters optimized for the hexagonite structure by CASTEP are given as $\mathbf{a} = 0.4772$ nm and $\mathbf{c} = 0.4129$ nm. Thus from this DFT-CASTEP optimization, the coordinates of the 10 C atoms in the hexagonite unit cell are given in Table 5.2. The density of hexagonite[4] is calculated to be 2.449 g/cm^3, it lies between the density of the 3-connected C allotrope, graphite, at 2.27 g/cm^3, and the density of the 4-connected C allotrope, diamond, at 3.56 g/cm^3, but is closer to graphite's density than to diamond's density.

From the CASTEP optimized lattice parameters we can calculate a simple diffraction pattern for the C-based hexagonite lattice (Warren 1990), and such a powder pattern is shown with Bragg spacings enumerated out to fourth order for the lattice in Table 5.3. For comparison, in Table 5.3, we report a set of nine Bragg reflections recorded from the synthetic C allotrope resulting from cold compression and collapse of oriented, powdered nanotubes at near 100 GPa pressure (Wang et al. 2004). It is believed that the powdered nanotubes have transformed into a novel

[4] The volume of the hexagonite unit cell was calculated with the formula $V = ((3)^{1/2}/2)(\mathbf{a}^2\mathbf{c})$, and the corresponding density of hexagonite, with 10 C atoms in the unit cell, was found to be 2.449 g/cm^3. The density of the hexagonal C polymorph reported by Wang et al. in 2004 is 32% greater than this calculated value, at 3.6 g/cm^3.

Table 5.2 CASTEP optimized fractional hexagonal crystallographic coordinates and lattice parameters of hexagonite

Atom#	x/a	y/b	z/c	a	c
1	2/3	1/3	0.1840	4.772 Å	4.129 Å
2	2/3	1/3	0.8160	4.772 Å	4.129 Å
3	1/3	2/3	0.1840	4.772 Å	4.129 Å
4	1/3	2/3	0.8160	4.772 Å	4.129 Å
5	1/2	1/2	0.3400	4.772 Å	4.129 Å
6	1/2	1/2	0.6600	4.772 Å	4.129 Å
7	0	1/2	0.3400	4.772 Å	4.129 Å
8	0	1/2	0.6600	4.772 Å	4.129 Å
9	1/2	0	0.3400	4.772 Å	4.129 Å
10	1/2	0	0.6600	4.772 Å	4.129 Å

hexagonal polymorph of C, which nonetheless possesses an unknown structure (Wang et al. 2004). This latter C allotrope is able to be quenched at room pressure, as well. The Bragg spacings in this cold-compressed sample of powdered C nanotubes, were monitored using the specialized technique of energy-dispersive-x-ray-diffraction (EDXRD) on a specially constructed high-pressure diamond anvil cell (DAC) synchrotron beamline at the Cornell High Energy Synchrotron Source (CHESS) in Ithaca, NY (Wang et al. 2004).

One can see in Table 5.3, that the nine reflections in the EDXRD dataset of the powdered C nanotube sample, as cold-compressed in the DAC to over 100 GPa, readily match to the set of Bragg spacings calculated theoretically from the optimized set of lattice parameters provided for the hexagonite lattice from the DFT-CASTEP algorithm (Segall et al. 2002). The average deviation over the nine reflections in the experimental dataset, matched to the calculated Bragg spacings of the optimized hexagonite lattice, is given by $\underline{\Delta x} = 0.003040$ nm/reflection, while the standard deviation over the nine matched reflections in the experimental dataset, is given by $\sigma_x = 0.003107$ nm/reflection.

Clearly all the data, over the nine matched reflections in the experimental dataset, have deviations that fit to within less than $3\sigma_x = 0.009321$ nm of the standard deviation of the dataset, as is reflected in Table 5.3. In a parallel comparison of the experimental dataset given here of the hexagonal C polymorph, with the commonly observed Bragg reflections from cubic diamond, it is important to point out that only the cubic diamond reflections (111), with a Bragg spacing of 0.2060 nm, and (220), with a Bragg spacing of 0.1261 nm, matched to the experimental set of reflections for the hexagonal C polymorph (Bucknum et al. 2005).

A caution should be made here with regard to the size of the dataset, with only nine reflections to fit, clearly the model proposed here of the hexagonite lattice, as an explanation of the experimentally obtained EDXRD synchrotron dataset for a proposed hexagonal polymorph of C, is not constrained as much by observation as one would like it to be. With only nine observations to work with, such a fit between experimental and theoretical diffraction data is little more than an educated guess.

Table 5.3 Observed diffraction data of proposed hexagonal C polymorph compared to calculated hexagonite pattern from CASTEP optimization

Calculated hexagonite reflections a = 0.4772 nm, c = 0.4129 nm		C polymorph reflections[a]	Absolute deviation per reflection[b]
(hkl)	d-spacing, nm	d-spacing, nm	Δd-spacing, nm
100	0.4133		
001	0.4129		
110	0.2386		
101	0.2921		
111	0.2066	0.2155	0.0089
200	0.2066	0.2155	0.0089
002	0.2065	0.2053	0.0012
102	0.1847		
120	0.1562		
201	0.1848		
211	0.1461	0.1495	0.0034
221	0.1146	0.1161	0.0015
212	0.1246	0.1248	0.0002
222	0.1033		
300	0.1378		
003	0.1376		
103	0.1306		
130	0.1146	0.1161	0.0015
301	0.1307		
311	0.1104		
331	0.07810		
313	0.08808		
333	0.06886		
203	0.1146	0.1161	0.0015
302	0.1146	0.1161	0.0015
320	0.09481		
223	0.09015		
232	0.08616		
332	0.07422		
323	0.07808		
321	0.09241		
312	0.1002		
213	0.1033		
104	0.1002		
401	0.1002		

[a] Wang et al. 2004

[b] Average deviation over the five reflections compared is 0.003040 nm/reflection

Still, the authors of the report (Wang et al. 2004) state that their fit to the data can be interpreted, (from the reflections that they have indexed in their dataset of five Bragg spacings from a hexagonal model with lattice parameters **a** = 0.249 nm and **c** = 0.412 nm) from rules of systematic absences among the hexagonal space groups, to a structure that lies in the hexagonal space group P-6 m2 (#190) (Wang et al. 2004). We offer here in response to this point, that the hexagonite lattice lies in the very closely related hexagonal symmetry space group P6/mmm (#191) (Bucknum and Castro 2006). It would therefore appear that the two sets of data, while only consisting of nine matches, are closely connected together by considerations of the potential symmetry of the unknown C phase.

5.5 Densities of Carbon Phases and Superhard Hexagonite

One area of disagreement in the current comparison of theoretical and experimental data, is in the density reported by Wang et al. (2004) in their compression study of the C nanotubes, and that reported for the candidate hexagonite structure from first principles theory. Experimentally, the density of the hexagonal C polymorph of unknown structure is given as 3.6 ± 0.2 g/cm^3, while the theoretically calculated value for the density of hexagonite is some 32% lower than this experimental value, at 2.449 g/cm^3. It is thus important here to note, that the density of the starting oriented, powdered nanotube material in this compression synthesis can be estimated to be lower than the known density of fullerite at 1.75 g/cm^3.

Wang et al. proposed that the C material of unknown structure should have a density in excess of the density of diamond, at 3.56 g/cm^3, based upon thermodynamic arguments. They reasoned that a novel allotrope of C would have collapsed to a denser structure than diamond, given the high pressures involved in the study, to well over 100 GPa (Wang et al. 2004). It is believed by the authors of the present study, however, that the inhomogeneous nature of the C material produced in the synchrotron study (Wang et al. 2004), in which amorphous C was present in the quenched DAC sample, along with the novel crystalline phase, suggests that kinetically stabilized products, such as the low density C material hexagonite,[5] may have been formed. The current study does not rule out such a kinetically favored crystalline C product from forming at such high pressures exceeding 100 GPa. It is thus physically reasonable to expect that the starting oriented, powdered nanotube material, at a density of less than that of fullerite at 1.75 g/cm^3, could quite possibly have collapsed to the hexagonite lattice at a density of 2.449 g/cm^3, in their study, if such a lattice was otherwise thermodynamically a stable phase up to 100 GPa.

It was reported in their study (Wang et al. 2004), that the hexagonal C allotrope produced under pressure, partially indented the diamond anvils in the high pressure cell used. It was conjectured by Wang et al. that the hexagonal, crystalline C phase, that was quenched from high pressure, was a superhard phase of C. They estimated

[5] On a stability scale at which diamond is at the 0 of energy, the glitter allotrope of C, described in [22], has an energy of formation of 0.5116 eV/C atom above that of diamond, by the CASTEP optimization, under the local density approximation (LDA). In contrast, the hexagonite allotrope of C, described herein, has an energy of formation of 0.5343 eV/C atom above that of diamond by the same computational method.

a bulk modulus of at least 447 GPa for the C form in the study. In this instance, it is worth pointing out that a semi-empirical estimate (Cohen 1994; Liu and Cohen 1989) of the bulk modulus of hexagonite, despite its low density of 2.449 g/cm^3, puts it at a value of $B_0 = 445$ GPa. This comes from a semi-empirical formula for bulk modulus in materials, developed by Cohen et al. (Cohen 1994; Liu and Cohen 1989) and is,

$$B_0 = \frac{1972 - 220I}{\langle d \rangle^{3.5}} \frac{\langle N_c \rangle}{4} \tag{5.2}$$

Here in formula 5.2, the parameter I represents the degree of ionicity of the bonding in the unit cell of a given material. For the C form of hexagonite this parameter is just 0, because the hexagonite structure is assumed to be a C allotrope in this instance. The parameter N_c is the averaged coordination number in the unit cell, this is just the connectivity in the lattice, p, which has a value of 3.4 for hexagonite, as was discussed above (Bucknum and Castro 2006). Finally, the parameter d is the weighted average bond distance in the unit cell. The C-C single bonds have a length of 0.1521 nm, while the C=C double bonds are 0.1326 nm in length, in the DFT-CASTEP hexagonite optimization. They average out to a distance of about 0.1460 nm over the unit cell. These unit cell parameters thus lead to a semi-empirical estimation of B_0 for hexagonite that is listed above. One can see that the bulk moduli from experiment and from theory are in good agreement with each other.

5.6 Conclusions

In this communication, a geometry optimization of a novel, hypothetical form of C called hexagonite (Karfunkel and Dressler 1992; Bucknum and Castro 2006) has been performed using a state-of-the-art DFT-based program called CASTEP (Segall et al. 2002). The lattice parameters for the hexagonal unit cell are given as $\mathbf{a} = 0.4772$ nm and $\mathbf{c} = 0.4129$ nm. The density of the C allotrope has been optimized to be at 2.449 g/cm^3. The coordinates for the 10 C atoms in the hexagonite unit cell are thus listed in Tables 5.1 and 5.2 above (Bucknum and Castro 2006).

From the optimized lattice parameters of hexagonite, a simple diffraction pattern has been calculated (Warren 1990), of the Bragg spacings in the crystalline material enumerated out to fourth order. From this optimized diffraction pattern, a comparison has been carried out over a set of nine Bragg reflections obtained from an experimental study in which oriented, powdered C nanotubes have been cold compressed in a diamond anvil cell (DAC), on a specially designed high pressure beam line (Wang et al. 2004). Comparison of the two datasets yields a close fit over

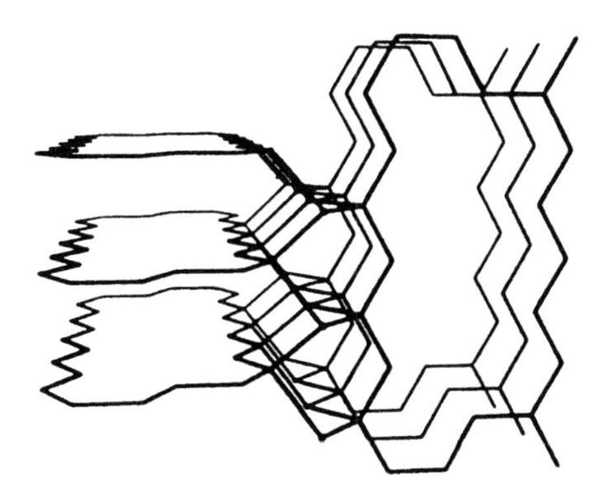

Fig. 5.6 Structure of a variety of the "ortho-" graphite-diamond hybrids

the nine experimentally determined x-ray reflections, with an average deviation given by $\underline{\Delta x} = 0.003040$ nm/reflection, and a standard deviation of $\sigma_x = 0.003107$ nm/reflection. It therefore appears possible, that the cold compressed, oriented, powdered nanotubes, in the high pressure study, have collapsed under compressive forces into the hexagonite lattice, which was first proposed as a likely structure for C to adopt by Karfunkel et al. (Karfunkel and Dressler 1992), and later refined by Bucknum et al. (Bucknum and Castro 2006) in 2006.

It should be emphasized here that the parent structure, called hexagonite, represents the first member of a possibly infinite family of related Catalan C allotropes, and it has possibly been synthesized in the experimental study cited here by Wang et al. (2004). And this infinite family of 3D network C structures, that are collectively described as Catalan networks, represent an interesting contrast to the infinity of Archimedean fullerene structures first identified in 1985 by Kroto et al. (1985). Collectively, the hexagonites and the fullerenes, being semi-regular structures, can be contrasted, on the one hand, with the graphite and diamond polytypes, which are regular (Platonic) structures of C, given by the Wells point symbols 6^3 and 6^6, respectively (Bucknum et al. 2005). Speculatively, on the other hand, these semi-regular structures can be contrasted, as well, with the Wellsean (or topologically demi-regular) tetragonal glitter network, which has been previously optimized and shown to have a stability close to that of hexagonite, and with the Wells point symbol given by $(6^2 8^4)(6^2 8)_2$ (Bucknum and Hoffmann 1994; Bucknum and Castro 2004). As well as other such demi-regular networks enumerated by Merz et al. (1987), Wells (1977 , 1979), and others in the references cited therein.

And finally, of course, the hexagonites can be speculatively compared directly to the infinite family of graphite-diamond hybrids proposed in 1994 by Balaban et al. (1994), which share the same Catalan Wells point symbol given by $(n, p) = (6, 3^{x/x+y})$, as described above. A couple of varieties of the graphite-diamond hybrids are shown in Figs. 5.6 and 5.7 (Balaban et al. 1994). It is important to note that the graphite-diamond hybrids collectively circumscribe several families of potential C structures, all of which possess orthorhombic symmetry, and which, to date, have not been structurally optimized in any of their various forms.

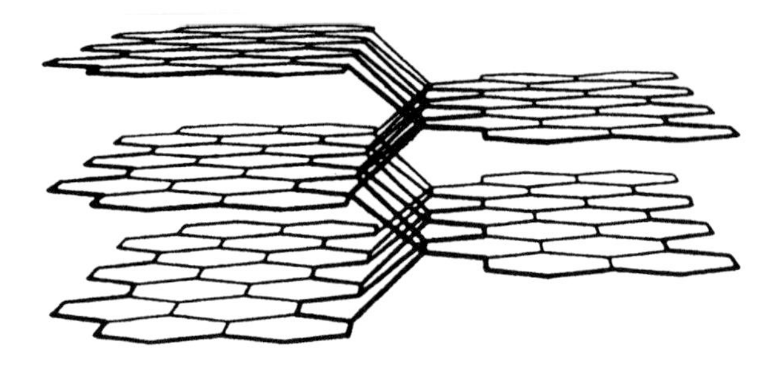

Fig. 5.7 Structure of a variety of the "para-" graphite-diamond hybrids

Acknowledgements MJB thanks his wife Hsi-cheng Shen for much love and patience in his work on C allotropy and the subtle structural issues of C. The authors wish to thank Norman Goldberg, PhD for producing the structural drawings of hexagonite while a post-doctoral associate in Professor Roald Hoffmann's theoretical chemistry group at Cornell University. The authors wish to thank Chris J. Pickard, PhD of the Theoretical Condensed Matter (TCM) Group at Cambridge University, for his great help in carrying out the DFT-CASTEP optimization calculations of the hexagonite structure. The authors wish to thank Roald Hoffmann for his suggestions in writing this manuscript. Finally, the authors wish to thank D.M.E. (Marian) Szebenyi, PhD at Cornell High Energy Synchrotron Source (CHESS) for helpful discussions of the symmetry aspects of hexagonite.

References

Balaban AT, Klein DJ, Folden CA (1994) Chem Phys Lett 217:266–270
Bucknum MJ, Castro EA (2004) J Math Chem 36(4):381–408
Bucknum MJ, Castro EA (2005) MATCH Commun Math Comput Chem 54:89–119
Bucknum MJ, Castro EA (2006) J Math Chem 39(3–4):611–628
Bucknum MJ, Castro EA (2009) J Math Chem 42(1):117–138
Bucknum MJ, Hoffmann R (1994) J Am Chem Soc 116:11456–11464
Bucknum MJ, Stamatin I, Castro EA (2005) Mol Phys 103(20):2707–2715
Burroughes JH, Jones CA, Friend RH (1988) Nature(London) 335:137–141
Cohen ML (1994) Solid State Commun 92(1–2):45–52
Cotton FA (1990) Chemical applications of group theory, 3rd edn. Wiley, New York, pp 166–172
Epstein AJ, Miller JS (1979) Sci Am 241:48–61
Hoffmann R (1963) J Chem Phys 39:1397–1407
Hoffmann R, Lipscomb WN (1962) J Chem Phys 37:2872–2878
Kaner RB, MacDiarmid AG (1988) Sci Am 258:60–72
Karfunkel HR, Dressler T (1992) J Am Chem Soc 114(7):2285–2288
Kroto HW, Heath JR, O'Brien SC, Curl RF, Smalley RE (1985) Nature(London) 318:162–163
Liu AY, Cohen ML (1989) Science 245:841–845
Merz KM Jr, Hoffmann R, Balaban AT (1987) J Am Chem Soc 109:6742–6751
Segall MD, Lindan PJD, Probert MJ, Pickard CJ, Hasnip PJ, Clark SJ, Payne MC (2002) J Phys Condens Matter 14(11):2717–2743
Wang Z, Zhao Y, Tait K, Liao X, Schiferl D, Zha C, Downs RT, Qian J, Zhu Y, Shen T (2004) P Natl A Sci (PNAS) 101(38):13699–13702
Warren BE (1990) *X-ray Diffraction*, 1st edition, Dover Publications, Inc., Mineola, NY: 21–22. The formula used to calculate Bragg spacings in the hexagonal crystal system of hexagonite is given in the book by B.E. Warren in the following format: $1/d_{hkl}^2 = (4/3)((h^2 + hk + k^2)/a^2) + l^2/c^2$

Wells AF (1977) Three dimensional nets and polyhedra, 1st edn. Wiley, New York, pp 1–150
Wells AF (1979) Further studies of three-dimensional nets, ACA monograph #8. ACA Press, Pittsburgh, pp 1–75
Whangbo MH, Hoffmann R (1978) J Am Chem Soc 100:6093–7002
Whangbo MH, Hoffmann R, Woodward RB (1979) P Roy Soc A 366:23–32
Wilcox CF Jr, Winstein S, McMillan WG (1960) J Am Chem Soc 82:5450–5453
Zimmerman HE, Paufler RM (1960) J Am Chem Soc 82:1514–1516

Chapter 6
Graph Drawing with Eigenvectors

István László[1], Ante Graovac[2,3,4], Tomaž Pisanski[5], and Dejan Plavšić[3]

Abstract The visualization of graphs describing molecular structures or other atomic arrangements is necessary in theoretical studying or examining nano structures of several atoms. In the present paper we review previous results obtained by drawing graphs with the help various matrices as the adjacency matrix, the Laplacian matrix and the Colin de Verdière matrix. In examples we show their possibilities and limits of applicability. We suggest a new matrix that reproduces well the same structures and those ones which were not drawn by the previous matrices.

6.1 Introduction

It happens very often in material science that only the topological arrangement of the atoms is given, and in order to perform some theoretical calculation or to understand the physical or chemical behavior of the compound, we must to perform some kind

[1] Department of Theoretical Physics, Institute of Physics, Budapest University of Technology and Economics, H-1521 Budapest, Hungary
e-mail: laszlo@eik.bme.hu

[2] Department of Chemistry, Faculty of Science, University of Split, Nikole Tesle 12, HR-21000 Split, Croatia

[3] NMR Center, The "Ruđer Bošković" Institute, HR-10002 Zagreb, Croatia
e-mail: Ante.Graovac@irb.hr; dplavsic@irb.hr

[4] IMC, University of Dubrovnik, Branitelja Dubrovnika 29, HR-20000 Dubrovnik, Croatia

[5] Department of Theoretical Computer Science, Institute of Mathematics, Physics and Mechanics, University of Ljubljana, Jadranska 19, SI-1000 Ljubljana, Slovenia
e-mail: tomaz.pisanski@fmf.uni-lj.si

M.V. Putz (ed.), *Carbon Bonding and Structures: Advances in Physics and Chemistry*, Carbon Materials: Chemistry and Physics 5, DOI 10.1007/978-94-007-1733-6_6,

of visualization of the structure under study. That is we have to generate the geometrical structure from the topological one. In these cases, however, the topological arrangement contains some information from the geometrical structure as well. Namely the neighboring atoms are close to each other in the Euclidean space R^3. We describe the topological structure with the help of a graph, where the vertices represent the atoms and the edges the pairs of neighboring atoms which are in most of the cases the chemical bonds. The recovering of geometry from topology correspond to embedding the graph into the Euclidean space R^3 or R^2. There are many methods for geometric representation and visualization of graphs (Di Battista et al. 1999; Fowler et al. 1995; Godsil and Royle 2001; Kaufmann and Wagner 2001; Koren 2005; László 2008; Lovász and Vesztergombi 1999; Pisanski et al. 1995; Tutte 1963), but few of them are using eigenvectors of adjacency matrices, Laplacians or other matrices of the graph. The spectra of graphs have many interesting properties (Biyikoglu et al. 2007; Colin de Verdière 1998; Godsil and Royle 2001; Trinajstić N 1992; van der Holst 1996) and here we shall present methods for drawing graphs with the help of eigenvectors.

First we shall present our basic notations and definitions, then drawing of spherical graphs and applications for fullerenes will be described. The graphs of fullerenes will be constructed with the help of the spiral conjecture (Fowler and Manolopoulos 1995). It will be shown that fullerenes can be represented properly with the help of three eigenvectors of the adjacency matrix. The next paragraphs will be devoted to toroidal structures, nanotubes and planar arrangements of atoms. We shall see that the three eigenvector method which was successful for fullerenes can not be applied for nanotori. It will be shown that drawing with four eigenvectors is applicable for nanotori. After this paragraph versions of the four eigenvectors procedure will be used for drawing nanotubes and planar structures. Before all of these we shall present a method for constructing graphs of nanotori nanotubes and planar structures including drawing Schlegel diagrams.

In the next paragraph in the frame work of our shape analysis we shall study the minimum number of eigenvectors of the adjacency matrix and the Laplacinan in order to obtain proper drawing of nonspherical structures. Before the conclusion we present a method for drawing fullerenes, straight, toroidal and helical nanotubes, and nanotube junctions with the help of only three eigenvectors of a matrix.

6.2 Basic Notions and Definitions

Although most of our results will be valid for a greater set of graphs if it will not be stated others, we understand molecular graphs under the notation $G(V, E)$ where V is the set of vertices and E is the set of edges. Molecular graphs are chemical graphs which represent the constitution of molecules (Trinajstić 1992) or any atomic arrangements. Thus in the set V of vertices each vertex $v \in V$ represent an atom in the structure under study and each edge $(u, v) \in E$ means that the atoms u and v are in some way related to each other, that is they are bounded or they are neighboring atoms. At the end of this Chapter we shell include second neighbors into the set of edges as well. The value $n = |V|$ equals to the number of vertices.

The identity matrix and the all-1 matrix, i.e., $J_{ij} = 1$will be denoted in order by $\mathbf{I}$ and $\mathbf{J}$. The matrix $\mathbf{A} = \mathbf{A}(G) = (a_{uv})$ will be the adjacency matrix of the graph $G(V, E)$ with the definition that $a_{uv} = 1$ if $(u, v) \in E$ and $a_{uv} = 0$ if $(u, v) \notin E$. If the graph$G$is a weighted graph, the values a_{uv} are equals to the corresponding weights. In molecular graphs there are not loops and thus the relation $a_{uu} = 0$ will be valid. Let $D = (d_{vv})$ be the diagonal matrix with $d_{vv} = \sum_{u:(u,v)\in E} a_{uv}$. The Laplacian matrix is defined to be $\mathbf{Q} = \mathbf{Q}(G) = \mathbf{Q}(\mathbf{A}) = \mathbf{D} - \mathbf{A}$. The $\lambda_i$eigenvalues of the adjacency matrix A and those of the Laplacian $\mathbf{Q}$ are numbered in descending and in ascending order respectively. From the definition of $\mathbf{Q}$ follows, that $\mathbf{Qc} = \mathbf{0}$ if the components of the eigenvector care the same, that is $c_i = \frac{1}{\sqrt{n}}$ for example.

We shall call an eigenvector c n-lobal, if in the graph $G(V, E)$ after deleting the vertices $i \in V$ if $c_i = 0$ and the edges $(i, j) \in E$ if the sign of c_i and c_j are different, the resulting graph will have n components. According to the corresponding sign there are thus positive and negative components. In our graph drawing procedures the bi-lobal eigenvectors will be the most important. For the nodal properties of graph Laplacians see the reference (Biyikoglu et al. 2004).

Under embedding a graph $G(V, E)$ into R^k we mean the a mapping

$$\tau : V(G) \rightarrow R^k \tag{6.1}$$

We will denote by τ_i the n-dimensional vector formed by taking the ith coordinate $\tau(u)_i$ of $\tau(u)$ for all $u \in V$. Thus τ_i is an n-dimensional vector indexed by the vertices of the graph $G(V, E)$. We shall use also the following notations $X = \tau_1$, $Y = \tau_2$ and $Z = \tau_3$, that is $(x_u, y_u, z_u) = \left(\tau(u)_1, \tau(u)_2, \tau(u)_3\right)$.

6.3 Drawing of Spherical Structures and Applications for Fullerenes

6.3.1 Topological Structure of Fullerenes

If the number of carbon atoms in a fullerene is $n = V$, then the number of edges is $E = \frac{3V}{2}$. From Euler formula $F - E + V = 2$ follows the number of faces as $F = 2 + \frac{V}{2}$. Also from the Euler formula follows that in fullerenes the number of pentagons is 12 and the number of hexagons is$\left(\frac{V}{2} - 10\right)$. Thus knowing the number

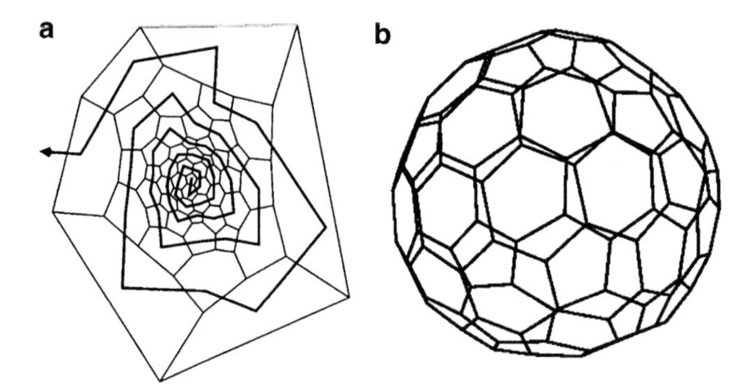

Fig. 6.1 The C_{100} fullerene which is described by the spiral code $(1,7,12,19,25,28,30,33,36,41,48,50)$. The spiral on the Schlegel diagram (**a**) and the picture of the fullerene obtained by the $\tau_1 = c^2$, $\tau_2 = c^3$and $\tau_3 = c^4$ bi-lobal eigenvectors of the adjacency matrix generated by the spiral code (**b**)

of carbon atoms in a fullerene we can generate the adjacency matrix of various fullerenes using the following conjecture (Fowler and Manolopoulos 1995):

> **Spiral conjecture:** *The surface of a fullerene polyhedron may be unwound in a continuous spiral strip of edge-sharing pentagons and hexagons such that each new face in the spiral after the second shares an edge with both (a) its immediate predecessor in the spiral and (b) the first face in the preceding spiral that still has an open edge.*

According to the spiral conjecture the spiral of the polyhedron is represented by a one-dimensional sequence of 5s and 6s which give the positions of pentagons and heptagons. Thus the list of the 12 pentagon serial number describe the topological structure of the fullerenes (Fig. 6.1).

6.3.2 Spherical Structures and Fullerenes

Hall suggested a k-dimensional quadratic placement algorithm for embedding a graph $G(V, E)$ to the Euclidean space R^k (Hall 1970). He defined the optimal embedding by minimizing the following energy function:

$$E_H(\tau) = \sum_{i=1}^{k} \tau_i^T \mathbf{Q} \tau_i \tag{6.2}$$

with the constraints $||\tau_i||^2 = \sum_{u=1}^{n} \tau(u)_i^2 = 1$. It was proved that this optimal embedding can be realized by the first k eigenvectors with non-zero eigenvalues of the Laplacian $\mathbf{Q}$. That is $\tau_i = \mathbf{c}^{i+1}$ for $i = 1, \ldots, k$.

The idea of using eigenvectors for drawing graphs was used first in chemical setting for molecular orbitals by Fowler and Manolopoulos (1995; Manolopoulos and Fowler 1992). They called it the topological coordinate method. In order to find

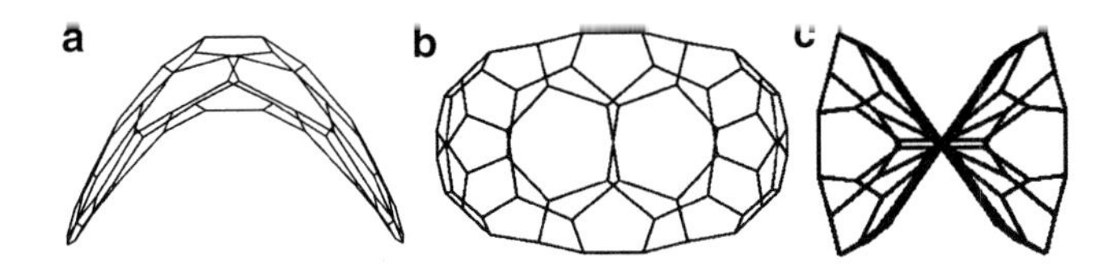

Fig. 6.2 Various drawing of a C_{60} fullerene isomer. The adjacency matrix was constructed from the $(1, 2, 3, 4, 5, 8, 25, 28, 29, 30, 31, 32)$ spiral code. Drawing by the eigenvectors c^2, c^3 and c^4 (**a**); c^2, c^4 and c^5 (**b**) c^4, c^5 and c^6 (**c**). All of these eigenvectors are bi-lobal except c^3 which is 3-lobal

the point group associated with a given fullerene molecular graph they wanted the Cartesian coordinates for each atom in the graph. Inspired by Stone's work (Stone 1981) they have found that the first few Hückel molecular orbitals of fullerenes invariably contain three bi-lobal eigenvectors which are discrete version of the continuous p_x, p_y and p_z orbitals. Since the p_x, p_y and p_z orbitals on a sphere are proportional to the x, y, and z coordinates, the three lowest bi-lobal eigenvectors c^{k_1}, c^{k_2} and c^{k_3} determine the (x_i, y_i, z_i) topological coordinates of the i-th atom by the relations:

$$x_i = S_1 c_i^{k_1} \tag{6.3}$$

$$y_i = S_2 c_i^{k_2} \tag{6.4}$$

$$z_i = S_3 c_i^{k_3} \tag{6.5}$$

with the scaling factors $S_\alpha = S_0$ or $S_\alpha = \dfrac{S_0}{\sqrt{\lambda_1 - \lambda_{k_\alpha}}}$ (Manolopoulos and Fowler 1992; Fowler and Manolopoulos 1995).

The most realistic picture of fullerenes can be found by the scaling factor $S_\alpha = \dfrac{S_0}{\sqrt{\lambda_1 - \lambda_{k_\alpha}}}$ (Manolopoulos and Fowler 1992; Fowler and Manolopoulos 1995; László 2004a; Pisanski and Shawe-Taylor 1993). As the fullerene graphs are regular, the bi-lobal eigenvectors can be obtained from the adjacency matrix **A** or from the Laplacian matrix **Q** as well. It was found that for most of the fullerenes the topological coordinates are generated by the eigenvectors c^2, c^3 and c^4 (Fig. 6.1), but there are some exceptions as well (Fig. 6.2).

Pisanski and Shawe-Taylor defined the optimal embedding of the weighted graph G(V, E) by minimizing the following energy function (Pisanski and Shawe-Taylor 1993, 2000):

$$E(\tau) = \sum_{(u,v)\in E} a_{uv} \|\tau(u) - \tau(v)\|^2 - \beta \sum_{(u,v)\notin E} \|\tau(u) - \tau(v)\|^2 \tag{6.6}$$

and it was subjected to the constraints
$\|\tau_i\| = 1$, $\tau_i^T c^1 = 0$ for i = 1,. . ., k
$\tau_i^T \tau_j = 0$ for $1 \le i < j \le k$, and β is a positive constant.

It was proven in Pisanski and Shawe-Taylor (1993, 2000) that the optimal embedding for this problem is given by $\tau_i = c^{i+1}$ for $i = 1, \ldots, k$, and the minimal value of $E(\tau)$ is

$$\sum_{l=2}^{k+1} \lambda_l - \beta nk. \quad (6.7)$$

Here the corresponding Laplacian $\mathbf{Q} = \mathbf{Q}(\mathbf{B})$ for the eigenvalues and eigenvectors was constructed from the matrix $\mathbf{B}$ with the matrix elements $b_{uv} = a_{uv} + \beta$ if $(u, v) \in E$ and $b_{uv} = 0$ otherwise. It was proved also that in the case where the graph is not weighted, the optimal embedding does not depend on the parameter β.

Lovász and Schrijver (1999) defined a symmetric $n \times n$ matrix $\mathbf{M}$ for the 3-connected planar graph $G(V, E)$ with the following properties:

(i) $\mathbf{M}$ has exactly one negative eigenvalue, of multiplicity 1;
(ii) for all $(u, v) \in E$ $m_{uv} < 0$ and if $u \neq v$ and $(u, v) \notin E$ $m_{uv} = 0$
(iii) $\mathbf{M}$ has rank $n - 3$.

They have proved that if we have a matrix $\mathbf{M}$ with the above mentioned conditions than the null space of $\mathbf{M}$ (the eigenvectors c^2, c^3 and c^4 of the eigenvalue $\lambda = 0$) gives a proper embedding of $G(V, E)$ in the sphere S^2as $\tau_i = c^{i+1}$ for $i = 1, \ldots, 3$, and $||\tau_i||^2 = 1$. Thus the relation $(x_u, y_u, z_u) = \left(\tau(u)_1, \tau(u)_2, \tau(u)_3\right)$ is valid for each vertex. It was also proved that this null space contains bi-lobal eigenvectors (van der Holst 1996). The matrix $\mathbf{M}$ is often called Colin de Verdière matrix in the scientific literature.

As an application the topological coordinate method was used to create a comprehensive catalogue of general fullerene isomers with up to 50 carbon atoms and isolated-pentagon isomers with up to 100 carbon atoms (Fowler and Manolopoulos 1995). It was used also for developing a graph-theoretical procedure for constructing chirality descriptors of fullerenes (Rassat et al. 2003).

6.4 Drawing of Toroidal and Planar Graphs

6.4.1 *Topological Structure of Periodic Systems*

The carbon nanotubes are imagined as rolled up graphene sheets and the atoms on the graphene sheet are described with the help of two unit cell vectors $a_1 = a\left(\sqrt{3}, 0\right)$ and $a_2 = a\left(\frac{\sqrt{3}}{2}, \frac{3}{2}\right)$. Here a is the inter-atomic distance and the centre of the hexagon (k, l) is given by the translation vector $\mathbf{R} = k\mathbf{a_1} + l\mathbf{a_2}$ with integers k and l. As in the graphene each unit cell contains two atoms with the serial numbers $r = 1$ and $r = 2$,

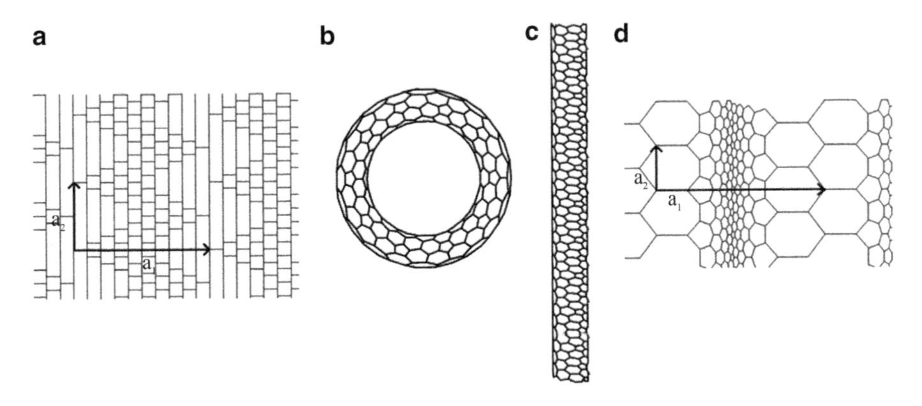

Fig. 6.3 The topological structure for periodic systems and drawing with four bi-lobal eigenvectors of the Laplacian. The unit cell with perpendicular unit vectors $\mathbf{a}_1$ and $\mathbf{a}_2$ which contains 60 vertices (**a**). Drawing the nanotorus of parameters $(\mathrm{m,n,p,q}) = (1,0,0,5)$ (**b**). Drawing the nanotube of parameters $(\mathrm{m,n,p,q}) = (1,-1,5,5)$ (**c**). The unit cell which was generated by four bi-lobal eigenvectors of the Laplacian (**d**)

the atoms of the graphene sheet can be parameterised by three integers as $(\mathrm{k,l,r})$. The hexagonal nanotubes are given with the help of the chiral vector $\mathbf{C}_\mathrm{h}$ as

$$\mathbf{C}_\mathrm{h} = \mathrm{m}\mathbf{a_1} + \mathrm{n}\mathbf{a_2} \tag{6.8}$$

where m and n are integers. The chiral vector and the multiple of the translation vector **T** defines a rectangle which is the super cell in our rolling up construction. By identifying those two parallel lines of this rectangle which are parallel also with **T** one can construct a cylinder. The vector **T** is perpendicular to the chiral vector C_h and defined as

$$\mathbf{T} = [-(2\mathrm{n} + \mathrm{m})\mathbf{a_1}, (2\mathrm{m} + \mathrm{n})\mathbf{a_2}]/\mathrm{d_R} \tag{6.9}$$

where d is the highest common divisor of $(\mathrm{m,n})$, and $\mathrm{d_R} = \mathrm{d}$ if $\mathrm{n} - \mathrm{m}$ is not a multiple of 3d and $\mathrm{d_R} = 3\mathrm{d}$ if $\mathrm{n} - \mathrm{m}$ is a multiple of $3d$ (Dresselhaus et al. 1996).

Keeping in mind the connectivity structure of the carbon atoms on the graphene sheet we can perform its affine transformation in such a way that the unit vectors $\mathbf{a}_1$ and $\mathbf{a}_2$ will be perpendicular and the unit cells have the same atoms with the same connectivity structure as before (Fig. 6.3a). Thus the rolling up algorithm can be generalised for any periodic structure, taking into account only the neighboring structure of the atoms. For the unit vectors $\mathbf{a}_1$ and $\mathbf{a}_2$ we suppose only that they are perpendicular to each other. The unit vector $\mathbf{a}_3 = \mathbf{a}_1 \times \mathbf{a}_2$ is their vectorial product and from definition follows that $\mathbf{a_i} \cdot \mathbf{a_j} = \delta_{\mathrm{ij}}$. Let k, l and r be three integers. Thus the vector $\mathrm{k}\mathbf{a}_1 + \mathrm{l}\mathbf{a}_2$ points to the unit cell $(\mathrm{k,l})$, and $(\mathrm{k,\ l,\ r})$ means the r-th atom in this unit cell in the same way as before in the case of polyhex structures. The neighbours of the atom $(\mathrm{k,\ l,\ r})$ can be in any of the following unit cells: $(\mathrm{k,\ l})$, $(\mathrm{k}+1,\mathrm{l})$, $(\mathrm{k}-1,\mathrm{l})$, $(\mathrm{k},\mathrm{l}+1)$, $(\mathrm{k},\mathrm{l}-1)$, $(\mathrm{k}+,\mathrm{l}+1)$, $(\mathrm{k}-1,\mathrm{l}-1)$, $(\mathrm{k}+1,\mathrm{l}-1)$ and $(\mathrm{k}-1,\mathrm{l}+1)$. These are numbered in the previous order with the type number t $= 0, 1, 2, 3, 4, 5, 6, 7, 8$ and 9. Thus for each atom (k, l, r) we can give the list of its neighbors with the corresponding (k', l', r') parameters. These are the data that will be used for constructing the adjacency matrix of the given structure. The super cell with a parallelogram of side vectors $\mathbf{b_1} = m\mathbf{a_1} + n\mathbf{a_2}$ and $\mathbf{b_2} = p\mathbf{a_1} + q\mathbf{a_2}$ is given with the integers m, n, p andq. It is easy to show that $\mathbf{b_1} \times \mathbf{b_2} = (mq - np)\mathbf{a_3}$. From these definitions follows that for the definition of the unit vectors $\mathbf{a_1}$ and $\mathbf{a_2}$ we do not need any coordinates. We shall use only that they are orthogonal unit vectors.

An atom of coordinates (k, l, r) is in the super cell of vectors $\mathbf{b_1} = m\mathbf{a_1} + n\mathbf{a_2}$ and $\mathbf{b_2} = p\mathbf{a_1} + q\mathbf{a_2}$ if with $\mathbf{c} = k\mathbf{a_1} + l\mathbf{a_2}$ the following relations are valid (László 2005):

$$(\mathbf{b_1} \times \mathbf{b_2}) \cdot \mathbf{a_3} = mq - np > 0 \tag{6.10}$$

$$(\mathbf{b_1} \times \mathbf{c}) \cdot \mathbf{a_3} = ml - nk \geq 0 \tag{6.11}$$

$$(\mathbf{c} \times \mathbf{b_2}) \cdot \mathbf{a_3} = kq - lp \geq 0 \tag{6.12}$$

$$\{(\mathbf{c} - \mathbf{b_2}) \times \mathbf{b_1}\} \cdot \mathbf{a_3} = (k - p)\,n - (l - q)m > 0 \tag{6.13}$$

$$\{\mathbf{b_2} \times (\mathbf{c} - \mathbf{b_1})\} \cdot \mathbf{a_3} = p(l - n) - q(k - m) > 0 \tag{6.14}$$

or

$$(\mathbf{b_1} \times \mathbf{b_2}) \cdot \mathbf{a_3} = mq - np < 0 \tag{6.15}$$

$$(\mathbf{b_1} \times \mathbf{c}) \cdot \mathbf{a_3} = ml - nk \leq 0 \tag{6.16}$$

$$(\mathbf{c} \times \mathbf{b_2}) \cdot \mathbf{a_3} = kq - lp \leq 0 \tag{6.17}$$

$$\{(\mathbf{c} - \mathbf{b_2}) \times \mathbf{b_1}\} \cdot \mathbf{a_3} = (k - p)n - (l - q)m < 0 \tag{6.18}$$

$$\{\mathbf{b_2} \times (\mathbf{c} - \mathbf{b_1})\} \cdot \mathbf{a_3} = p(l - n) - q(k - m) < 0 \tag{6.19}$$

In the rolling up process the connectivity graph $G = (V, E)$ of a nanotube is constructed by identifying the two $\mathbf{b_2}$ sides of the super cell (they are parallel with the vector $\mathbf{b_2}$). The atoms (k_1, l_1, r_1) and (k_2, l_2, r_2) are identified if $r_1 = r_2$ and for the vectors $\mathbf{c_1} = k_1\mathbf{a_1} + l_1\mathbf{a_2}$ and $\mathbf{c_2} = k_2\mathbf{a_1} + l_2\mathbf{a_2}$ the following relation is valid:

$$\mathbf{c_2} - \mathbf{c_1} = u\,\mathbf{b_1} \tag{6.20}$$

where uis an integer in the range $-1 \leq u \leq 1$.

The graph $G = (V, E)$ of a torus is constructed in a similar way, but each opposite sides of type $\mathbf{b_1}$ and $\mathbf{b_2}$ are identified and the condition for identifying the two atoms (k_1, l_1, r_1) and (k_2, l_2, r_2) is $r_1 = r_2$. That is

$$\mathbf{c_2} - \mathbf{c_1} = u\,\mathbf{b_1} + v\,\mathbf{b_2} \tag{6.21}$$

with integers u and v in the ranges $-1 \leq u \leq 1$ and $-1 \leq v \leq 1$.

Fig. 6.4 A nanotube which was drawn by three eigenvectors c^2, c^3 and c^{opt} of the adjacency matrix

After constructing the graph $G = (V, E)$ for the structure under study we can construct its adjacency matrix A and its Laplacian Q as well. From the Eqs. 6.10–6.21 follows that we do not need the special Descartes coordinates of unit cell atoms for the determination of the connectivity graph $G = (V, E)$. The initial data contain only the integers m, n, p, q and the list of neighbours for each unit cell atoms. A neighbouring atom is described by two numbers the serial number r and the type number t.

6.4.2 Topological Coordinates for Nano Tori

After generating the graph $G = (V, E)$ of a torus one can construct its adjacency matrix **A** or its Laplacian **Q**. In Graovac et al. (2000) the triplet, c^2, c^3 and c^{opt} of the adjacency matrix eigenvecrors of a 3-valent toroidal structure was used for generating Descartes coordinates for nanotori. The second and third eigenvectors are c^2 and c^3. The eigenvector c^{opt} was selected in a way to obtain optimal drawing. The toroidal structures obtained from three eigenvectors usually are distorted or flattened in some way (Fig. 6.4) (Graovac et al. 2000; László et al. 2001). These problems can be solved using four eigenvectors of the adjacency matrix (László et al. 2001). In this method the position of a point on the surface of a torus is given as the sum of two vectors R and r, where the vector R is in the xy plane and r is in the planes perpendicular to the plane of R. These two vectors are two dimensional planar vectors, each of them can be described by two bi-lobal eigenvectors. If c^{k_1}, c^{k_2}, c^{k_3} and $c^{k_4}$are for bi-lobal eigenvectors of the adjacency matrix A then using basic geometrical construction the (x_i, y_i, z_i) topological coordinates of the i-th atom on the torus is given by the relations (László et al. 2001; László and Rassat 2003):

$$x_i = S_1 c_i^{k_1}(1 + S_4 c_i^{k_4}) \tag{6.22}$$

$$y_i = S_2 c_i^{k_2}(1 + S_4 c_i^{k_4}) \tag{6.23}$$

$$z_i = S_3 c_i^{k_3}. \tag{6.24}$$

A torus which was drawn using these relations can be seen in Fig. 6.3b.

6.4.3 *Topological Coordinates for Nanotubes*

Although nanotubes can be constructed easily with the help of rolling up the graphene sheet, it is interesting to pose the question whether the topological coordinate method can be used to construct nanotubes as well (László 2004a). For the nanotube let us suppose periodic boundary condition and construct the graph $G = (V, E)$ of the corresponding torus. Using the four bi-lobal eigenvectors $\mathbf{c}^{k_1}$, $\mathbf{c}^{k_2}$, $\mathbf{c}^{k_3}$ and $\mathbf{c}^{k_4}$ of the adjacency matrix of this graph, we can generate the Descartes coordinates of a torus and after cutting it by a plane and transforming it into a tube we obtain the following relations for the atomic coordinates of the nanotube (László and Rassat 2003; László 2004b, 2005):

$$x_i = S_3 c_i^{k_3}, \tag{6.25}$$

$$y_i = S_4 c_i^{k_4}, \tag{6.26}$$

$$\mathrm{z_i} = \mathrm{R}\ \arccos\left(\mathrm{S_1 c_i^{k_1}}/R\right)\quad \textit{if}\ \ \mathrm{c_i^{k_2}} \geq 0 \tag{6.27}$$

$$\mathrm{z_i} = \mathrm{R}\left(2\pi - \arccos\left(\mathrm{S_1 c_i^{k_1}}/R\right)\right)\quad \textit{if}\ \ \mathrm{c_i^{k_2}} < 0 \tag{6.28}$$

Here the radius R governs the size of the torus and it is the average of the $\mathrm{R_i}$ radii of atomic coordinates. The vectors $\mathbf{c}^{k_1}$, $\mathbf{c}^{k_2}$, $\mathbf{c}^{k_3}$ and $\mathbf{c}^{k_4}$ are the four bi-lobal eigenvectors of the adjacency matrix $G = (V, E)$ characterizing the torus with the parameters $(\mathrm{m, n, p, q})$. From this construction follows, that the obtained tube has the same $(\mathrm{m, n, p, q})$ parameters (Fig. 6.3c).

6.4.4 *Topological Coordinates for Planar Structures*

In the previous paragraph we obtained the nanotube topological coordinates by cutting and transforming a torus into a tube. If we cut the tube we can transform it into a planar structure. Thus here we start once more with an appropriate graph $G = (V, E)$ of a torus, calculate its topological coordinates with the help of four bi-lobal eigenvectors of the adjacency matrix and we transform it into a planar structure using simple geometrical transformations. The topological coordinates for the planar graph are the followings (László and Rassat 2003; László 2004b, 2005):

$$x_i = r \arccos\left(S_4 c_i^{k_4}/r\right) \text{ if } c_i^{k_3} \geq 0, \tag{6.29}$$

$$x_i = -r \arccos\left(S_4 c_i^{k_4}/r\right) \text{ if } c_i^{k_3} < 0, \tag{6.30}$$

$$y_i = 0, \tag{6.31}$$

$$z_i = R \arccos\left(S_1 c_i^{k_{14}}/R\right) \text{ if } c_i^{k_2} \geq 0, \tag{6.32}$$

and

$$z_i = R\left(2\pi - \arccos\left(S_1 c_i^{k_1}/R\right)\right) \text{ if } c_i^{k_2} < 0 \tag{6.33}$$

The radii R and r average values of R_i and r_i determine the size and aspect ration of the auxiliary torus, and $\mathbf{c}^{k_1}$, $\mathbf{c}^{k_2}$, $\mathbf{c}^{k_3}$, $\mathbf{c}^{k_4}$ are four bi-lobal eigenvectors of the corresponding toroidal adjacency matrix (Fig 6.3d).

From the Eqs. 6.29–6.33 it could be thought that these formula are valid only for planar structures having two dimensional translation symmetry. As it was shown in László (2005) the method works in each cases where the given planar graph $G' = (V', E')$is a subgraph of the graph $G = (V, E)$having a two dimensional translation symmetry and in Eqs. 6.29–6.33 only the vertices of the graph $G' = (V', E')$ are calculated. In László (2005) we used this method for calculating Schlegel diagrams of fullerenes.

6.5 Shape Analysis of Carbon Nanotubes, Nanotori and Nanotube Junctions

Until now our structures under study were spherical or circular or they were generated from these structures. The fullerenes are three dimensional spheres and the tori can be imagined as the direct product of two circles. It was found that the sphere in d-dimensions can be described with the help d bi-lobal eigenvectors of the Laplacian matrix or in the case of regular graphs of the adjacency matrix. The question arrives what drawings will be obtained using d bi-lobal eigenvectors of structures in d dimension, or increasing the number of eigenvectors how many eigenvectors are needed for a satisfactory drawing?

In Figs. 6.4–6.6 we present the drawings obtained using three bi-lobal eigenvectors of the adjacency matrix and the Laplacian matrix of the corresponding

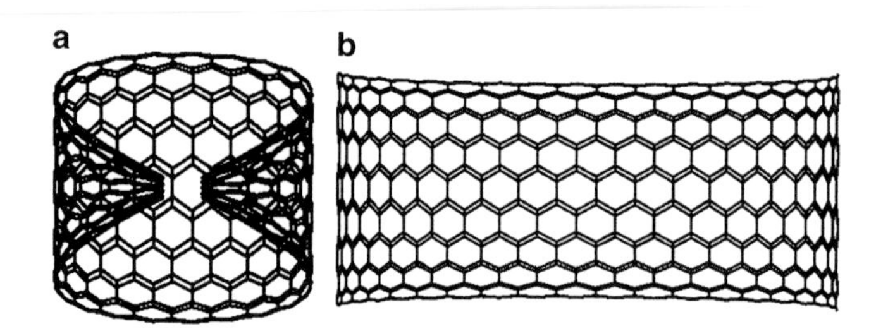

Fig. 6.5 Drawing of nanotube using three bi-lobal eigenvectors of the adjacency matrix (**a**) and of the Laplacian (**b**)

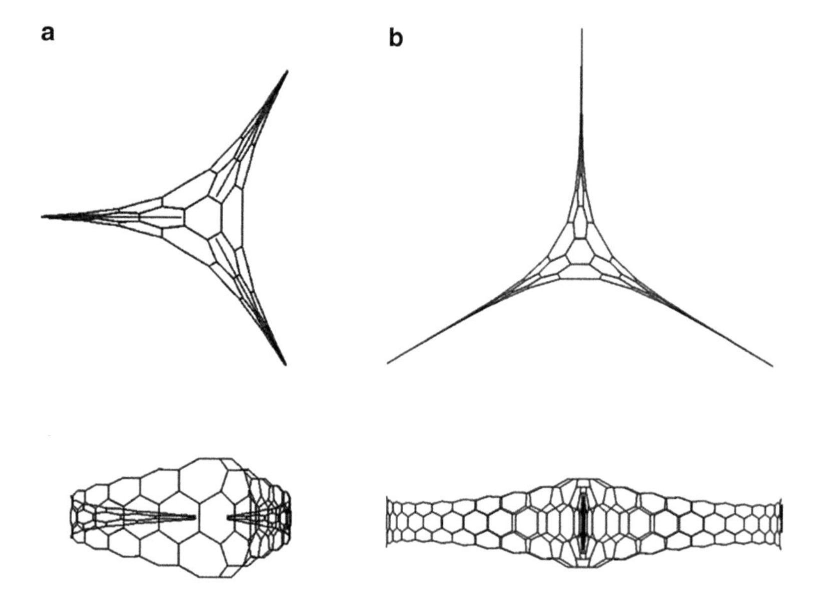

Fig. 6.6 Top and side view of nanotube junctions obtained by three bi-lobal eigenvectors of the adjacency matrix (**a**) and of the Laplacian (**b**)

graphs describing nanotubes nanotori and nanotube junctions. In each cases we can note some kind of flattening. The nanotube ends are spiked and they are turning back in the case of adjacency matrices. This phenomenon shows that for non-regular graphs the eigenvectors of the Laplacian matrix give better drawings than those of the adjacency matrix.

Let us suppose that we have calculated the Descartes coordinates (x_i, y_i, z_i) of the atoms in a non-spherical structure of *n* atoms. Thus **X, Y** and **Z** are n-dimensional vectors containing the x, y and z coordinates of the atoms in order. Let us suppose further that the centre of mass of the molecule is in the origin of the coordinate system and the eigenvectors of its tensor of inertia are showing to the direction of the x, y and z axis.

With the help of the following scalar products

$$\alpha_{Xk} = Xc^k,\ \alpha_{Yk} = Yc^k \text{ and } \alpha_{Zk} = Zc^k \tag{6.34}$$

the atomic coordinates can be written as

$$X = \sum_{k=1}^{n} \alpha_{Xk} c^k,\ Y = \sum_{k=1}^{n} \alpha_{Yk} c^k \text{ and } Z = \sum_{k=1}^{n} \alpha_{Zk} c^k. \tag{6.35}$$

Here $\mathbf{c}_k$ is the eigenvector of the Laplacian $\mathbf{L}$ and the corresponding eigenvalues λ_k are ordered in increasing order. We say that the weights of the eigenvector $\mathbf{c}_k$ in $\mathbf{X}$, $\mathbf{Y}$ and $\mathbf{Z}$ are in order $|\alpha_{Xk}|^2$, $|\alpha_{Yk}|^2$ and $|\alpha_{Zk}|^2$.

The measure of the convergence using only $m \le n$ terms in the summation can be described with the following notations:

$$X_i^{(m)} = \sum_{k=1}^{m} \alpha_{Xk} c_i^k,\ Y_i^{(m)} = \sum_{k=1}^{m} \alpha_{Yk} c_i^k \text{ and } Z_i^{(m)} = \sum_{k=1}^{m} \alpha_{Zk} c_i^k \tag{6.36}$$

and

$$R = (X, Y, Z),\ R^{(m)} = \left(X^{(m)}, Y^{(m)}, Z^{(m)}\right). \tag{6.37}$$

The convergence of the structure is quantified as follows:

$$\left|X - X^{(m)}\right| = \frac{1}{n}\left(\sum_{i=1}^{n}\left(X_i - X_i^{(m)}\right)^2\right)^{\frac{1}{2}} \tag{6.38}$$

$$\left|Y - Y^{(m)}\right| = \frac{1}{n}\left(\sum_{i=1}^{n}\left(Y_i - Y_i^{(m)}\right)^2\right)^{\frac{1}{2}} \tag{6.39}$$

$$\left|Z - Z^{(m)}\right| = \frac{1}{n}\left(\sum_{i=1}^{n}\left(Z_i - Z_i^{(m)}\right)^2\right)^{\frac{1}{2}} \tag{6.40}$$

$$\left|R - R^{(m)}\right| = \frac{1}{n}\sum_{i=1}^{n}\left(\left(X_i - X_i^{(m)}\right)^2 + \left(Y - Y_i^{(m)}\right)^2 + \left(Z - Z_i^{(m)}\right)^2\right)^{\frac{1}{2}} \tag{6.41}$$

In Graovac et al. (2008b) we obtained that only few of the coefficients $|\alpha_{Xk}|^2$, $|\alpha_{Yk}|^2$ and $|\alpha_{Zk}|^2$ are significantly greater than the others. Thus we made such kind of summations similar to the Eq. 6.36 which contained only the significant $|\alpha_{Xk}|^2$, $|\alpha_{Yk}|^2$ and $|\alpha_{Zk}|^2$ terms. In these summations we have found that three bi-lobal

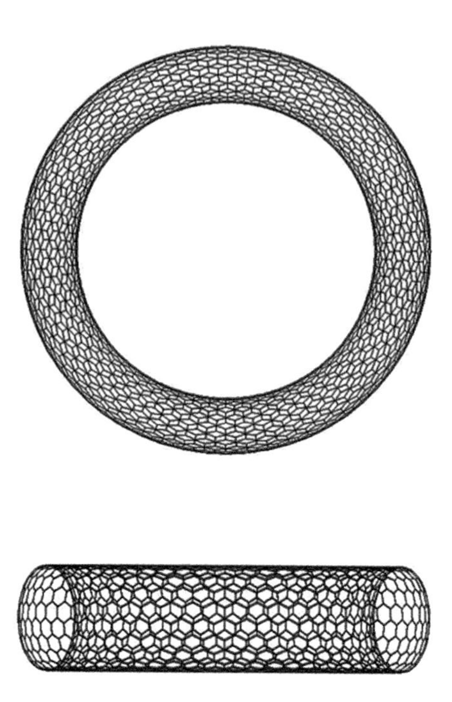

Fig. 6.7 Drawing of nanotorus using ten eigenvectors of the Laplacian. *Top view* and *side view*

eigenvectors of the Laplacian reproduce well the nanotubes without any spiky phenomenon at the ends (Fig. 6.5b). For nanotori we had to use 8–10 eigenvectors of the Laplacian and most of them were bi-lobal. One representative structure is in the Fig. 6.7.

The details of our nanotube junction analysis can be found in Graovac et al. (2008a).

We have found that the greatest absolute values of α_{X2}, α_{Y3} and α_{Z7} are attributed to the three bi-lobal eigenvectors . If the number of eigenfunctions (m) in Eq. 6.36 is smaller than 7 the picture of the structure can be described as a planar or a curved two dimensional surface (Fig. 6.8). The eigenvectors $\mathbf{c}_4$, $\mathbf{c}_5$and $\mathbf{c}_6$ are 4, 4 and 5 –lobal but they have relatively small weight in **Z**. Although the eigenvectors $\mathbf{c}_8$, $\mathbf{c}_9$, $\mathbf{c}_{10}$, $\mathbf{c}_{11}$ and $\mathbf{c}_{12}$ have relatively small weight in **Z** and their lobality is from 3 to 5 they are important in eliminating the spiky features at the tube ends (Fig. 6.8). We can obtain no flattened structure only if m is greater than the serial number of any bi-lobal eigenfunction. The weight of the eigenvectors $\mathbf{c}_{13}$ and $\mathbf{c}_{14}$ is also significant in the vector **Z**. If m is greater than 16, practically there are no appreciated changing in the picture of the structure.

In Graovac et al. (2008a) we have investigated altogether 11 junctions and obtained three-dimensional structures only if all the three bi-lobal eigenfunctions were included in the summations. We had to take into consideration some other eigenvectors for eliminating the spiky behaviours at the end of the tubes. We have found in each case a value for m which gave satisfactory coordinates that can be used for initial coordinates in molecular mechanics calculations.

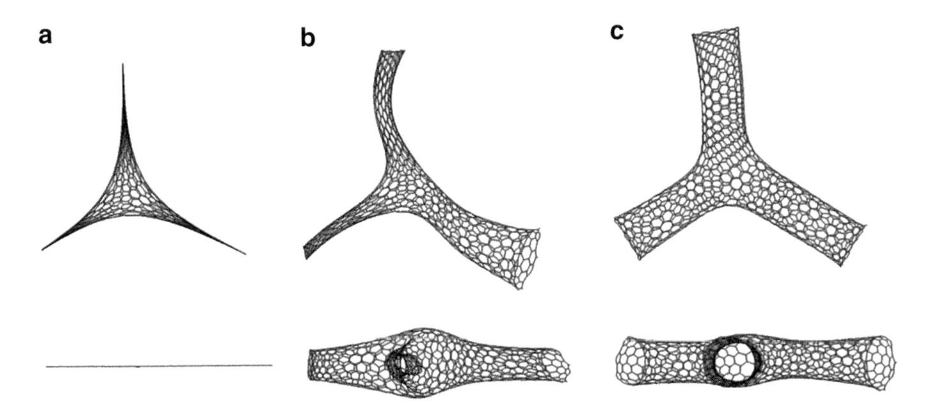

Fig. 6.8 Shape analysis of a nanotube junction. *Top* and *side view* of the structure obtained by 3 (**a**), 9 (**b**) and 16 (**c**) eigenvectors of the Laplacian

6.6 Toward Topological Coordinates of any Molecular Arrangement

6.6.1 Embedding with Three Eigenvectors into R^3

In the previous paragraph we have seen that the applicability of three eigenvectors is restricted only to spherical structures and nanotubes. Here we shall show that there exist such kind of matrix or matrices which can reproduce practically exactly the (x_i, y_i, z_i) Descartes-coordinates with the help of three eigenvectors. Let us suppose that we have the exact (x_i, y_i, z_i) coordinates of the atoms. In this case, as the number of atoms is n, the values of **X**, **Y** and **Z** are n-dimensional vectors containing the x, y and z coordinates of the atoms in order. We suppose further, as before that the centre of mass of the molecule is in the origin and the molecule is directed in such a way that the eigenvectors of its tensor of inertia are showing to the directions of the x, y and z axis. Thus we have the following relations:

$$\sum_{i=1}^{n} x_i = 0, \sum_{i=1}^{n} y_i = 0, \sum_{i=1}^{n} z_i = 0 \tag{6.42}$$

and

$$\mathbf{XY} = \sum_{i=1}^{n} x_i y_i = 0, \mathbf{YZ} = \sum_{i=1}^{n} y_i z_i = 0, \mathbf{ZX} = \sum_{i=1}^{n} z_i x_i = 0 \tag{6.43}$$

If we define the vector **U** of n dimension with the relations $u_i = \frac{1}{\sqrt{n}}$ from Eq. 6.42 follows that

$$\mathbf{UX} = 0, \mathbf{UY} = 0, \mathbf{UZ} = 0 \tag{6.44}$$

Equations 6.43 come from the conditions that the off-diagonal matrix elements of the tensor of inertia are zero because of the special position of the molecule.

Let us suppose that the total energy

$$E(r) = E(r_{12}, r_{21}, ...r_{ij}, r_{ji}...) \tag{6.45}$$

depends only on the inter-atomic distances

$$r_{ij} = \left((x_i - x_j)^2 + (y_i - y_j)^2 + (z_i - z_j)^2 \right)^{\frac{1}{2}} \tag{6.46}$$

Here we suppose further that total energy $E(r)$ depends on r_{ij} and r_{ji} in a symmetric way. If this is not the case, we substitute r_{ij} or r_{ji} by the value $r_{ij} = r_{ji} = (r_{ij} + r_{ij})/2$.

In the followings we shall use the relations:

$$\frac{dr_{ij}}{dx_i} = \frac{x_i - x_j}{r_{ij}}, \frac{dr_{ij}}{dy_i} = \frac{y_i - y_j}{r_{ij}}, \frac{dr_{ij}}{dz_i} = \frac{z_i - z_j}{r_{ij}} \tag{6.47}$$

$$\frac{dr_{ij}}{dx_j} = \frac{x_j - x_i}{r_{ij}}, \frac{dr_{ij}}{dy_j} = \frac{y_j - y_i}{r_{ij}}, \frac{dr_{ij}}{dz_j} = \frac{z_j - z_i}{r_{ij}} \tag{6.48}$$

As the gradients give the forces $F_i = (F_{x_i}, F_{y_i}, F_{z_i})$ acting on the i-th atom, we have:

$$\frac{\partial E(r)}{\partial x_i} = -F_{x_i}, \frac{\partial E(r)}{\partial y_i} = -F_{y_i}, \frac{\partial E(r)}{\partial z_i} = -F_{z_i} \tag{6.49}$$

Applying the above mentioned relations we obtain.

$$\frac{\partial E}{\partial x_i} = \sum_{j=1}^{n} w_{ij} x_j = -F_{x_i} \tag{6.50}$$

$$\frac{\partial E(r)}{\partial y_i} = \sum_{j=1}^{n} w_{ij} y_j = -F_{y_i} \tag{6.51}$$

$$\frac{\partial E}{\partial z_i} = \sum_{j=1}^{n} w_{ij} z_j = -F_{z_i} \tag{6.52}$$

where

$$w_{ij} = -\frac{\partial E(r)}{r_{ij} \partial r_{ij}} - \frac{\partial E(r)}{r_{ji} \partial r_{ji}} \tag{6.53}$$

and

$$w_{ii} = \sum_{j \neq i}^{n} \left(\frac{\partial E(r)}{r_{ij} \partial r_{ij}} + \frac{\partial E(r)}{r_{ji} \partial r_{ji}} \right) = -\sum_{j \neq i}^{n} w_{ij} \tag{6.54}$$

That is

$$\mathbf{WX} = -\mathbf{F_x},\ \mathbf{WY} = -\mathbf{F_y}\ \mathbf{WZ} = -\mathbf{F_z} \tag{6.55}$$

If the atoms are in the equilibrium positions, **X, Y** and **Z** can be seen as eigenvectors of **W** with zero eigenvalue, that is

$$\mathbf{WX} = \mathbf{0},\ \mathbf{WY} = \mathbf{0},\ \mathbf{WZ} = \mathbf{0} \tag{6.56}$$

From the relations of Eqs. 6.53 and 6.54 follows that

$$\mathbf{WU} = \mathbf{0} \tag{6.57}$$

with $u_i = \frac{1}{\sqrt{n}}$.

If the centre of mass of the molecule is in the origin and the molecule is directed in such a way that the eigenvectors of its tensor of inertia are showing to the directions of the x, y and z axis, from the Eqs. 6.43, 6.44, 6.56, 6.57 follows that **X**, **Y**, **Z** and **U** are orthogonal eigenvectors of the matrix **W**. That is

$$\mathbf{X} = S_x \mathbf{C^x},\ Y = S_y \mathbf{C^y},\ Z = S_z \mathbf{C^z} \tag{6.58}$$

where $\mathbf{C^x}$, $\mathbf{C^y}$, $\mathbf{C^z}$ and U are orthogonal and normalized eigenvectors of **W** with zero eigenvalue and S_x, S_y and $S_z$are appropriate scaling factors.

The question arises if we have any orthogonal and normalized eigenvectors $\mathbf{A^x}$, $\mathbf{A^y}$, $\mathbf{A^z}$ and **U** of **W** with zero eigenvalue are there any appropriate scaling factors S_x, S_y and S_z for obtaining the Descartes coordinates with a relation,

$$\mathbf{X} = S_x \mathbf{A^x},\ Y = S_y \mathbf{A^y},\ Z = S_z \mathbf{A^z} \tag{6.59}$$

If the number of eigenvectors with zero eigenvalue is four, the answer is yes, but in Eq. 6.59 we obtain a rotation of the molecule as the vectors $\mathbf{A^x}$, $\mathbf{A^y}$, $\mathbf{A^z}$ can be obtained as linear combination of the vectors $\mathbf{C^x}$, $\mathbf{C^y}$, $\mathbf{C^z}$. If the vectors $\mathbf{C^x}$, $\mathbf{C^y}$, $\mathbf{C^z}$ are mixed with the vector **U** it means arbitrary translation and a rotation of the molecule. As the vector **U** is known it can be easily subtracted from the linear combinations in the case of mixings.

Usually the first neighbour distances in a molecule do not determine the positions of the atoms but the full structure can be describe if we know the second neighbour distances as well. From this follows that if the edges of a molecular graph $G = (V, E)$correspond to the first and second neighbours of a molecule, the matrix **W** can be generated from a total energy E(r) which depends only on the

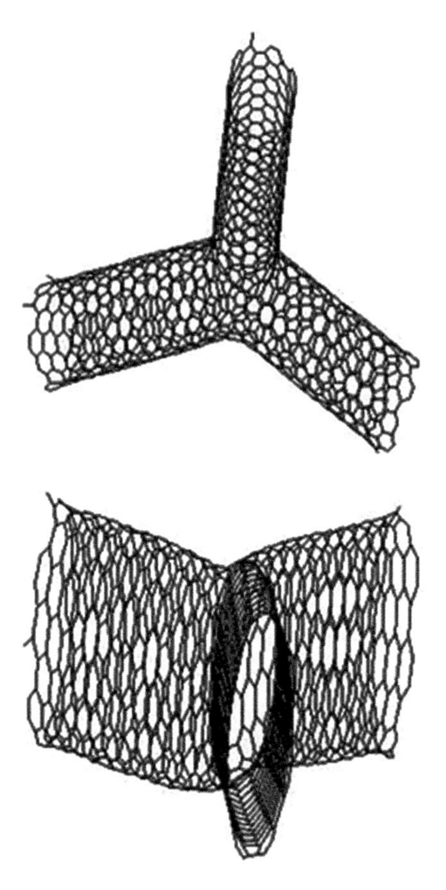

Fig. 6.9 Coordinates of a nanotube junction of the tubes (9,6), (8,7) and (10,5) obtained by three eigenvectors of the matrix W (*top view* and *side view*)

first and second neighbors of the molecule. If the dimension of the null space of **W** is four than this null space contains three eigenvectors which give a proper embedding of $G = (V, E)$ into R^3.

6.6.2 *Examples for Embedding with Three Eigenvectors into* R^3

We tested our ideas for several structures. Here we present our results obtained for nanotube junctions nano tori and helical nanotubes. We calculated the interatomic interactions with the help of the Brenner potential (Brenner 1990) and harmonic potentials as well. In the Brenner potential there are first neighbour and second neighbour interactions.

The matrix **W** has non-zero matrix elements only for the first and the second neighbours. We calculated the equilibrium position of the carbon atoms in a nanotube junction, in a torus and in a helical structure. In each cases we could reconstruct the original coordinates with the help of the relations of Eq. 6.59, see Figs. 6.9–6.11. In each cases we used $S_x = S_y = S_z$. The final values for the parameters S_x, S_y and S_z can be obtained from scaling three independent distance to given values.

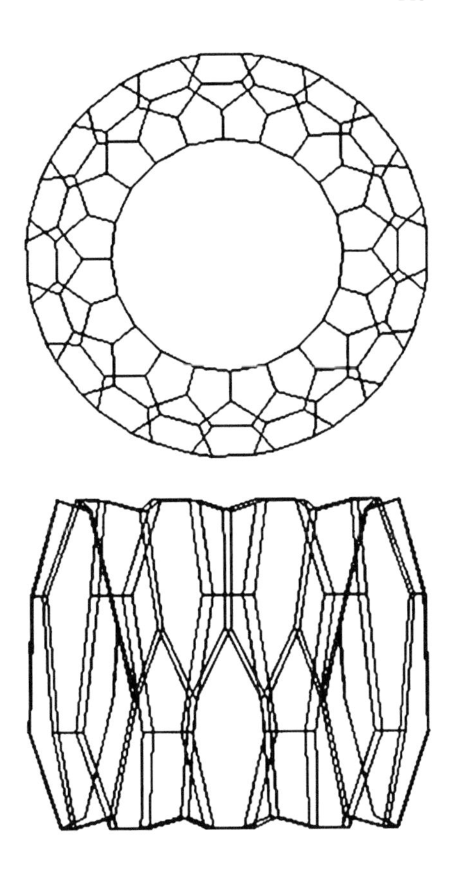

Fig. 6.10 Coordinates of a torus obtained by three eigenvectors of the matrix W (*top view* and *side view*)

6.7 Conclusions

We have examined the possibilities of drawing graphs with eigenvectors of the adjacency matrix **A** the Laplacian **L** and the Colin de Verdière matrix **M**. We have suggested the matrix **W** for drawing not only spherical graphs with eigenvectors. For these symmetric matrices the off-diagonal matrix element of $(u, v) \notin E$ are equals to zero. The absolute value of other off-diagonal matrix element for matrices **A** and **L** are equals to 1. In matrices **L** and **W** the diagonal matrix elements are calculated in such a way that the sum of matrix elements be equal to zero in each row. The off-diagonal matrix elements of **M** and **W** are determined using special kind of conditions. We have shown that in each cases where the adjacency matrix was applicable the Laplacian matrix gave good drawings as well. We have found also that the Laplacian could be used even for nanotubes where the adjacency matrix was not applicable. It was demonstrated that three eigenvectors of the matrix **W** produced good drawing in each cases under study in this paper. The drawback of **W** is that at present there is not a simple algorithm for its construction. In this work we could generate it using an energy minimalization algorithm with the help of a

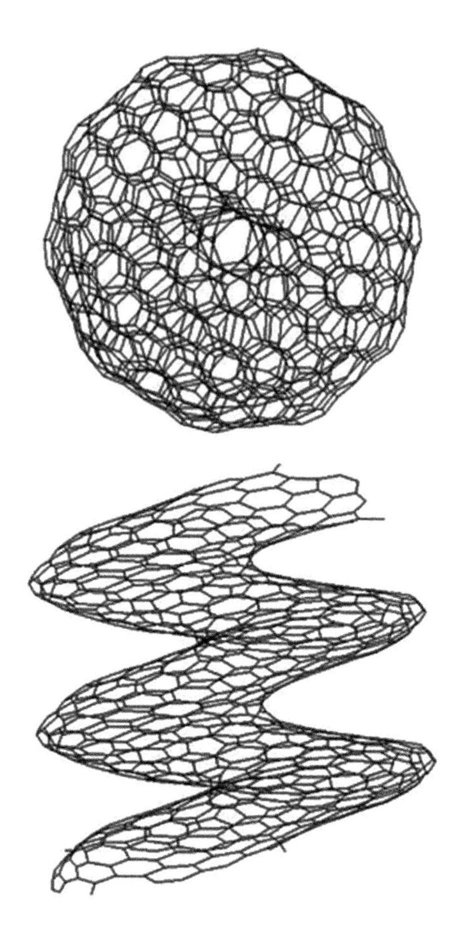

Fig. 6.11 Coordinates of a helix obtained by three eigenvectors of the matrix W (*top view* and *side view*)

conjugate gradient algorithm. There is a hope, however, that using appropriate approximations for the matrix elements of W a method can be found for constructing topological coordinates of complicated non-spherical structures as well.

Acknowledgements I. László thanks for the supports of grants TAMOP-4.2.1/B-09/1/KONV-2010-0003, TAMOP-4.2.1/B-09/1/KMR-2010-0002 and for the support obtained in the frame work of bi-lateral agreement between the Croatian Academy of Science and Art and the Hungarian Academy of Sciences.

References

Biyikoglu T, Hordijk W, Leydold J, Pisanski T, Stadler PF (2004) Linear Algebra Appl 390:155–174

Biyikoglu T, Leydold J, Stadler PF (2007) Laplacian eigenvectors of graphs. Perron-Frobenius and Faber-Krahn type theorems, LNM1915. Springer, Berlin/Heidelberg

Brenner DW (1990) Phys Rev B 42:9458–9471

Colin de Verdière Y (1998) Spectres de graphes, cours spécialisés 4. Société Mathématique de France, Paris
Di Battista G, Eades P, Tamassia R, Tollis IG (1999) Graph drawing: algorithms for the visualization of graphs. Prentice Hall, Upper Saddle River
Dresselhaus MS, Dresselhaus G, Eklund PC (1996) Science of fullerenes and carbon anotubes: their properties and applications. Academic, New York/London
Fowler PW, Manolopulos DE (1995) An atlas of fullerenes. Clarendon, Oxford
Fowler PW, Pisanski T, Shaw-Taylor J (1995) In Tamassia R, Tollis EG (eds) Graph drawing. Lecture notes in computer science, vol 894. Springer-Verlag, Berlin
Godsil CD, Royle GF (2001) Algebraic graph theory. Springer, Heidelberg
Graovac A, Plavšić D, Kaufman M, Pisanski T, Kirby EC (2000) J Chem Phys 113:1925–1931
Graovac A, Lászlo I, Plavšić D, Pisanski T (2008a) MATCH Commun Math Comput Chem 60:917–926
Graovac A, László I, Pisanski T, Plavšić D (2008b) Int J Chem Model 1:355–362
Hall KM (1970) Manage Sci 17:219–229
Kaufmann M, Wagner D (eds) (2001) Drawing graphs. Methods and models, LNCS 2025. Springer-Verlag, Germany
Koren Y (2005) Comput Math Appl 49:1867–1888
László I (2004a) Carbon 42:983–986
László I (2004b) In Buzaneva E, Scharff P (eds) Frontiers of multifunctional integrated nanosystems. NATO science series, II. Mathematics, physics and chemistry. Kluwer Academic Publishers, Dordrecht, Boston, London. Vol 152, 11
László I (2005) In: Diudea MV (ed) Nanostructures: novel architecture. Nova Science, New York, pp 193–202
László I (2008) In: Blank V, Kulnitskiy B (eds) Carbon nanotubes and related structures. Research Singpost, Kerala, pp 121–146
László I, Rassat A (2003) J Chem Inf Comput Sci 43:519–524
László I, Rassat A, Fowler PW, Graovac A (2001) Chem Phys Lett 342:369–374
Lovász L, Schrijver A (1999) Ann Inst Fourier (Grenoble) 49:1017–1026
Lovász L, Vesztergombi K (1999) In Halász L, Lovász L, Simonovits M, T Sós V (eds) Paul Erdős and his mathematics. Bolyai Society – Springer Verlag. Berlin, Heidelberg, New York
Manolopoulos DE, Fowler PW (1992) J Chem Phys 96:7603–7614
Pisanski T, Shawe-Taylor JS (1993) In Technical report CSD-TR-93-20, Royal Holloway, University of London, Department of Computer Science, Egham, Surrey TW200EX, England
Pisanski T, Shawe-Taylor JS (2000) J Chem Inf Comput Sci 40:567–571
Pisanski T, Plestenjak B, Graovac A (1995) Croat Chim Acta 68:283–292
Rassat A, László I, Fowler PW (2003) Chem Eur J 9:644–650
Stone AJ (1981) Inorg Chem 20:563–571
Trinajstić N (1992) Chemical graph theory. CRC Press/ Boca Raton/ Ann Arbor, London/Tokyo
Tutte WT (1963) Proc Lond Math Soc 13:743–768
van der Holst H (1996) Topological and spectral graph characterizations. Ph.D. Thesis, University of Amsterdam, Amsterdam

Chapter 7
Applications of Chemical Graph Theory to Organic Molecules

Lionello Pogliani[1]

Abstract The full combinatorial search algorithm and the greedy search algorithm have been used on a set of molecular connectivity indices, five experimental parameters and the molar mass to extract the best descriptor for 12 properties of a set of organic solvents. The quality of the descriptions has been compared with a 'zero-level' description based on random numbers. The optimal molecular connectivity indices of the best descriptors obtained with the full combinatorial search and belonging to different configurations have been used to derive super-descriptors. These last types of descriptors achieve an impressive description for the melting points. It has also been found that few properties can advantageously be described with full combinatorial descriptors that include random indices, the so-called semi-random descriptors. The model of a set of randomized properties with super-descriptors and with random numbers has been tried to further check the validity of the Topliss-Costello rule. The full combinatorial search algorithm is quite effective in finding the best descriptor, while it highlights the advantages to work with a selected choice of indices belonging to different configurations, which encode different hydrogen and core electron contributions. The forward search greedy algorithm is nevertheless able to find some no bad descriptors. Both search algorithms underline the descriptive importance of the experimental parameters. Previous conclusions on the contributions of the hydrogen atoms to obtain an optimal model are confirmed as well as the need of a more stringent limit for the q^2 statistics, and for the Topliss-Costello rule.

[1] Dipartimento di Chimica, Università della Calabria,
via P. Bucci, 87036 Rende (CS), Italy
e-mail: lionp@unical.it

M.V. Putz (ed.), *Carbon Bonding and Structures: Advances in Physics and Chemistry*, Carbon Materials: Chemistry and Physics 5, DOI 10.1007/978-94-007-1733-6_7,

7.1 Introduction

Recent studies (Pogliani 2010, 2011) describe the satisfactory QSPR model of twelve properties of a highly heterogeneous class of organic solvents done with two different model strategies to find out the best descriptor. The first model strategy uses the *greedy* algorithm, a forward step search algorithm, which finds the best multi-index descriptor by adding the next best index while keeping constant the previous ones. The second model strategy, the full combinatorial algorithm, spans the entire combinatorial space (normally from millions to billions of descriptors) made by the molecular connectivity (MC) indices to find the best multi-index descriptors. The greedy algorithm, even if it uses a drastically reduced number of combinations (here around hundreds), a drawback that all inclusion stepwise methods share, finds, nevertheless, rather satisfactory descriptors. Clearly, it is not evident how far from optimal the found greedy descriptors are. The full combinatorial algorithm, working on the entire combinatorial space is, instead, able to find out the best overall descriptor. The model of the 12 properties with both model strategies with 30 molecular connectivity indices to which the molar mass (M) was added is here reviewed and deepened. For the boiling and melting points as well as for the dipole moment three *ad hoc* parameters were introduced, AH^b, AH^m, and ϕ (0, 1), respectively, to take care of the hydrogen bond problem and/or of other structural problems. To improve the model also semi-empirical descriptors were used, i.e., multilinear descriptors that include experimental parameters as indices. The experimental parameters that were used as indices were those with the greatest number of points: T_b, the boiling temperature, T_m, the melting temperature, ε, the dielectric constant, d, the density, and RI, the refractive index of the organic solvents. The model of the surface tension with the greedy algorithm can, for instance, only be achieved by the aid of these five experimental parameters. The full combinatorial algorithm confirms the utility of this methodology, as many properties can satisfactorily be modeled only with semiempirical descriptors made of molecular connectivity indices plus experimental parameters.

Let us remind that while the number of six-index combinations from a set of 36 indices (30 molecular connectivity indices plus five empirical indices, plus M) entails many millions of combinations, the number of greedy combinations entails only hundreds of combinations to find out a six-index descriptor. In the present chapter an overview will also be given on the model quality of semi-random combinations, i.e., multilinear descriptors made of graph-theoretical and experimental indices plus random indices. These last type of indices were chosen from two sets of 38 different subsets of random indices, and in some cases they show an interesting model quality.

7.2 Method

The valence delta number, δ^v, is the basic parameter not only for the valence molecular connectivity indices (χ^v) but also for the *I*- and *E*-State indices ($\psi_{E,I}$: *E* means electrotopological, and *I* intrinsic), as *I* and *S* are δ^v-dependent indices, in fact (Kier and Hall 1999):

$$I = (\delta^v + 1)/\delta, S = I + \Sigma\Delta I, \text{with } \Delta I = \left(I_i - I_j\right)/r^2{}_{ij} \tag{7.1}$$

r_{ij} counts the atoms in the minimum path length separating atoms *i* and *j*, which equals the graph distance, $d_{ij} + 1$; $\Sigma\Delta I$ incorporates the information about the influence of the remainder of the molecular environment, and, as it can be negative, *S* can also be negative for some atoms. To avoid imaginary ψ_E values (see later on), every *S* value has to be rescaled (Pogliani 2002; Garcia-Domenech et al. 2008). Throughout the present model the rescaling value is 6.611. The δ^v number defined in the following Eq. 7.2 encodes the core electrons, by the aid of two parameters belonging to complete graphs, *p* and *r*. It encodes also the depleted hydrogen atoms, by the aid of the perturbation parameter, f_δ (Garcia-Domenech et al. 2008; Pogliani 2005a, b, 2006, 2007, 2010),

$$\delta^v = \frac{(q + f_\delta^n)\delta^v(ps)}{(p \cdot r + 1)} \tag{7.2}$$

Here, $\delta^v(ps)$ is the valence of a vertex in a chemical pseudograph. This type of graph also known as general graph allows multiple bonds and self-connections (or loops). Normally, in chemical graph theory simple graphs (with no multiple bonds and loops) and pseudographs are hydrogen-depleted (or suppressed, HS). Parameters *p* is the order of a complete graph, K_p, and *r* is its regularity ($r = p - 1$). A complete graph is a graph where every pair of its vertices is adjacent (Trinajstić 1992). A K_p is always *r*-regular, i.e., all its vertices have the same regularity, *r*. The first order complete graph, K_1, is just a vertex and it is usually been used to encode second row atoms, and especially the carbon atom. Parameter *q* in relation 7.2 is two-values, i.e., $q = 1$ or *p*, where *p* can be either odd-valued ($p = 1, 3, 5, 7,\ldots$) or sequential-valued ($p = 1, 2, 3, 4,\ldots$). Four representations for δ^v are thus possible: K_p-(*p-odd*) for $q = 1$, and $p = odd$; K_p-(*p-seq*) for $q = 1$ and $p = seq$ (sequential); K_p-(*pp-odd*) for $q = p$ and $p = odd$, and K_p-(*pp-seq*) for $q = p$ and $p = seq$. To keep the number of MC indices under control only two representations are chosen: the K_p-(*p-odd*) representation with the smallest δ^v values ($p = 1$, 3, and 5 and $q = 1$), and the K_p-(*pp-odd*) representation with the second largest δ^v values, that is, the ($p = 1$, 3, and 5 and $q = p$). The f_δ fractional perturbation parameter encodes the depleted hydrogen atoms, and is defined in the following way,

$$f_\delta = [\delta^v{}_m(ps) - \delta^v(ps)]/\delta^v{}_m(ps) = 1 - \delta^v(ps)/\delta^v{}_m(ps) = n_H/\delta^v{}_m(ps) \tag{7.3}$$

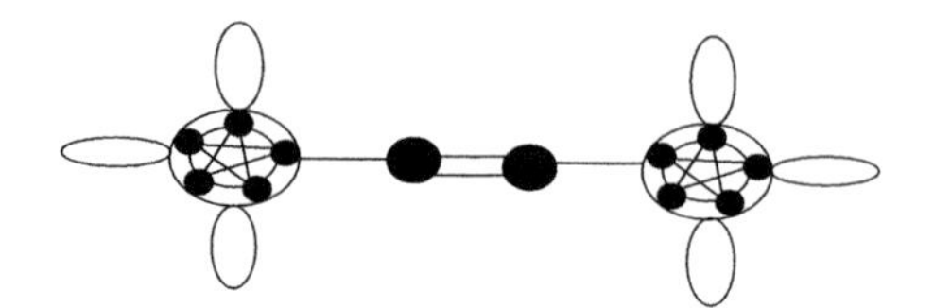

Fig. 7.1 The HS pseudograph plus complete graph of Br-CH = CH-Br (K_1 for C and K_5 for Br)

$\delta^{v}_{m}(ps)$ is the maximal $\delta^{v}(ps)$ value a heteroatom (a vertex) can have in a hydrogen depleted chemical pseudograph when all bonded hydrogen atoms are substituted by heteroatoms, and n_H equals the number of hydrogen atoms bonded to a heteroatom. For completely substituted heteroatoms, $f_\delta = 0$ as $\delta^{v}_{m}(ps) = \delta^{v}(ps)$ (i.e., $n_H = 0$). It should be remarked that in hydrocarbons $\delta^{v}(ps) = \delta$, which is the delta number in simple chemical graphs with no multiple bonds and loops. In this case: $\delta^{v} = (1+f_\delta{}^{n})\delta$ (for $p = 1$). For quaternary carbons $f_\delta = 0$ and $\delta^{v} = \delta$.

Exponent n in f_δ quantifies the importance of the perturbation, i.e., the higher the n values the lower the importance of the perturbation. Different values for n give rise to different sets of indices, where in each set n is constant, and, consequently, the corresponding δ^{v} is constant. In this study, $n = -1, -0.5, -0.1, 0.1, 0.5, 1, 2, 5, 8, 50$. The greedy study does not consider the two n values $n = -0.1$ and 0.1. From now on a set of indices, which has been derived either with a K_p(*p-odd*) or with a K_p(*pp-odd*) representation for the core electrons and with a particular type of $f_\delta{}^{n}$ hydrogen perturbation will be considered to belong either to the K_p(*p-odd*) / $f_\delta{}^{n}$ or to the K_p(*pp-odd*) / $f_\delta{}^{n}$ configuration.

Figure 7.1 shows the HS pseudograph plus complete graph for Br-CH = CH-Br. The δ^{v} values for the bromine and carbon atoms in this compound are the following: $\delta^{v}(\mathrm{Br}) = 7/21$, and $\delta^{v}(\mathrm{C}) = 3.1875$, where, the core electrons of Br have been encoded with K_5 odd complete graphs ($p = 5$, $q = 1$, and $n_H = 0$), while the two carbon atoms have been encoded with K_1 vertices (or complete graphs) and their valence delta values have been calculated with a $f_\delta{}^{2}$ hydrogen perturbation ($n_H = 1$, $\delta^{v}_{m}(ps) = 4$). The δ values for the bromine and carbon atoms of the HS simple graph (•—•—•—•) of this same compound are $\delta(\mathrm{Br}) = 1$, and $\delta(\mathrm{C}) = 2$. The $\delta^{v}(ps)$ values of the HS pseudograph of compound in Fig. 7.1, and whose vertices are undistinguishable K_1 vertices are: $\delta^{v}(psBr) = 7$, and $\delta^{v}(psC) = 3$.

The molecular connectivity indices used in this study are:

$$\{\chi\} = \{D, {}^{0}\chi, {}^{1}\chi, \chi_t, D^{v}, {}^{0}\chi^{v}, {}^{1}\chi^{v}, \chi^{v}{}_{t}\},$$
$$\{\psi\} = \{{}^{S}\psi_1, {}^{0}\psi_1, {}^{1}\psi_1, {}^{T}\psi_1, {}^{S}\psi_E, {}^{0}\psi_E, {}^{1}\psi_E, {}^{T}\psi_E\},$$
$$\{\beta_d\} = \{{}^{0}\chi_d, {}^{1}\chi_d, {}^{1}\chi_s, {}^{0}\chi^{v}{}_{d}, {}^{1}\chi^{v}{}_{d}, {}^{1}\chi^{v}{}_{s}, {}^{0}\psi_{1d}, {}^{1}\psi_{1d}, {}^{1}\psi_{1s}, {}^{0}\psi_{Ed}, {}^{1}\psi_{Ed}, {}^{1}\psi_{Es}\}, \ \Delta, \ \Sigma \qquad (7.4)$$

The definition of these 30 indices is shown in Tables 7.1 and 7.2, where the Δ index encodes the number of electronegative atoms (n_{EA}), while the Σ index encodes the sum of the *S*-State index for the electronegative atoms, *N*, *O*, *F*, *Cl*, *Br* ($\langle S_{EA} \rangle$ is the average value for a specific type of atom, sulphur has not been considered as an electronegative atom). The rescaling procedure brings about

Table 7.1 Definition of the *MCI* indices used in this study[a]

MCI $\{\chi\}$	*Dual MCI* $\{\beta_d\}$
$D = \Sigma_i\delta_i$	${}^0\chi_d = (-0.5)^N\Pi_i(\delta_i)$
${}^0\chi = \Sigma(\delta_i)^{-0.5}$	${}^1\chi_d = (-0.5)^{(N+\mu-1)}\Pi(\delta_i + \delta_j)$
${}^1\chi = \Sigma(\delta_i\delta_j)^{-0.5}$	${}^1\chi_s = \Pi(\delta_i + \delta_j)^{-0.5}$
$\chi_t = (\Pi\,\delta_i)^{-0.5}$	$\Delta = \Sigma_{EA}n_{EA}$

[a] N is the number of atoms, ij means σ bond, μ is the cyclomatic number. Replacing δ with δ^v and I with S other eight valence χ^v indices and eight ψ_E indices are obtained for a total of 28 χ-ψ MC indices. Δ and Σ are special indices (see paragraph after Eq. 7.4)

Table 7.2 Definition of the *pseudo-MCI* indices used in this study[a]

pMCI $\{\psi\}$	*Dual pMCI* $\{\beta_d\}$
${}^S\psi_I = \Sigma_iI_i$	${}^0\psi_{Id} = (-0.5)^N\Pi_i(I_i)$
${}^0\psi_I = \Sigma(I_i)^{-0.5}$	${}^1\psi_{Id} = (-0.5)^{(N+\mu-1)}\Pi\,(I_i + I_j)$
${}^1\psi_I = \Sigma\,(I_iI_j)^{-0.5}$	${}^1\psi_{Is} = \Pi\,(I_i + I_j)^{-0.5}$
${}^T\psi_I = (\Pi\,I_i)^{-0.5}$	$\Sigma = \Sigma_{EA} < S_{EA} >$

[a] N is the number of atoms, ij means σ bond, μ is the cyclomatic number. Replacing δ with δ^v and I with S other eight valence χ^v indices and eight ψ_E indices are obtained for a total of 28 χ-ψ MC indices. Δ and Σ are special indices (see paragraph after Eq. 7.4)

that ${}^S\psi_I \neq {}^S\psi_E$, while, normally, the electrotopological state with no rescaling procedure concept implies $\Sigma_iS_i = \Sigma_iI_i$. In relation 7.4 $\{\chi\}$ is the subset of eight molecular connectivity indices (*MCI*); $\{\psi\}$ is the subset of eight molecular pseudoconnectivity indices (*pMCI*), and $\{\beta_d\}$ the subset of 12 dual connectivity and pseudoconnectivity indices [*Dual MCI* and *Dual pMCI*] (Pogliani 2010; Garcia-Domenech et al. 2008). All these indices are considered molecular connectivity indices, as, formally, they originated from Randić's molecular connectivity index (Randić 1975; Kier and Hall 1986; Todeschini and Consonni 2009) further developed by Kier and Hall (1986). Indices $\{D, {}^0\chi, {}^1\chi, \chi_t, {}^0\chi_d, {}^1\chi_d, {}^1\chi_s, \Delta\}$ are independent of the hydrogen content of the solvent molecules as well as from the complete graph representation for the core electrons. As already told to these 30 indices, the molar mass, M, and the five experimental parameters $\{T_b, T_m, \varepsilon, d, RI\}$ will be added and this brings the number of indices used for model purposes to 36. The huge number of possible combinations that these indices give rise should be multiplied by 20 as there are two representations for the core electrons: K_p-(*p-odd*) and K_p-(*pp-odd*) and for each representation there are ten different values for $f_\delta{}^n$. The full combinatorial space for the best descriptor has been searched with the help of Statistica 6.0 of Statsoft Inc.

The best descriptor, which is normally chosen with a least-squares procedure, is characterized by the lowest standard deviation of the estimates, s, and highest correlation coefficient, r. The minimum requirement is that at least $r > 0.84$ for a ratio N° data/N° indices ≥ 5 (Topliss-Costello rule, Topliss and Costello 1972). The best descriptor usually has a high Fischer ratio, F, which is monotonically related to r for constant number N of data points, and constant number of variables, v.

It is usual practice in model studies to mention F statistics, a practice some workers in the field deplore, and for this reason it will be left out. It is also practice to mention r^2 instead of r, which can easily be compared with q^2 (next paragraph). The deviation, s_i, of the regression parameters, c_i, of the multilinear relationships should also be rather low (see Randić 1994 for a discussion on this topic). Observed vs. calculated plots for all models should more often than not be given as sometimes 'good' statistics are contradicted by rather poor plots (Garcia-Domenech 2008; Pogliani and de Julian-Ortiz 2004; Besalu et al. 2006a, b, 2007; de Julian-Ortiz et al. 2010). The prediction q^2 coefficient, used to check the validity of the leave-one-out (or Jackknife) method will also be given. This method belongs to the wide family of cross-validation methods, and it is a (usually) computer intensive method to estimate parameters, and/or to gauge uncertainty in the estimates. Each observation is removed one at a time and during the removal it is assumed that the descriptor does not change. The prediction coefficient q^2 equals (SD- $PRESS$)/SD, where $SD = \Sigma(y_i - < y >)^2$ is the squared deviation of the observed value from their mean, and $PRESS = (y_i - y_{iloo})^2$, where y_{iloo} is a predicted value of the studied property where the prediction has been made by the leave-one-out method (Todeschini and Consonni 2009; Carbó-Dorca et al. 2000). There is no common accord about the minimum q^2 value for the acceptance of a model. It has been suggested that $q^2 > 0.6$, but the present study will show that his is a too optimistic choice. The leave-one-out method has some drawbacks with small data sets and with strong clusterization, which is here rarely the case (Eriksson et al. 2000; Martens and Dardenne 1998). The model quality of the optimal descriptor will also be tested leaving out some compounds (those with '°' in Table 7.3) in a sort of *back-training* test. Strong outliers, whenever advantageous, will be excluded from the model. Outliers are a few data that have a large undesirable effect on the model and for this reason they have to be removed. Outliers that should be as few as possible are not necessarily outside the chemical domain but depend, as shown in ref. 1, on the QSPR model employed.

The 'zero-level' model will here be searched among two 38 sets, *r1-r38*, and *rd1-rd38* of 0–1 valued random numbers. Furthermore, a combinatorial space made of the *r1-r38* and then of *rd1-rd38* sets of random indices plus the given MC indices, the five experimental parameters, and M is also searched for the best semi-random descriptor. The resulting descriptors could also be called either semi-random descriptors or *mc-exp-rn*-descriptors. Here, *mc* means molecular connectivity, *exp* means experimental and *rn* means random. Whenever it is not computationally feasible only the indices of the best descriptors will be added to the *r1-r38* and/or *rd1-rd38* random sets. Random numbers have been obtained with the algorithm of the 2003 Microsoft Excel electronic sheet. A second random test will be performed with randomized properties, where the values of the properties have been randomized among the different compounds.

The correlation parameters c_i of the multilinear regressions as well their errors s_i will be collected into a vector, $\boldsymbol{C} = [c_1(s_1)\ c_2,(s_2), \ldots, c_n(s_n), c_0(s_0)]$. This formalism allows to read vector $\boldsymbol{C}$ as the ordered list of regression values of a property $\boldsymbol{P}$ with respect to the vector of the MC indices, $\boldsymbol{\chi} = (\chi_1, \chi_2, \ldots, \chi_n, \chi_0 \equiv 1)$.

Table 7.3 Two properties of organic solvents plus the molar mass M (g·mol^{-1}): boiling points, T_b (K, in parenthesis AH^b values), and melting points, T_m (K, in parenthesis AH^m values)

Solvents	M	T_b	T_m
(°)Acetone (56)[a]	58.1	329	179
(°)Acetonitrile (61)	41.05	355 *(1)*	225 *(1)*
Benzene (37)	78.1	353	278 *(1)*
Benzonitrile (18)	103.1	461	260
1-Butanol (44)	74.1	391 *(1)*	183 *(1)*
(°)2-Butanone (49)	72.1	353	186
Butyl Acetate (13)	116.2	398	195 *(−1)*
CS_2 (38)	76.1	319	161
CCl_4 (5)	153.8	350	250 *(1)*
Cl-Benzene (15)	112.6	405	228
1Cl-Butane (24)	92.6	351	150
$CHCl_3$ (11)	119.4	334	210
Cyclohexane (34)	84.2	354	280 *(1)*
(°)Cyclopentane (50)	70.1	323	179 *(1)*
1,2-diCl-Benzene (6)	147.0	453	257
1,2-diCl-Ethane (21)	98.95	356	238
diCl-Methane (33)	84.9	313	176
N,N-diM-Acetamide (29)	87.1	438 *(1)*	253 *(1)*
N,N-diM-Formamide (45)	73.1	426 *(1)*	212 *(1)*
1,4-Dioxane (27)	88.1	374	285
Ether (41)	74.1	308	157
Ethyl acetate (26)	88.1	350	189 *(−1)*
(°)Ethyl alcohol (59)	46.1	351 *(1)*	143
Heptane (19)	100.2	371	182
Hexane (31)	86.2	342	178
2-Methoxyethanol (40)	76.1	398 *(1)*	188
(°)Methyl alcohol (62)	32.0	338 *(1)*	175 *(1)*
(°)2-Methylbutane (47)	72.15	303	–
4-Me-2-Pentanone (20)	100.2	391	193
2-Me-1-Propanol (43)	74.1	381 *(1)*	165 *(1)*
2-Me-2-Propanol (42)	74.1	356 *(1)*	298 *(2)*
DMSO (36)	78.1	462 *(2)*	292 *(2)*
(°)Nitromethane (52)	61.0	374	244
1-Octanol (9)	130.2	469 *(1)*	258 *(1)*
(°)Pentane (46)	72.15	309	143
3-Pentanone (32)	86.1	375	233
(°)1-Propanol (54)	60.1	370 *(1)*	146 *(1)*
(°)2-Propanol (53)	60.1	356 *(1)*	184 *(1)*
Pyridine (35)	79.1	388	231
tetraCl-Ethylene (4)	165.8	394	251
(°)tetra-Hydrofuran (48)	72.1	340	165
Toluene (25)	92.1	384	180
1,1,2triCl,triFEthane (1)	187.4	321	238
2,2,4-triMe-Pentane (14)	114.2	372	166 *(−1)*
o-Xylene (17)	106.2	417	249

(continued)

Table 7.3 (continued)

Solvents	M	T_b	T_m
p-Xylene (16)	106.2	411	286
(°)Acetic acid (55)	60.05	391 *(1)*	290 *(1)*
Decaline (7)	138.2	465	230
diBr-Methane (2)	173.8	370	221 *(1)*
1,2-diCl-Ethylen(Z) (63)	96.9	334	193
(°)1,2-diCl-Ethylen(E) (22)	96.9	321	223
1,1-diCl-Ethylen (23)	96.9	305	151 *(−1)*
Dimethoxymethane (39)	76.1	315	168 *(−1)*
(°)Dimethylether (58)	46.1	249 *(−0.5)*	134 *(−0.5)*
Ethylen Carbonate (28)	88.1	511 *(2)*	310 *(2)*
(°)Formamide (60)	45.0	484 *(2)*	276 *(2)*
(°)Methylchloride (57)	50.5	249 *(−0.5)*	175 *(−0.5)*
Morpholine (30)	87.1	402 *(1)*	270 *(1)*
Quinoline (10)	129.2	510	258
(°)SO_2 (51)	64.1	263 *(−1)*	200
2,2-tetraCl-Ethane (3)	167.8	419	229
tetraMe-Urea (12)	116.2	450	272
triCl-Ethylen (8)	131.4	360	200

(°) Left out Compounds to build the training set
[a] Enumeration has to do with a randomization test performed on property values (see Sect.7.4.14)

The property can thus be computed ($P = \mathbf{C}\cdot\boldsymbol{\chi}$) as the scalar product of the row vector $\boldsymbol{C} = (c_1, c_2, \ldots, c_n, c_0)$ with the column $\boldsymbol{\chi}$ vector.

Correlations among the indices of a multilinear descriptor will be given to check, in the case of strong correlations, about the possibility of a loss of meaning of the structure–property relationship. Two indices are considered strongly correlated if $r > 0.98$ (Mihalić et al. 1992). In the case of strong correlation it might be advantageous to work with orthogonalized regressions, which can be obtained with the Randić's stepwise orthogonalization method (Randić 1991a, b). It has been suggested that even strongly correlated indices are not that detrimental (Peterangelo and Seybold 2004).

7.3 Results

Tables 7.3–7.6 show the experimental values of the 12 properties (for the origin of these values see Pogliani 2010). Tables 7.7 and 7.8 shows the best results obtained with the *greedy* algorithm, and they are either pure MCI or semi-empirical descriptions (MCI plus experimental indices). Superscripts on the left and right side of a combination means the type of configuration, for instance, superscript *fl* means that the (δ^v-based) valence MC indices have been obtained with the f_δ^l hydrogen perturbation, while superscript *pp-odd* on the right means that these

Table 7.4 Four properties of organic: refractive index, *RI* (20 °C) density, *d* (at 20°C±5°C relative to water at 4°C, g/cc), dielectric constant, ε, and FlashPoint, *FP* (K)

Solvents	*RI*	*d*	ε	*FP*
Acetone	1.359	0.791	20.7	256
Acetonitrile	1.344	0.786	37.5	278
Benzene	1.501	0.84	2.3	262
Benzonitrile	1.528	1.010	25.2	344
1-Butanol	1.399	0.810	17.1	308
2-Butanone	1.379	0.805	18.5	270
Butyl Acetate	1.394	0.882	5.0	295
CS_2	1.627	1.266	2.6	240
CCl_4	1.460	1.594	2.2	–
Cl-Benzene	1.524	1.107	5.6	296
1Cl-Butane	1.4024	0.886	7.4	267
$CHCl_3$	1.446	1.492	4.8	–
Cyclohexane	1.426	0.779	2.0	255
Cyclopentane	1.400	0.751	2.0	236
1,2-diCl-Benzene	1.551	1.306	9.9	338
1,2-diCl-Ethane	1.444	1.256	10.4	288
diCl-Methane	1.424	1.325	9.1	–
N,N-diM-Acetamide	1.438	0.937	37.8	343
N,N-diM-Formamide	1.431	0.944	36.7	330
1,4-Dioxane	1.422	1.034	2.2	285
Ether	1.353	0.708	4.3	233
Ethyl acetate	1.372	0.902	6.0	270
Ethyl alcohol	1.360	0.785	24.3	281
Heptane	1.387	0.684	1.9	272
Hexane	1.375	0.659	1.9	250
2-Methoxyethanol	1.402	0.965	16.0	319
Methyl alcohol	1.329	0.791	32.7	284
2-Methylbutane	1.354	0.620	1.8	217
4-Me-2-Pentanone	1.396	0.800	13.1	286
2-Me-1-Propanol	1.396	0.803	17.7	310
2-Me-2-Propanol	1.387	0.786	10.9	277
DMSO	1.479	1.101	46.7	368
Nitromethane	1.382	1.127	35.9	308
1-Octanol	1.429	0.827	10.3	354
Pentane	1.358	0.626	1.8	224
3-Pentanone	1.392	0.853	17.0	279
1-Propanol	1.384	0.804	20.1	288
2-Propanol	1.377	0.785	18.3	295
Pyridine	1.510	0.978	12.3	293
tetraCl-Ethylene	1.506	1.623	2.3	–
tetra-Hydrofuran	1.407	0.886	7.6	256
Toluene (25)	1.496	0.867	2.4	277
1,1,2triCl,triFEthane	1.358	1.575	2.4	–
2,2,4-triMe-Pentane	1.391	0.692	1.9	266
o-Xylene	1.505	0.870	2.6	305

(continued)

Table 7.4 (continued)

Solvents	*RI*	*d*	ε	*FP*
p-Xylene	1.495	0.866	2.3	300
Acetic acid	1.372	1.049	6.15	–
Decaline	1.476	0.879	2.2	–
diBr-Methane	2.497	1.542	7.8	–
1,2-diCl-Ethylen(Z)	1.449	1.284	9.2	–
1,2-diCl-Ethylen(E)	1.446	1.255	2.1	–
1,1-diCl-Ethylen	1.425	1.213	4.7	–
Dimethoxymethane	1.356	0.866	2.7	–
Dimethylether	–	–	5.0	–
Ethylen Carbonate	1.425	1.321	89.6	–
Formamide	1.448	1.133	109	–
Methylchloride	1.339	0.916	12.6	–
Morpholine	1.457	1.005	7.3	–
Quinoline	1.629	1.098	9.0	–
SO_2	–	1.434	17.6	–
2,2-tetraCl-Ethane	1.487	1.578	8.2	–
tetraMe-Urea	1.449	0.969	23.1	–
triCl-Ethylen	1.480	1.476	3.4	–

Table 7.5 Three properties of organic solvents: viscosity, η (Cpoise, 20 C; at 25 C, at 15 C), surface tension, γ (mN/m at 25 C). Cutoff UV values, *UV* (nm)

Solvents	η	γ	*UV*
Acetone	0.32	23.46	330
Acetonitrile	0.37	28.66	190
Benzene	0.65	28.22	280
Benzonitrile	1.24[1]	38.79	–
1-Butanol	2.95	24.93	215
2-Butanone	0.40	23.97	330
Butyl Acetate	0.73	24.88	254
CS_2	0.37	31.58	380
CCl_4	0.97	26.43	263
Cl-Benzene	0.80	32.99	287
1Cl-Butane	0.35	23.18	225
$CHCl_3$	0.57	26.67	245
Cyclohexane	1.00	24.65	200
Cyclopentane	0.47	21.88	200
1,2-diCl-Benzene	1.32	–	295
1,2-diCl-Ethane	0.79	31.86	225
diCl-Methane	0.44	27.20	235
N,N-diM-Acetamide	–	–	268
N,N-diM-Formamide	0.92	–	268
1,4-Dioxane	1.54	32.75	215
Ether	0.24	16.95	215
Ethyl acetate	0.45	23.39	260
Ethyl alcohol	1.20	21.97	210

(continued)

Table 7.5 (continued)

Solvents	η	γ	*UV*
Heptane	–	19.65	200
Hexane	0.33	17.89	200
2-Methoxyethanol	1.72	30.84	220
Methyl alcohol	0.60	22.07	205
4-Me-2-Pentanone	–	–	334
2-Me-2-Propanol	–	19.96	–
DMSO	2.24	42.92	268
Nitromethane	0.67	36.53	380
1-Octanol	10.6^2	27.10	–
Pentane	0.23	15.49	200
3-Pentanone	–	24.74	–
1-Propanol	2.26	23.32	210
2-Propanol	2.30	20.93	210
Pyridine	0.94	36.56	305
Tetra-Cl-Ethylene	0.90	–	
tetra-Hydrofuran	0.55	–	215
Toluene	0.59	27.93	285
1,1,2triCl,triFEthane	0.69	–	230
2,2,4-triMe-Pentane	0.50	–	215
o-Xylene	0.81	29.76	–
p-Xylene	0.65	28.01	–
Acetic acid	–	27.10	–
diBr-Methane	–	39.05	–
Formamide	–	57.03	–
Quinoline	–	42.59	–
2,2-tetraCl-Ethane	–	35.58	–

Table 7.6 Three properties of organic solvents: dipole moments, μ (debye, 1D = 10^{-18} esu cm = 3.3356 10^{-3} cm), magnetic susceptibility, $-\chi \cdot 10^6$, (*MS*, emu mol^{-1}, 1 emu = 1 cm^3, temperatures cover a range from 15°C to 32°C), and Elutropic value, *EV* = (silica)

Solvents	μ	*MS*	*EV*
Acetone	2.88	0.46	0.43
Acetonitrile	3.92	0.534	0.50
Benzene	0	0.699	0.27
2-Butanone	–	–	0.39
CS_2	0	0.532	–
CCl_4	0	0.691	0.14
$CHCl_3$	1.01	0.740	0.31
Cyclohexane	0	0.627	0.03
Cyclopentane	–	0.629	–
1,2-diCl-Benzene	2.50	0.748	–
1,2-diCl-Ethane	1.75	–	–
diCl-Methane	1.60	0.733	0.32
N,N-diM-Acetamide	3.8	–	–
N,N-diM-Formamide	3.86	–	–

(continued)

Table 7.6 (continued)

Solvents	μ	*MS*	*EV*
1,4-Dioxane	0.45	0.606	–
Ether	1.15	–	0.29
Ethyl acetate	1.8	0.554	0.45
Ethyl alcohol	1.69	0.575	–
Heptane	–	–	0.00
Hexane	–	–	0.00
Methyl alcohol	1.70	0.530	0.73
2-Me-1-Propanol	–	0.534	–
2-Me-2-Propanol	1.66	–	–
DMSO	3.96	–	–
Nitromethane	3.46	0.391	–
Pentane	–	–	0.00
2-Propanol	–	–	0.63
Pyridine	2.2	0.611	0.55
Tetra-Cl-Ethylene	–	0.802	–
tetra-Hydrofuran	1.75	–	0.35
Toluene	0.36	0.618	0.22
1,1,2triCl,triFEthane	–	–	0.02
2,2,4-triMe-Pentane	–	–	0.01
Acetic acid	1.2	0.551	–
Decaline	–	0.681	–
diBr-Methane	1.43	0.935	–
1,2-diCl-Ethylen(Z)	1.90	0.679	–
1,2-diCl-Ethylen(E)	0	0.638	–
1,1-diCl-Ethylen	1.34	0.635	–
Dimethoxymethane	–	0.611	–
Ethylen-Carbonate	4.91	–	–
Formamide	3.73	0.551	–
Methylchloride	1.87	–	–
Morpholine	–	0.631	–
Quinoline	2.2	0.729	–
SO_2	1.6	–	–
2,2-tetraCl-Ethane	1.3	0.856	–
tetraMe-Urea	3.47	0.634	–
triCl-Ethylen	–	0.734	–

indices have been obtained with the $K_p(pp\text{-}odd)$ representation for the core electrons, where p = odd and $q = p$.

The full combinatorial algorithm confirms the validity of the *ad hoc* parameters, AH^b, AH^m, and $\phi(0, 1)$ introduced with the use of the greedy algorithm to model T_b and T_m and μ, respectively. The rationale for these three *ad hoc* parameters will be explained in the next section. The descriptions obtained with the full combinatorial method, either pure MCI or semiempirical, are collected in Tables 7.9 and 7.10.

The 'zero-level' descriptors obtained with the full combinatorial method are collected in Tables 7.11 and 7.12, while the best full combinatorial semi-random descriptors are collected in Tables 7.13 and 7.14.

Table 7.7 The best descriptors for the 12 properties (P) of the organic solvents obtained with the greedy algorithm

P	Descriptor
T_b	${}^{f1}\{D^v, {}^0\psi_{Id}, \varepsilon, M, AH^b\}^{pp-odd}$
T_m	${}^{f8}\{D^v, {}^0\chi_d, {}^0\psi_{Ed}, {}^1\chi, {}^0\psi_{Id}, AH^m\}^{pp-odd}$
ε	${}^{f8}\{\Sigma^3/M^{1.7}, T_b, {}^1\chi_s\}^{p-odd}$
d	${}^{f8}\{{}^0\chi^v/M, {}^T\psi_I/M, \Delta, T_b\}^{pp-odd}$
RI	${}^{f5}\{\chi_t^v, D^v, {}^1\psi_{Is}, {}^S\psi_I, {}^1\chi_s\}^{p-odd}$
FP	${}^{f-0.5}\{T_b, {}^0\psi_E, \Delta, {}^S\psi_I, {}^1\chi^v\}^{p-odd}$
η	${}^{f-0.5}\{{}^0\psi_{Ed}, {}^1\psi_{Ed}, \Sigma, {}^0\psi_I\}^{p-odd}$
γ	$\{T_b, d, \varepsilon, RI, M\}$
UV	${}^{f50}\{1/RI, ({}^S\psi_I / M)^2, (1/\eta)^{0.5}, ({}^T\psi_I / M)^2\}^{pp-odd}$
μ	${}^{f50}\{\phi(\Sigma \cdot T_b)/M^{0.5}, \phi(\Sigma /M)^{2.3}, \phi(\boldsymbol{\varepsilon} / (\Sigma + 5))^{0.6}, \phi(\Sigma \cdot {}^0\psi_{Id})^{0.4}\}^{pp-odd}$
$-\chi \cdot 10^6$	${}^{f50}\{(M + \varepsilon^{0.5} + T_b^{0.7}), ({}^S\psi_I + 5RI), ({}^0\chi_d^v / M\varepsilon), ({}^1\psi_{Ed} / \varepsilon)\}^{pp-odd}$
EV	${}^{f0.5}\{{}^1\chi^v, T_b, \Sigma, RI\}^{p-odd}$

Table 7.8 The statistics of descriptors of Table 7.7

P	N	r^2	s	q^2
T_b	63	0.913	17.8	0.826
T_m	62	0.742	25	0.654
ε	62	0.903	5.3	0.867
d	62	0.973	0.04	0.965
RI	61	0.922	0.04	0.861
FP	41	0.985	4.7	0.979
η	39	0.930	0.5	0.839
γ	40	0.916	2.5	0.847
UV	33	0.892	17.8	0.854
μ	34	0.903	0.4	0.866
$-\chi \cdot 10^6$	32	0.886	0.04	0.854
EV	20	0.920	0.07	0.820

Table 7.9 The best descriptors, either pure MCI or semiempirical descriptors, for the 12 properties (P) of the organic solvents obtained with the full combinatorial algorithm

P	Descriptor
T_b	${}^{f50}\{\varepsilon, AH^b, M, D, {}^1\psi_E, \Sigma\}^{pp-odd}$
T_m	${}^{f-0.5}\{T_b, AH^m, \chi_t, {}^0\chi_d, {}^1\chi_d, {}^1\psi_E, \Delta, \Sigma\}^{p-odd}$
ε	${}^{f8}\{T_b, {}^0\psi_E, T_{\Sigma/M}\}^{p-odd}$
d	${}^{f50}\{D/M, {}^S\psi_I/M, {}^0\psi_E/M, \Delta\}^{pp-odd}$
RI	${}^{f5}\{M, D, {}^1\chi_s, \chi_t^v, {}^0\psi_I, {}^1\psi_{Is}, \Delta\}^{p-odd}$
FP	${}^{f-0.5}\{T_b, d, RI, {}^S\psi_I, {}^S\psi_E\}^{p-odd}$
η	${}^{f-0.5}\{T_b, \varepsilon, {}^1\psi_{Id}, {}^0\psi_{Ed}, \Sigma\}^{p-odd}$
γ	${}^{f1}\{\varepsilon, d, RI, \chi_t^v, {}^T\psi_I\}^{p-odd}$
UV	${}^{f1}\{RI, AH^b, {}^0\psi_I, {}^S\psi_E, {}^0\psi_E\}^{p-odd}$
μ	${}^{f0.5}\{\phi\varepsilon, \phi{}^1\chi, \phi D^v, \phi\Sigma, \phi T_{\Sigma/M}\}^{pp-odd}$
$-\chi \cdot 10^6$	${}^{f50}\{M, \chi_t, {}^0\chi^v, {}^1\chi_s^v\}^{p-odd}$
EV	${}^{f0.5}\{{}^1\chi^v, {}^0\psi_I, {}^0\psi_E, \Sigma\}^{pp-odd}$

Table 7.10 The statistics of the best descriptors of Table 7.9

P	N	r^2	s	q^2
T_b	63	0.975	13.4	0.933
T_m	62	0.799	22.0	0.698
ε	62	0.916	4.9	0.902
d	62	0.975	0.4	0.969
RI	61	0.944	0.04	0.915
FP	41	0.984	4.7	0.979
η	39	0.963	0.3	0.863
γ	41	0.959	1.8	0.915
UV	33	0.876	19.0	0.802
μ	34	0.926	0.4	0.818
$-\chi \cdot 10^6$	32	0.848	0.05	0.789
EV	20	0.949	0.06	0.906

Table 7.11 The 'zero-level' (*Zl*) descriptors for the 12 properties (*P*) of the organic solvents obtained with the full combinatorial algorithm and with the two sets of random indices, *r1*-*r38* and *rd1*-*rd38*

P	Zl-Descriptor
T_b	{*rd2*, *rd6*, *rd9*, *rd24*, *rd34*, *rd37*}
T_m	{*r1*, *r23*, *r27*, *r29*, *r30*, *r31*, *r35*, *r38*}
ε	{*rd9*, *rd14*, *rd16*, *rd20*, *rd27*, *rd32*, *rd34*, *rd37*}
d	{*rd21*, *rd36*, *r1*, *r34*}
RI	{*r1*, *r5*, *r13*, *r20*, *r30*, *r35*, *r36*}
FP	{*rd6*, *rd11*, *rd16*, *rd19*, *rd20*}
η	{*rd8*, *rd28*, *rd34*, *r9*, *r20*}
γ	{*rd27*, *r10*, *r27*, *r30*, *r35*}
UV	{*rd11*, *r13*, *r21*, *r27*, *r38*}
μ	{ϕ*r17*, ϕ*r21*, ϕ*r23*, ϕ*r26*, ϕ*r34*}
$-\chi \cdot 10^6$	{*rd7*, *rd16*, *rd21*, *r25*}
EV	{*rd14*, *rd22*, *rd25*, *r30*}

Table 7.12 The statistics of the random descriptors of Table 7.11

P	N	r^2	s	q^2
T_b	63	0.410	45	0.255
T_m	62	0.427	37	0.252
ε	62	0.479	13	0.265
d	62	0.336	0.2	0.213
RI	61	0.329	0.13	0.14
FP	41	0.702	20	0.605
η	39	0.567	1.2	0.01
γ	41	0.677	4.9	0.545
UV	33	0.649	33	0.480
μ	34	0.774	0.7	0.639
$-\chi \cdot 10^6$	32	0.569	0.1	0.379
EV	20	0.865	0.1	0.751

Table 7.13 The semi-random descriptors (*Sr*) for the 12 properties (*P*) of the organic solvents obtained with the full combinatorial algorithm

P	Sr-Descriptor
T_b	$\{\varepsilon, AH^b, M, D, rd14, rd25\}$
T_m	No good descriptor found
ε	$^{f8}\{T_b, {}^0\psi_E, T_{\Sigma/M}, r21, r32\}^{p-odd}$
d	$^{f50}\{D/M, {}^S\psi_I/M, {}^0\psi_E/M, rd31\}^{pp-odd}$
RI	$^{f5}\{D, {}^1\chi_s, \chi_t^v, {}^0\psi_I, {}^1\psi_{Is}, rd6, rd25\}^{p-odd}$
FP	$^{f-0.5}\{T_b, D, {}^S\psi_I, \Delta, rd38\}^{p-odd}$
η	$^{f-0.5}\{T_b, \varepsilon, {}^0\psi_{Ed}, \Sigma, r32\}^{p-odd}$
γ	$^{f1}\{\varepsilon, d, RI, \chi_t^v, rd33\}^{p-odd}$
UV	$\{RI, rd11, rd34, r21, r38\}$
μ	$^{f0.5}\{\phi\varepsilon, \phi\Sigma, \phi T_{\Sigma/M}, \phi rd10, \phi rd34\}^{pp-odd}$
$-\chi \cdot 10^6$	$^{f50}\{M, \chi_t, {}^T\psi_I, r30\}^{p-odd}$
EV	$^{f0.5}\{{}^1\chi^v, rd1, rd10, rd25\}^{p-odd}$

Table 7.14 The statistics of the semi-random descriptors of Table 7.13

P	N	r^2	s	q^2
T_b	63	0.901	19	0.858
ε	62	0.939	4.3	0.923
d	62	0.960	0.06	0.951
RI	61	0.941	0.04	0.924
FP	41	0.987	4.2	0.983
η	39	0.956	0.4	0.791
γ	41	0.948	2.0	0.822
UV	33	0.767	27	0.645
μ	34	0.953	0.3	0.908
$-\chi \cdot 10^6$	32	0.887	0.04	0.842
EV	20	0.895	0.1	0.812

The observed vs. calculated plots have been obtained with the given correlation vector ***C*** of the optimal descriptor of Table 7.9 (unless otherwise stated). The derivation of the semi-random descriptors, due to the limits of the our PC, could not be searched with the two sets of random indices plus all MC indices plus the five experimental parameter and *M*. They have been calculated in a more compact way, which is explained in each property's paragraph. The super-descriptors have been obtained with a full search over the set of indices of the optimal descriptors for the 12 properties (see the corresponding paragraph).

7.4 Discussion

In the following discussion the model of each property will be discussed in detail together with their corresponding observed vs. calculated and residual plots. The rationale for the *ad hoc* indices used for the boiling and melting temperature data as well as for the dipole moment data will be made clear. In this section also it will also be discussed how the space for the super-descriptors has been built and how to obtain a particular model of random properties.

7.4.1 Boiling Points, T_b

The best $K_p(pp\text{-}odd)\,/\,f_\delta{}^1$ greedy descriptor can be seen in Tables 7.7 and 7.8 and the best full combinatorial $K_p(pp\text{-}odd)\,/\,f_\delta{}^{50}$ descriptor is shown in Tables 7.9 and 7.10 (here only ${}^1\psi_E$ and Σ are δ^v- or configuration-dependent). The full combinatorial method, achieves to find also a good five-index semi-empirical descriptor,

$$f^{50}\{\varepsilon, AH^b, M,\ D,\ {}^1\psi_E\}^{pp-odd} : N = 63,\ r^2 = 0.922,\ s = 14.8,\ q^2 = 0.918$$

The dielectric constant is the only experimental parameter that can here be used, as it is the only property that has $N = 63$. Comparison between Tables 7.8 and 7.10 shows the noticeable improvement achieved by the full combinatorial method over the greedy method, especially at the q^2 level. The correlation vector of descriptor (Table 7.9) used to model this property and to obtain Fig. 7.2 is:

$$C = [0.88\,(0.1),\ 44.1\,(4.2),\ 0.68\,(0.07), 26.1(1.9), \\ -295(31), 2.27(0.6), 195.2(7)]$$

The best ‘zero-level’ random description is shown in Tables 7.11 and 7.12, the reader should notice its very bad s and q^2.

The strongest correlated indices instead are: $r(D, {}^1\psi_E) = 0.97$ and $r(\varepsilon, AH^b) = 0.73$, while T_b shows the strongest correlation with index D: $r(T_b, D) = 0.71$. The largest correlation between two random variables is around 0.22 and between a random variable and T_b is 0.33. Poor correlations will not be given anymore. Notice that from $n = -0.5$ till $n = 50$ the quality of the description normally goes up and down, even if the superior quality of $K_p(pp\text{-}odd)$ is always respected. Such a trend is detected also with the other properties.

The search for a five-index *mc-exp-rn*-description first with *r1*-*r38* and then with *rd1*-*rd38* plus ε and *MCI*- $K_p(pp\text{-}odd)\,/\,f_\delta{}^{50}$ indices finds exactly the same first five-index descriptor we already found. The search for a six-index descriptor extends over hundreds of millions of combinations and is well outside the range of our PC. If the search for a six-index descriptor is, instead, restricted to *r1*-*r38* plus *rd1*-*rd38* plus the indices of descriptor, $^{f50}\{\varepsilon, AH^b, M, D, {}^1\psi_E, \Sigma\}^{pp-od}$, we land again on the same six-index descriptor. The search for a six-index *sr*-descriptor is now done with *r1*-*r38* and *rd1*-*rd38* plus the four indices $\{\varepsilon, AH^b, M, D\}$ (their quality is: $r^2 = 0.870$, $s = 21$, $q^2 = 0.826$). The resulting descriptor can be found in Tables 7.13 and 7.14. The improvement brought about by the random indices is minimal but detectable.

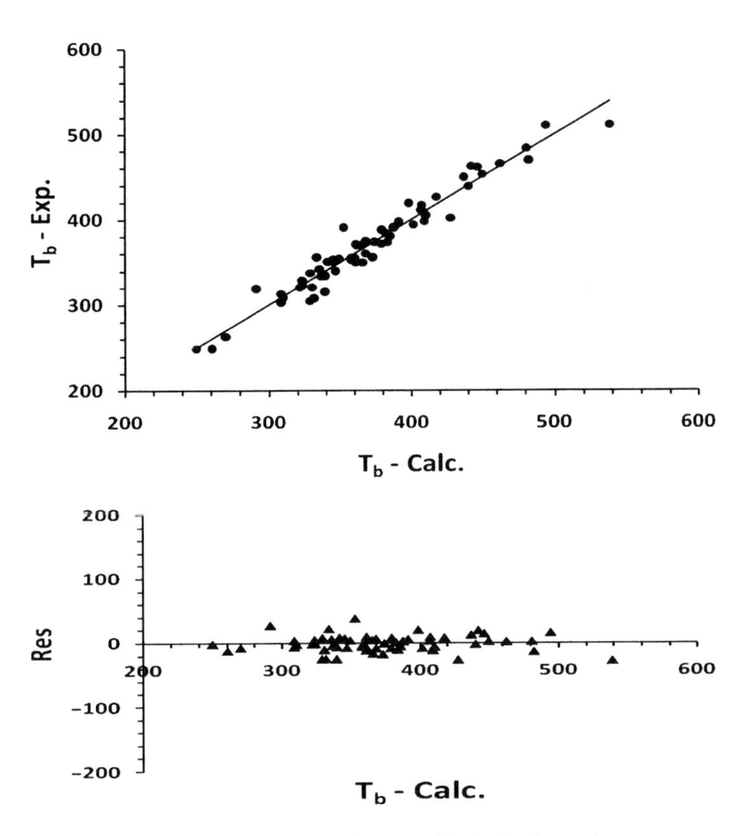

Fig. 7.2 The Exp./Calc. plot of T_b together with its residual plot (*bottom*)

If items (°) in Table 7.3 are left out the statistical quality of the set with $N = 45$ points with the given semiempirical descriptor (1), 'zero-level' descriptor (2) and semi-random descriptor (3) are:

$$(1)\ r^2 = 0.939,\ s = 14,\ q^2 = 0.802$$
$$(2)\ r^2 = 0.575,\ s = 37,\ q^2 = 0.412$$
$$(3)\ r^2 = 0.871,\ s = 20,\ q^2 = 0.732$$

The values of the '*ad hoc*' AH^b parameter (Table 7.3) have been derived (Pogliani 2010) keeping in mind that the lack of a detailed information about hydrogen bonds, steric, dipolar, and van der Waals interactions obliges to find out, to obtain a good model of the property, a parameter that takes into account such factors. Many solvents have –OH, –NH or $-NH_2$ groups which can undergo hydrogen bonding. Moreover, DMSO and SO_2 are rather small molecules (as well as CH_3–Cl and diMe-Ether) but while DMSO has a relatively high boiling point, SO_2, CH_3–Cl and diMe-Ether have a quite low boiling point. Now, while DMSO has a quite high ε and μ values (see Table 3A, B), the other three compounds show low ε and μ values. Formamide can undergo hydrogen bonding and has quite high ε, and μ values. Also the rather high μ and ε values of N,N-diMe-Acetamide, N,N-diMe-Formamide, Nitromethane, and Acetonitrile cannot be overlooked. The considerations here done suggest to use the following values for the 'ad hoc' (AH^b) descriptor: for compounds with a –OH and –NH groups, a $AH = 1$ is used, for EthylenCarbonate, instead, a $AH = 2$ is used. This compound not only can undergo hydrogen bond but has three oxygen atoms and a double bond and its ε and μ values are quite consistent. For the small molecule Formamide with a $-NH_2$ group (and a O group) and high ε and μ values a $AH = 2$ is used, while for Acetonitrile, Nitromethane, N,N-diMe-Acetamide and N,N-diMe-Formamide a $AH = 1$ is used. For DMSO a $AH = 2$ is used, for SO_2 a $AH = -1$ is used, and for CH_3–Cl and diMe-Ether a $AH = -0.5$ is used. For all other compounds a $AH = 0$ is used.

7.4.2 *Melting Points, T_m*

To improve the model of this property a fine-tuning of the AH^b into a AH^m parameter has here been done on the following grounds (Pogliani 2010): (*i*) cyclic compounds, with no atoms other than carbon/hydrogen atoms, have $AH^m = 1$ (Benzene, Cyclohexane, Cyclopentane), (*ii*) for spherical or nearly spherical molecules $AH^m = 1$ (diBr-methane, CCl_4), for 2-Me-2-Propanol $AH^m = 2$ (1 for hydrogen bond and 1 for sphericity), (*iii*) linear asymmetric molecules around a central C or CH_2 group $AH^m = -1$ (Butyl Acetate, 1,1-diCl-Ethylen, Ethyl Acetate, 2,2,4-triMe-Pentane, Dimethoxymethane), (*iv*) SO_2 as well as 2-Methoxyethanol, nitromethane, and Ethyl alcohol $AH^m = 0$. For the last compounds the hydrogen bond contributes 1 and the asymmetric conjecture contributes -1. The set of AH^m values is shown in Table 7.3, T_m column.

The greedy description is shown in Tables 7.7 and 7.8. There are two similar not impressive six-index full combinatorial descriptors with quite different perturbations for the hydrogen atoms ($f_\delta^{\cdot -0.5}$ and $f_\delta^{\cdot 50}$) of which the following has a slightly better q^2 value (here only $\chi^v{}_t$ is configuration-dependent), and which can be compared with the greedy descriptor of Tables 7.7 and 7.8.

$$f^{50}\{AH^m, D, \chi_t, {}^0\chi_d, {}^1\chi_d, \chi^v{}_t\}^{pp-odd} : N = 62, r^2 = 0.757, \; s = 24, q^2 = 0.689$$

The full combinatorial algorithm finds, nonetheless, an improved semiempirical descriptor with eight indices, which is shown in Tables 7.9 and 7.10, and which belongs to the $K_p(p\text{-}odd)/f_\delta^{-0.5}$ configuration (for ${}^1\psi_E$ and Σ only). Its correlation vector is:

$$C = [0.21(0.08), 34.1(5.2), 63.9(26), 32.5(6.5), 0.014(0.003), 109(25), 7.15(2.6), \\ -1.72(0.4), 43.2(35)]$$

Figure 7.3 shows the obs/calc and residual plots for T_m. The strongest correlated indices are: $r(\chi_t, {}^1\psi_E) = 0.83$, and the strongest correlation for T_m is: $r(T_m, T_b) = 0.64$.

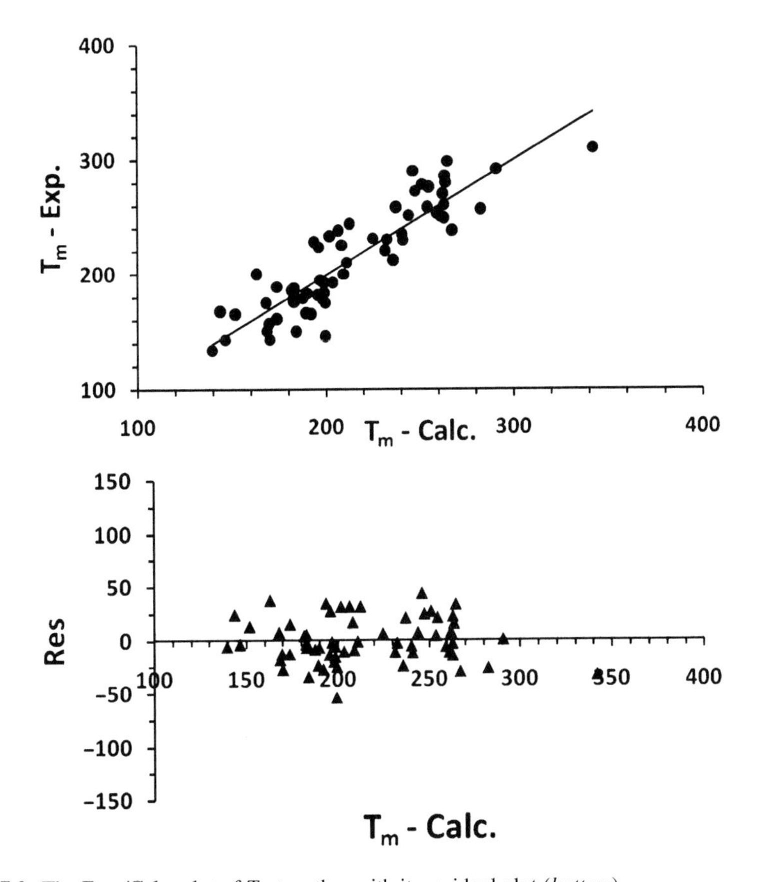

Fig. 7.3 The Exp./Calc. plot of T_m together with its residual plot (*bottom*)

The random descriptor of Tables 7.11 and 7.12 is quite bad. Due to the limits of our PC the search for a semi-random descriptor is restricted to a six-index descriptor with the *r1-r38* and then *rd1-rd38* plus $^{f50}\{AH^m, D, \chi_t, {}^0\chi_d, {}^1\chi_d, \chi^v{}_t\}^{pp\text{-odd}}$ indices. No good descriptor is here found. If items (°) in Table 7.3 are left out, the training quality of the semiempirical descriptor (1) and 'zero-level' descriptor (2) with $N = 45$ are:

$$(1)\ r^2 = 0.939,\ s = 14,\ q^2 = 0.802$$
$$(2)\ r^2 = 0.575,\ s = 37,\ q^2 = 0.412$$

7.4.3 Dielectric Constant, ε

A rather good eight-index semiempirical combination (remind that only T_b has $N = 63$) belonging to the $K_p(p\text{-}odd)\ /\ f_\delta{}^{-0.5}$ configuration (only for ${}^0\chi^v, {}^0\psi_I, {}^1\psi_I, {}^0\psi_E$, and ${}^1\psi_E$) is here found with the full combinatorial algorithm,

$$^{f-0.5}\{T_b, M, {}^0\chi, {}^0\chi^v, {}^0\psi_1, {}^1\psi_1, {}^0\psi_E, {}^1\psi_E\}^{p-odd} : N = 63, r^2 = 0.891,\ s = 6.7,\ q^2 = 0.812$$

The *greedy* algorithm without ethylencarbonate, which is a strong outlier (Tables 7.7 and 7.8), was unable to find such a descriptor. Without the strong outlier ethylencarbonate the full combinatorial search finds the following improved semiempirical combination (last five indices belong to the K_p(*p-odd*) / $f_\delta^{-0.5}$ configuration), which outperforms the 'bad' zero-level description shown in Tables 7.11 and 7.12.

$$^{f-0.5}\{T_b, {}^0\chi, {}^1\chi, D^v, \chi_t^v, {}^0\psi_{1d}, {}^0\psi_E, {}^1\psi_E\}^{p-odd} : N = 62, r^2 = 0.914,\ s = 5.2,\ q^2 = 0.835$$

Both search algorithms find a highly compact semiempirical combination shown in Tables 7.7 and 7.8 (greedy) and 7.9 and 7.10 (full combinatorial). Here, descriptor $T_{\Sigma/M} = \Sigma^3/M^{1.7}$ [Σ and ${}^0\psi_E$ belong to K_p(*p-odd*) / $f_\delta{}^8$ configuration]. The correlation vector of descriptor in Table 7.9 is,

$$C = [0.11(0.02), -8.22(1.5), 25.4(1.8), -19.9(4.8)]$$

The model plots obtained with this correlation vector are given in Fig. 7.4.

The full combinatorial technique finds also an improved semiempirical combinations with five indices (see later on), which include T_b and $T_{\Sigma/M}$, but we prefer the more economical three-index combination. The improvement relatively to the *greedy* algorithm is mainly in q^2. The strongest correlations are: $r(T_b, {}^0\psi_E) = 0.60$, the strongest correlation of the dielectric constant is, $r(\varepsilon, T_{\Sigma/M}) = 0.93$. Notice that the $T_{\Sigma/M}$ term encodes the information about the molar mass but also about the overall E-State index of the electronegative atoms. It reflects the charge distribution due to these atoms, normalized to the molar mass, a no trivial choice.

The search for a three-index *mc-exp-rn*-description (with no Ethylencarbonate) with *r1-r38* plus *rd1-rd38* plus the *MCI*- K_p(*p-odd*) / $f_\delta{}^8$ indices ($T_{\Sigma/M}$ inclusive) finds the already found three-index descriptor. The search for a five-index descriptor done with the previous three-index descriptor plus the random indices finds the optimal semi-random description of Tables 7.13 and 7.14. Had we started with the indices of the three-index descriptor and added the random indices, we would have done a good guess. These results are illuminating about the potential uses of random numbers.

If items (°) in Table 7.3 are left out the training quality of the semiempirical descriptor (1), 'zero-level' descriptor (2) and semi-random descriptor (3, a quite good description) with $N = 44$ are:

$$(1)\ r^2 = 0.833,\ s = 20,\ q^2 = 0.684$$
$$(2)\ r^2 = 0.569,\ s = 7.5,\ q^2 = 0.390$$
$$(3)\ r^2 = 0.897,\ s = 3.5,\ q^2 = 0.854$$

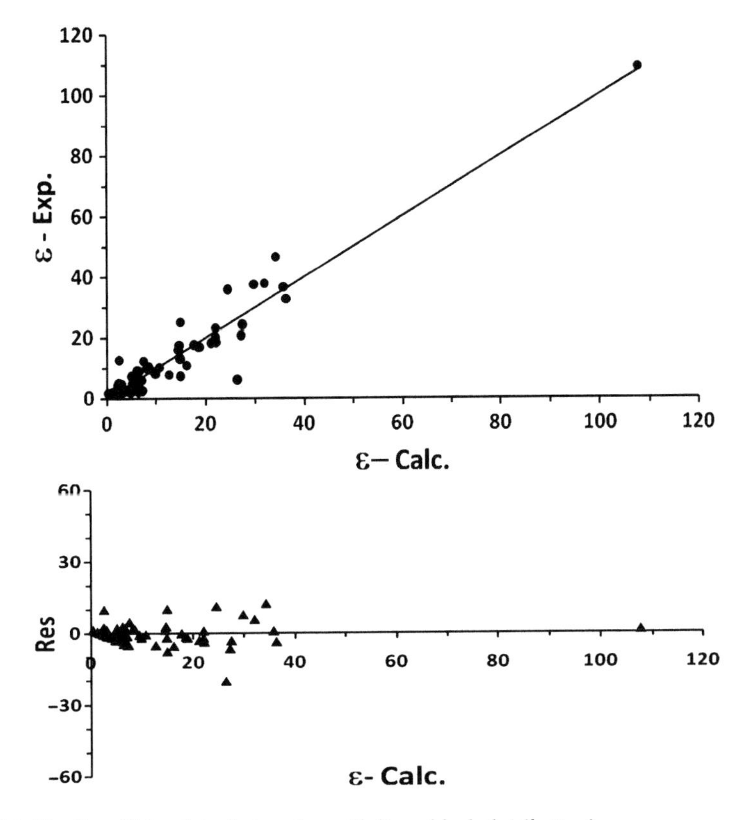

Fig. 7.4 The Exp./Calc. plot of ε together with its residual plot (*bottom*)

7.4.4 Density, d

The full combinatorial search (Tables 7.9 and 7.10) confirms the superior model quality of *M* normalized indices, plus *Δ* index, already found with the greedy search (Tables 7.7 and 7.8), even if it does not need any empirical parameters to improve the model quality. The zero-level description shown in Tables 7.11 and 7.12 is not at all satisfactory. The correlation vector for the $K_p(pp\text{-}odd)/f_\delta^{50}$-descriptor shown in Table 7.9, with whom Fig. 7.5 has been obtained, is,

$$C = [10.9(0.5), 1.72(0.2), -109(4.5), 0.05(0.01), 1.79(0.04)]$$

The strongest correlations are: $r(D/M, {}^0\psi_E/M) = 0.92$, $r(D/M, \Delta) = 0.58$, $r(d, {}^0\psi_E/M) = 0.81$, $r(d, \Delta) = 0.78$.

The search for a *mc-exp-rn*-description with *r1-r38* and then *rd1-rd38* indices plus the *M*-normalized MC indices finds exactly the same optimal four-index descriptor. A space made of all random indices plus ${}^{f50}\{D/M, {}^S\psi_I/M, {}^0\psi_E/M\}^{pp-odd}$

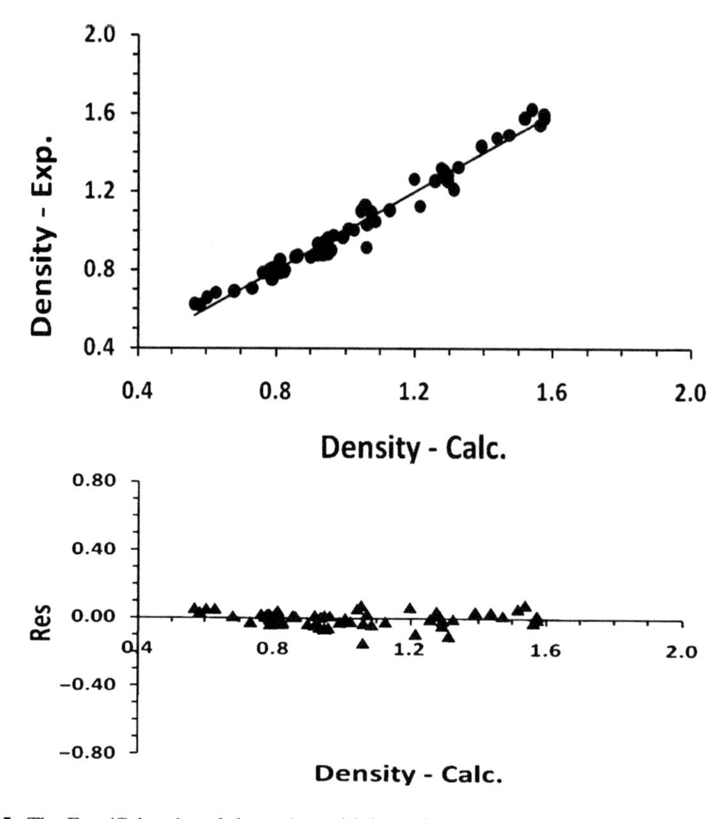

Fig. 7.5 The Exp./Calc. plot of *d* together with its residual plot (*bottom*)

indices ($r^2 = 0.936$, $s = 0.07$, $q^2 = 0.925$) finds the semi-random descriptor of Tables 7.13 and 7.14.

If items (°) in Table 7.3 are left out the training quality of the semiempirical descriptor (1), 'zero-level' descriptor (2) and semi-random descriptor (3) with $N = 45$ are:

$$(1)\ r^2 = 0.969,\ s = 0.05,\ q^2 = 0.960$$
$$(2)\ r^2 = 0.407,\ s = 0.23,\ q^2 = 0.248$$
$$(3)\ r^2 = 0.969,\ s = 0.05,\ q^2 = 0.960$$

7.4.5 Refractive Index, RI

A pure MCI five-index descriptor belonging to the $K_p(p\text{-}odd)/f_\delta^5$ configuration (for indices ${}^0\chi^v$, χ_t^v, and ${}^1\psi_{Is}$) with a very good q^2 value is found with the full combinatorial technique, which can be compared with the greedy descriptor of Tables 7.6–7.8,

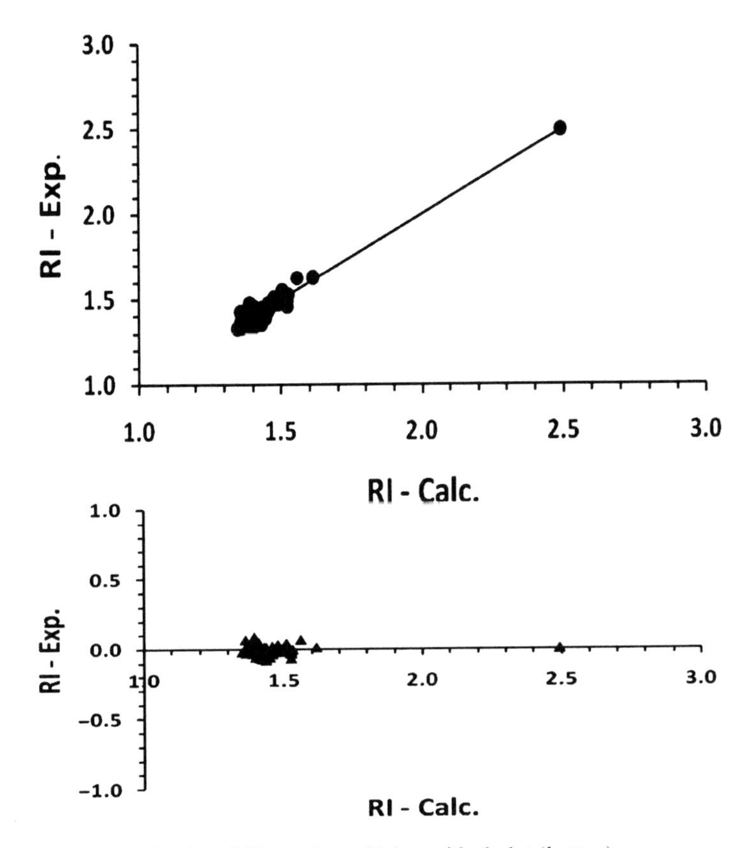

Fig. 7.6 The Exp./Calc. plot of *RI* together with its residual plot (*bottom*)

$$^{f5}\{D,{}^{1}\chi_s,{}^{0}\chi^{v},\chi_t^{v},{}^{1}\psi_{1s}\}^{p-odd}:\ N=61,\ r^2=0.927,\ s=0.04,\ q^2=0.910$$

The full combinatorial technique (Tables 7.9 and 7.10) finds an even better seven-index descriptor, belonging to the same configuration (for indices, χ_t^{v}, ${}^{0}\psi_I$, and ${}^{1}\psi_{Is}$). With the following correlation vector *RI* plots of Fig. 7.6 have been obtained,

$$C=[0.0019(0.0006),0.054(0.006),0.88(0.15),0.82(0.05),-0.17(0.02),\\ -2.73(0.3),-0.043(0.01),1.41(0.03)]$$

The best ‘zero-level’ description is shown in Tables 7.11 and 7.12, and it is a no ‘amazing’ description.

The strongest correlations are: $r(D,{}^{0}\psi_I)=0.97$, $r({}^{1}\chi_s,{}^{0}\psi_{Is})=0.93$, $r(\chi_t^{v},{}^{1}\psi_{Is})=0.79$, $r(RI,\chi_t^{v})=0.69$.

The search for a five-index *mc-exp-rn*-description throughout *r1*-*r38* and then *rd1*-*rd38* plus the MC- $K_p(p\text{-}odd)/f_\delta^5$ indices and *M* finds the same five-index descriptor already found. This suggest that also the seven-index descriptor should be the same, even if, due the huge number of combinations, the search is outside our possibility. If the search for a seven-index semirandom descriptor is done with the five indices of descriptor $^{f\,5}\{M, D, {}^1\chi_s, \chi_t^v, {}^0\psi_I, {}^I\psi_{Is}\}^{p-odd}$ ($N = 61$, $r^2 = 0.887$, $s = 0.05$, $q^2 = 0.792$) plus *r1-r38* and then *rd1-rd38* we obtain the best q^2 description shown in Tables 7.13 and 7.14.

If items (°) in Table 7.3 are left out the training quality of the graph-theoretical/ semiempirical descriptor (1), 'zero-level' descriptor (2) and semi-random descriptor (3, a quite good descriptor) with $N = 45$ are:

$$(1)\ r^2 = 0.957,\ s = 0.04,\ q^2 = 0.878$$
$$(2)\ r^2 = 0.403,\ s = 0.14,\ q^2 = 0.203$$
$$(3)\ r^2 = 0.954,\ s = 0.04,\ q^2 = 0.911$$

7.4.6 *Flash Point, FP*

It is here reminded that *FP* (in Kelvin) is the lowest temperature at which there is enough fuel vapour to ignite (T_b is here expected to play a role). It is here possible to use as indices parameters T_b, ε, d, and *RI*, but no T_m as it has no data for 2-methylbutane.

The greedy descriptor is shown in Tables 7.7 and 7.8. The full combinatorial search algorithm (Tables 7.9 and 7.10) finds many similar semiempirical descriptors throughout the many types of configurations, among which the descriptor of Tables 7.9 and 7.10 has been chosen. The correlation vector to obtain Fig. 7.7 is,

$$C = [0.85(0.03), 34.6(7.9), -95.6(18), 1.71(0.2), 1.06(0.1), 95.2(22)]$$

The strongest correlations are: $r({}^S\psi_E, {}^S\psi_I) = 0.84$, $r(d, RI) = 0.70$, $r(T_b, {}^S\psi_E) = 0.64$, and for *FP*, $r(FP, T_b) = 0.93$, $r(FP, {}^S\psi_I) = 0.53$.

The 'zero-level' descriptor, with *poor s* and q^2, is in Tables 7.11 and 7.12.

The search for a *mc-exp-rn*-description with *r1-r38* and then *rd1-rd38* plus T_b, d, *RI* and the set of MC indices finds the descriptor of Tables 7.13 and 7.14 (only ${}^S\psi_I$ is configuration-dependent). Notice that $^{f\,-\,0.5}\{T_b, D, {}^S\psi_I, \Delta\}^{p\text{-}odd}$ is a quite good descriptor with: $r^2 = 0.982$, $s = 5.0$, $q^2 = 0.976$. If we are not too severe the following semi-random descriptor searched among *r1- r38* plus *rd1-rd38* and $\{T_b, d, RI\}$, ($r^2 = 0.931$, $s = 9.5$, $q^2 = 0.914$), could be considered quite satisfactory,

$$\{T_b, d, RI, rd2, rd32\} : N = 41, r^2 = 0.956, s = 7.8, q^2 = 0.943$$

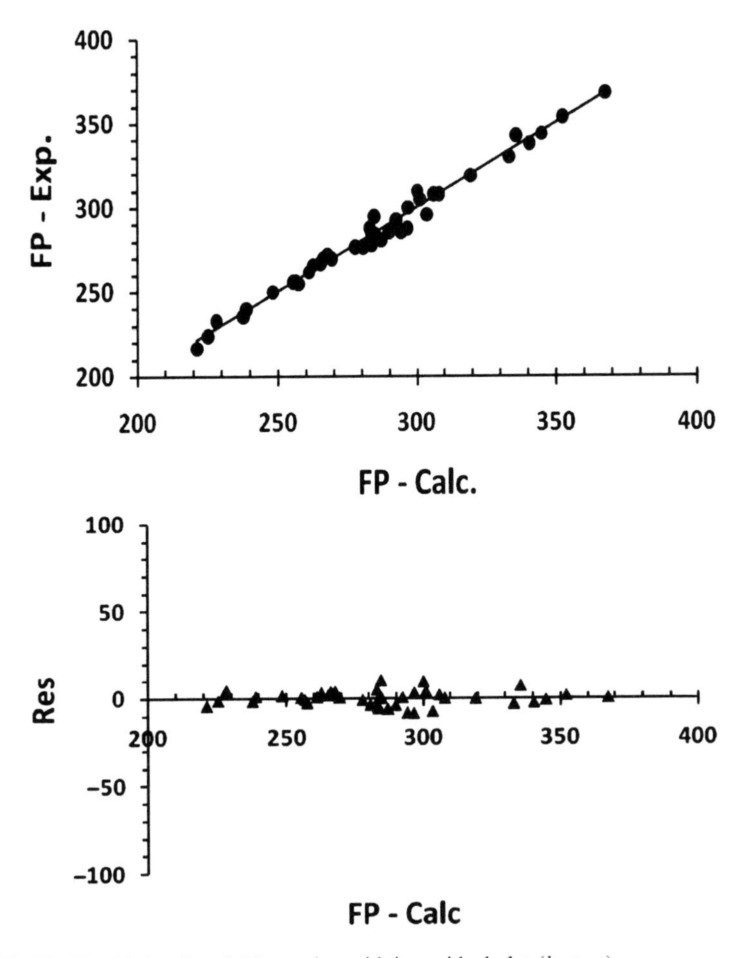

Fig. 7.7 The Exp./Calc. plot of *FP* together with its residual plot (*bottom*)

If items (°) in Table 7.3 are left out the training quality of the graph-theoretical/semiempirical descriptor (1), 'zero-level' descriptor (2) and semi-random descriptor (3, a quite good description) with $N = 29$ are:

$$(1)\ r^2 = 0.985,\ s = 4.3,\ q^2 = 0.981$$
$$(2)\ r^2 = 0.769,\ s = 18,\ q^2 = 0.633$$
$$(3)\ r^2 = 0.989,\ s = 4.0,\ q^2 = 0.982$$

7.4.7 Viscosity, η

This property has a rather good greedy descriptor (Tables 7.7 and 7.8). The full combinatorial technique (Tables 7.9 and 7.10) finds a very good semiempirical

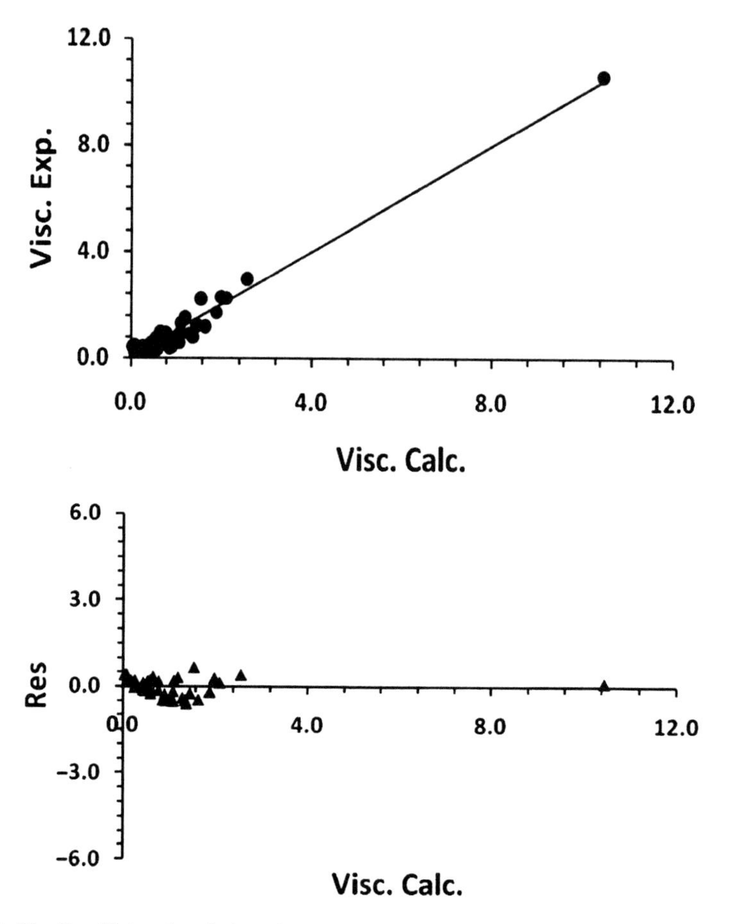

Fig. 7.8 The Exp./Calc. plot of viscosity, η, and its residual plot (*bottom*)

five-index descriptor belonging to the same configuration of the greedy descriptor. The best 'zero-level' has very poor s and q^2 statistics (Tables 7.11 and 7.12). The correlation vector of descriptor in Table 7.9 used to derive plots of Fig. 7.8, is,

$$C = [0.014(0.002), -0.029(0.006), 0.25 \cdot 10^{-4}(6 \cdot 10^{-6}), \\ -0.37 \cdot 10^{-5}(10^{-7}), 0.07(0.01), -4.23(0.6)]$$

The strongest correlations are: $r(\varepsilon,\Sigma) = 0.50$, $r(T_b, {}^1\psi_{Id}) = 0.45$, $r(\eta, {}^0\psi_{Ed}) = 0.82$, and $r(\eta, \Sigma) = 0.58$.

The search for a *mc-exp-rn*-description with *r1-r38* and then *rd1-rd38* plus T_b, ε, and the set of MC-$K_p(p\text{-}odd)/f_\delta^{-0.5}$ indices finds exactly the already found q^2 five-index descriptor. A search among *r1- r38* and *rd1-rd38* plus the five indices of the optimal descriptor finds again this same descriptor. The descriptor is now chosen among *r1- r38* plus *rd1-rd38* plus ${}^0\psi_{Ed}$, Σ, T_b and ε. The full search lands on the descriptor of Tables 7.13 and 7.14, which, relatively to the optimal descriptor of Table 7.9 has a worse q^2 statistics.

If items (°) in Table 7.3 are left out the training quality of the graph-theoretical/semiempirical descriptor (1), 'zero-level' descriptor (2) and semi-random descriptor (3) with $N = 28$, are:

$$(1)\ r^2 = 0.972,\ s = 0.35,\ q^2 = 0.786$$
$$(2)\ r^2 = 0.675,\ s = 1.2,\ q^2 = 0.02$$
$$(3)\ r^2 = 0.963,\ s = 0.4,\ q^2 = 0.718$$

7.4.8 Surface Tension, γ

To achieve a satisfactory description the greedy algorithm requires the exclusion of CH_2Br_2 compound and uses only experimental parameters (Tables 7.7 and 7.8). The full combinatorial technique (Tables 7.9 and 7.10) finds a very good $K_p(p\text{-}odd)/f_\delta{}^1$ semiempirical descriptor, with no need to exclude CH_2Br_2. The 'zero-level' description is shown in Tables 7.11 and 7.12 and shows poor s and q^2 statistics. The correlation vector for the descriptor of Table 7.9, with whom Fig. 7.9 has been obtained, is,

$$C= [0.27(0.01), 21.0(1.4), 48.8(3.5), -42.2(4.5), 39.4(9.1), -64.7(5.4)]$$

The strongest correlations are: $r(\chi_t{}^v, {}^T\psi_I) = 0.92$, $r(\chi_t{}^v, RI) = 0.80$, $r(\chi_t{}^v, d) = 0.63$, $r(\gamma, \varepsilon) = 0.60$, and $r(\gamma, d) = 0.56$. Concerning the experimental indices notice that solvents with high dielectric constants and high density have usually high surface tension.

The search for a *mc-exp-rn*-description with *r1-r38* and then *rd1-rd38* plus ε, *d*, *RI*, and the set of MC- $K_p(p\text{-}odd)/f_\delta{}^1$ indices finds the same optimal five-index descriptor already found. Instead, the *mc-exp-r*-descriptor searched among *r1- r38* plus *rd1-rd38* plus ε, *d*, *RI*, and $\chi_t{}^v$, finds the semi-random descriptor of Tables 7.13 and 7.14, with a worse q^2 quality [$^{f1}\{\varepsilon, d, RI, \chi_t{}^v\}^{p-odd}$: $N = 41$, $r^2 = 0.937$, $s = 2.2$, $q^2 = 0.721$].

If items (°) in Table 7.3 are left out the training quality of the graph-theoretical/semiempirical descriptor (1), 'zero-level' descriptor (2) and semi-random descriptor (3, quite poor q^2) with $N = 29$ are:

$$(1)\ r^2 = 0.948,\ s = 1.7,\ q^2 = 0.916$$
$$(2)\ r^2 = 0.572,\ s = 5.0,\ q^2 = 0.283$$
$$(3)\ r^2 = 0.909,\ s = 2.3,\ q^2 = 0.355$$

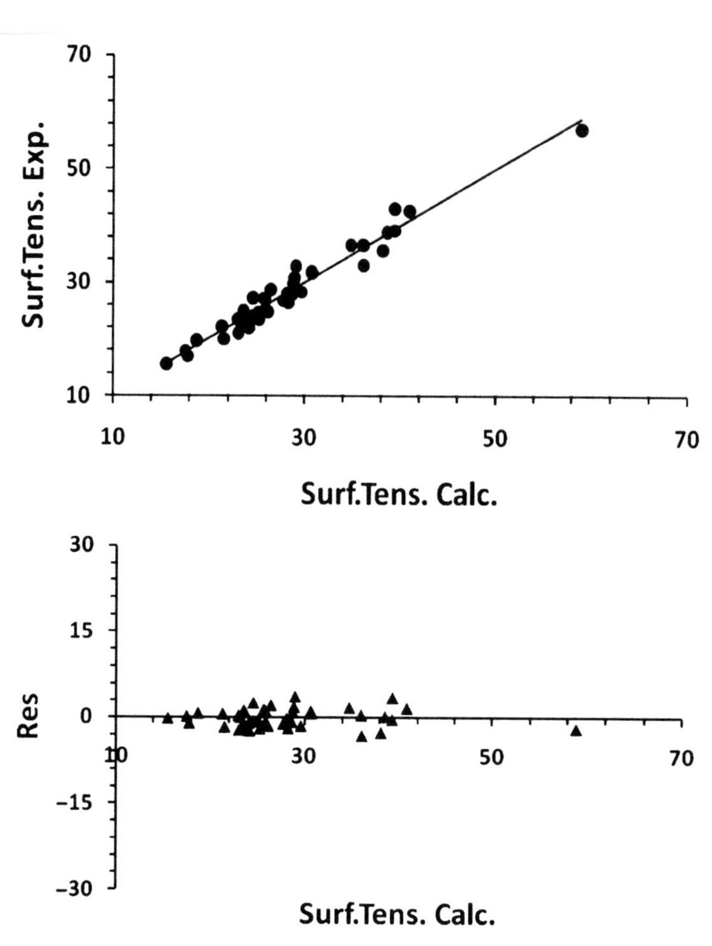

Fig. 7.9 The Exp./Calc. plot of γ, and its residual plot (*bottom*)

7.4.9 Cutoff UV Values, UV

Paralleling the greedy description (Tables 7.7 and 7.8) the full combinatorial search also needs to exclude the four compounds as strong outliers, heptane, 4-Me-2-pentanone, N,N-diMe-acetamide, and acetonitrile. The good semiempirical descriptor (Tables 7.9 and 7.10) has no need of any composite indices and M. The 'zero-level' description of Tables 7.11 and 7.12 has poor s and q^2 statistics. The correlation vector of descriptor of Table 7.9, with whom Fig. 7.10 has been obtained, is,

$$C = [736.820(62), -106.109(13), 500.009(79), 39.7165(4.7), -2065.97(277), \\ -785.973(88)]$$

The strongest correlations are: $r({}^{0}\psi_I, {}^{0}\psi_E) = 0.98$, $r({}^{S}\psi_E, {}^{0}\psi_E) = 0.96$, $r({}^{0}\psi_I, {}^{S}\psi_E) = 0.89$, $r({}^{0}\psi_I, AH^b) = 0.52$, and $r(UV, RI) = 0.53$.

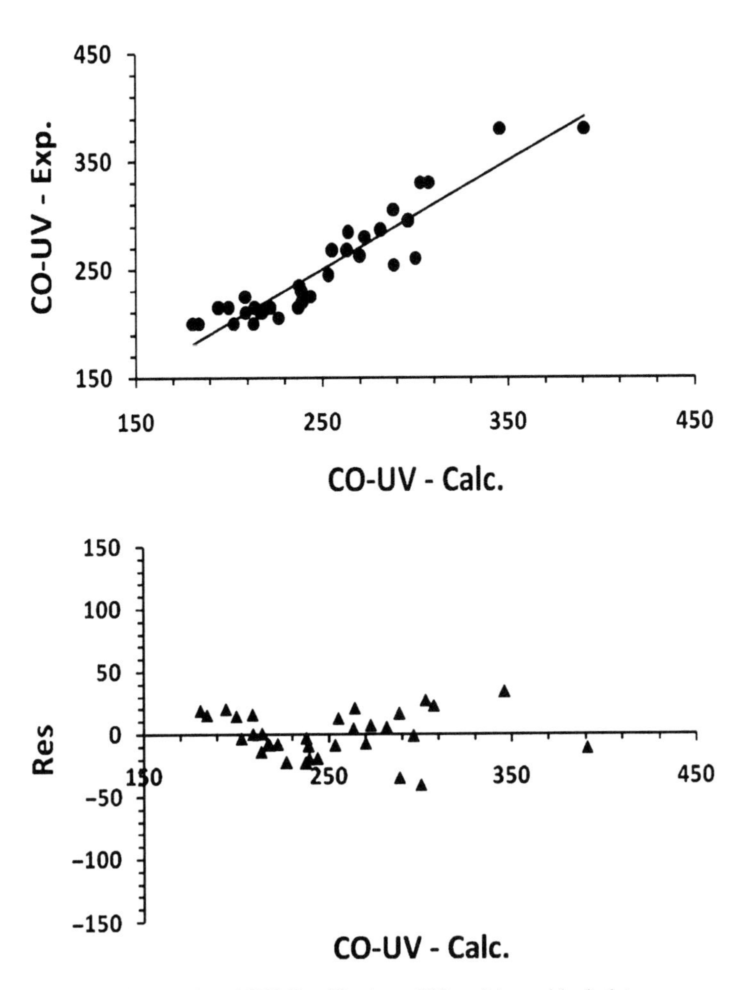

Fig. 7.10 The Exp./Calc. plot of UV Cutoff values, *UV*, and its residual plot

The search for a *mc-exp-rn*-descriptor that uses *r1-r38* and then *rd1-rd38* plus *RI*, AH^b and the MC- $K_p(p\text{-}odd)/\ f_\delta^{\ 1}$ indices (no heptane, 4-Me-2-pentanone, N,N-diMe-acetamide, and acetonitrile) finds exactly the optimal five-index descriptor already found. Limiting the search to *r1-r38* plus *rd1-rd38* plus *RI*, AH^b, ${}^0\psi_I$, and ${}^S\psi_E$, the mediocre q^2 descriptor of Tables 7.13 and 7.14 is found.

The training tests are now done with a four-index descriptor only (no last index) to obey N° Data/N° indices $\geq$ 5. If items (°) in Table 7.3 are left out the training quality of the graph-theoretical/semiempirical descriptor (1), 'zero-level' descriptor (2) and semi-random descriptor (3) with $N = 23$ are:

$$(1)\ r^2 = 0.879,\ s = 16,\ q^2 = 0.783$$

$$(2)\ r^2 = 0.325,\ s = 38,\ q^2 = 0.115$$

$$(3)\ r^2 = 0.825,\ s = 20,\ q^2 = 0.643$$

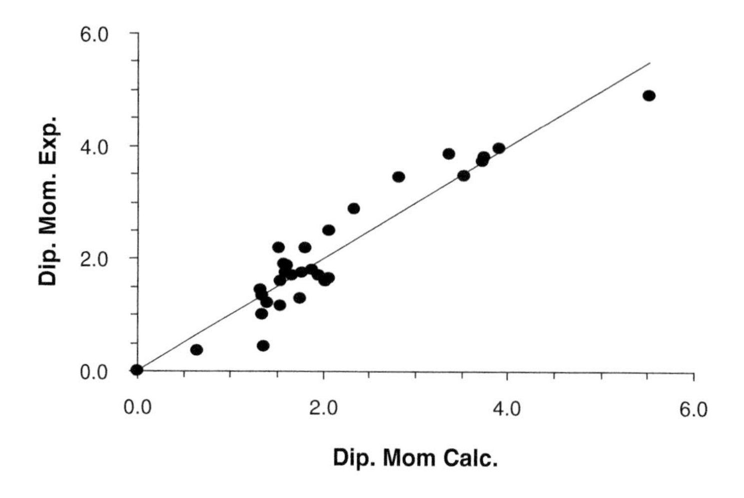

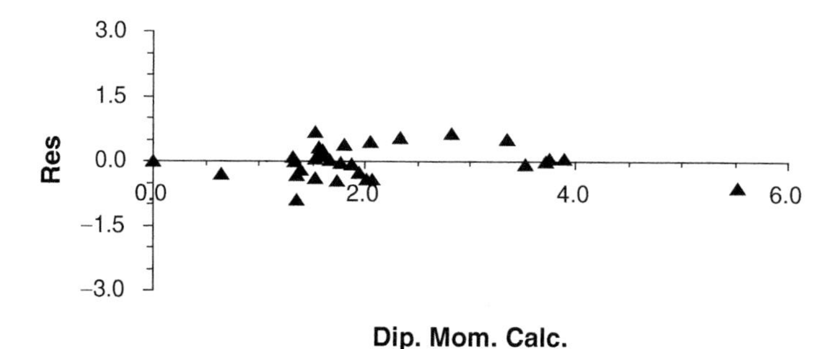

Fig. 7.11 The Exp./Calc. plot of μ, and its residual plot (*bottom*)

7.4.10 Dipole Moment, μ

The greedy (Tables 7.7 and 7.8) and complete combinatorial search algorithm (Tables 7.9 and 7.10) consider acetonitrile a strong outlier and take advantage of the two-valued $\phi(0, 1)$ symmetry parameter that zeroes all indices whose properties have $\mu = 0$, while leaves them unchanged if $\mu \neq 0$, i.e., for $\mu = 0$, $\phi{\cdot}\chi = 0$, while for $\mu \neq 0$, $\phi{\cdot}\chi = \chi$. The optimal full combinatorial descriptor with an improved *r* but is shown in Tables 7.9 and 7.10, while in Tables 7.11 and 7.12 is shown the corresponding 'zero-level' description, which seems not excessively bad even if the aid of the non-random ϕ *ad hoc* index plays a no minor role. The correlation vector of descriptor in Table 7.9, with whom Fig. 7.11 has been obtained, is,

$$C= [0.048(0.004), 0.64(0.3), -0.05(0.03), 0.19(0.03), -0.31(0.05), -0.13(0.1_5)]$$

The best non-ϕ descriptor, and the corresponding non-ϕ 'zero-level' descriptor are (this last quite unsatisfactory),

$$^{f2}\{\varepsilon, AH^m, \Delta, \Sigma, T_{\Sigma/M}\}^{pp-odd} : N = 34, r^2 = 0.896, s = 0.5, q^2 = 0.767$$
$$\{rd9, rd14, rd16, rd24, rd28\} : N = 34, r^2 = 0.516, s = 1.0, q^2 = 0.326$$

The correlations are: $r(\phi^1\chi, \phi D^v) = 0.97$, $r(\phi\varepsilon, \phi T_{\Sigma/M}) = 0.71$, $r(\phi\Sigma, \phi T_{\Sigma/M}) = 0.70$, $r(\phi\Sigma, \phi\varepsilon) = 0.69$, $r(\mu, \phi\varepsilon) = 0.78$, and $r(\mu, \phi\Sigma) = 0.72$. The positive role of the dielectric constant and of Σ, the overall E-State index for the electronegative atoms, and of $T_{\Sigma/M}$ ($= \Sigma^3/M^{1.7}$) is not unexpected.

The search encompassing a combinatorial space made of the $\phi r1$-$\phi r38$ and then $\phi rd1$-$\phi rd38$ plus ϕMCI and $\phi\varepsilon$ (no acetonitrile) finds the *mc-exp-rn*-descriptor of Tables 7.13 and 7.14, which is even better (and easier) than the *greedy* highly convoluted descriptor (only $\phi\Sigma$ and $\phi T_{\Sigma/M}$ are configuration-dependent).

Correlations among the normal and random indices and among μ and *rd10* and *rd34* are less than 0.4, as it is normally the case with the other properties also. Anyway, notice that the quality of $^{f0.5}\{\phi\varepsilon, \phi\Sigma, \phi T_{\Sigma/M}\}^{pp\text{-}odd}$ is: $N = 34$, $r^2 = 0.903$, $s = 0.4$, $q^2 = 0.771$. This means that the two random indices bring a striking contribution to q^2.

The training tests are now done with a four-index descriptor only (no last index) to obey N° Data/N° indices ≥ 5. If items (°) in Table 7.3 are left out the training quality of the graph-theoretical/semiempirical descriptor (1), 'zero-level' descriptor (2) and semi-random descriptor (3, quite good) with $N = 24$ are:

(1) $r^2 = 0.928$, $s = 0.4$, $q^2 = 0.742$
(2) $r^2 = 0.596$, $s = 1.0$, $q^2 = 0.422$
(3) $r^2 = 0.960$, $s = 0.3$, $q^2 = 0.915$

7.4.11 Magnetic Susceptibility, MS, $-\chi \cdot 10^6$

The greedy descriptor of Tables 7.7 and 7.8 is quite good but also quite convoluted. The best normal non-convoluted descriptor is the graph-theoretical descriptor of Tables 7.9 and 7.10. The description here needs M as a helpful index, which is a choice physically grounded as this property is really M-dependent. In this descriptor only $^0\chi^v$ and $^1\chi_s^v$ are configuration-dependent. With growing number of indices q^2 worsens. The 'zero-level' description has very poor q^2 (Tables 7.11 and 7.12).

The best correlations are: $r(\chi_t, {}^0\chi^v) = 0.854$, $r(\chi_t, {}^1\chi_s^v) = 0.810$, $r(-\chi \cdot 10^6, M) = 0.839$. This last strong value with M is physically grounded.

The search for a good four-index semi-random descriptor done among *MCI* plus M, plus *r1-r38* and then *rd1-rd38* finds the optimal *mc-exp-rn*-descriptor in Tables 7.13 and 7.14. This semi-random descriptor is statistically similar to the highly convoluted greedy descriptor but formally more down-to-earth. Without random index *r30* the statistics are: $r^2 = 0.788$, $s = 0.05$, $q^2 = 0.729$ (here only $^T\psi_I$ is

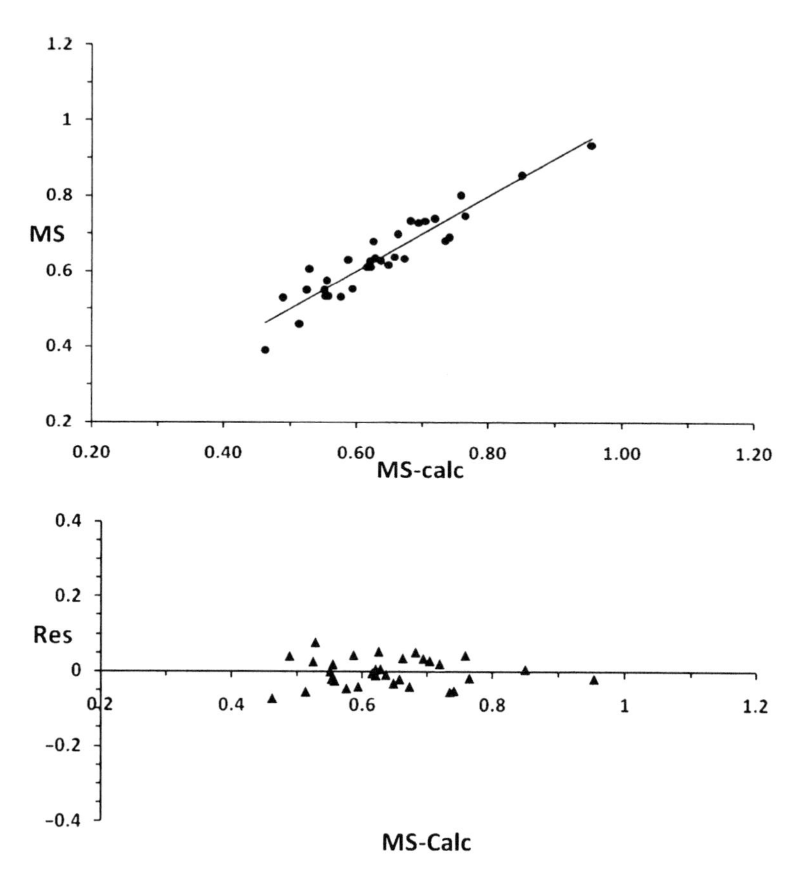

Fig. 7.12 The Exp./Calc. plot of MS, and its residual plot (*bottom*)

configuration-dependent). The correlation vector for the semi-random descriptor of Table 7.13 has been used to obtain Fig. 7.12,

$$C = [0.0019(0.0002), -0.13(0.04), 0.36(0.07), 0.14(0.03), 0.037(0.03)]$$

If items (°) in Table 7.3 are left out the training quality of the graph-theoretical/semiempirical descriptor (1), 'zero-level' descriptor (2) and semi-random descriptor (3, quite good description) with $N = 23$ are:

$$(1)\ r^2 = 0.850,\ s = 0.04,\ q^2 = 0.758$$
$$(2)\ r^2 = 0.483,\ s = 0.08,\ q^2 = 0.123$$
$$(3)\ r^2 = 0.858,\ s = 0.04,\ q^2 = 0.777.$$

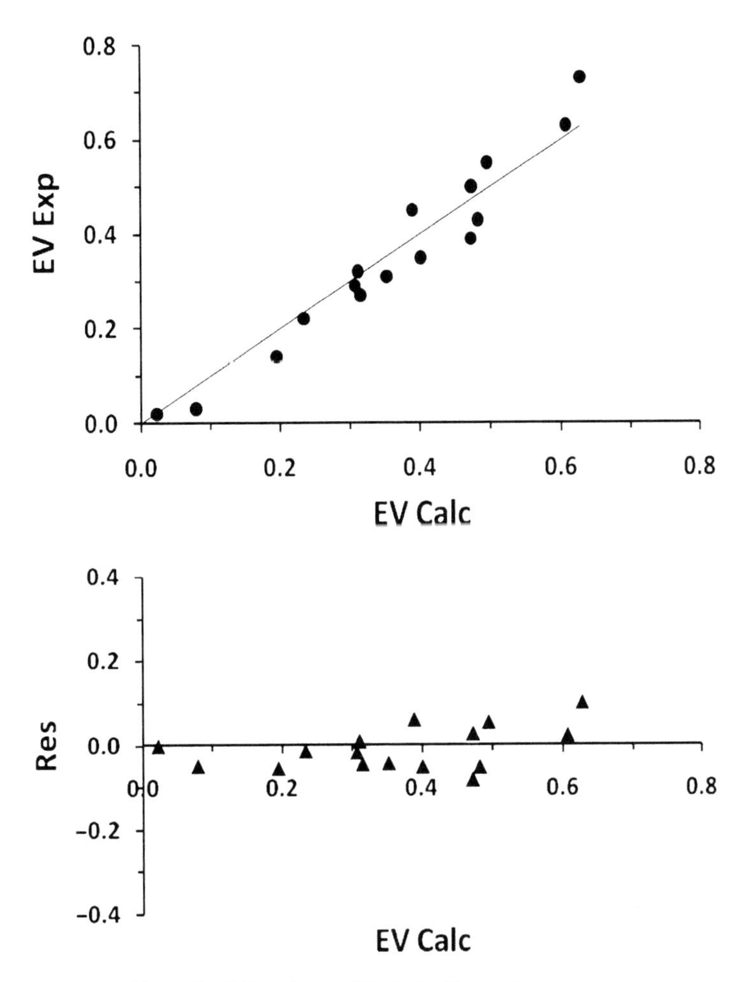

Fig. 7.13 The Exp./Calc. plot *EV*, and its residual plot (*bottom*)

7.4.12 Elutropic Values, EV

The greedy descriptor is shown in Tables 7.7 and 7.8. The full combinatorial descriptor belonging to the $K_p(pp\text{-}odd)/f_\delta^{0.5}$ configuration is shown in Tables 7.9 and 7.10. In Tables 7.11 and 7.12 is shown the 'zero-level' descriptor. The correlation vector of descriptor in Table 7.9, with whom Fig. 7.13 has been obtained, is,

$$C = [-2.24(0.2), 1.99(0.2), -2.02(0.2), 0.05(0.01), 0.08(0.07)]$$

It should here be noticed that we are bordering the Topliss-Costello rule for what concerns $N°$ data/$N°$ indices ≥ 5, and for this reason the training test is not considered (not enough data). The 'zero-level' descriptor, even if it is quite good, differs consistently at the q^2 level from the optimal descriptor. The strongest correlations are: $r({}^1\chi^v, {}^0\psi_I) = 0.99$, $r({}^0\psi_I, {}^0\psi_E) = 0.97$, $r({}^1\chi^v, {}^0\psi_E) = 0.94$, $r({}^1\chi^v, \Sigma) = 0.49$, $r(EV, {}^1\chi^v) = 0.77$, $r(EV, {}^0\psi_I) = 0.72$, $r(EV, {}^0\psi_E) = 0.70$.

The attempt to model this property with a four-index descriptor searched among the sets of MC indices plus the *r1-r38* and then *rd1-rd38* random numbers gave no new results (the previous optimal descriptor of Tale 7.9 was confirmed). Nevertheless, if the search is restricted to *r1-r38* plus *rd1-rd38* plus ${}^1\chi^v$, and ${}^0\psi_I$, it is possible to obtain the rather good semi-random descriptor (here only ${}^1\chi^v$ is configuration-dependent) of Tables 7.13 and 7.14, but its q^2 is not as good as the q^2 of the optimal descriptor,

Notice that the quality of ${}^{f\,0.5}\{{}^1\chi^v\}^{\text{p-odd}}$ is: $r = 0.707$, $s = 0.1$, $q^2 = 0.661$. The 'zero-level' description let us guess that giving up the Topliss-Costello rule (N° data / N° indices $\geq$ 5) with a descriptor encompassing seven random indices it should be possible to reach a more than satisfactory random model, even at the q^2 level, as it is the case with the following random descriptor, which, neverthe-less, continue to have a lower q^2 quality than the previous descriptor with four indices,

$$\{rd3, rd10, rd15, rd22, rd25, rd32, r37\} : N = 20, r^2 = 0.954, s = 0.06, q^2 = 0.856$$

7.4.13 The Super-Descriptors

All different MC or empirical indices of the full normal (non semi-random) descriptors of Tables 7.9 and 7.10 are now joined together to form a new space of indices, kind of super-indices, which will be used for a full combinatorial search of the best super-descriptors in a kind of configuration interaction of best indices. This super-descriptor space gave no remarkable results with the *greedy* algorithm, but it does find improved descriptions for the following four properties.

7.4.13.1 Melting Points, T_m

The following is, probably, the best description ever obtained of a set of 62 melting points with graph-theoretical methods, and we would dare to say, with any theoret-ical method,

$$\begin{aligned}&\{AH^m, {}^0\chi_d, D^v(f^{0.5} - ppo), {}^1\psi_E(f^{-0.5} - po), {}^0\psi_{Ed}(f^{-0.5} - po),\\&\Sigma(f^{-0.5} - po), \Sigma(f^{0.5} - ppo), {}^1\psi_E(f^{50} - ppo)\}:\\&\qquad N = 62, r^2 = 0.852, s = 19, q^2 = 0.804\end{aligned}$$

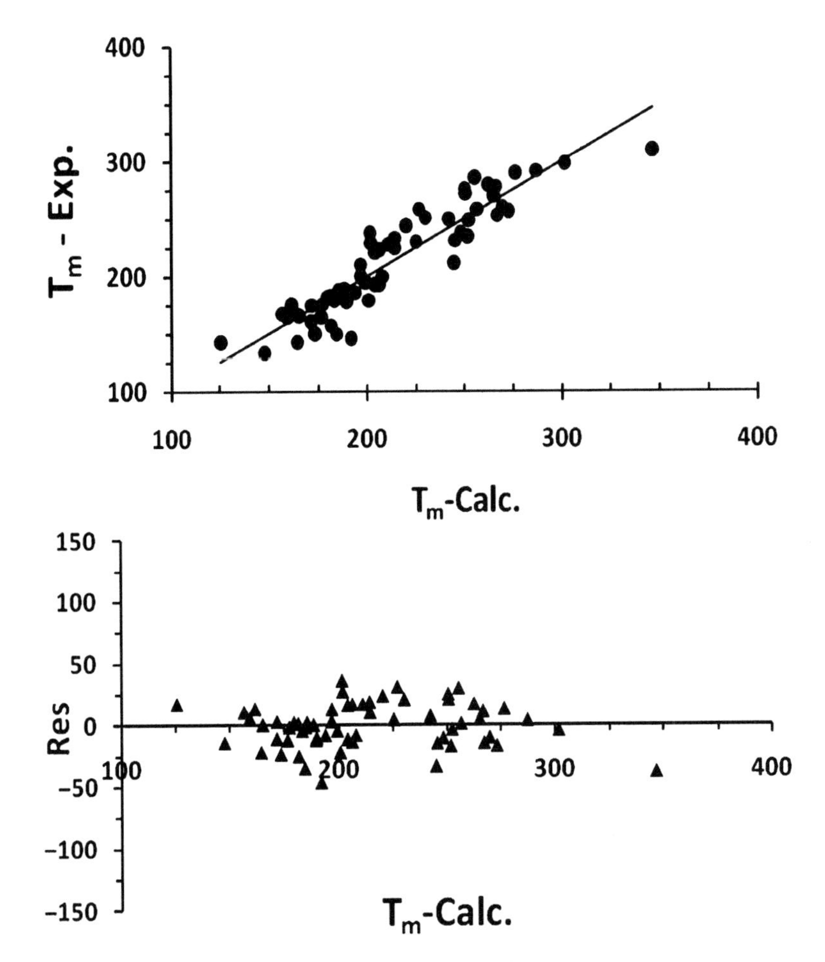

Fig. 7.14 The Exp./Calc. plot of T_m and its residual plot (*bottom*)

$$C = [40.3(3.9), 36.0(5.9), 3.75(0.9), 286(45), -1 \cdot 10^{-5}(10^{-6}), \\ -3.90(0.5), 4.14(1.1), -288(49), 142.7(8.0)]$$

Figure 7.14 shows the model plots for this super-description of T_m, obtained with the previous correlation vector and which should be compared with plots of Fig. 7.3. The correlations are: $r[^1\psi_E(f^{0.5} - po), {}^1\psi_E(f^{50} - ppo)] = 0.97$, $r[D^v(f^{0.5} - ppo), {}^1\psi_E(f^{50} - ppo)] = 0.94$, $r[^1\psi_E(f^{-0.5} - po), D^v(f^{0.5} - ppo)] = 0.91$, $r[\Sigma(f^{-0.5} - po), \Sigma(f^{0.5} - ppo)] = 0.81$, $r[^0\chi_d, {}^0\psi_{Ed}(f^{-0.5} - po)] = 0.66$, $r[T_m, AH^m] = 0.51$. Leaving out (°) items in Table 7.3 we obtain: $N = 45$, $r^2 = 0.853$, $s = 18.7$ $q^2 = 0.760$.

The search of a seven-index semi-random super-descriptor with *r1-r38* or *rd1-rd38* plus the super-indices is beyond the power of our PC. A search for a seven-index semi-random super-descriptor with the eight indices of the super descriptor plus the *r1-r38* and then the *rd1-rd38* indices confirmed the seven super-indices while excluded the super-index $\Sigma(f^{0.5} - ppo)$. This last seven-index super-descriptor has: $N = 62$, $r^2 = 0.810$, $s = 21$, $q^2 = 0.750$. Another test for a seven super-semi-random descriptor with the first five indices of the super-descriptor plus all the random indices gave no positive results.

7.4.13.2 Flash Points, FP

No better super-descriptor is here found, but the search throughout a combinatorial space made of T_b, the 32 super-indices plus the *r1-r38* and then *rd1-rd38* finds a five-index *mc-exp-rn*-super-descriptor, which improves slightly in s and q^2 over the previous semi-random descriptor,

$$\{T_b, \Delta, \Sigma(f^{-0.5} - po), {}^S\psi_E(f^1 - po), rd8\} : N = 41, r^2 = 0.989, s = 4.0, q^2 = 0.984$$

Excluding *rd8* the description of the remaining four-index descriptor is quite good: $r^2 = 0.982$, $s = 5.0$, $q^2 = 0.975$, i.e., *rd8* brings about a rather small but noticeable improvement. The important interrelations are: $r[{}^S\psi_E(f^{\,1} - po), T_b] = 0.62$, $r[{}^S\psi_E(f^{\,1} - po), rd8] = -0.53$, $r[FP, T_b] = 0.93$, $r[FP, \Delta] = 0.50$.

Leaving out (°) items we have: $N = 29$, $r^2 = 0.988$, $s = 4.2$, $q^2 = 0.980$.

7.4.13.3 Cutoff UV Values, UV

Excluding the four strong outliers, heptane, 4-Me-2-pentanone, N,N-diMe-acetamide, and acetonitrile, the following improved super-descriptor can be found

$$\{RI, AH^b, {}^1\chi_s, {}^S\psi_1/M(f^{50} - ppo), D^v(f^{0.5} - ppo)\} : N = 33, r^2 = 0.904, s = 17, q^2 = 0.807$$

Correlations are: $r[{}^1\chi_s, D^v(f^{\,0.5} - ppo)] = 0.66$, $r[AH^b, {}^S\psi_1/M(f^{\,50} - ppo)] = 0.66$, $r[UV, RI] = 0.53$.

Leaving out (°) items in Table 7.3 and the last index in the previous descriptor, to obey the Toplis-Costello rule, the quality of the model is (with none of the four outliers): $N = 23$, $r^2 = 0.902$, $s = 15$, $q^2 = 0.798$.

The entire combinatorial space made of RI, AH^b, the 32 super-indices plus the *r1-r38* and then *rd1-rd38* random sets has been searched and no interesting results have been found. The given five-index super-descriptor is by far the best descriptor. The search with the five indices of the previous descriptor plus the two sets of random indices finds also no different descriptor. The search with the first four indices of the previous descriptor (r^2 =0.733, s = 28, q^2 = 0.425) plus the random indices finds the following semi-random descriptor,

$$\{RI, AH^b, {}^S\psi_1/M(f^{50} - ppo), rd18, r8\} : N = 33, r^2 = 0.830, s = 23, q^2 = 0.667$$

7.4.13.4 Dipole Moment, μ

No best super-descriptor can be found (remind: no acetonitrile!). Instead, a quite good semi-random super-descriptor is found by the aid of a search with a space made of ϕ-super-indices (inclusive $\phi\varepsilon$ and ϕM) plus $\phi r1$-$\phi r38$ and then $\phi rd1$-$\phi rd38$ random sets, and it is,

$$\{\phi\varepsilon, \phi\Delta, \phi\Sigma(f^{50} - ppo), \phi T_{\Sigma/M}(f^{0.5} - ppo), \phi rd5\}\, N = 34, r^2 = 0.960,$$
$$s = 0.3, q^2 = 0.914$$

Without *rd5* the statistics of the remaining super-descriptor is: r^2 = 0.929, s = 0.3, q^2 = 0.883. Correlations are: $r[\phi T_{\Sigma/M}\ (f^{0.5} - ppo), \phi\varepsilon]$ = 0.71, $r[\phi\varepsilon, \phi\Sigma(f^{50} - ppo)]$ = 0.62, $r[\phi T_{\Sigma/M}\ (f^{0.5} - ppo), \phi\Sigma(f^{50} - ppo)]$ = 0.59, $r[\mu, \phi\Sigma(f^{50} - ppo)]$ = 0.80, $r[\mu, \phi\varepsilon]$ = 0.78. All other correlations are less than 0.5.
Without items (°) we obtain: N = 24, r^2 = 0.967, s = 0.3, q^2 = 0.909.

7.4.14 A Randomized Model

This is another sort of random test for the 12 properties. Our properties have been randomized following a procedure outlined elsewhere (Pogliani 2010). Compounds have sequentially been numbered following their molar mass M, and from a hat were chosen the 7° (acetone), 2° (Acetonitrile), 26°(benzene) 37° (Ethylacetate), 1° (CH_3-OH), and 27° (DMSO) molar mass. The 12 properties of Acetone, Acetonitrile, Benzene, Ethylacetate, Methyl alcohol, and the first three properties of DMSO, for a total of N = 63 points were then sequentially assigned to compounds sorted out from a hat, and the assignment is shown in Table 7.3. The number in parenthesis in this Table means, for instance, that T_b of acetone has been assigned to 1,1,2triCl,triFEthane (1), T_m of acetone to diBr-methane (2), *RI* of acetone to 2,2-tetraCl-Ethane (3), to tetraMe-Urea (12) is assigned the EV value of Acetone (0.43), and to Butylacetate (13) is assigned T_b of Acetonitrile, and so on. The full combinatorial search model was performed with the indices used to find out the super-descriptors (here we depart from Pogliani 2010 where normal indices have been used). Four different data sets of randomized properties were studied (N = 63, 32, 16, and 8, always deleting the bottom compounds numbered in parenthesis in

Table 3) with seven-, five-, three-, and two-index index super-descriptor to keep an eye to the importance of the Topliss-Costello rule. In the following lines are the results,

$$\{\varepsilon, {}^0\psi_{Ed}(f^{-0.5}-po), {}^0\psi_1(f^{0.5}-ppo), {}^T\psi_1(f^1-po), {}^0\psi_E(f^8-po), {}^S\psi_1/M(f^{50}-ppo), {}^0\psi_E/M(f^{50}-ppo)\}:$$
$$N=63, r^2=0.33, s=119, q^2=0.20$$

$$\{T_m, D^v(f^{0.5}-ppo), {}^1\chi_s{}^v(f^{50}-po), {}^0\psi_{Ed}(f^{-0.5}-po), \Sigma(f^{-0.5}-po)\}:$$
$$N=32, r^2=0.54, s=100, q^2=0.35$$

$$\{d, D^v(f^{0.5}-ppo), D/M\}: N=16, r^2=0.750, s=80, q^2=0.619, F=12$$
$$\{{}^S\psi_1(f^{-0.5}-po), {}^0\psi_E(f^8-po)\}: N=8, r^2=0.849\ \ s=61, q^2=0.707$$

The last model of eight randomized properties does not obey the Topliss-Costello rule anymore, while the model with $N = 16$ borders this rule ($r > 0.84$ for $N°$ data/$N°$ indices ≥ 5). The model of $N = 8$ compounds with just an index is quite negative. The reader should remind that the values of the properties are completely random.

In the following is the best model of the randomized properties with *r1-r38* and *rd1-rd38* random numbers. The quality of this seven-random-index descriptor is not far from the quality of the previous seven index super-descriptor, especially at the q^2 level,

$$\{r1, r9, r23, r30, r31, r32, r36\}: N=63,\ r^2=0.41,\ s=112,\ q^2=0.23$$

This similarity suggests that the randomized properties are unable to differentiate between random numbers and meaningful indices.

7.5 Conclusions

An optimal model for the 12 properties has here been achieved and for some properties the model is even outstanding. Other and more specific conclusions deserve out attention, and the first one follows from the use of random and semi-random descriptions of Tables 7.11–7.14, respectively. The 'zero-level' descriptors show their main weakness at the s and q^2 level. The random tests of the *EV* property confirms in a clear way the importance of the Topliss-Costello rule. The very good model of *EV* by a semi-random descriptor with three random indices borders the limit of five for the ratio *N°data/N°indices*. The importance of this rule and of the q^2 statistics is confirmed by the 'zero-level' descriptors for *EV* and *FP*, as well as by the last property randomization test (the cases with $N = 16$ and $N = 8$). These studies suggest that the lowest value for q^2 should be placed around 0.75, and that a more stringent Topliss-Costello rule should be established, i.e., $r^2 > 0.8$ (and not r) for a ratio N° data/N° indices ≥ 6. The critical importance of these last statistical parameters (N° data/N° indices, r^2, and q^2) has already been underlined by other authors (Golbraikh and Tropsha 2002; Hawkins et al. 2003; Hawkins 2004) while the importance of the s statistics has already been stressed (Randić 1991a, b, 1994).

Once the new lowest value for $q^2 \geq 0.75$ is accepted, the semi-random model of *UV* (see Tables 7.13 and 7.14) should be rejected, while the semi-random model for the viscosity, η, should be considered with a critical eye. The other nine good to optimal semi-random model studies, instead, confirm the importance for a semi-random description to have at disposition several good MC indices or/and experimental parameters that by themselves could give rise to a satisfactory description, otherwise random indices do not work. If the positive side of a semi-random description is that it allows an economy of the number of MC indices to be used for a model, the negative side is that it is not possible to know from scratch which MC indices or experimental parameters are the best ones for the random indices. Four properties are optimally modeled with semi-random descriptors: the refractive index, *RI*, the flash point, *FP*, the dipole moment, μ, and the magnetic susceptibility, $-\chi\cdot 10^6$. The use of two sets of random numbers *r1-r38* and *rd1-rd38* shows that the more the number of random indices the highest is the possibility to find random indices that fit together with MC indices or experimental parameters. Similar conclusions about random descriptions indices, that did not concern semi-random descriptions, were recently drawn (Katritzky et al. 2008).

The very good model of the 12 properties inclusive of the corresponding 'back-training' test (excluding compounds with "°" in Table 7.3) either with normal, semiempirical, semi-random, and super-descriptors is good news for the full combinatorial algorithm. This algorithm confirms the utility of the experimental parameters and *M* as well as of the *ad hoc* parameters AH^b, AH^m, and ϕ. Due to the huge number of combinations of the full combinatorial space we should expect that the configuration of the full combinatorial descriptors be different from the configuration of the *greedy* descriptor. Looking at our best descriptors in the corresponding Tables we notice that things are not exactly that way. Properties, ε, d, *RI*, and η, share the same (nearly the same for d) configuration with both search algorithms, and this is good news for the *greedy* algorithm. Assuming q^2 as the critical statistics for a model the full combinatorial algorithm does an optimal job with T_b (the *greedy* algorithm could not find a better descriptor with higher number of indices), T_m (excellent description with a super-descriptor), ε, *RI*, η, γ, μ (the last two with super-descriptors), and *EV*. Concerning *UV* the full combinatorial descriptor has not the same q^2 of the greedy descriptor but it is formally much simpler. The full combinatorial algorithm also does a better job than the *greedy* algorithm with d and with *FP* (with a super-descriptor). The tendency to look for the help of a semiempirical descriptor is the same with both search algorithms (nine properties). The *greedy* algorithm is the only that chooses a pure empirical descriptor for the surface tension, γ. Normally, the full combinatorial algorithm does not need the help, to achieve an optimal model, of any highly convoluted descriptor. The full combinatorial algorithm normally chooses descriptors with a strong hydrogen perturbation, i.e., $f_\delta{}^{n}$ with $n = -0.5$, 0.5, and 1 and prefers also a K_p(*p-odd*) encoding for the core electrons. The full combinatorial algorithm prefers to compensate a strong hydrogen contribution with a weaker electron core contribution. The model of the properties with the highest and similar number N of compounds (T_b, T_m, ε, d, and RI) confirms that the hydrogen perturbation is mainly property-dependent. All in all the *greedy* algorithm even if it covers only a tiny fraction of the full combinatorial space achieves to detect some good descriptors, even if some are quite convoluted.

The advantage to work with indices belonging to different configurations is emphasized by possibility to derive super-descriptors (*Sd*) for model purposes, i.e., combinations of indices of the best descriptors (super-indices) of the 12 properties in a kind of configuration interaction procedure performed with the full combinatorial algorithm. Super-descriptors are, in fact, able to consistently improve the model of T_m, and, a bit less consistently, to improve the model of FP and μ.

Last but not least, throughout the model of the 12 properties the pseudoconnectivity indices together with the electrotopological index, Σ, which encode specific information on the electronic environment, are the indices that are mainly responsible for the quality of the model.

References

Besalu E, de Julian-Ortiz JV, Iglesias M, Pogliani L (2006a) J Math Chem 39:475

Besalu E, de Julian-Ortiz JV, Pogliani L (2006b) MATCH Commun Math Comput Chem 55:281

Besalu E, de Julian.Ortiz JV, Pogliani L (2007) J Chem Inf Model 47:751

Carbó-Dorca R, Robert D, Amat L, Girones X, Besalu E (2000) Molecular quantum similarity in QSAR and drug design. Springer Verlag, Berlin

de Julián-Ortiz JV, Pogliani L, Besalu E (2010) J Chem Educ 87:994

Eriksson L, Johansson E, Muller M, Wold S (2000) J Chemometr 14:599

Garcia-Domenech R, Galvez J, de Julian-Ortiz JV, Pogliani L (2008) Chem Rev 108:1127

Golbraikh A, Tropsha A (2002) J Mol Graph Model 20:269

Hawkins DM (2004) J Chem Inf Comp Sci 44:1

Hawkins DM, Basak SC, Mills DJ (2003) Chem Inf Comp Sci 43:579

Katritzky AR, Dobchev DA, Slavov S, Karelson M (2008) J Chem Inf Model 48:2207

Kier LB, Hall LH (1986) Molecular connectivity in structure-activity analysis. Wiley, New York

Kier LB, Hall LH (1999) Molecular structure description. The electrotopological state. Academic, New York

Martens HA, Dardenne P (1998) Chemom Intell Lab Syst 44:99

Mihalić Z, Nikolić S, Trinajstić N (1992) J Chem Inf Comput Sci 32:28

Peterangelo SC, Seybold PG (2004) Int J Quantum Chem 96:1

Pogliani L (2002) J Chem Inf Comput Sci 42:1028

Pogliani L (2005a) Int J Quantum Chem 102:38

Pogliani L (2005b) New J Chem 29:1082

Pogliani L (2006) J Comput Chem 27:868

Pogliani L (2007) J Pharm Sci 96:1856
Pogliani L (2010) J Comput Chem 31:295
Pogliani L (2011) MATCH Commun Math Comput Chem 65:347
Pogliani L, de Julian-Ortiz JV (2004) Chem Phys Lett 393:327
Randić M (1975) J Am Chem Soc 97:6609
Randić M (1991a) J Comput Chem 12:970
Randić M (1991b) New J Chem 15:517
Randić M (1994) Int J Quantum Chem: Quantum Biol Symp 21:215
Todeschini R, Consonni V (2009) Molecular descriptors for chemical informatics, vol 1, Alphabetical Listing. Wiley-VCH, Weinheim
Topliss JG, Costello RJ (1972) J Med Chem 15:1066
Trinajstić N (1992) Chemical graph theory. CRC Press, Boca Raton

Chapter 8
Structural Approach to Aromaticity and Local Aromaticity in Conjugated Polycyclic Systems

Alexandru T. Balaban[1] **and Milan Randić**[2]

Abstract After an introductory brief discussion of the important and much debated concept of aromaticity, we elaborate on a recently proposed scheme for the partition of π-electrons of Kekuléan polycyclic conjugated hydrocarbons to individual rings, so that by summing π-electrons within individual rings of polycyclic conjugated hydrocarbons one obtains the total number of π-electrons in a molecule. We discuss separately for various types of hydrocarbons the π-Electron Content of individual rings, to be denoted as EC. In particular we summarize the EC results that we published in a series of papers which included: benzenoid catafusenes, benzenoid coronafusenes, benzenoid perifusenes, alternant and nonalternant conjugated hydrocarbons. Finally we return to Clar structures of benzenoid hydrocarbons as well as Clar structures of a selection of non-benzenoid alternant hydrocarbons like biphenylene, and point to a significant difference in distribution of π-electrons when instead of using all Kekulé valence structures one focuses on the subset of Kekulé valence structures indicated by Clar's model.

8.1 Aromaticity

Aromaticity is a cornerstone of chemistry, accounting for more than three quarters of all known and recorded substances (about 60 million). It was discussed recently in *Chemical Reviews*: a special issue published in 2001 reviewed most of the recent developments in aromaticity (Schleyer 2001); other relevant reviews have been published soon afterwards: one on polycyclic aromatic hydrocarbons

[1] Texas A&M University at Galveston, MARS, 5007 Avenue U, Galveston, TX 77551, USA
e-mail: balabana@tamug.edu

[2] National Institute of Chemistry, P.O. Box 3430, 1001 Ljubljana, Slovenia
e-mail: mrandic@msn.com

M.V. Putz (ed.), *Carbon Bonding and Structures: Advances in Physics and Chemistry*,
Carbon Materials: Chemistry and Physics 5, DOI 10.1007/978-94-007-1733-6_8,

(Randić 2003), and three other ones on aromatic heterocycles (Balaban et al. 2004; Balaban 2009, 2010). In addition to other reviews (Balaban 1980; Klein 1992; Kikuchi 1997), aromaticity was the topic of many books, from which a few ones are cited here (Balaban et al. 1987; Clar 1972; Harvey 1991, 1997; Minkin et al. 1994).

Although aromaticity is a fuzzy concept (a fact that has generated many discussions – even an outrageous and unrealistic proposal to abandon this concept), aromaticity remains a useful, albeit multidimensional phenomenon, in the sense that it displays in different situations different manifestations. There are many "hyphenated" types of aromaticity in addition to the well-known in-plane-aromaticity, such as anti-aromaticity, homo-aromaticity, hetero-aromaticity, sigma-aromaticity, 3D-aromaticity, spherical aromaticity, etc. Historically, the principal names associated with aromaticity are *Michael Faraday* who discovered benzene in 1820; *August Kekulé* who, having advocated the tetravalency of carbon in 1857, proposed in 1865 the well-known cyclic formula with continuous conjugation, arguing that such a structural criterion should be taken as the basis for defining aromaticity; alternative structural formulas for benzene proposed by *J. Dewar* and *A. Ladenburg* correspond to strained valence-isomeric hydrocarbons that could be synthesized in the 1960s; *Emil Erlenmeyer* who advanced in 1866 the analogous naphthalene formula, and advocated that reactivity based on substitution (being preferred to addition) should be the aromaticity criterion; *Victor Meyer* who recognized the aromaticity of thiophene that he had discovered serendipitously in 1863; *Eugen Bamberger* who extended ideas of six centric bonds to five-membered heterocycles; *Wilhelm Körner* who assigned structures to many poly-substituted benzene derivatives on the basis of his "absolute" method based on the topology of poly-substitution (Körner 1867); *Johannes Thiele* who hypothesized that cyclic continuous conjugation was the necessary and sufficient conjugation for aromatic character; *Richard Willstätter* who disproved Thiele's hypothesis based on his unsuccessful attempts to synthesize cyclobutadiene and on his study of cyclooctatetraene; *Robert Robinson* (Armit and Robinson 1925) who advanced the idea of the "aromatic sextet of electrons", although *Crocker* (1922; see also Balaban et al. 2005) had published in 1922 in *J. Am. Chem. Soc.* another article, before the 1925 Armit–Robinson paper, with the same idea, which had already been adumbrated by Thomson (Thomson 1921); *Pauling* (1931, 1940; Pauling and Wheland 1933) who explained aromatic stability via his resonance theory and hybridization concept; *Hückel* (1931, 1937, 1938, 1940, 1975; Hückel and Hückel 1932) and who applied early quantum-chemical calculations demonstrating that monocyclic planar aromatic systems have a marked stabilization when they possess $4n + 2$ π-electrons but are destabilized when they have $4n$ π-electrons ($n = 1, 2$, etc.), predicting the aromaticity of tropylium; Dewar (1967, 1969; Dewar and de Llano 1969) who confirmed this prediction by proposing tropolone structures and who established a better scale than Hückel's for aromatic stabilization and destabilization; and *Clar* (1964, 1972) who assigned a special meaning to the "aromatic sextet circle" that had been introduced by Robinson, arguing that "Clar formulas" should never have such circles in adjacent condensed rings. Three recent papers (Balaban 2004; Gutman 1982; Gutman and Cyvin 1989) explain the rules ("how to do and how not to do") about writing Clar formulas. For an alternative simple notation proposed for benzenoids that does not contradict Clar's model, one may see Figs. 133 and 138 (the latter reproduced on the cover of the September 1993 issue of *Chemical Reviews*) in an ample review (Randić 2003).

The extraordinary stability of benzene derivatives that is responsible for their substitution reactions also explains why the eutectic liquid mixture of *meta*- and *para*-terphenyl is so resistant to gamma radiation that it can be used as a neutron moderator and thermal transfer fluid in "organic-cooled nuclear reactors". Also, this stability accounts for the ubiquity of the strongly carcinogenic benzo[*a*]pyrene that is formed in many combustion processes and is found in exhaust gases and in cigarette smoke. Its carcinogenicity is due to a paradoxical effect of living cells' defense mechanism involving oxidation of aromatic hydrocarbons, but in this case the epoxy-diol formed at the "bay region" leads to an extra-stabilized carbocationic intermediate which reacts with the DNA nucleophilic bases, playing havoc with the double strand of DNA (Harvey 1991).

8.2 Classification of Polycyclic Conjugated Hydrocarbons

Conventionally, throughout this review, all benzenoid systems will be drawn such that two bonds in each benzenoid ring are shown as vertical line segments.

When two rings share a bond, the system is called "condensed" or "fused". There are two main types of *substances* that are polycyclic benzenoids with condensed rings, namely planar portions of the graphene honeycomb lattice, and helicene- or cyclophane-like non-planar aggregates with benzenoid rings that on flattening would become superimposed. Both these types are real ensembles of zillions of molecules with identical structures. Even the large peri-condensed benzenoids prepared by Müllen and coworkers (Watson et al. 2001) deserve to be called "*substances*" because they are monodisperse (ignoring the random distribution of stable isotopes ^{13}C and ^{2}H admixed with the more abundant ^{12}C and ^{1}H). By contrast, nanotubes, nanocones, and other similar carbon aggregates are *mixtures with various degrees of dispersity* like most of the natural and synthetic polymers that consist of variable numbers of monomer units (only proteins and polynucleotides are monodisperse natural polymers).

Polycyclic aromatic hydrocarbons with continuous conjugation have only sp^2-hybridized carbon atoms, and can be cata-condensed, peri-condensed, or corona-condensed. Benzenoids in the first class (also called catafusenes) have no carbon atoms common to more than two benzenoid rings, whereas the second class, perifusenes, have one or more carbon atoms common to three benzenoid rings. An alternative and simpler way to characterize the above three classes is based on the *dualists* of benzenoid hydrocarbons which are obtained by representing each benzene ring in a polycyclic benzenoid system as a vertex, and by connecting vertices of adjacent (condensed) rings sharing a CC bond. To be more precise, one places vertices in the molecular plane at the centers of the rings and then one connects them by edges which are perpendicular to the underlying CC bonds of

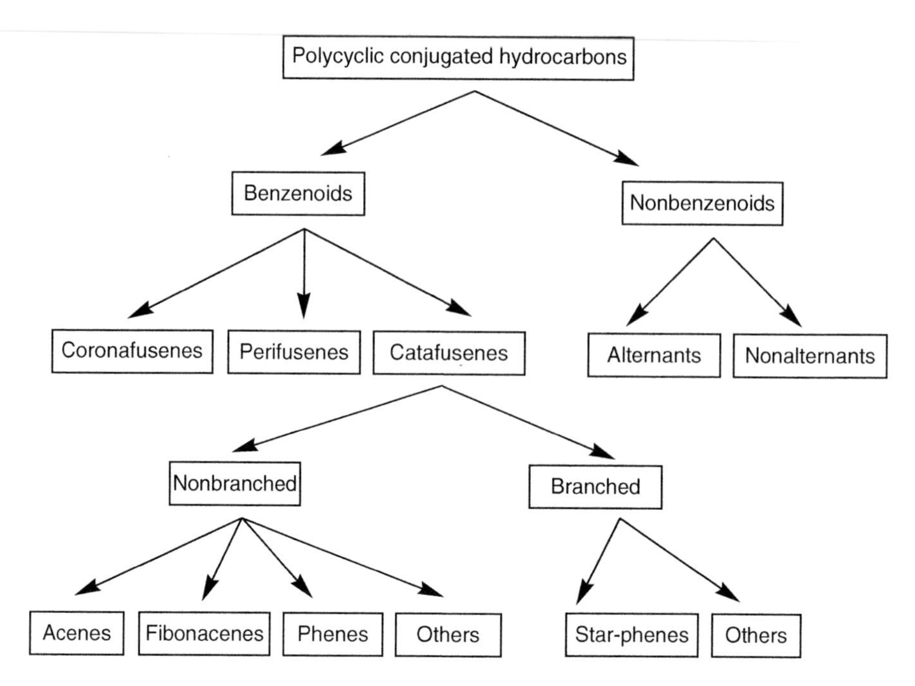

Fig. 8.1 Classification of conjugated polycyclic hydrocarbons

benzene rings. The so-constructed dualists differ from the standard dual graphs of geometry and graph theory in two ways: (1) One does not consider the outside space as a face, as is the case with graph-theoretical duals of polyhedra; and (2) One firmly embeds the dualist graph over the regular hexagonal (graphite-like) grid. Thus, bond angles in dualist graphs do matter, unlike the situation for ordinary graphs. In Fig. 8.1 we show a chart that summarizes the classification of polycyclic conjugated hydrocarbons.

In Fig. 8.2 we have illustrated dualist graphs for a selection of smaller benzenoid hydrocarbons. As one can see, dualist graphs of catafusenes, shown in the first two lines of Fig. 8.2, are trees (that is, they do not contain cycles); dualist graphs of perifusenes contain three-membered rings, and dualist graphs of coronafusenes have larger rings (Balaban and Harary 1968; Balaban 1969). Perifusenes or coronafusenes include also such systems having cata-fused branches. However, when the dualist graph of a benzenoid has both three-membered and larger rings, the respective "peri + corona-fusene" could be considered to belong to a special class, but so far no such compound has apparently been reported.

In the case of catafusenes, the approach based on dualist graph allows a simple coding of structures, which is based on the following convention: one starts from a terminal vertex of the longest non-branched chain in the dualist and assigns digits 0, 1, or 2 to vertices on this chain; if the next vertex continues on a straight-line (angle 180°), then one uses digit 0. When one arrives at a vertex that is off the straight line, that is at angles 60° or −60° for any left or right bending, respectively, one uses

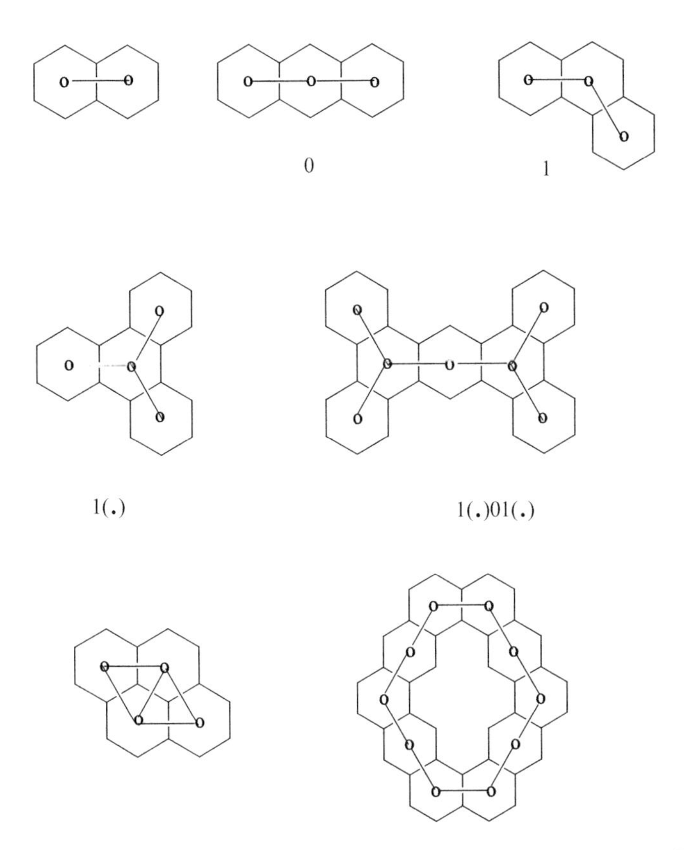

Fig. 8.2 Examples of benzenoids and their dualist graphs: trees for non-branched catafusenes (*top three benzenoids*), branched catafusenes (*middle row*), a perifusene and a coronafusene (*bottom row*)

digits 1 or 2 to denote the two alternatives (once this choice is made, e. g. left-kink = 1 and right-kink = 2, it must be kept throughout the length of the dualist graph). For non-branched catafusenes, among all possible choices of 1 or 2 digits, or of which one from the two terminal vertices for the start of the coding, the canonical code is the one corresponding to the smallest number. For branched catafusenes the main chain of rings is the one with the largest number of condensed rings (just as in the IUPAC system for naming alkanes); a similar convention applies, with the difference that a branch is included in brackets after the digit corresponding to the ring at the branching point, and a one-vertex branch is symbolized by a dot in the brackets (.). In one of the following sections it will be evident how this coding system (which we illustrate in Fig. 8.2 on a few smaller catafusenes) operates. It will be seen that catafusenes whose codes differ only by interchanging digits 1 and 2 have the same number of Kekulé structures and set of EC values (see below what EC stands for); they are called *isoarithmic* (Balaban and Tomescu 1983). Isoarithmic systems also exist for perifusenes that have cata-condensed isoarithmic appendages.

In alternant hydrocarbons, carbon atoms can be divided into two sets (starred and non-starred) such that an atom of one set is connected only to atoms of the other set. The necessary and sufficient condition for a hydrocarbon to be alternant is to have only even-membered rings. Benzenoid hydrocarbons have only six-membered rings and are therefore alternant or bipartite.

The following notation will be used throughout this review: the number of carbon atoms, hence also of π-electrons, will be denoted by $n = D/2$, the number of double bonds by D, the number of rings by R, and the numbers of Kekulé structures by K. For benzenoid catafusenes, $n = 4R + 2$, and their formulas are $C_{4R+2}H_{2R+6}$, or alternatively $C_{2R-4}(CH)_{2R+6}$, which immediately shows the number (2R – 4) of internal or quaternary carbon atoms participating in the fusion of rings. While clearly all benzenoids are alternant hydrocarbons, not all alternants are benzenoid (e. g. biphenylene).

In the present paper we will discuss only Kekuléan structures, which are systems that must have at least one Kekulé valence structure, i. e. in all systems covered by this review, all carbon atoms must have a C=C double bond and two single bonds (one of which is a C–C bond and the other may be either a C–C or a C–H bond).

8.3 Local Aromaticity of Polycyclic Benzenoid Hydrocarbons

It may look surprising, but the local aromaticity of individual rings of polycyclic conjugated hydrocarbons has received little attention if any, till Eric Clar (a giant in the synthetic organic chemistry of benzenoid hydrocarbons) drew attention to local aromaticity. Eric Clar obtained an astounding number of 122 benzenoid hydrocarbons, starting from the year 1924 when he synthesized pentaphene till 1981 when he synthesized his last benzenoid: dinaphtho[8,1,2-*abc*:2′,1′,8′-*klm*] coronene. For a list of all these benzenoid hydrocarbons see Table 55 in the recent review on the aromaticity of conjugated polycyclic hydrocarbons (Randić 2003). To honor E. Clar's memory and his achievements, we have decided deliberately to use the abbreviation EC for the π-**E**lectron **C**ontent of individual rings, because it also coincides with the initials of **E**ric **C**lar! By contrast, we may observe that the conceptually related quantity, the Pauling bond order, which of course relates to individual characters of CC bonds rather than rings, has been around since the 1930s. However, in 1967 Oskar E. Polansky reconsidered the empirical approach of Eric Clar and searched for a possible quantum chemical justification of the notion of Clar that some rings in polycyclic systems show considerable aromaticity while other rings are devoid of comparable local aromatic character. In a seminal paper with Derflinger (Polansky and Derflinger 1967) using the then widely popular HMO approach, they were able to show that Hückel molecular orbitals imply a non-uniform distribution of the "local aromatic character," and moreover that the variations of ring aromaticity derived from HMO strongly parallel Clar's empirical rules for distribution of π-aromatic sextets to some rings, while designating some other rings as "empty" (of π-aromatic sextets) and referring to intermediate cases as

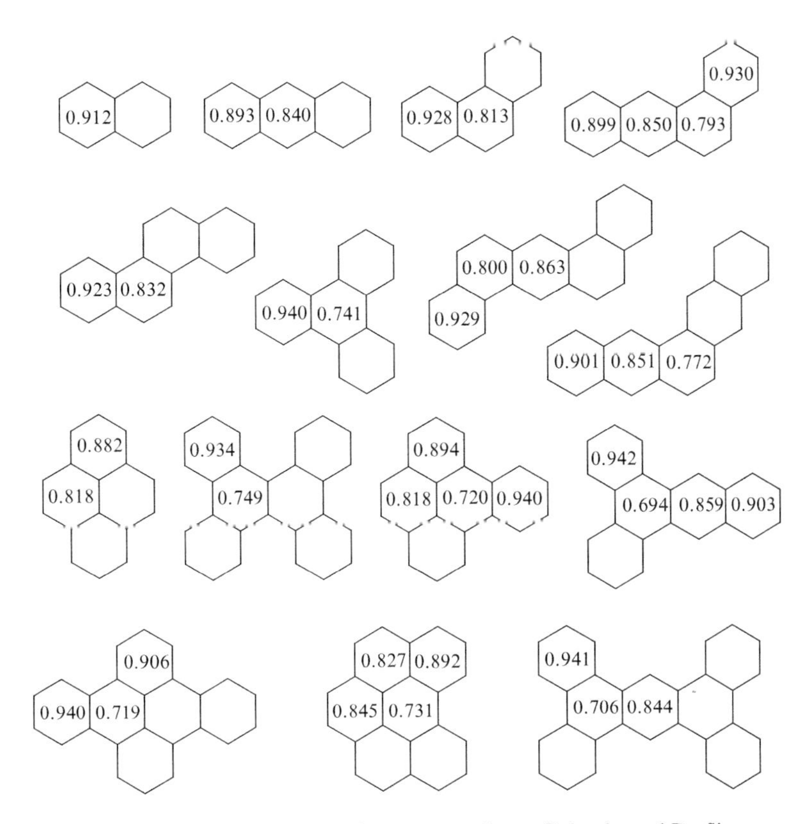

Fig. 8.3 Local aromaticity values for catafusenes according to Polansky and Derflinger

"migrating" π sextets. In Fig. 8.3 we show for a selection of 15 benzenoid hydrocarbons the local aromaticity indices as reported by Polansky and coworkers (Polansky and Derflinger 1967; Monev et al. 1981).

Let us reformulate the rules that Clar proposed for the construction of benzenoid structures depicting π-sextets, structures that today we refer to as Clar structures:

Inscribe as many as possible circles (that signify π-aromatic sextets) in individual benzene rings such that circles do not appear in adjacent rings and that carbon atoms which are not involved in these sextets allow completion of assignment of CC double bonds for the rest of the structure. Thus a benzenoid ring must have either a Clar sextet circle, or 0, 1, or 2 C=C double bonds.

In Fig. 8.4 we have illustrated on the same 15 smaller benzenoids as in Fig. 8.3 their Clar valence structures, which although being essentially of a qualitative nature show a clear parallelism with the quantitative results of Polansky and Derflinger.

Observe that the characterization of Clar structures given above, which can be viewed as a geometric "definition" of Clar structures, has been proposed by Clar purely on the basis of his intuition and abundant experimental experience, based on electronic absorption spectra (and later also on NMR spectra)! Observe also that

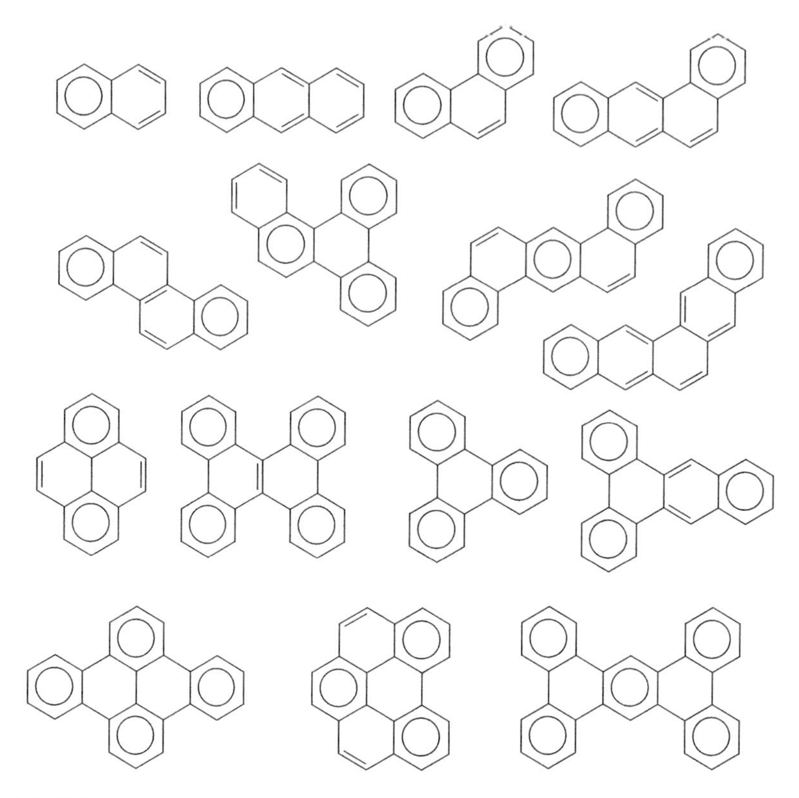

Fig. 8.4 Valence structures according to Clar for several benzenoids

the above "Construction Rule" has no immediately apparent connection either with Kekulé valence structures or with Quantum Chemistry whatsoever, yet there is some "hidden" connection, as we will see later with both Kekulé valence structures and Quantum Chemistry! In addition, let us point out that most of the 15 benzenoids in Fig. 8.4 have a unique Clar structure, with the exceptions (the 1st, 2nd, 4th, 5th, 6th and 12th) being considered by Clar to have "migrating sextets" along an acenic portion.

The paper by Polansky and Derflinger must have awakened certain circles of chemists, who since the early 1970s started to investigate the "local benzenoid" character of rings in various benzenoids. Several authors developed various approaches for the characterization of "local aromaticity" which included proposals of many numerical indices for characterizing local aromaticity (Kruszewski 1971, 1980; Kruszewski and Krygowski 1972; Krygowski and Cyrański 1996, 1998, 2001; Krygowski et al. 1995, 1996; Cyrański et al. 2000; Cyrański and Krygowski 1996, 1998; Schleyer et al. 1996, 2001; Aido and Hosoya 1980; Aihara 1977, 1982, 1987, 2003; Gutman and Bosanac 1977; Herndon and Ellzey 1974; Jiang et al. 1993; Zhu and Jiang 1992; Li and Jiang 1995; Moyano and Paniagua 1991; Poater et al. 2003a, b; Randić 1974, 1975, 1980; El-Basil and Randić 1987; Sakurai et al. 1986;

Moran et al. 2003; Patchovskii and Thiel 2000; Suresh and Gadre 1999; Tarko and Filip 2003; Matta and Hernández-Trujillo 2003; Oonishi et al. 1992; Mandado 2006). For convenience, these different approaches to local aromaticity can be classified as:

(1) Experimental criteria
(2) Classical Modeling criteria
(3) Quantum Chemical criteria
(4) Graph Theoretical criteria

There is substantial overlap between some of these approaches: the experimental criteria such as bond lengths determined for instance by X-ray diffracto metry have served in several classical modeling criteria. Krygowski's harmonic oscillator model and Schleyer's nucleus-independent chemical shifts, based on classical physics principles, can be considered to belong to class (2). A whole recent issue of *Chemical Reviews* debated problems connected with aromaticity, including local aromaticity (Schleyer 2001).

Recently, Randić advanced an approach that strictly speaking and surprisingly does not belong to any of the four classes listed above. But what is even more surprising is that this (rather simple and one could say elegant) approach could have been "discovered" long ago – but was not! Its only ingredients are the Kekulé valence structures, which have been around for quite some time and have been intensively used in the early days of Valence Bond Theory. What this approach implies is a full partitioning of all n π-electrons which are associated with the set of C=C double bonds in individual benzene rings of each individual Kekulé valence structure. The approach is general and applies equally to non benzenoid polycyclic conjugated hydrocarbons. The partition is based on a "natural" convention that in a Kekulé structure of a polycyclic conjugated hydrocarbon, a shared C=C double bond contributes one π-electron to that ring's "share" or "part" (the π-electron content of each ring) whereas a non-shared C=C double bond contributes with two π-electrons (Balaban and Randić 2004a, b; Randić 2003, 2004; Randić and Balaban 2004). By assuming equal weights for all Kekulé structures and by averaging the number of π-electrons assigned to each ring, one obtains a partition of all π-electrons to the rings of the polycyclic conjugated hydrocarbon. We may classify this approach as the last additional class:

(5) Structural criteria

We present this structural approach in Fig. 8.5 for the four Kekulé valence structures of anthracene. As we can see from Fig. 8.5, the electron count for a terminal benzene ring in this acene ranges from four to six π-electrons, the total being 19, and the average 19/4 or 4.75. The electron count for the central ring of anthracene is either four or five π-electrons, the total being 18, and the average is 18/4 or 4.50. When all this information is combined we obtain the numerical (or algebraic) Kekulé structure of anthracene. Observe that now for the first time benzenoid systems (like the anthracene considered here) are represented by a **single** (numerical or algebraic) valence structure! It will be discussed

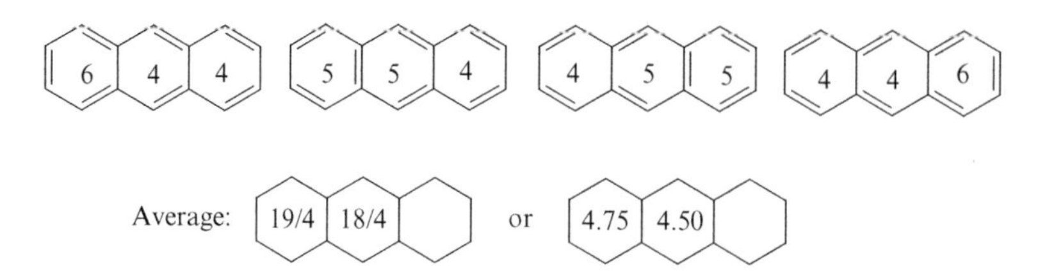

Fig. 8.5 *Upper row*: the four Kekulé (geometric) structures for anthracene with their numerical (algebraic) counterparts inscribed in each ring. Each of the four geometric Kekulé structures has a unique partition of π-electrons or "algebraic Kekulé structure": 6,4,4; 5,5,4; 4,5,5; and 4,4,6. *Lower row*: the EC values for the marginal and central rings of anthracene

below that in many cases there is a one-to-one correspondence between individual geometric Kekulé structures and numerical (algebraic) Kekulé structures of polycyclic conjugated hydrocarbons. The averaged numerical (algebraic) value of the valence (Kekulé) structures will be denoted by EC, the π-electron content (averaged) of the polycyclic conjugated hydrocarbon.

For smaller benzenoid systems it is not difficult to examine all Kekulé valence structures, as we did here for anthracene. Of course, this is a relative view about what a *small benzenoid* is. This is how Linus Pauling described the difference between smaller and larger benzenoids (Pauling 1980):

A few minutes suffice to draw the four unexcited structures for anthracene, the five for phenanthrene and the six for pyrene ... an hour or two might be needed for the 110 structures of tetrabenzoheptacene.

Clearly everyone can spend a few minutes of his/her time to draw various Kekulé valence structures of smaller benzenoids, but few will have the patience to draw hundreds of structures – particularly because today's computers can supply such structures if needed. So an alternative and more efficient route to examine the π-electron content of larger benzenoids is needed, and is possible. This alternative procedure for finding the values for the partition of π-electrons in individual rings makes use of Pauling bond orders, which can be computed without much difficulty even for larger benzenoid hydrocarbons (see the next paragraph).

In a series of papers we have investigated and developed further this new concept for partitioning π-electrons to individual benzenoid rings (Balaban and Randić 2004a, b, c, 2005, 2007, 2008a, b; Randić 2003, 2004; Randić and Balaban 2004, 2006, 2008). Also, the partition of π-electrons in rings of polycyclic benzenoids was discussed in joint papers with Gutman and several coworkers (Gutman et al. 2004). We may add here that a fast way to obtain Pauling bond orders for rather large benzenoid systems can be obtained using the HMO approach and the connection between the calculation of the Coulson bond orders and the Pauling bond orders, unexpectedly discovered and described by Platt (1953) and by Ham and Ruedenberg (1958). By using a similar procedure extended to Coulson bond orders, Gutman and coworkers developed a parallel approach.

The effect of nitrogen heteroatoms on partitions was studied in collaboration with Gutman and his coworkers (Balaban et al. 2007, 2008a; Gutman 2006): in catafusenes, replacing a CH group in an α-position (i. e. next to a shared bond) by a nitrogen heteroatom reduces the EC value of this ring, whereas a similar replacement in a β-position (i. e. next to α) raises the EC value.

Gutman had defined the "energy content" of rings in conjugated polycyclic hydrocarbons (Gutman 2005; Gutman et al. 2004a, b, 2005b, 2007). Further collaborations involved the "energy effect" (*ef*) caused by linear, angular, or geminal benzo-annelation of a marginal benzenoid ring (i. e. changes in the EC or the *ef* value in the ring next to this marginal ring on going from naphthalene to anthracene, phenanthrene, or triphenylene, respectively): quantitative rules can be formulated in terms of the numbers of the linearly, angularly, or geminally annelated rings (Balaban et al. 2009, 2010c, 2011; Đurđević et al. 2009; Gutman et al. 2009). If this ring next to the marginal ring is five-membered (as in acenaphthylene or fluoranthene) or six-membered (as the central ring in anthracene), then linear benzo-annelation decreases the EC or *ef* values, whereas linear or geminal benzo-annelation increases these values. For annelations involving fusing the four-membered ring of benzocyclobutadiene (BCBD-annelation), the opposite effect is obtained.

An alternative energy content of rings in benzenoids and non-benzenoids) can be obtained by partitioning the expression for molecular RE (resonance energy) expressed in terms of R_1, R_2, R_3, conjugated circuits (CCs) to individual rings. For example in the four resonance structures of anthracene, a marginal ring participates in two 6-membered CCs (R_1), one 10-membered CC (R_2), and one 14-membered CC (R_3), whereas the central ring participates in $2R_1$ and $2R_2$. After substitution of values for parameters R_1, R_2, R_3 anthracene RE $= (6R_1 + 4 R_2 + 2 R_3)/6$ one obtains numerical "energy contents" which are at the level of Pariser-Parr-Pople method rather than the HMO level.

A further extension involves looking at atom-based electron content, denoted by *ec* (adding to the two types of partitionings defining EC a third case of carbon atoms common to three rings, when only a third of an electron is assigned to such an atom). This involves considering the effect of nearest-neighboring benzenoid rings: from zero to six such rings there are 13 possibilities, leading to ten situations of a benzene ring and to the following set of ten *ec* values: 6.00, 5.00, 4.33, 4.00, 3.67, 3.33, 3.00, 2.67, 2.33, and 2.00. If the corresponding *ec* is subtracted from EC, one obtains the "electron excess" which can be positive or negative (Gutman 2004, 2005, 2009; Gutman and Bosanac 1977; Gutman et al. 1993, 2007).

In most figures in this article we shall inscribe the π-electron content (EC) in each ring, but we leave symmetry-equivalent rings without EC values, as in Fig. 8.3. In Figs. 8.6–8.8 one can see several examples that will be discussed in detail in the following sections of this review. Since for simplicity the numbering of compounds in each figure starts with boldface **1**, we will use the notation **X**/Y to label compound **X** in Fig. Y.

A final average of EC values over all rings then provides a *global aromaticity index* per benzenoid ring. We recall that all catafusenes have the molecular formula $C_{4R+2}H_{2R+6}$ and that each of the $n = 4R + 2$ carbon atoms contributes with one

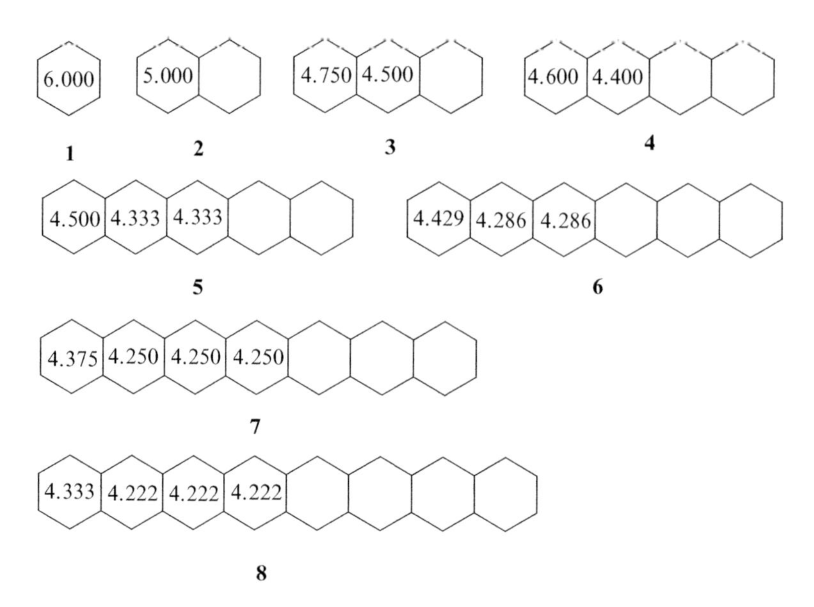

Fig. 8.6 Acenes and their EC values

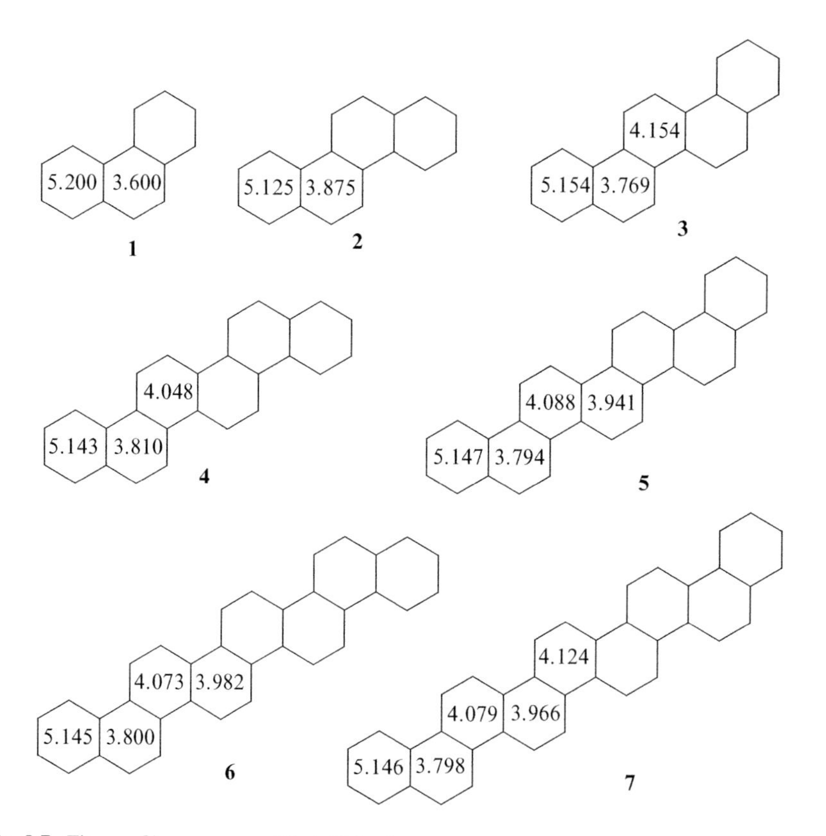

Fig. 8.7 Zig-zag fibonacenes and their EC values

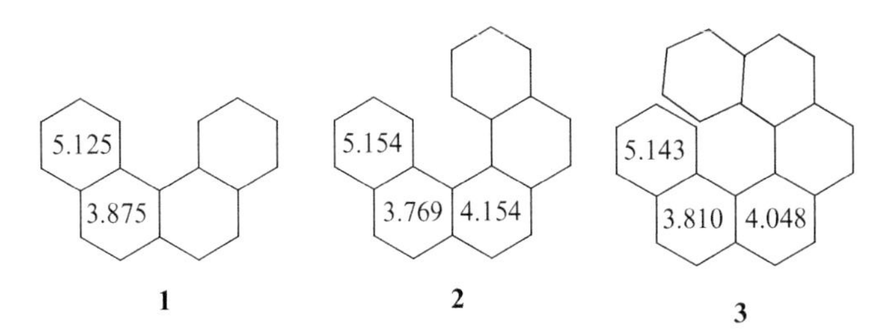

Fig. 8.8 Helicenes with 4, 5, and 6 benzenoid rings and their EC values

π-electron to the molecular orbitals. Thus, benzene has a *global index* of six π-electrons and naphthalene has five π-electrons per ring, in agreement with the following crude formula, which will be discussed at the end of the paragraph (*i*) about acenes, namely EC = n/R = 4 + 2/R.

8.4 Cata-Condensed Benzenoids (Benzenoid Catafusenes)

We present separately two classes and several subclasses of benzenoid catafusenes, although they were examined together in the first part of the series entitled "Partitioning of π-electrons in rings of polycyclic conjugated hydrocarbons. Part 1. Catacondensed benzenoids" (Randić and Balaban 2004). The class of non-branched catafusenes includes four subclasses: (i) *Acenes*, whose dualist graphs form a straight line; (ii) *Fibonacenes*, in which every internal ring is a kink ring and which therefore have no "anthracenic subgraphs," hence their code has no 0 digit. The dualist graphs of the family phenanthrene, chrysene, picene, fulminene, etc. are represented as zigzag lines with code 1212121.... Helicenes on the other hand have dualist graphs with code 11111...; and in general a catacondensed non-branched benzenoid described by any combination of digits 1 and 2 represents a fibonacene; all fibonacenes with the same number R of rings are isoarithmic; (iii) *Phenes*, which we will define below; (iv) *Other* non-branched catafusenes with less or no symmetry; We also have the class of branched catafusenes with the subclass of *star-phenes* (v) and *other* branched catafusenes (vi), seen in Fig. 8.1.

In Table 8.1 we illustrate the coding of catafusenes for benzenoids with five and six condensed rings.

(i) *Acenes*. In Fig. 8.6 we have illustrated smaller members of this family, which actually terminates with R = 7, heptacene being till now the longest linearly fused benzenoid system known. It is thermally and photochemically unstable and reacts with oxygen having a diradicalic character. However, hexaphenyl-heptacene and derivatives are stable in the absence of oxygen (Chun et al. 2008; Kaur et al. 2009). A persistent nonacene substituted with ten *para-tert*-butylphenyl-thio groups was recently described (Kaur et al. 2010). Pentacene derivatives are organic semiconductors and have found numerous applications

Table 8.1 Codes [and formula **numbers** in square brackets, when available] of isoarithmic (iso-Kekulean) catacondensed systems with five and six benzenoid rings

Code	Isoarithmic benzenoids
Non-branched benzenoids	
Catafusenes with five condensed rings	
000 [**4**/6]	1
001 [**6**/11]	1
010 [**1**/9]	1
011 [**8**/11], 012 [**7**/11]	2
101 [**2**/11], 102 [**3**/11]	2
111 [**2**/8], 112 [**4**/11], 121 [**3**/7]	3
Total	*10*
Catafusenes with six condensed rings	
0000 [**6**/6]	1
0001	1
0010	1
0011, 0012	2
0101, 0102	2
0110, 0120	2
0111, 0112, 0121, 0122	4
1001, 1002	2
1011, 1012, 1021, 1022	4
1111 [**3**/8], 1112, 1121, 1122, 1212 [**4**/7], 1221	6
Total	*25*
Branched benzenoids	
Catafusenes with five condensed rings	
01(.) [**2**/12]	1
11(.)	1
Total	*2*
Catafusenes with six condensed rings	
001(.)	1
011(.)	1
01(.)0	1
01(.)1, 01(.)2	2
101(.)	1
111(.), 121(.)	2
11(.)1, 11(.)2, 12(.)1	3
1(.)1(.)	1
Total	*12*

as organic field-effect transistors, organic light-emitting diodes and organic photovoltaics.

For acenes, $K = \mathrm{R} + 1$, but this formula does not give a hint about the instability of long acenes. If we write the expressions for the resonance energy, RE, of acenes in terms of the contributions of the smallest conjugated circuits R_1, R_2, and R_3 with 3, 5, and 7 double bonds, respectively, we obtain for naphthalene, anthracene, tetracene, pentacene, . . .:

Benzene	$(2\ R_1)/2$
Naphthalene	$(4\ R_1 + 2\ R_2)/3$
Anthracene	$(6\ R_1 + 4\ R_2 + 2\ R_3)/4$
Tetracene	$(8\ R_1 + 6\ R_2 + 4\ R_3)/5$
Pentacene	$(10\ R_1 + 8\ R_2 + 6\ R_3)/6...$

which upon substituting for $R_1 = 0.815$ eV, $R_2 = 0.302$ eV and $R_3 = 0.085$ eV affords 1.323, 1.600, 1.766, and 1.876, respectively. This is a rather slow increase of RE with the molecular size. By the time we reach the systems with eight and nine linearly fused rings we have for RE: $(16\ R_1 + 14\ R_2 + 12\ R_3)/9$ and $(18\ R_1 + 16\ R_2 + 14\ R_3)/10$ giving numerical values 2.033 and 2.069 eV, which is about half of the typical RE for benzenoid hydrocarbons having eight and nine fused benzene rings.

Hess and Schaad (1980) have argued convincingly that a better measure for the relative stability of benzenoid hydrocarbons would be REPE (that is, the resonance energy per π-electron, which is more nearly a size-independent quantity) than RE (which typically increases with molecular size). The REPE values for naphthalene, anthracene, tetracene, and pentacene are: 0.132, 0.114, 0.098 and 0.085 eV respectively. As we see, with increasing acene size the REPE values steadily decrease. For comparison, the REPE value for benzene is 0.136 eV, which interestingly and significantly is not the largest REPE value. The largest REPE values are found in benzenoid hydrocarbons known as "fully benzenoid hydrocarbons" (Clar 1972), all-benzenoid hydrocarbons (Gutman and Cyvin 1989), "sextet-resonant benzenoids" (Dias 1987; Balaban and Schmalz 2006), or "claromatics" (Balaban and Klein 2009) which are compounds in which all benzene rings are either represented by aromatic π-sextets, or are the so-called "empty" rings, in the nomenclature of Clar. Rings of such fully aromatic benzenoids in Clar formulas are either empty or with a sextet circle, and then no ring has one or two C=C double bonds. The REPE values for eight-ring and nine-ring linearly fused benzenoid hydrocarbons are 0.060 eV and 0.054 V, which apparently set the bounds for stable benzenoid hydrocarbons.

As we can see from Fig. 8.6, the π-electron ring partition discriminates only two types of rings in acenes: terminal and non-terminal rings. Terminal rings have a π-electron content EC = (n + 5)/K or (4R + 7)/(R + 1), while all non-terminal rings have EC = (n + 4)/K or (4R + 6)/(R + 1).

In Fig. 8.6 the EC values of anthracene, tetracene, pentacene, and hexacene (**3**/6 – **6**/6) are presented. The global aromaticity indices of anthracene, tetracene, and pentacene are 4.67, 4.50, and 4.40 π-electrons per ring, respectively, decreasing asymptotically towards four π-electrons per ring. This asymptotic value is actually common to all catafusenes when the number R of rings increases to infinity. The same value results by remembering that all catafusenes with R rings have n = 4R + 2 carbon atoms and n = 4R + 2 π-electrons, so that per ring one has EC = n/R = 4 + 2/R π-electrons; which tends towards 4 when R increases indefinitely. By the same token, for perifusenes that are increasing indefinitely towards graphene, the asymptotic EC value is 3.

(ii) *Fibonacenes*. The name fibonacene apparently appeared for the first time in the mathematical literature (Anderson 1986; Balaban 1989) to embrace families of benzenoids for which the numbers of Kekulé structures obey the relationship $K_{n+1} = K_n + K_{n-1}$, where the subscripts $n + 1$, n, and $n - 1$ relate to successive members of the family, where this relationship represents the recursive formula for Fibonacci numbers. The most common fibonacene family is one involving benzene, naphthalene, phenanthrene, chrysene, picene, fulminene, etc., whose K values are 2, 3, 5, 8, 13, 21, etc., and whose dualists form a series of zigzag lines. It was shown earlier that all fibonacenes with the same number R of rings that differ only by any permutation of digits 1 and 2 have the same set of EC values, i. e. are isoarithmic, because their codes differ only by interchanging digits 1 and 2, without changing their position. The smallest isoarithmic benzenoids are chrysene or benzo[*a*]phenanthrene (**2**/7) coded as 12, and benzo[*c*]phenanthrene or [4]helicene (**1**/8) coded as 11, whose dualist graphs correspond to hydrogen-depleted *transoid* and *cisoid* forms of butadiene, respectively, in which there is a 1:1 correspondence between individual Kekulé valence structures that have the same distributions of CC single and CC double bonds in their common substructure – the phenanthrene fragment. In Fig. 8.7 we present the EC values for phenanthrene, chrysene, picene, and fulminene (**1**/7 – **4**/7). Exactly the same EC values are observed for 4-, 5-, and 6-helicene (**1**/8 – **3**/8) and other related fibonacenes of Fig. 8.8.

It is interesting to compare isomeric catafusenic benzenoids such as anthracene and phenanthrene: their 14 π-electrons are partitioned differently (4.75 + 4.50 + 4.75, and 5.2 + 3.6 + 5.2, respectively) and this difference accounts for their different properties. Of course, the propensity of anthracene to react with dienophiles and the unsaturated behavior of carbons 9 and 10 in both these catafusenes has subtler explanations.

In a joint paper of the present authors with Gutman and Kiss-Tóth (Gutman et al. 2005a), it was proved that in fibonacenes the EC values of rings numbered 1, 3, 5,. . . from the marginal rings towards the center decrease steadily, whereas the EC values of rings numbered 2,4,6,. . . increase steadily; the asymptotic values are 5.146 for EC(1), 3.798 for EC(2), and 4.000 π-electrons for the central ring(s). The global aromaticity index for phenanthrene, chrysene, picene, and fulminene is exactly the same as for their acenic isomers (4 + 2/R), i. e., 4.67, 4.50, 4.40, and 4.33, respectively, also decreasing asymptotically towards four π-electrons per ring. If one is referring to the above-mentioned *global aromaticity index*, instead of global π-electron ring density index, one is assuming a strong correlation between the local π-electron density and the local aromaticity, but whether these two values parallel one another and to what degree may still be an open question.

The high chemical reactivity of higher acenes towards oxygen or dienophiles contrasts strongly with the stability of fibonacenes, which are stable no matter how many rings they have. This difference in behavior is in good agreement with Clar's theory: acenes can have only one Clar sextet ring which can migrate along the linear chain, but a fibonacene with R rings has R/2 or (R + 1)/2 Clar sextet rings.

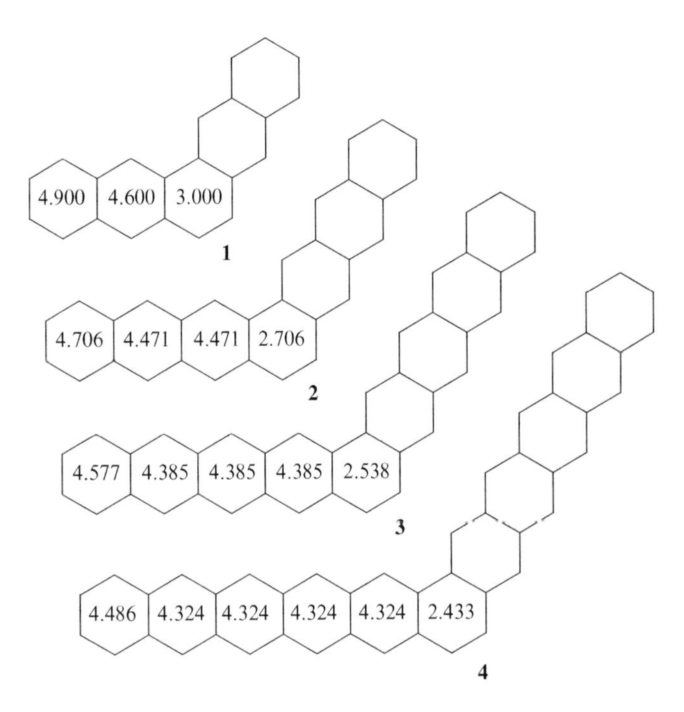

Fig. 8.9 Phenes and their EC values

Chemical stability towards heat or oxidation in air correlates with the number of Clar sextet rings. On the other hand, the opto-electronic properties of medium-sized acenes such as pentacene have led to many more applications than any fibonacene.

(iii) *Phenes*. These cata-condensed benzenoids shown in Fig. 8.9 have two equally long linear branches of fused benzene rings at an angle of 120°. The first member of this family is phenanthrene and the next member is *pentaphene* (**1/9**). The benzenoid rings in linear branches belonging to the "migrating" sextets have intermediate values for EC, while the ring at the kink has the lowest EC value. In Fig. 8.9 we show EC values for the first members of this series and in Fig. 8.10 we illustrate an alternative way of obtaining EC values on pentaphene, which has $K = 10$ Kekulé structures. The approach is based on examination of all possible assignments of C=C double bonds in different rings. The number of occurrences of each ring type is given by the italic underlined number of Kekulé structures of the fragment whose double bonds remain unassigned. It will be observed that on summing up the italic underlined numbers for each ring type, one obtains $K = 10$, and that the ratio between the final sums and this K value affords the EC values for each type of ring in pentaphene.

There are only three EC values (in decreasing order) for any member of the phene class: the terminal, intermediate, and central rings; all EC values

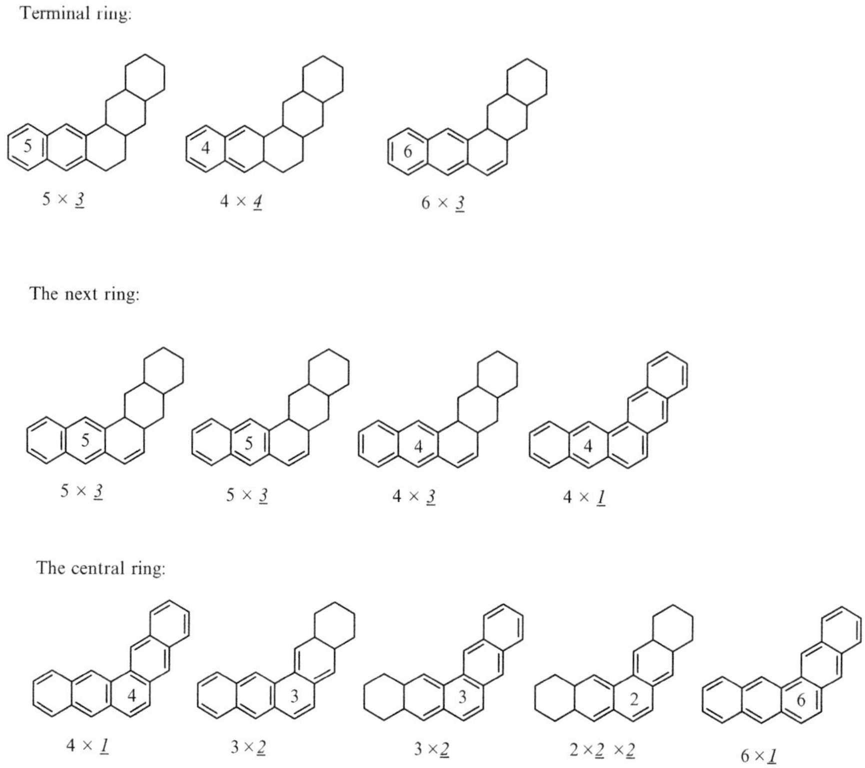

Terminal ring:

5 × 3 4 × 4 6 × 3

The next ring:

5 × 3 5 × 3 4 × 3 4 × 1

The central ring:

4 × 1 3 × 2 3 × 2 2 × 2 × 2 6 × 1

Fig. 8.10 An alternative procedure for obtaining EC values for rings in pentaphene

decrease as *n* increases, because their sum total must amount to the number *n* of π-electrons. However, with increasing *n*, the difference between EC values of terminal and intermediate rings decreases, but the difference between EC values of intermediate and central rings increases.

(iv) *Other non-branched catafusenes,* with less or no symmetry. The smallest member in this subset is benz[*a*]anthracene (**1**/11) with four rings, all symmetry-non-equivalent. There are several members of this class having five rings, almost all of them shown in Fig. 8.11. It will be observed that in addition to the non-symmetrical benzenoids benz[*a*]anthracene (**1**/11, coded as 01) and benz[*a*]tetracene (**6**/11, code 001), there are three pairs of isoarithmic (or iso-Kekuléan) benzenoids in this class: the pair dibenzo[*a,j*]anthracene (**2**/11, code 101) and dibenzo[*a,h*]anthracene (**3**/11, code 102), the pair benzo[*b*]chrysene (**7**/11, code 012), and naphtho[1,2-*a*]anthracene (**8**/11, code 011), and the pair of fibonacenes benzo[c]chrysene (**4**/11, code 112) and picene (**5**/11, code 121). The last structures, benzo[*c*]chrysene and picene, are isoarithmic to other fibonacenes with five benzenoid rings such as [5]helicenes (code 111), because all their rings except the two terminal rings are "kink" rings.

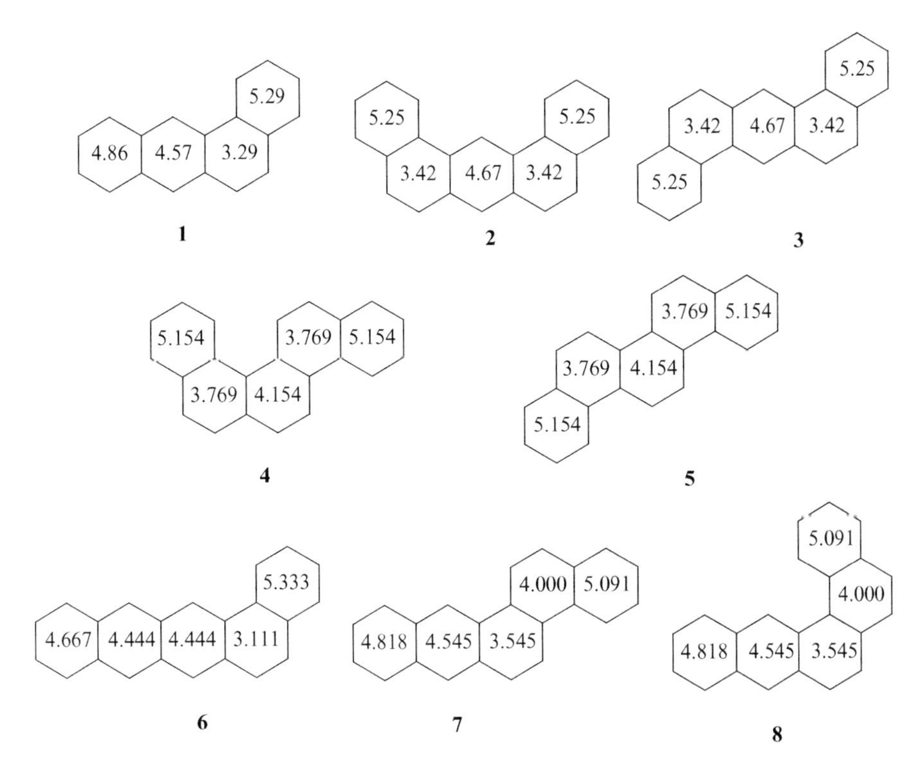

Fig. 8.11 Non-branched catafusenes having low symmetry, with their EC values

(v) *Branched catafusenes*. The smallest branched catacondensed benzenoid is triphenylene (**1**/12) whose Clar formula has π-electron sextet circles in the three terminal rings, leaving the central, the so-called "empty" ring quite depleted in π-electrons. In Fig. 8.12 we have depicted several additional branched benzenoids, including dibenz[*a,c*]anthracene (**2**/12) and dibenzo[*a,c*]phenanthrene (**5**/12). It appears that as the lengths of the acenic branches increase, the EC value of the dibenzo rings increases and the EC value of the ring adjacent to them decreases. In dibenz[*a,c*]acenes **1**/12 to **4**/12, tribenzacenes **5**/12 to **7**/12, and tetrabenzacenes **8**/12 to **10**/12 (corresponding to the three rows of Fig. 8.12) the effect of the longer acenic parts is to squeeze more efficiently the π-electrons from the "marginal acenic' ring into the adjacent benzenoid rings.

In the lower part of Fig. 8.12 we included several smaller benzenoid hydrocarbons having two branching centers, the smallest being tetrabenzo[*a, c,f,h*]naphthalene or dibenzo[*g,p*]chrysene (**8**/12). In all cases the terminal rings have the highest EC values, the rings associated with the "migrating" π-sextets have intermediate EC values, the rings with a single fixed C=C double bond have still smaller values, and finally the marginal rings of the acenic portion adjacent to three other rings have the smallest EC values.

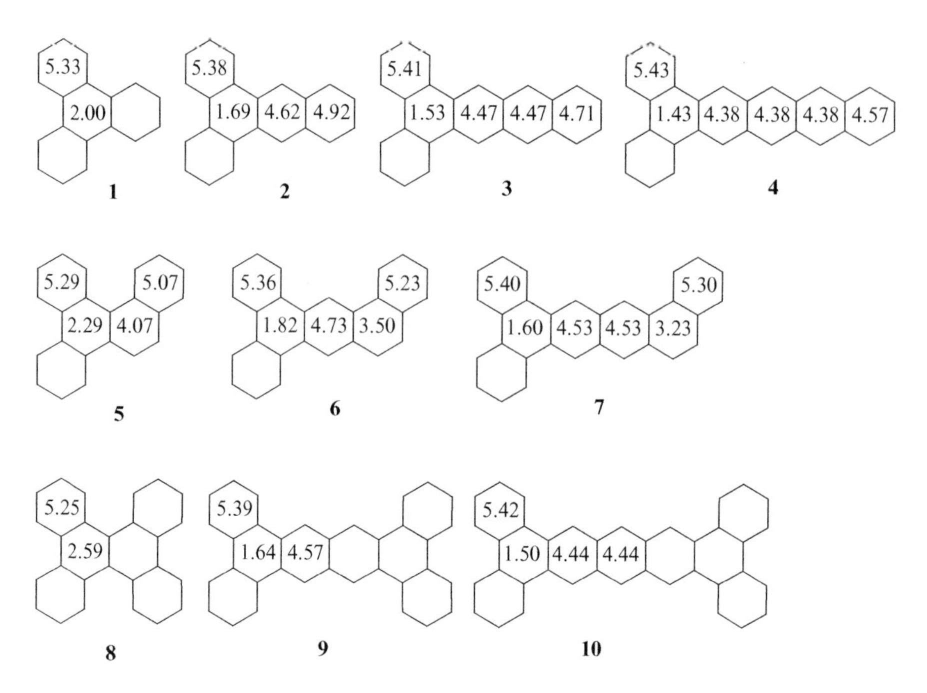

Fig. 8.12 Branched catafusenes and their EC values

(vi) *Star-phenes.* These benzenoids represent a particular case of branched catafusenes having a single branching point and three branches. They may have either three acenic branches of equal length (**1**/13 to **3**/13) or two equally long branches and a shorter branch. The latter type was analyzed in detail in Part 1 of the above-mentioned series. We have included in Fig. 8.13 other catafusenes with three symmetrical branches (**4**/13, **5**/13); the last benzenoid can be considered to be a 'second-generation dendrimeric catafusene'. In all branched structures in Figs. 8.12 and 8.13 the benzenoid ring at the branching point(s) has a low EC value, which agrees with previously made observations concerning benzenoid rings at "kink" points.

8.5 Corona-Condensed Benzenoids

The best known and most investigated among the corona-condensed benzenoids is kekulene, the first structure in Fig. 8.14, which was synthesized in 1983 (Staab and Diederich 1983). Aihara showed that it is not superaromatic (Aihara 1992, 2008). It has only two distinct types of rings, the "kink" rings at the corner of the hexagon of hexagons, and the central rings in the "sides" of the hexagon of hexagons, which correspond to Clar sextets. The EC value for the "kink" hexagon is 3.3, which is close to the value found for the "kink" hexagon in phenanthrene (**1**/7),

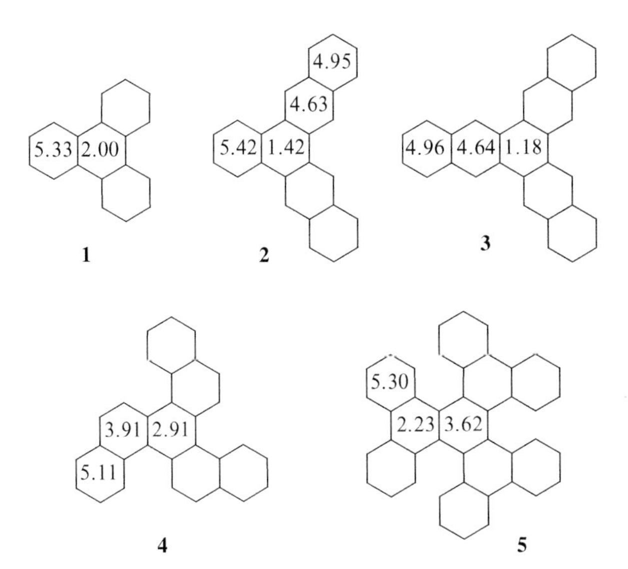

Fig. 8.13 Star-phenes and related branched catafusenes with their EC values

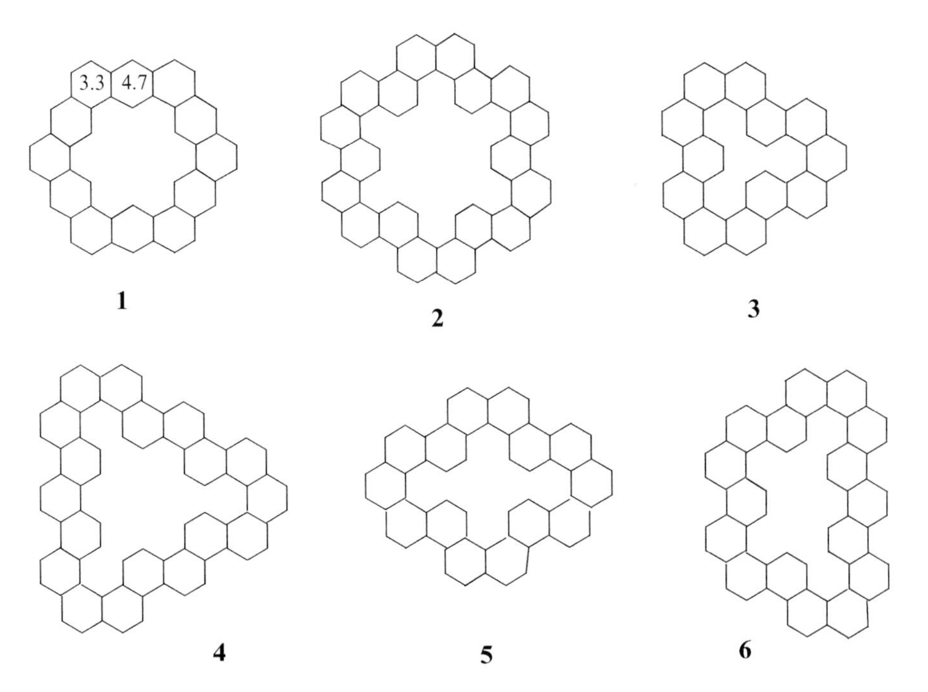

Fig. 8.14 Benzenoid coronoids: kekulene and its EC values, followed by fibonacenic coronoids with EC = 4 for all their rings

benzanthracene (**1**/11) and dibenz[*a,c*]anthracene (**2**/12), which have one fixed C=C double bond.

The remaining five corona-condensed benzenoids of Fig. 8.14 represent a special subclass of corona-condensed compounds in that they have every benzene ring on their periphery being a "kink" ring, and because of this we will refer to them as fibonacenic coronoids. Although till now only a few benzenoid coronoids have been experimentally obtained, the second part of the above-mentioned series entitled "Partitioning of π-electrons in rings of polycyclic benzenoid hydrocarbons. Part 2 Catacondensed coronoids" (Balaban and Randić 2004a) analyzed in detail such systems. A remarkable feature of fibonacenic coronoids is the fact that such systems are among the few benzenoids with more than two benzene rings where all rings have the same EC value, which in this case is four π-electrons per ring.

8.6 Peri-Condensed Benzenoids

Peri-condensed benzenoid hydrocarbons or perifusenes can have one or more "inner" carbon atoms common to three rings, that is, their dualist graph has three-membered cycles. As a well-known rule, Kekuléan alternant systems must have even numbers of inner carbon atoms, whereas Kekuléan non-alternants may have either even or odd numbers of inner carbon atoms. In "Part 3. Perifusenes" (Balaban and Randić 2004c) all possible benzenoid perifusenes having four, five and six benzenoid rings were displayed and examined. In Fig. 8.15 we show some of them.

The smallest and simplest perifusenes are pyrene and perylene (compounds **1**/15 and (**2**/15), respectively), which may appear similar but structurally speaking are quite different. From the viewpoint of Clar's π-sextet model, both molecules have two π-sextets, but in pyrene the sextets are in fixed locations while in perylene we have "migrating" sextets. The central ring in perylene has "essentially" single CC bonds (that is, the two CC bonds connecting the two naphthalene fragments are single bonds in all Kekulé valence structures). Despite these differences we find coincidentally equal EC values of the remotest ring from the center in both systems. However, since the bonds connecting the two naphthalenic moieties in perylene are essentially single bonds, the central ring is quite poor in π-electrons. The same is true for coronene (**7**/15). On fusing an extra benzenoid ring to one bond of perylene, the naphthalenic rings on the opposite side keep their initial EC values (4.667) because the inter-naphthalenic bonds in perylene are essentially single bonds (compounds **11**/15 and **12**/15). This value when compared to EC of naphthalene (5.000) is somewhat smaller, because now two CC bonds are shared with the central ring. The situation is somewhat similar with the "anthracene" and "phenanthrene" fragments in **11**/15 and **12**/15, the rings adjacent to the central ring have smaller values than the corresponding ring values in anthracene and phenanthrene, but the added benzene ring has larger values, in one case the EC value 4.75 π-electrons, and in the other case 5.20 π-electrons, respectively. It is interesting to observe that these are precisely the values found for the terminal rings of anthracene and phenanthrene, and this ought not to be surprising because the added

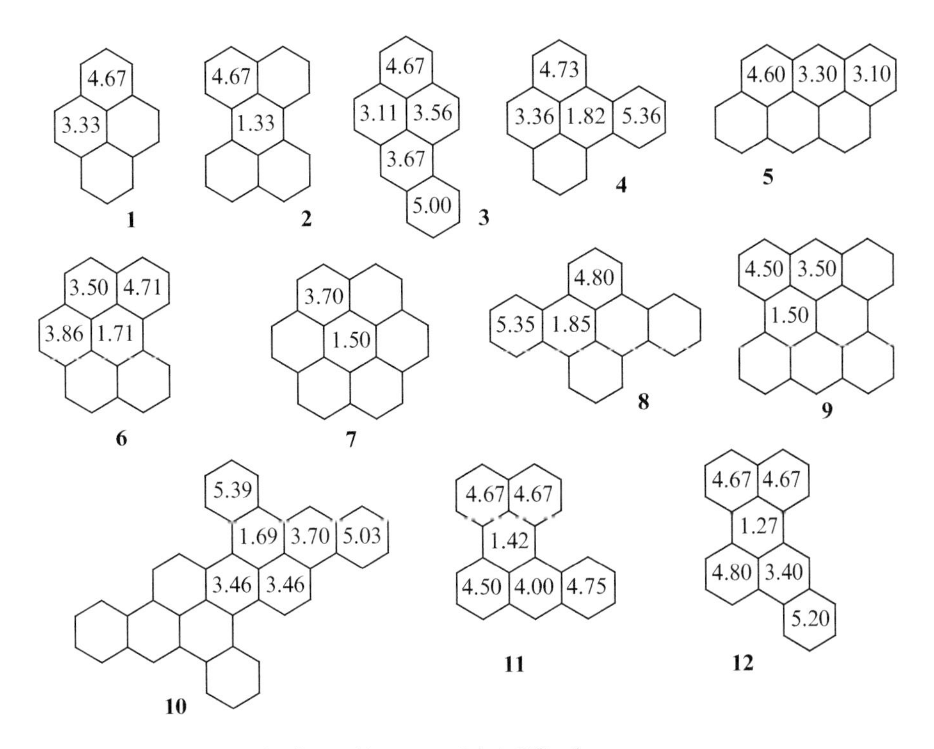

Fig. 8.15 Benzenoid Kekuléan perifusenes and their EC values

ring shares its π-electrons only with the adjacent ring, just as in the case of anthracene and phenanthrene. We will encounter again these same values later on.

Part 3 of the above-mentioned series displayed benzenoid perifusenes having four, five, and six benzenoid rings, and Fig. 8.15 displays some of them. In Fig. 8.15 we include also tribenzo[*a,h,rst*]phenanthra[1,2,10-*cde*]pentaphene (**10**/15) with 11 benzenoid rings (Krygowski et al. 1995; Oonishi et al. 1992) which will be discussed in more detail in another section of this review.

8.7 Benzenoids with More Than One Geometric Kekulé Structure Corresponding to the Same Algebraic Kekulé Structure

In this section we no longer look at EC values, which are averaged over all possible Kekulé structures, but at individual geometric Kekulé structures, and examine the corresponding individual algebraic values. In turns out that for many polycyclic conjugated hydrocarbons there is one-to-one correspondence between geometric and algebraic Kekulé structures. In particular this is always true for catacondensed benzenoids, as has been proven mathematically in a paper by Gutman and coworkers

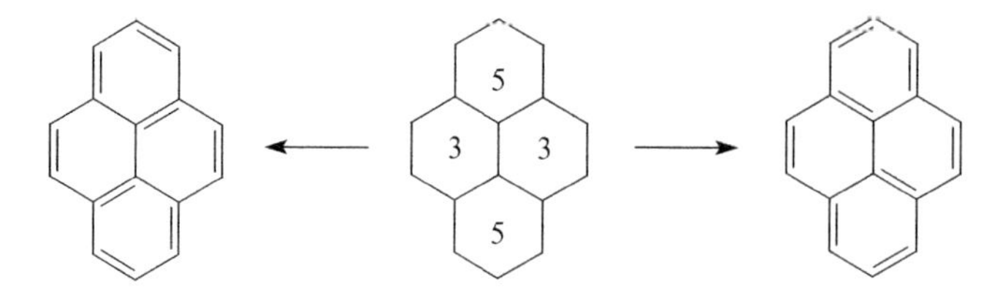

Fig. 8.16 For pyrene, two different geometric Kekulé structures (from the six possible Kekulé structures) correspond to the one and the same algebraic Kekulé structure shown between them

(Gutman et al. 2004). However, there are a few exceptions, and they are mentioned in this section, but for more details readers should consult Part 4 of the above-mentioned series entitled "Partitioning of π-electrons in rings of polycyclic benzenoid hydrocarbons. Part 4. Benzenoids with more than one geometric Kekulé structure corresponding to the same algebraic Kekulé structure" (Vukičević et al. 2004). The most obvious case is benzene itself, which has two (symmetry equivalent) Kekulé valence structures, having necessarily the same π-electron ring partition.

Whereas all benzenoid catafusenes and most benzenoid perifusenes show a one-to-one correspondence between geometric and algebraic Kekulé structures (as shown in Fig. 8.5 for anthracene), some perifusenes are exceptions, the smallest being pyrene. As can be seen from Fig. 8.16, there are two pyrene Kekulé structures (albeit symmetry related) that have identical π-electron ring partitions for all rings. A one-to-one correspondence between the "geometric" and "numerical" (or "algebraic") individual Kekulé valence structures signifies that no loss of information is associated with the numerical Kekulé valence structures, which therefore allow a full reconstruction of Kekulé valence structure from the "algebraic" form. In the case of pyrene, however, there is no longer a one-to-one correspondence between the two types of Kekulé valence structures, but nevertheless there is no loss of information accompanying the "numerical" Kekulé valence structures. That is, in the case of pyrene a single numerical structure yields two solutions.

If one condenses other benzenoid moieties on both sides of pyrene, the resulting systems also show the same kind of degeneracy that we have seen in pyrene, as shown on the three algebraic Kekulé structures of dibenzopyrene in Fig. 8.17. Interestingly, it is not necessary for the moieties attached to the left and right side of pyrene to be identical, as shown in the next lower part of Fig. 8.17. All these types of degeneracy were examined in Part 4 of the above-mentioned series (Vukičević et al. 2004). This observation can be further generalized to the class of benzenoids schematically represented by the diagram at the bottom of Fig. 8.17, which is discussed again for non-alternant systems that present a similar degeneracy.

In Fig. 8.18 at the top we show two special structures among 200 geometric Kekulé structures of kekulene that correspond to one and the same algebraic Kekulé structure. Again these two Kekulé structures are symmetry related, but in this case all rings have the same EC value, even though the distribution of C=C double bonds in different rings is different. In the lower part of Fig. 8.18 we show a fourfold degeneracy of a perifusene combining pyrene and perylene moieties.

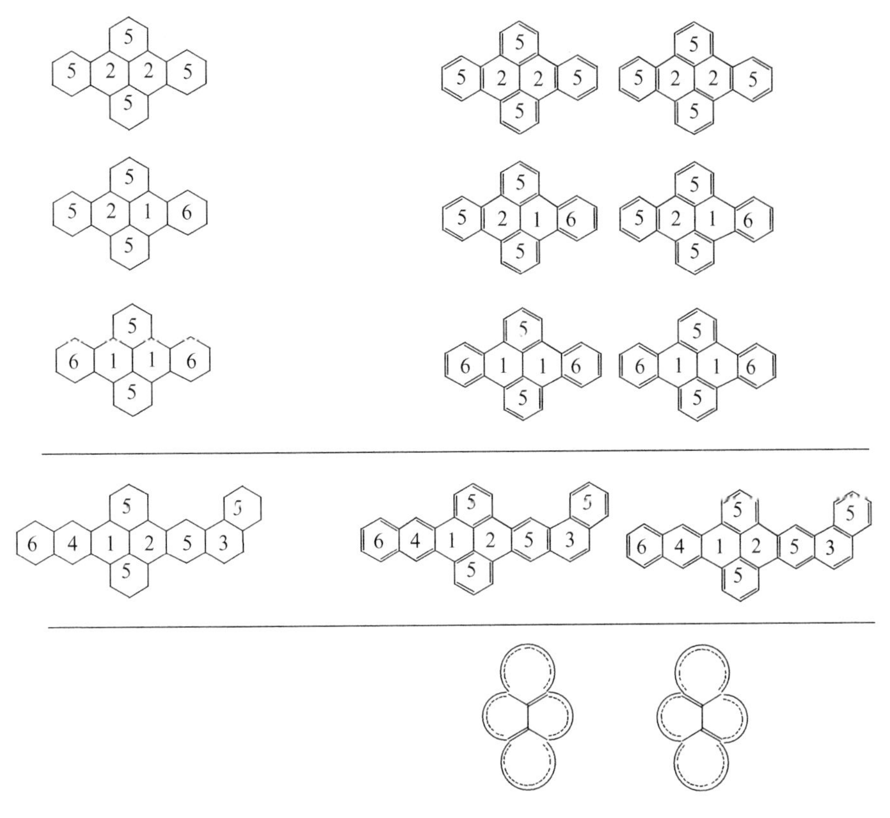

Fig. 8.17 The first three rows display some of the Kekulé structures of one and the same benzenoid perifusene: one and the same algebraic Kekulé structure on the *left* corresponds to the pair of geometric Kekulé structures on the *right*. The fourth row shows that a similar degeneracy occurs even when the moieties condensed to the *left*- and *right*-hand bonds of pyrene are different. The *bottom line* shows the generalized feature of degeneracy that holds even for nonalternants

8.8 Alternant Non-benzenoids

The EC values of alternant non-benzenoids have not been discussed till now. We shall limit the discussion to cata-fused non-benzenoids having two or three condensed rings with ring sizes equal to 4, 6, or 8, and to peri-condensed non-benzenoids with the above ring sizes having at most four rings. Figure 8.19 contains structures of systems that are discussed in this section.

Observe that among the 35 structures of Fig. 8.19 the first five are bicyclic. As we see the systems with two condensed rings have integer EC values. Recall that both rings of naphthalene also have integer EC values. The EC $= 5$ π-electrons in naphthalene is the average of three resonance (Kekulé) structures, two of which are symmetry related having partitions 4:6 and 6:4 and the Fries structure with a C=C double bond in the middle with partition 5:5. It was postulated by Fries that among

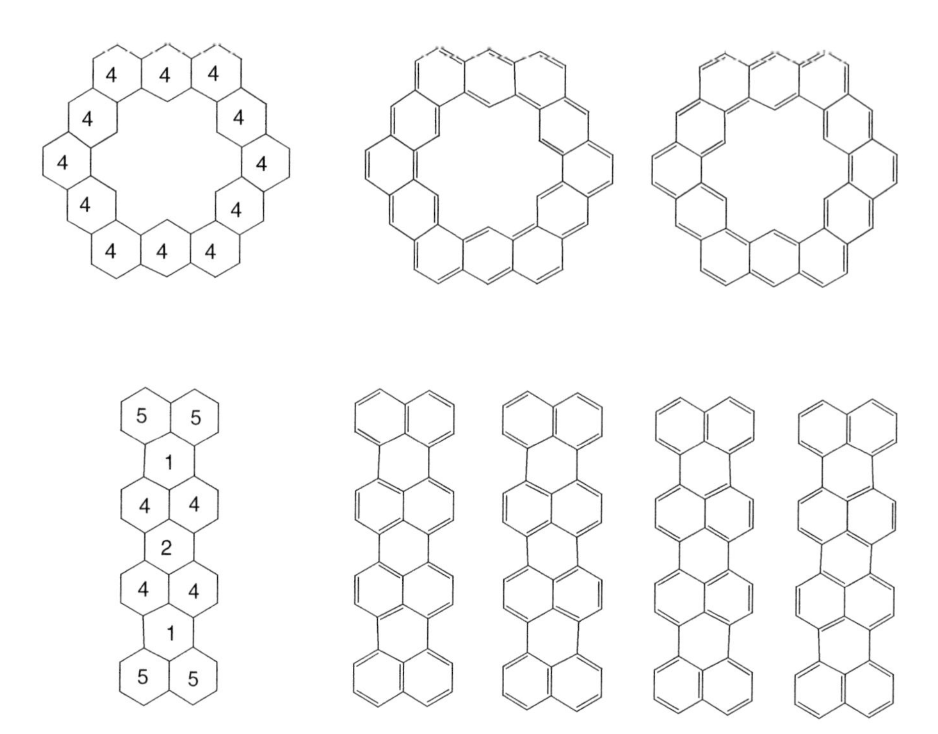

Fig. 8.18 *Top row*: one algebraic Kekulé structure (*left*) corresponds (*right*) to a pair of geometric Kekulé structures of kekulene (a coronafusene) – note that these are not EC counts, and that there are many other geometric Kekulé structures. *Bottom row*: one algebraic Kekulé structure (*left*) corresponds to four geometric Kekulé structures (*right*) for the perifusene with 11 benzenoid rings combining pyrene and perylene moieties (*left*)

all Kekulé resonance structures, the most important one(s) is (are) that (those) with a maximal number of benzenoid rings with three double bonds (Fries 1927). An algorithm for finding such Fries structures is available (Ciesielski et al. 2010). But observe that the average of the first two structures gives 5:5, thus the partition of the Fries resonance structures give also the overall partition for a molecule as a whole. By contrast, all systems with three cata-condensed rings discussed here have non-integer partitions. Interestingly, on fusing a benzenoid ring to one bond of a bicyclic system such as **4**/19 one obtains one of the following two situations: (i) when the extra benzenoid ring is fused to a double bond of the "Fries-type" structure, its EC value is 5.2 π-electrons, the ring to which it is fused loses 1.4 π-electrons, and the last ring gains 0.2 π-electrons; (ii) when the extra benzenoid ring is fused to a single bond of the "Fries-type" Kekulé structure, its EC value is 4.75 π-electrons, the ring to which it is fused loses 0.5 π-electrons, and the last ring loses 0.25 π-electrons. This is a general situation, which was encountered also in the annelation of naphthalene yielding either phenanthrene (**1**/7) – case(i) –, or anthracene (**3**/6) – case (ii); and in the annelation of perylene to form either **12**/15 or **11**/15, corresponding to the above two cases, respectively. Even more

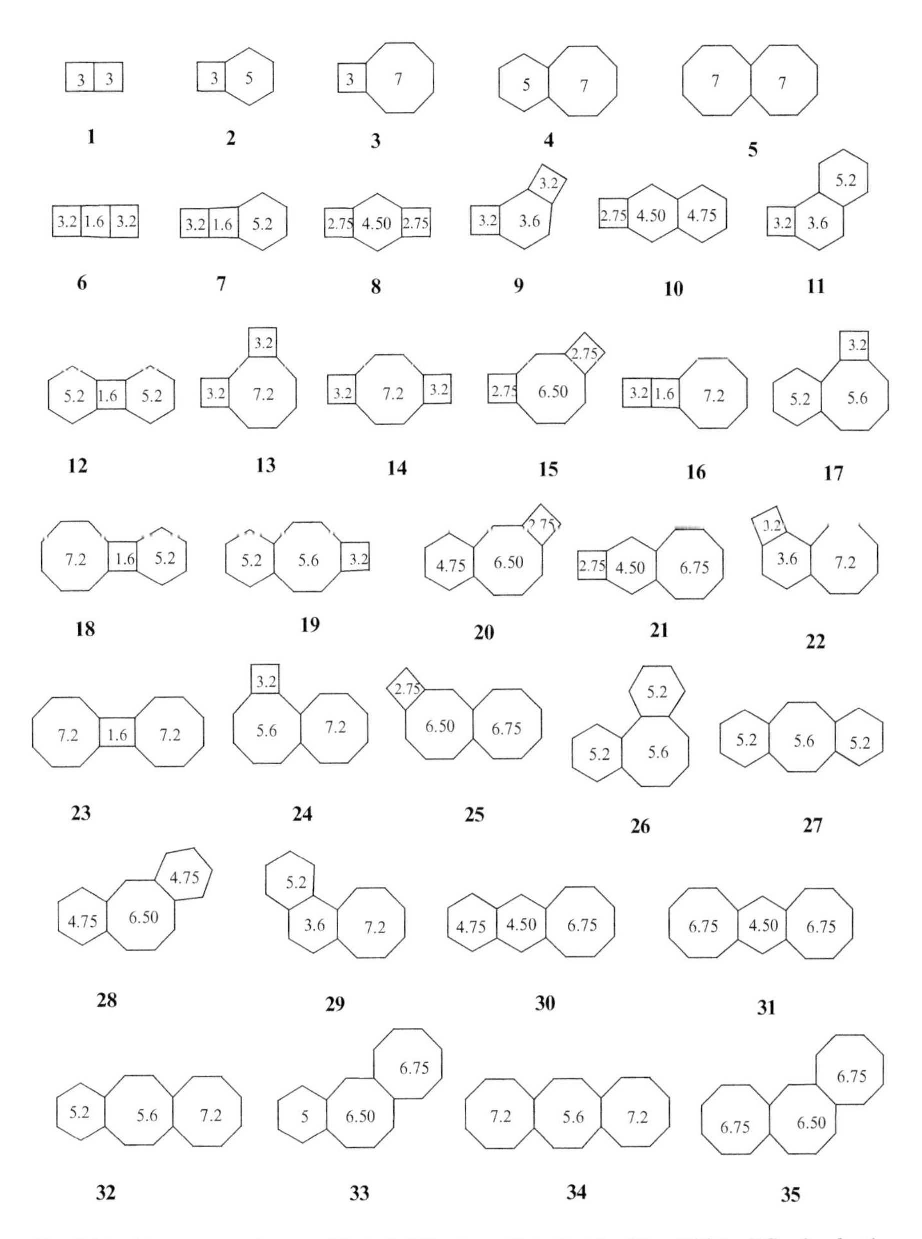

Fig. 8.19 Alternant non-benzenoids their EC values. Note that for **13** and **14** the EC value for the eight-membered ring is 5.6. and not 7.2

generally, on inspecting Fig. 8.19, one may see that the same scenario is valid for annelating a polycyclic system with definite single/double bonds by fusing an even-membered ring (having, instead of 5 as for a six-membered ring, an EC value in the bicyclic system, denoted by V) on one of the bonds of the bicyclic system: the extra ring has after annelation an EC value of either V + 0.2, or V – 0.25, whereas the two other rings have exactly the values discussed in the two cases mentioned above.

8.9 Non-alternant Conjugated Hydrocarbons

Part 5 of the series was entitled "Partitioning of π-electrons in rings of polycyclic conjugated hydrocarbons. 5. Nonalternant compounds" (Balaban and Randić 2004b) and discussed in detail non-alternant systems, providoing a (presumed) complete list of all 189 possible systems with two, three, or four condensed rings having ring sizes equal to 5, 6, or 7. Table 8.2 contains data about all these systems, but we include here formulas only for a small part of these structures (the absent structures are indicated by a long dash in Table 8.2, and/or may be found by difference from the columns "Compound" and "Isomer count"). The first two columns in this Table indicate ring sizes and their sum. Whereas cata-condensed systems have no internal vertices and no odd-numbered sums of ring sizes, peri-condensed systems have either even-membered or odd-numbered sums of ring sizes. In the latter case, the system may be tricyclic with one internal vertex (and three odd-numbered ring sizes or one odd-numbered and two even-numbered ring sizes) or tetracyclic (with one internal vertex, one or three odd-membered rings and the remaining rings with even ring sizes). In the former case (peri-condensed systems with even-numbered sums of ring sizes) there are two internal vertices, and 0, 2, or 4 odd-membered ring sizes.

In the present review we shall not aim at such exhaustive enumerations of all possible systems, but we shall expand the ring sizes to include also four- and eight-membered rings, which were also included in the preceding section. Structures of the systems that will be discussed in this section are presented in Fig. 8.20 (bicyclic and tricyclic cata-condensed systems), Fig. 8.21 (tetracyclic cata-condensed systems) and Fig. 8.22 (peri-condensed systems).

It was noted in Part 5 (Balaban and Randić 2004b) that among the 189 nonalternants (cata- and peri-condensed systems) with two, three, or four condensed rings having ring sizes equal to 5, 6, or 7, more than half had integer EC values. In fact all tricicyclic catacondensed systems considered have integer π-electron partitions. A simple rule accounts for the EC values of Fig. 8.20 that have ring sizes ranging from 4 to 8 sp^2-hybridized carbon atoms: for catacondensed tricyclic systems composed of r, s, and t-membered rings, the EC values are $r - 1$, $s - 2$, and $t - 1$, respectively.

The cata-condensed systems with three rings of sizes 4–8 presented in Fig. 8.20 reveal two interesting aspects: (i) all of them possess integer EC values, namely for terminal rings EC = R – 1, and for rings condensed to two other rings EC = R – 2; Another interesting observation, derived from the preceding one, is that in some cases presented in Fig. 8.20, rings differing in size have the same EC value. It is not difficult

Table 8.2 Nonalternant Kekulénoids with two, three, or four 5-, 6-, and 7-membered rings. A hyphen in the column 'internal vertices' means 'not applicable'. A long line in the column 'compound' indicates that no examples are illustrated in this review

		Catafused nonalternant systems			Perifused nonalternant systems		
Ring size	Sum of size	Internal vertices	Compound	Isomer count	Internal vertices	Compound	Isomer count
5,5	10	0	**1**/20	1[a]	–	None	0
5,7	12	0	**2**/20	1[a,b]	–	None	0
7,7	14	0	**3**/20	1[a]	–	None	0
5,5,5	15	–	None	0	1	**1**/22	1
5,5,6	16	0	**10–12**/20	3[a]	–	None	0
5,5,7	17	–	None	0	1	**5**/22	1
5,6,6	17	–	None	0	1	**4**/22	1
5,6,7	18	0	**13–17**/20	5[a,b]	–	None	0
5,7,7	19	–	None	0	1	**2**/22	1
6,6,7	19	–	None	0	1	**6**/22	1
6,7,7	20	0	–	4[a]	–	None	0
5,5,5,5	20	0	–	2[a]	2	**7**/22	1[a]
7,7,7	21	–	None	0	1	**3**/22	1
5,5,5,6	21	–	None	0	1	–	2
5,5,5,7	22	0	–	7[a]	2	–	2[a]
5,5,6,6	22	0	**1**,**2**,**8**, etc./21	13	2	**8–10**/22	3
5,5,6,7	23	–	None	0	1	–	11
5,6,6,6	23	–	None	0	1	–	4
5,5,7,7	24	0	–	18[a,b]	2	–	3[a]
5,6,6,7	24	0	**3-7**,**10**, etc./21	25[b]	2	**11**,**12**/22	4
5,6,7,7	25	–	None	0	1	**13-16**, etc./22	14
6,6,6,7	25	–	None	0	1	–	5
5,7,7,7	26	0	–	14[a]	2	–	2[a]
6,6,7,7	26	0	–	20	2	–	3
6,7,7,7	27	–	None	0	1	–	6
7,7,7,7	28	0	–	8[a]	2	–	1[a]

[a] All isomers have integer partitions
[b] Aromatic systems

to see that this happens only when a terminal r-membered ring with odd ring size r is connected to a middle $(r + 1)$-membered ring, namely for systems having two rings of different sizes with the same EC value: **5**, **9–11**, **14–17**, **22–25**, all of Fig. 8.20. Moreover, in some cases three cata-condensed rings share the same size, as was the case with **10**/20 and **11**/20. A similar situation with four rings can be observed for compounds **11**/21 and **7**/22, but in the last case it involves a peri-condensed system. It is not difficult to see that any number of cata-condensed rings can share one EC value if the ring sizes are $r - 1, r, r, \ldots, r, r - 1$, where r is odd. Finally, one may observe that several tetracyclic systems such as pyracylene (**8**/22) and acepleiadylene (**11**/22) have integer partitioning of π-electrons to their four rings.

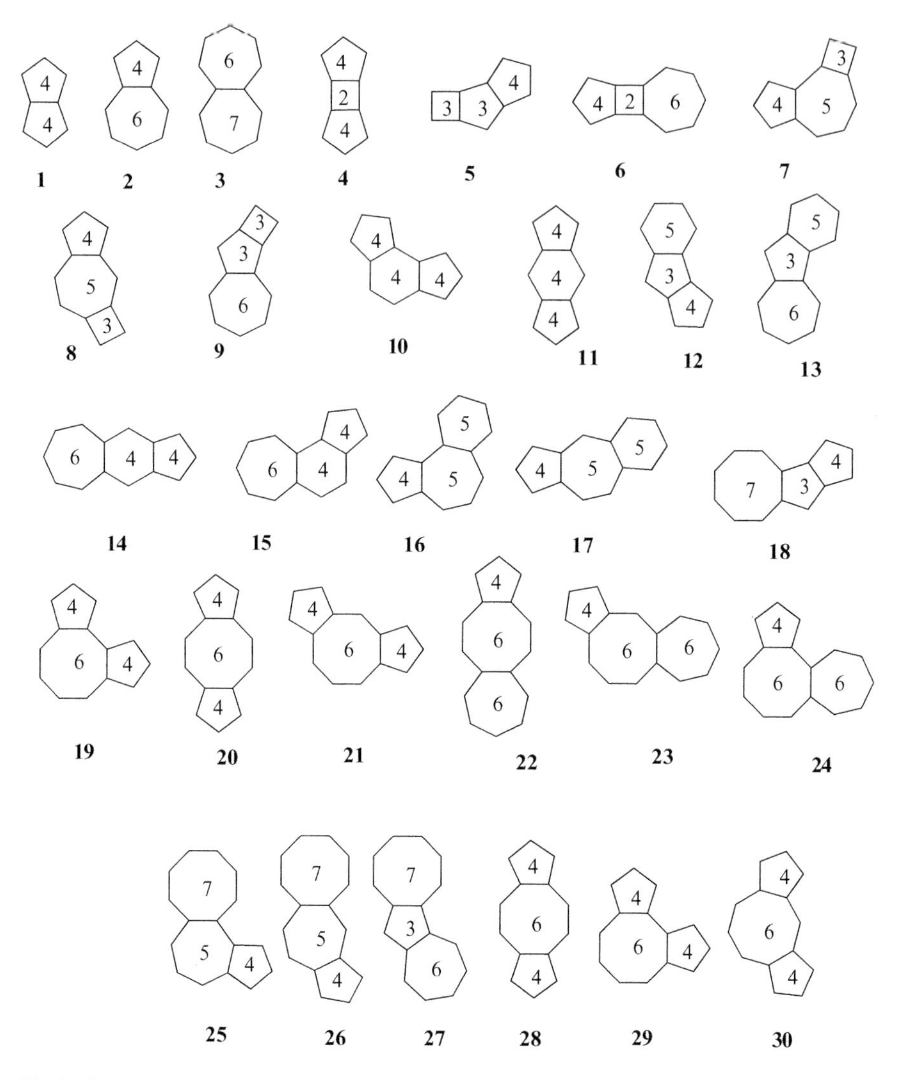

Fig. 8.20 Non-alternant bicyclic and tricyclic cata-condensed conjugated hydrocarbons with their EC values

8.10 Comparison with Other Methods for Estimating Local Aromaticity of Rings in Polycyclic Benzenoids

We continue this review with a discussion of "local aromaticities" estimated by EC values in comparison with other methods. Among the many published approaches to this problem, pioneered by Oskar Polansky (Polansky and Derflinger 1967; Monev et al. 1981) which were presented in Part 6 of our series, entitled "Partitioning of π-electrons in rings of polycyclic conjugated hydrocarbons.

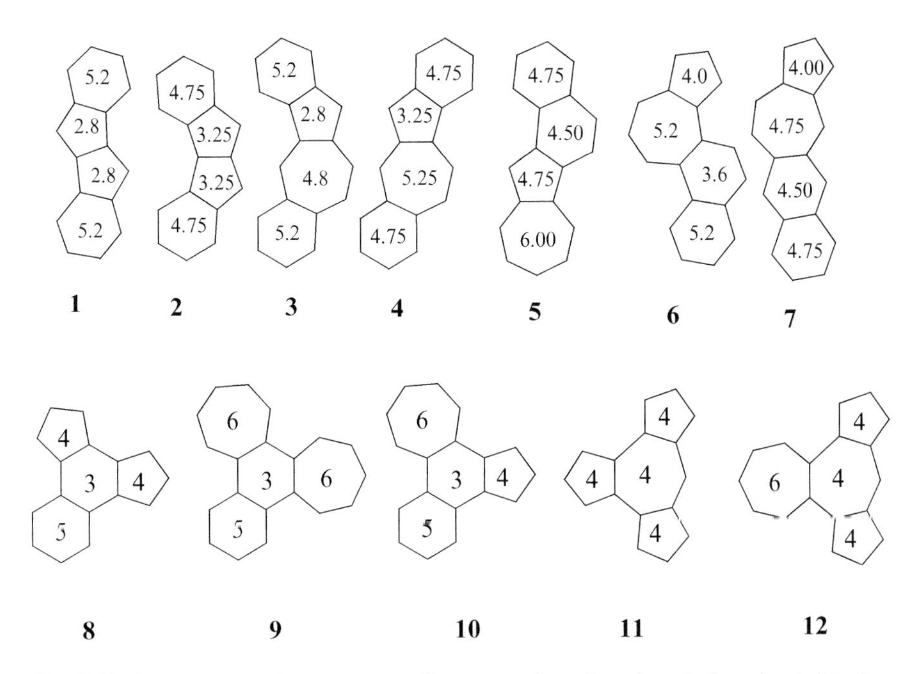

Fig. 8.21 *Top row*: non-alternant tetracyclic cata-condensed nonbranched conjugated hydrocarbons with their EC values. *Bottom row*: non-alternant tetracyclic cata-condensed branched conjugated hydrocarbons with integer EC values

Part 6. Comparison with other methods for estimating the local aromaticity of rings in polycyclic benzenoids" (Balaban and Randić 2005) we shall confine our discussion in this review only to a few other methods: Randić's graph-theoretical method (Randić 1975), Schleyer's nuclear-independent chemical shifts over the ring center at a distance of y Ångstroms from the molecular plane, NICS(y), where $y = 0$ or 1 Å (Chen et al. 2005; Schleyer et al. 1996, 2001) and the harmonic oscillator model of aromaticity (HOMA) values introduced by Kruszewski et al. (Kruszewski 1971, 1980; Kruszewski and Krygowski 1972; Krygowski and Cyrański 1996, 1998, 2001; Krygowski et al. 1995, 1996; Cyrański et al. 2000; Cyrański and Krygowski 1996, 1998). A fair correlation exists between EC values and HOMA values, and a somewhat lower correlation with NICS(1) values (readers should consult Part 6 for details). Here we will present in detail only an unpublished comparison with one index (indicated in Table 8.3 as GT) for the graph-theoretical local aromaticity of benzenoid rings in polycyclic aromatic hydrocarbons published in 1975 by Randić (1975). This index is easily calculated as the ratio between the number of Kekulé structures in which a particular benzenoid ring has three conjugated double bonds and the total number of Kekulé structures of that polycyclic aromatic hydrocarbon.

For 24 cata- and perifusenes with 3 to 7 benzenoid rings, whose rings are labeled with capital letters starting with marginal rings belonging to acenic portions, Table 8.3 presents local EC and GT indices. It can be seen that isoarithmic systems such as zigzag catafusenes and helicenes (fibonacenes with the same number R of rings)

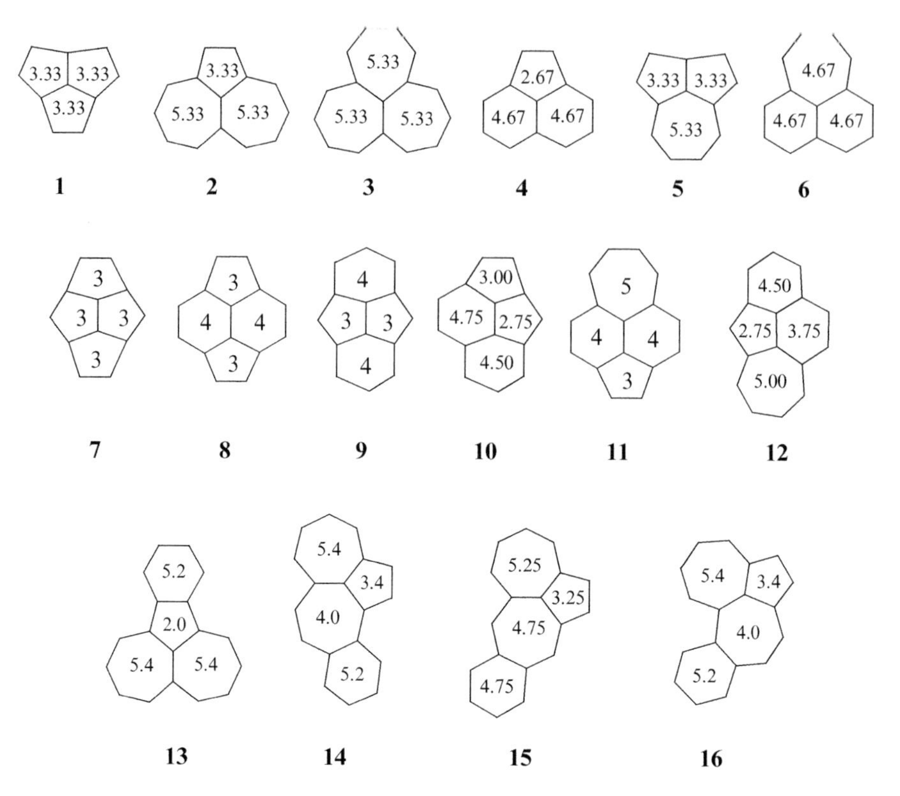

Fig. 8.22 Non-alternant tricyclic and tetracyclic peri-condensed conjugated hydrocarbons with their EC values

have identical EC values and very similar GT values. The points in the plot GT versus EC (Fig. 8.23) reveal three linear correlations. The rightmost trendline corresponds to marginal rings (labeled A) in acenes or fibonacenes; its equation is:

$$\mathrm{GT} = 0.664\ \mathrm{EC} - 2.652, \text{and the correlation coefficient is } r^2 = 0.9999.$$

The middle trendline with the highest slope corresponds to rings next to marginal rings in acenes; its equation is:

$$\mathrm{GT} = 0.986\ \mathrm{EC} - 3.936, \text{and the correlation coefficient is } r^2 = 0.9998.$$

The leftmost trendline with the lowest slope groups together corresponds to kink rings in catafusenes; its equation is:

$$\mathrm{GT} = 0.377\ \mathrm{EC} - 0.955, \text{and the correlation coefficient is } r^2 = 0.997.$$

In a very recent paper (Balaban and Mallion 2011) correlations between three local aromaticity indices for benzenoids, namely EC values, topological ring currents (Mallion 2008), and six-center delocalization-indices obtained by

Table 8.3 Local aromaticities of benzenoids (EC and GT values)

Benzenoid	Ring	EC	GT	Benzenoid	Ring	EC	GT
Anthracene	A	4.750	0.500	Triphenylene	A	4.920	0.615
3/6	B	4.500	0.500	**1**/12	B	4.620	0.615
Naphthacene	A	4.600	0.400	Coronene	A	4.860	0.571
(tetracene), **4**/6	B	4.400	0.400	**7**/15	B	4.570	0.571
Pentacene	A	4.500	0.333	Benz[*a*]-	A	3.290	0.286
5/6	B	4.330	0.333	anthracene	B	5.290	0.857
	C	4.330	0.333	**1**/11	C	5.250	0.833
Hexacene	A	4.420	0.286		D	3.420	0.333
6/6	B	4.280	0.286	Dibenz[*a*,*j*]-	A	4.670	0.667
	C	4.280	0.286	anthracene	B	5.250	0.833
Heptacene	A	4.375	0.250	**2**/11	C	3.420	0.333
8/6	B	4.250	0.250	Dibenz[*a*,*h*]-	A	4.670	0.667
	C	4.250	0.250	anthracene	B	4.900	0.600
	D	4.250	0.250	**3**/11	C	4.600	0.600
Phenanthrene	A	5.200	0.800	Fulminene	A	5.143	0.762
1/7	B	3.600	0.400	**4**/7	B	3.810	0.476
Chrysene	A	5.125	0.750		C	4.048	0.571
2/7	B	3.875	0.500	Zig-zag	A	5.147	0.765
Picene	A	5.154	0.769	[7]fibonacene	B	3.794	0.471
3/7, **5**/11	B	3.769	0.462	**5**/7 and	C	4.088	0.588
	C	4.154	0.615	heptahelicene	D	3.941	0.529

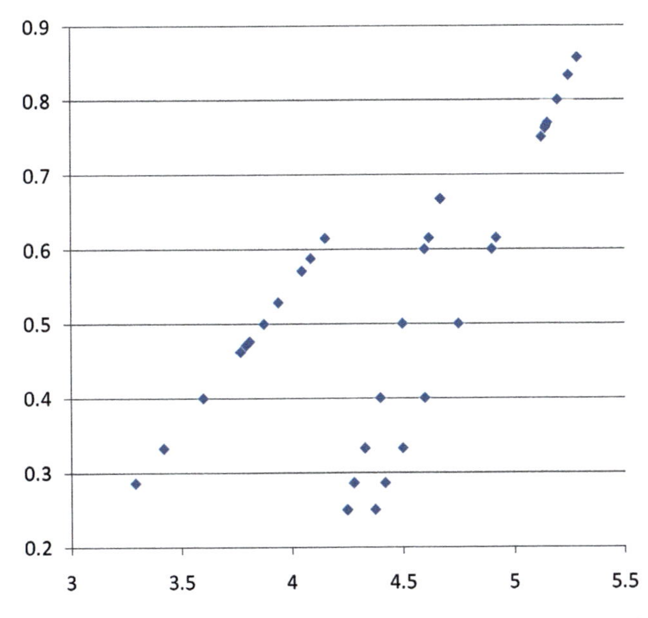

Fig. 8.23 Plot of the correlation between GT values and EC values for the benzenoids from Table 8.3

Table 8.4 Local aromaticity indices of compound **10**/15 (EC, GT and HOMA)

Ring	EC	GT	HOMA	CLAR
A	5.39	0.929	0.726	1.000
B	1.69	0.143	0.670	0.000
C	3.70	0.486	0.701	0.375
D	5.03	0.668	0.725	0.625
E	3.46	0.514	0.694	0.500
F	3.46	0.400	0.700	0.250

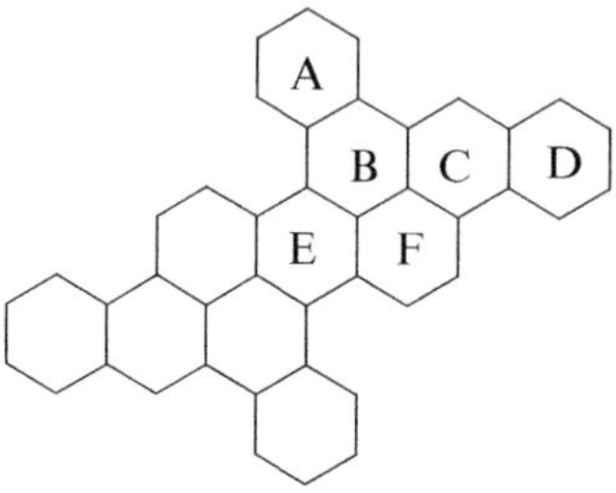

A B C D E F

10/15

quantum-chemical computations (Mandado et al. 2006) showed similar linear equations depending on the topology of the ring.

For one and the same benzenoid with 11 rings of six different types, namely tribenzo[*a,h,rst*]phenanthra[1,2,10-*cde*]pentaphene (**10**/15) with 11 benzenoid rings (Krygowski et al. 1995; Oonishi et al. 1992) the data in Table 8.4 indicate fair correlations between EC and various local aromaticity descriptors, albeit with only six points. Thus for EC and GT, $r^2 = 0.926$); for EC and HOMA, $r^2 = 0.986$; for EC and the "algebraic Clar structure" index (Randić 2011), $r^2 = 0.864$.

8.11 Ring Partition of π-Electrons for Clar's Fully Benzenoid Systems ('Claromatic Benzenoids')

Clar's intuitive approach in the characterization of local properties of benzenoids is necessarily qualitative in nature (Clar 1972). In Clar's approach individual benzenoid rings in polycyclic systems can be classified into three classes, namely as 'aromatic sextets', 'migrating sextets', and 'empty rings'. Fully benzenoid (claromatic) systems are those that have only 'aromatic sextets' or 'empty rings', the former of course being more interesting as the sites of possible chemical activity. It seems therefore of interest to examine closely individual rings in "fully benzenoid" systems to see what kind of variation there is between different classes of rings.

In Fig. 8.24 we show 11 smaller *fully benzenoid* (*claromatic*) systems that include triphenylene as the smallest fully benzenoid molecule (**1**/23), and hexabenzocoronene (**6**/23); the latter was described by Clar as an unusually stable compound which "resisted" the measurement of its melting point because in

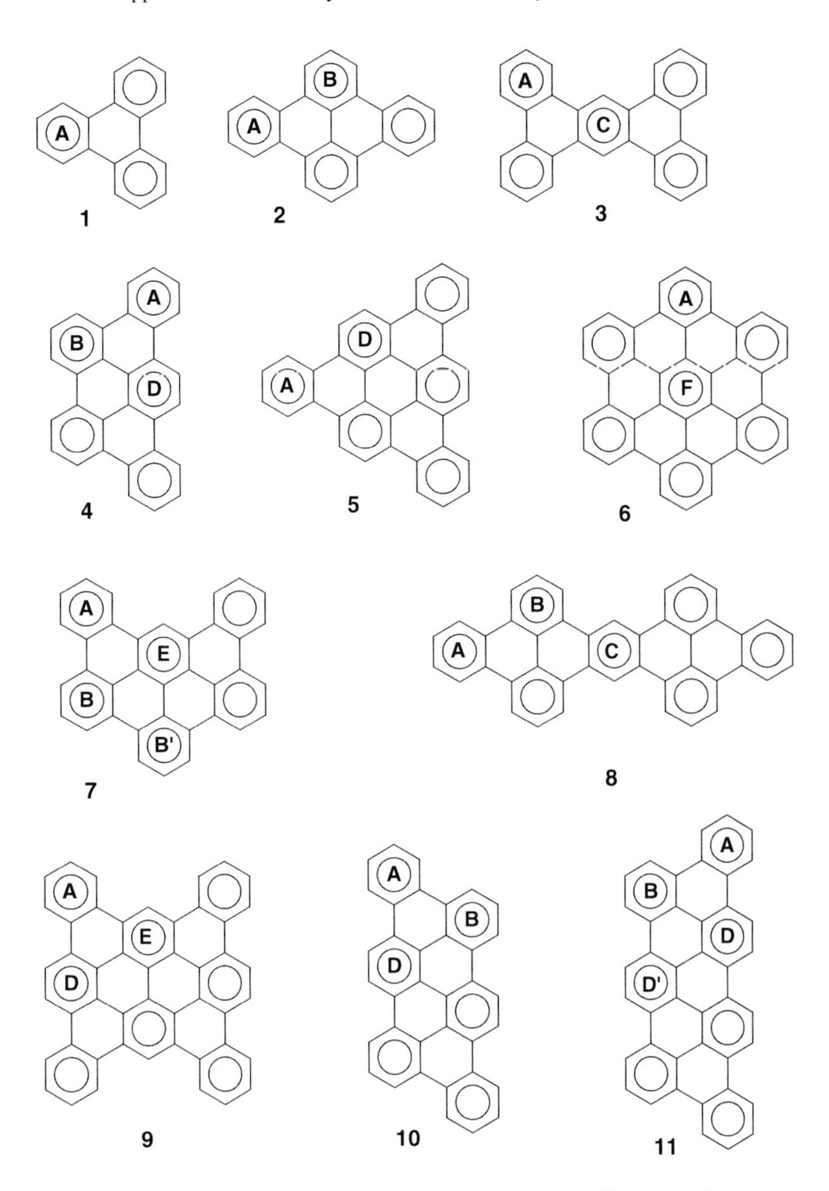

Fig. 8.24 Several symmetrical claromatic benzenoids with labeled Clar sextet rings

the attempt to do this, the glass tube melted before the compound did! We use the letter A for an "exposed" benzenoid ring having one common CC bond with the rest of the molecule, letters B, B′ and C for benzenoid rings having two CC bonds common to the rest of the molecule, letter D for benzenoid rings having three CC bonds common to the rest of the molecule, E for benzenoid rings having four CC bonds common to the rest of the molecule, and F for interior benzenoid rings having all CC bonds common to the rest of the molecule. In Table 8.5 we have

Table 8.5 Data for fully benzenoid (claromatic) compounds from Fig. 8.24

Comp.	Ring	EC	GT
1	A	48/9 = 5.333	8/9 = 0.889
2	A	107/20 = 5.350	18/20 = 0.900
	B	96/20 = 4.800	16/20 = 0.800
3	A	214/40 = 5.350	36/40 = 0.900
	B	192/40 = 4.800	32/40 = 0.800
4	A	240/45 = 5.333	40/45 = 0.889
	B	216/45 = 4.800	36/45 = 0.800
	D	186/45 = 4.133	32/45 = 0.711
5	A	551/104 = 5.298	90/104 = 0.865
	D	425/104 = 4.087	72/104 = 0.692
6	A	1200/250 = 4.800	200/250 = 0.800
	F	600/250 = 2.400	128/250 = 0.512
7	A	535/100 = 5.350	90/100 = 0.900
	B	480/100 = 4.800	80/100 = 0.800
	B′	480/100 = 4.800	80/100 = 0.800
	E	360/100 = 3.600	64/100 = 0.640
8	A	1059/198 = 5.348	178/198 = 0.899
	B	916/198 = 4.808	160/198 = 0.808
	C	882/198 = 4.828	162/198 = 0.818
9	A	2770/520 = 5.327	460/520 = 0.885
	D	1904/520 = 4.031	360/520 = 0.692
	E	1856/520 = 3.285	320/520 = 0.615
10	A	539/101 = 5.337	90/101 = 0.891
	B	484/101 = 4.792	80/101 = 0.792
	D	419/101 = 4.149	72/101 = 0.713
11	A	1211/227 = 5.335	202/227 = 0.890
	B	1088/227 = 4.793	180/227 = 0.793
	D	918/227 = 4.044	160/227 = 0.705
	D′	938/227 = 4.132	162/227 = 0.714

listed the EC values and the graph theoretical ring indices GT for the symmetry non-equivalent benzenoid rings qualifying as 'aromatic or Clar sextets' of these claromatic systems.

We see from Table 8.5 that the highest π-content belongs to rings A, which has over 5.300 π-electrons per ring. Most of the B-rings and C rings have values around 4.800, the smallest value belongs to **5**/5. As rings become more "deeply" attached to the rest of the molecule, the π-ring content decreases to around 4.000 and finally drops below 4.000 for rings that are in the interior part of the molecule.

In the last column of Table 8.5 we give the graph theoretical local aromaticity ring index GT [70] based on the count of Kekulé valence structures in which a ring has three double and three single CC bonds (that is, it appears as one of the Kekulé structures of benzene). There is very good correlation ($r^2 = 0.985$, Fig. 8.25) between the two indices, i. e. the π-electron ring partition (EC) and the count of 'benzenic' ring fragments with three double bonds in a larger benzenoid for individual benzenoid rings. This is very interesting and clearly illustrates that

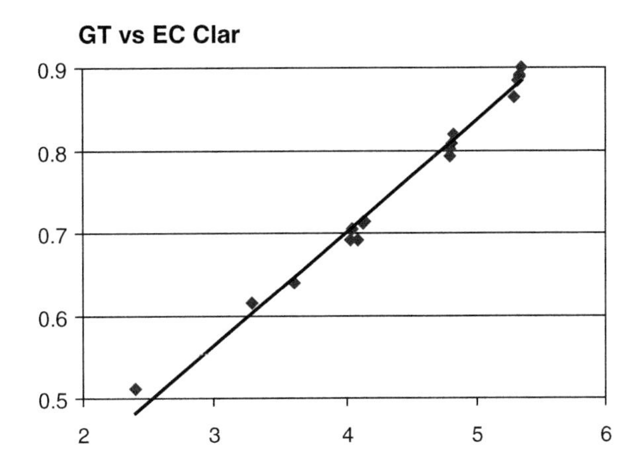

Fig. 8.25 Plot of the correlation between the GT index and the π-electron partition (*EC value*) for the Clar sextet rings of benzenoids from Fig. 8.24

Kekulé valence structures do involve more information than one might superficially anticipate. It appears that the fate of Kekulé valence structures followed the proverbial: "throwing away the baby with dirty linen" – because so much information innate to Kekulé valence structures has apparently been overlooked and would have never been recognized, were it not for the rise of Chemical Graph Theory and the application of Discrete Mathematics to chemical structure problems.

Interesting correlations between the π-electron ring partition (EC) and Gutman's energy effect for bipartite conjugated hydrocarbons (including, in addition to benzenoids, biphenylene and related compounds) were recently found (Balaban et al. 2010c). Gutman's observation of the 'PCP Rule' was extended to the effect of outer benzenoid rings twice removed from a central ring (the PCP Rule involved a central five-membered ring). Remarkably, the correlations could be expressed in simple numerical terms for just three types of outer benzenoid rings, depending on their condensation: linear, angular, or geminal (Balaban et al. 2010b).

8.12 r-Sequences and Signatures

On considering all *K* Kekulé structures of a benzenoid with *n* carbon atoms and *n* π-electrons, in addition to partitioning these π-electrons as discussed till now, it is possible to ascribe them in a few other different ways to individual rings leading to several types of local aromaticity accounts.

For sextet-resonant benzenoids (called "claromatic") there is a "winner takes all" solution, in which the Clar sextet rings get six π-electrons each, and the remaining rings remain "empty", as seen in the left-hand structure of Fig. 8.26; its π-electron

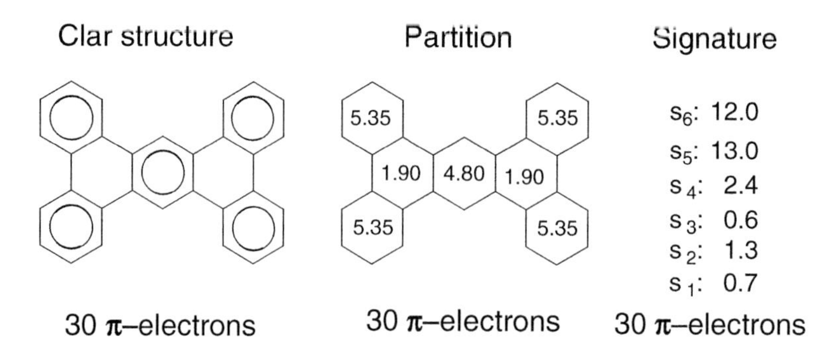

Fig. 8.26 Three ways of accounting for the π-electrons in a 7-ring claromatic perifusene. From *left to right*: five Clar sextets ("winner takes all"); partitions (note that the Clar sextets correspond to the highest partitions); and signatures

partition is shown in the middle structure, and one sees that Clar-sextet rings correspond to the highest EC values (Balaban and Randić 2007, 2008a, b; Balaban et al. 2008; Pompe et al. 2008; Randić and Balaban 2006, 2008). Finally, the same 30 π-electrons may be accounted for by what we called *the signature* of the benzenoid (Balaban and Randić 2008; Randić and Balaban 2008), which is explained in the following.

Let us have a look at Fig. 8.27 showing for each ring of a heptaperifusene, from left to right, the number q_i of Kekulé structures with $i = 6, 5, 4, 3, 2, 1, 0$ π-electrons according to the partition convention discussed earlier. Sums for each row are equal to the number K of Kekulé structures, $\Sigma_i\, q_i = K$. Then partitions of electrons to individual rings are given by $\mathrm{EC} = i \times q_i/K$. Of course, the sum of all partitions for a given benzenoid is the number n of carbon atoms and π-electrons. There is a close correspondence between Clar sextet rings and the rings with the highest partitions, but even for claromatic benzenoids there are no longer extreme values of EC = 6 and EC = 0.

The last row of numbers for columns with various i values, printed in boldface characters, is the column sum (C_i), and indicates how many times there are i π-electrons in any of the R rings for all Kekulé structures. The sum of all these numbers (also printed in boldface characters under the partition values) is $\Sigma_i\, C_i = \mathrm{R}K$. The integer C_i numbers allow us to group together the π-electrons in a different way: divide by K to obtain the r_i sequence, and multiply by i to obtain the signature s_i. Of course, on summing all signatures for a benzenoid we obtain again the numbers of π-electrons, as shown in Fig. 8.26. For four hexaperifusenes, Fig. 8.28 presents their r_i-sequences, partitions, and Clar structures.

All non-branched catafusenes have $C_\mathbf{0} = C_1 = 0$, and $s_1 = 0$. There are some interesting correlations among r_6 and r_5, illustrating the grouping of catafusenes into several classes with common structural features of their dualists, as discussed in detail in two side-by-side papers (Balaban and Randić 2008; Randić and Balaban 2008). Whenever a benzenoid ring in catafusenes or perifusenes has at least three

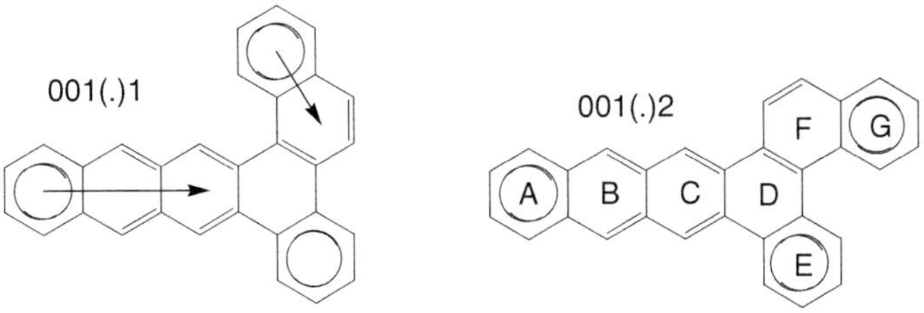

ring	*partit.*	*6-ring*	*5-ring*	*4-ring*	*3-ring*	*2-ring*	*1-ring*	*0-ring*
A	4.6923	6	6	14	0	0	0	0
B	4.4615	0	12	14	0	0	0	0
C	4.4615	0	12	14	0	0	0	0
D	1.7692	2	0	0	2	9	10	3
E	5.3846	12	12	2	0	0	0	0
F	4.1923	8	0	8	9	1	0	0
G	5.0385	9	9	8	0	0	0	0
$K = 26$	**182**	**37**	**51**	**60**	**11**	**10**	**10**	**3**

r-sequence	r_6	r_5	r_4	r_3	r_2	r_1	r_0
	1.4231	1.9615	2.3077	0.4231	0.3846	0.3846	0.1154

signature	s_6	s_5	s_4	s_3	s_2	s_1
	8.5386	9.8075	9.2308	1.2693	0.7692	0.3846

Fig. 8.27 Partitions, the r-sequence, and the signature (s-sequence) for both isoarithmic heptacatafusenes shown above

surrounding benzenoid rings as in triphenylene, this benzenoid has a nonzero r_0 value. A combination of several r_i values may serve as a yardstick for benzenoids, allowing an ordering or at least a partial ordering for isomeric systems (Pompe et al. 2008).

8.13 Biphenyl-Type Conjugation

Aromatic benzenoid hydrocarbons are important raw materials for the chemical industry, allowing the manufacture of plastics, fibers, strong textiles and a large variety of other products. However, aromatic benzenoid hydrocarbons as such do not have many uses, especially since some of them (benzene, benzanthracene, benzopyrene) are proved carcinogens. One of the few uses involves a mixture of *meta-* and *para-*terphenyl which has such a high thermal and radiolytical stability that it can be used as moderator and heat transfer fluid in "organic-moderator"

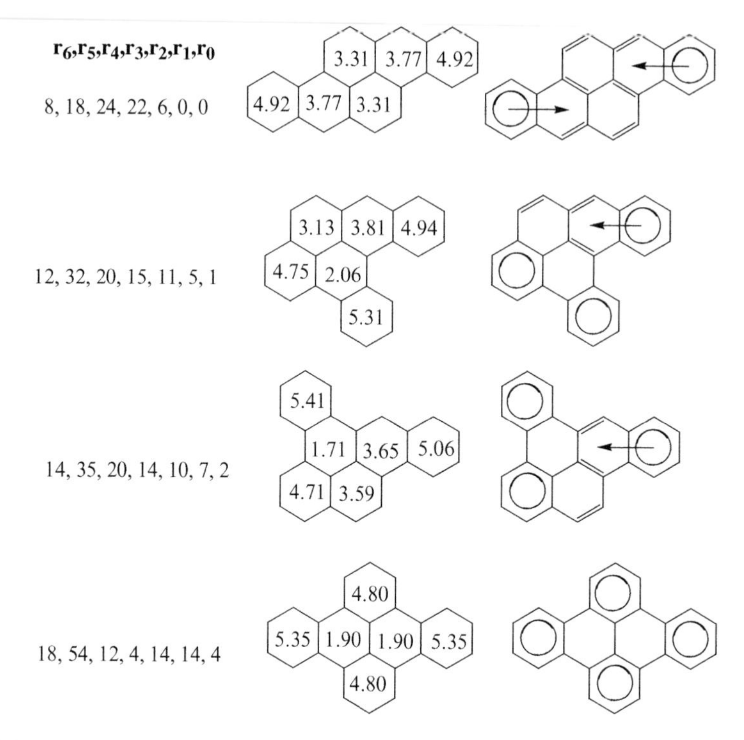

Fig. 8.28 From *left to right*: r_i sequences, π-electron partitions, and Clar structures of four hexaperifusenes (*top to bottom*) with 2, 3, 3, and 4 Clar sextet rings illustrating the fact that the Clar sextet rings have the highest partitions, and the "empty rings" have the lowest partitions among all isomeric structures

nuclear reactors. Indeed, biphenyl-type conjugation is also the cause for the high stability of various polychlorobiphenyls which now (like the freons which are also very stable) cause environmental problems.

Looking at Clar structures of polycyclic benzenoids, one sees that Clar sextet rings benefit always from biphenyl-type conjugation: triphenylene, the smallest claromatic structure, is derived from *ortho*-terphenyl by intramolecular dehydrogenation. Structures presented in Fig. 8.29 make it evident that *meta*- and *para*-terphenyl conjugation contributes to the stability of systems having closely-situated Clar sextet rings. With two exceptions to be discussed below, all structures in Fig. 8.29 reveal that the highest EC values correspond to Clar sextets. The first three rows contain isoarithmic structures, namely a pair of symmetrical catafusenes **1**/29 and **2**/29, a pair of nonsymmetrical perifusenes **6**/29 and **7**/29, and a triplet of symmetrical catafusenes (**3**/29 to **5**/29). The symmetrical (**8**/29) and

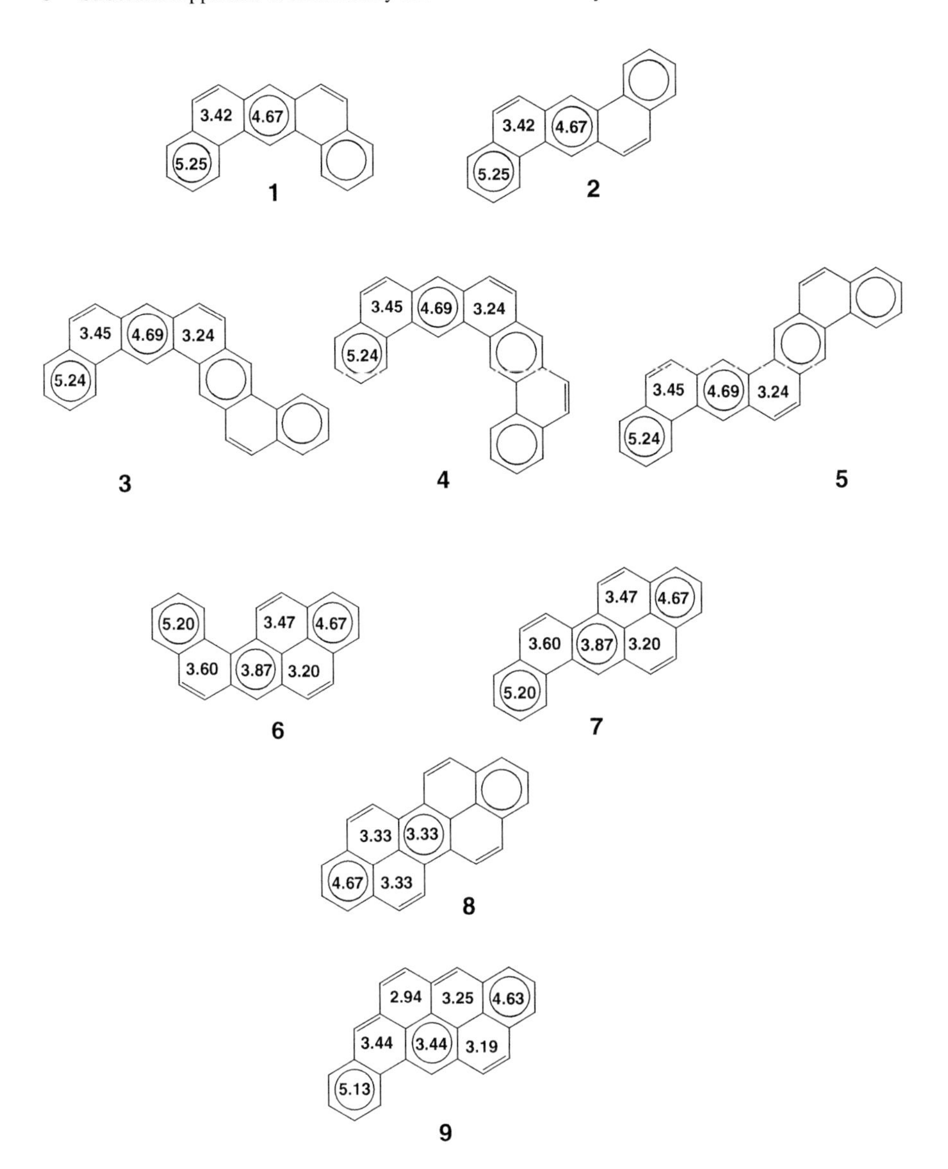

Fig. 8.29 Several benzenoids with Clar structures and EC values

nonsymmetrical perifusene (**9**/29) happen to have a Clar sextet ring with the same EC value as a ring that does not claim a Clar sextet; such situations are seldom encountered.

8.14 Concluding Remarks

In concluding this analysis of mathematical properties of individual Kekulé valence structures we should not be complacent and self-satisfied that we have come to the "end of the road" concerning Kekulé valence structures. The recent extension of this analysis of Kekulé valence structures to non-benzenoid structures has already hinted to some areas for future explorations that may be of interest. For instance, while in the case of benzenoid hydrocarbons it has been recently recognized that the inverse problem of Clar structures (that is, the mathematical characterization of Clar's structures) can be solved by selecting only the Kekulé valence structures with the highest "degree of freedom" (df), this however does not extend to non-benzenoid systems. For example, in the case of C_{60} (buckminsterfullerene) only a few of the Kekulé valence structures contribute to the Clar structure, namely: those having the degree of freedom df = 10 (one structure), df = 9 (two structures) and df = 8 (19 structures out of 32) (Randić 2002). Similarly, while the "resonance graph" of benzenoid hydrocarbons is always "connected" (i.e. all Kekulé structures form a single connected graph), this is not the case for nonbenzenoids. Thus already in the case of biphenylene, one structure from the five Kekulé structures has no single conjugated circuit R_1, and this cannot be part of the resonance graph. In our view, these differences in mathematical properties of benzenoids and non-benzenoids deserve close attention. It seems promising to focus attention not to all Kekulé valence structures but to a subset of "important" Kekulé valence structures. One such subset is the set of Kekulé valence structures contributing to the Clar structure – and we are looking into this matter. Another subset is even larger: the set of Kekulé valence structures that constitute the "connected" resonance graph.

Klein and coworkers (Flocke et al. 1998; Wu et al. 2003) have shown that in the case of C_{60}, out of 12,500 Kekulé valence structures, less than half make the dominant contribution to molecular resonance energy (RE), namely about 99.9% of RE, and these all form a connected resonance subgraph. It appears of considerable interest to find out when one must restrict attention to these subsets of Kekulé valence structures, how they reflect upon the π-electron ring partition and on the graph theoretical benzenoid ring characterization. In other words, it will be interesting to examine how well some qualified subsets of Kekulé valence structures can characterize benzenoid and non-benzenoid hydrocarbons, possibly including fullerenes. This is a task that appears worthy of attention. That this has not been hitherto considered may be due primarily to the preconceived idea that such 'refinement' of analysis of Kekulé valence structures will not make any difference in the case of benzenoid hydrocarbons, although it may be significant. However, one may expect that in the case of non-benzenoids we may see more visible aspects of the modified view on Kekulé valence structures. That this may be expected has already been reflected in the consideration of linearly and angularly fused higher homologues of biphenylene – the angular members of which, if one takes into account all Kekulé valence structures would be less and less stable as the number of fused biphenylene units increases – but Vollhardt has been able to synthesize the "unstable" systems" (Dosche et al. 2002; Han et al. 2002; Miljanić and Vollhardt 2006) thus showing that the model based on all-Kekulé valence structures is deficient. However, when one restricts attention to Kekulé structures with higher degree of freedom one can see that there is no substantial difference in the stability of linear and angular higher biphenylenes (Randić 2003; Miličević et al. 2004; Trinajstić et al. 1991).

Although the present discussion was centered on benzenoids, π-electron partitions of heterocyclic compounds have also been studied, adding a new dimension to the findings presented in the preceding pages (Balaban et al. 2007, 2008). Finally, it was interesting to investigate π-electron partitions of polyhedral carbon aggregates (Balaban et al. 2008c). One can conclude that "*the more we discover, the more diverse are the directions into the unknown, awaiting exploration* (Balaban, 2011)".

References

Aido M, Hosoya H (1980) Tetrahedron 36:1317–1326
Aihara J (1977) J Am Chem Soc 99:2048–2053
Aihara J (1982) Pure Appl Chem 54:1115–1128
Aihara J (1987) Bull Chem Soc Jpn 60:3581–3584
Aihara J (1992) J Am Chem Soc 114:865–868
Aihara J (2003) J Phys Chem A 107:11553–11557
Aihara J (2008) J Phys Chem A 112:4382–4385
Anderson PG (1986) In: Philippou AN, Bergum PG, Horadam AF (eds) Fibonacci numbers and their applications. Riedel, Dordrecht, p 2
Armit JV, Robinson R (1925) J Chem Soc 127:1604–1618
Balaban AT (1969) Tetrahedron 25:2949–2956
Balaban AT (1980) Pure Appl Chem 52:1409–1429
Balaban AT (1989) MATCH Commum Math Comput Chem 24:29–38
Balaban AT (2004) Polycyclic Aromat Compd 24:83–89
Balaban AT (2009) In: Krygowski TM, Cyranski MK (eds) Aromaticity in heterocyclic compounds. Springer, New York, pp 203–246
Balaban AT (2010) In: Katritzky AR (ed) Advances of heterocyclic chemistry, vol 99. Elsevier, Amsterdam, pp 61–105
Balaban AT (2011) as Guest Editor for a special issue on Aromaticity. Open Org Chem J (in press)
Balaban AT, Harary F (1968) Tetrahedron 24:2505–2516
Balaban AT, Klein DJ (2009) J Phys Chem C 113:19123–19133
Balaban AT, Mallion RB (2011) Croat Chem Acta (in press)
Balaban AT, Randić M (2004a) J Chem Inf Comput Sci 44:50–59
Balaban AT, Randić M (2004b) J Chem Inf Comput Sci 44:1701–1707
Balaban AT, Randić M (2004c) New J Chem 28:800–806
Balaban AT, Randić M (2005) J Math Chem 37:443–453
Balaban AT, Randić M (2007) J Univers Comput Sci 13:1514–1539
Balaban AT, Randić M (2008a) New J Chem 32:1071–1078
Balaban AT, Randić M (2008b) Int J Quantum Chem 108:898–926
Balaban AT, Schmalz TG (2006) J Chem Inf Model 46:1563–1579
Balaban AT, Tomescu I (1983) MATCH Commum Math Comput Chem 14:155–182

Balaban AT, Banciu M, Ciorba V (1987) Annulenes, benzo-, hetero-, homo-derivatives, and their valence isomers, vol 1. CRC Press, Boca Raton, p 5
Balaban AT, Oniciu D, Katritzky AR (2004) Chem Rev 104:2777–2812
Balaban AT, Schleyer PvR, Rzepa HS (2005) Chem Rev 105:3436–3447
Balaban AT, Furtula B, Gutman I, Kovačević R (2007) Polycycl Aromat Comp 27:51–63
Balaban AT, Gutman I, Stanković S (2008a) Polycycl Aromat Comp 28:85–97
Balaban AT, Pompe M, Randić M (2008b) J Phys Chem A 112:4148–4157
Balaban AT, Randić M, Vukičević D (2008c) J Math Chem 43:773–779
Balaban AT, Đurđević J, Gutman I (2009) Polycycl Aromat Comp 29:185–208
Balaban AT, Bean DE, Fowler PW (2010a) Acta Chem Slov 57:507–512
Balaban AT, Dickens TK, Gutman I, Mallion RB (2010b) Croat Chem Acta 83:209–215
Balaban AT, Đurđević J, Gutman I, Jeremić S, Radenković S (2010c) J Phys Chem A 114:5870–5877
Balaban AT, Gutman I, Jeremić S, Đurdević J (2011) Monatsh Chem 142:53–57
Chen Z, Wannere CS, Corminboeuf C, Puchta R, Schleyer PvR (2005) Chem Rev 105:3842–3888
Chun D, Cheng Y, Wudl F (2008) Angew Chem Int Ed Engl 47:8380–8385
Ciesielski A, Krygowski TM, Cyrański MK (2010) Symmetry 2:1390–1400
Clar E (1964) Polycyclic hydrocarbons, vol 1. Academic, New York
Clar E (1972) The aromatic sextet. Wiley, London
Crocker EC (1922) J Am Chem Soc 44:1618–1630
Cyrański MK, Krygowski TM (1996) J Chem Inf Comput Sci 36:1142–1145
Cyrański MK, Krygowski TM (1998) Tetrahedron 54:14919–14924
Cyrański MK, Stępień BT, Krygowski TM (2000) Tetrahedron 56:9663–9667
Dewar MJS (1967) In: Aromaticity: An international symposium held at Sheffield on 6th–8th July 1966, The Chemical Society, London, pp 177–215
Dewar MJS (1969) The molecular orbital theory of organic chemistry. McGraw-Hill, New York
Dewar MJS, de Llano C (1969) J Am Chem Soc 91:789–795
Dias JR (1987) Handbook of polycyclic hydrocarbons. Elsevier, Amsterdam
Dosche C, Löhmannröben H-G, Bieser A, Dosa PI, Han S, Iwamoto M, Schleifenbaum A, Vollhardt KPC (2002) Phys Chem Chem Phys 4:2156–2161
Đurđević J, Gutman I, Terzić J, Balaban AT (2009) Polycycl Aromat Comp 29:90–102
El-Basil S, Randić M (1987) J Math Chem 1:281–307
Flocke N, Schmalz TG, Klein DJ (1998) J Chem Phys 109:873–880
Fries K (1927) Justus Liebigs Ann Chem 454:121
Gutman I (1982) Bull Soc Chim Beograd 47:453–471
Gutman I (2004) Indian J Chem 43A:1615–1618
Gutman I (2005) Monatsh Chem 136:1055–1069
Gutman I (2006) MATCH Commun Math Comput Chem 56:345–356
Gutman I (2009) In: Mathematical methods and modelling for students of chemistry and biology. Graovac A, Gutman I, Vukičević D (eds), Hum, Zagreb, p 13
Gutman I, Bosanac S (1977) Tetrahedron 33:1809–1812
Gutman I, Cyvin SJ (1989) Introduction to the theory of benzenoid hydrocarbons. Springer, Berlin
Gutman I, Petrović V, Mohar B (1993) Chem Phys Lett 203:378–382
Gutman I, Morikawa T, Narita S (2004a) Z Naturforsch 59a:295–298
Gutman I, Vukičević D, Graovac A, Randić M (2004b) J Chem Inf Comput Sci 44:296–299
Gutman I, Balaban AT, Randić M, Kiss-Tóth C (2005a) Z Naturforsch 60a:171–176
Gutman I, Gojak S, Turković N, Furtula B (2005b) MATCH Commum Math Comput Chem 53:139–145
Gutman I, Randić M, Balaban AT, Furtula B, Vučković V (2005c) Polycycl Aromat Comp 25:215–226
Gutman I, Stanković S, Đurđević J, Furtula B (2007) J Chem Inf Model 47:776–781
Gutman I, Đurđević J, Balaban AT (2009) Polycycl Aromat Comp 29:3–11
Ham NS, Ruedenberg K (1958) J Chem Phys 29:1215–1229

Han S, Anderson DR, Bond AD, Chu HV, Disch RL, Holmes D, Schulman JM, Teat SJ, Vollhardt KPC, Whitener GD (2002) Angew Chem Int Ed 2002(41):3227–3230
Harvey RG (1991) Polycyclic aromatic hydrocarbons: chemistry and carcinogenicity. Cambridge University Press, Cambridge
Harvey RG (1997) Polycyclic aromatic hydrocarbons. Wiley-VCH, New York
Herndon WC, Ellzey ML (1974) J Am Chem Soc 96:6631–6642
Hess BA, Schaad LJ (1980) Pure Appl Chem 52:1471–1494
Hückel E (1931) Z Phys 70:204–286
Hückel E (1937) Z Elektrochem Angew Phys Chem 43:752–758–827–849
Hückel E (1938) Z Phys 72:310–337
Hückel E (1940) Grundzüge der Theorie ungesättigter und aromatischer Verbindungen. Verlag Chemie, Berlin, p 71
Hückel E (1975) Ein Gelehrtesleben. Verlag Chemie, Weinheim
Hückel E, Hückel W (1932) Nature 129:937–938
Jiang Y, Zhu H, Wang G (1993) J Mol Struct 297:327–335
Kaur I, Stein NN, Kopreski RP, Miller GP (2009) J Am Chem Soc 131:3424–3425
Kaur I, Jazdzyk M, Stein NN, Prusevich P, Miller GP (2010) J Am Chem Soc 132:1261–1263
Kikuchi S (1997) J Chem Educ 74:194–201
Klein DJ (1992) J Chem Educ 69:691–694
Körner W (1867) Bull Acad Roy Belg 24:166
Kruszewski J (1971) Acta Chim Lodz 16:77–82
Kruszewski J (1980) Pure Appl Chem 52:1525–1540
Kruszewski J, Krygowski TM (1972) Tetrahedron Lett 13:3839–3842
Krygowski TM, Cyrański MK (1996) Tetrahedron 52:1713–1722
Krygowski TM, Cyrański MK (1998) In: Párkányi C (ed) Theoretical organic chemistry. Elsevier, New York, pp 153–187
Krygowski TM, Cyrański MK (2001) Chem Rev 101:1385–1419
Krygowski TM, Ciesielski A, Bird CW, Kotschy A (1995) J Chem Inf Comput Sci 35:203–210
Krygowski TM, Cyrański MK, Ciesielski A, Świrska B, Leszczyński P (1996) J Chem Inf Comput Sci 36:1135–1141
Li S, Jiang Y (1995) J Am Chem Soc 117:8401–8406
Mallion RB (2008) Croat Chem Acta 81:227–246
Mandado M, Bultinck P, González-Moa MJ, Mosquera RA (2006) Chem Phys Lett 433:5–9
Matta CF, Hernández-Trujillo J (2003) J Phys Chem A 107:7496–7504
Miličević A, Nikolić S, Trinajstić N (2004) J Chem Inf Comput Sci 44:415–412
Miljanić OS, Vollhardt KPC (2006) In: Haley M, Tykwinski R (eds) Carbon-rich compounds. Wiley-VCH, Weinheim, p 140
Minkin VI, Glukhovtsev MN, Simkin BY (1994) Aromaticity and antiaromaticy: electronic and structural aspects, Wiley-VCH, Weinheim
Monev V, Fratev F, Polansky OE, Mehlhorn A (1981) Tetrahedron 37:1187–1191
Moran D, Stahl F, Bettinger HF, Schaeffer HF III, Schleyer PvR (2003) J Am Chem Soc 125:6746–6752
Moyano A, Paniagua J-C (1991) J Org Chem 56:1858–1866
Oonishi L, Ohshima S, Fujisawa S, Aoki J, Ohashi Y, Krygowski TM (1992) J Mol Struct 265:283–292
Patchovskii S, Thiel W (2000) J Mol Model 6:67–75
Pauling L (1931) J Am Chem Soc 53:3225–3237
Pauling L (1940) The nature of the chemical bond and the structure of molecules and crystals; an introduction to modern structural chemistry. Cornell University Press, Ithaca
Pauling L (1980) Acta Crystalllogr B 36:1898–1901
Pauling L, Wheland GW (1933) J Chem Phys 1:362–374
Platt JR (1953) J Chem Phys 21:1597–1600
Poater J, Fradera X, Duran M, Solà M (2003a) Chem Eur J 9:1113–1122

Poater J, Fradera X, Duran M, Solà M (2003b) Chem Eur J 9:400–406
Polansky OE, Derflinger G (1967) Int J Quantum Chem 1:379–401
Pompe M, Randić M, Balaban AT (2008) J Phys Chem A 112:11769–11776
Randić M (1974) Tetrahedron 30:2067–2074
Randić M (1975) Tetrahedron 31:1477–1481
Randić M (1980) Pure Appl Chem 52:1587–1596
Randić M (2002) In: Sen KD (ed) Reviews of modern quantum chemistry. A celebration of the contributions of Robert G. Parr, vol 1. World Scientific, New Jersey, pp 204–239
Randić M (2003) Chem Rev 103:3449–3605
Randić M (2004) J Chem Inf Comput Sci 44:365–372
Randić M (2011) Acta Chem Slov 58 (in press)
Randić M, Balaban AT (2004) Polycycl Aromat Comp 24:173–193
Randić M, Balaban AT (2006) J Chem Inf Model 46:57–64
Randić M, Balaban AT (2008) Int J Quantum Chem 108:865–897
Sakurai K, Kitaura K, Nishimoto K (1986) Theor Chim Acta 69:23–34
Schleyer PvR (2001) as Guest Editor. Chem Rev 101(5) (May issue) pp 1115–1566
Schleyer PvR, Maerker C, Dransfeld A, Jiao H, Hommes NJRvE (1996) J Am Chem Soc 118:6317–6318
Schleyer PvR, Manoharan M, Jiao H, Stahl F (2001) Org Lett 3:3643–3646
Staab HA, Diederich F (1983) Chem Ber 116:3487–3503
Suresh CH, Gadre SR (1999) J Org Chem 64:2505–2512
Tarko L, Filip P (2003) Rev Roum Chim 48:745–758
Thomson JJ (1921) Philos Mag 41:510–544
Trinajstić N, Schmalz TG, Živković TP, Nikolić S, Hite GE, Klein DJ, Seitz WA (1991) New J Chem 15:27–31
Vukičević D, Randić M, Balaban AT (2004) J Math Chem 36:271–279
Watson MD, Fechtenkötter A, Müllen K (2001) Chem Rev 101:1267–1300
Wu J, Schmalz TG, Klein DJ (2003) J Chem Phys 119:11011–11016
Zhu H, Jiang Y (1992) Chem Phys Lett 193:446–450

Chapter 9
Coding and Ordering Benzenoids and Their Kekulé Structures

Bono Lučić[1], Ante Miličević[2], Sonja Nikolić[1], and Nenad Trinajstić[1]

Abstract The coding and ordering of benzenoids and their Kekulé structures is discussed. The Wiswesser coding and the binary boundary coding systems are delineated and illustrated. The Wiswesser codes are also used to order cata- and peri-condensed benzenoids. The concept of numeric Kekulé structure is used for coding and ordering Kekulé structures of cata-condensed benzenoids. In the case of peri-condensed benzenoids the numeric code of Kekulé structures is not discriminative enough since there are Kekulé structures with identical numeric codes. In that case an additional code is needed, and here we suggest the perimeter code.

9.1 Introduction

Benzenoid hydrocarbons or simply benzenoids have been continuously studied by experimentalists and theoreticians since the discovery of benzene (Faraday 1825; Clar 1941; Randić 2003). We have been interested for a long time in the mathematical and combinatorial properties of benzenoids (e.g., Dewar and Trinajstić 1970; Gutman and Trinajstić 1976; Graovac et al. 1980; Živković et al. 1981, 1995; von Knop et al. 1983; Randić et al. 1987,1988; Živković and Trinajstić 1987; Klein and Trinajstić 1990; Nikolić et al. 2006; Vukičević and Trinajstić 2007; Lučić et al. 2009). One of the topics of our interest is concerned with coding and ordering benzenoids and their Kekulé structures. For example, Randić, Balaban and their co-workers published a number of papers on coding Kekulé structures *via* partitioning of π-electrons in various polycyclic conjugated systems and related heterosystems

[1] The Rugjer Bošković Institute, Bijenička 54, P.O.B. 180, HR-10 002 Zagreb, Croatia
e-mail: lucic@irb.hr; sonja@irb.hr; trina@irb.hr

[2] The Institute for Medical Research and Occupational Health, Ksaverskac. 2, P.O.B. 291, HR-10 002 Zagreb, Croatia
e-mail: antem@imi.hr

M.V. Putz (ed.), *Carbon Bonding and Structures: Advances in Physics and Chemistry*, Carbon Materials: Chemistry and Physics 5, DOI 10.1007/978-94-007-1733-6_9,

(e.g., Randić 2004; Randić and Balaban 2004; Balaban et al. 2007; Pompe et al. 2008; Klein 2010). In the present report, we will utilize some of their ideas and also use a less-known work of Wiswesser and his co-workers on coding benzenoids (Henson et al. 1975).

William Joseph Wiswesser (1914–1989), Bill for friends and collegues, is well-known for his work on the line-formula notation (Wiswesser 1954). A year before he died one of us (NT) talked to Bill Wiswesser and during the conversation Bill mentioned his work on coding benzenoids. Since that time we have been engaged in coding, generating and enumerating benzenoids (Balaban et al. 1987; von Knop et al. 1983, 1984a, b, 1990), and we were interested in Wiswesser's work on coding benzenoids. Some of our work on characterization, generation and enumeration of benzenoids and some other classes of chemical structures, such as, alkanes, alkenes, alkynes, alkyl radicals, polyenes, aza-benzenoids, was summarized in our two monographs entitled *Computer Generation of Certain Classes of Molecules* (von Knop et al. 1985) and *Computational Chemical Graph Theory* (Trinajstić et al. 1991). In the second book, we briefly mentioned Wiswesser coding system of benzenoids. This second book of ours was also dedicated to Bill Wiswesser, who died of heart problems in 1989. When NT talked to him in spring of 1988, Bill said he recently returned from hospital where he underwent yet another open-heart operation. Incidentally, a few years later NT had also open-heart surgery.

We will use in this report the term polyhexes for benzenoids to emphasize that we deal only with the carbon skeleton of benzenoids schematically presented as ensembles of regular hexagons. Polyhexes are mathematical structures which can be obtained by any combination of regular hexagons such that any two of its hexagons have exactly one common edge or are disjoint (e.g., Trinajstić 1990, 1992). In mathematical literature different names are given to polyhexes, such as, for example, hexagonal animals (Harary and Palmer 1973). We will use the names of benzenoids according to the *Handbook of Polycyclic Hydrocarbons* (Dias 1987).

9.2 The Wiswesser Code

Wiswesser and his co-workers (Henson et al. 1975) introduced the simplest and certainly the most elegant compact code for polyhexes in the following manner. They introduced the tilted Cartesian coordinate system in which the y axis is at the angle of 60° to the x axis. This was, of course, dictated by the hexagonal structure of benzene. A network can be placed in the xy plane and a polyhex can be placed onto this network in such a way that the x,y intersections appear in the centre of each of its hexagon. In Fig. 9.1, the tilted Cartesian coordinate system is given with the tilted network in the xy plane and a polyhex modeling the carbon skeleton of dibenzo[*b*,*g*]phenanthrene.

The position of each hexagon is determined by the x,y coordinates fixing the position of its center denoted by dot in the figure. The Wiswesser code is then simply defined by the y-values. The rule for setting up the Wiswesser code

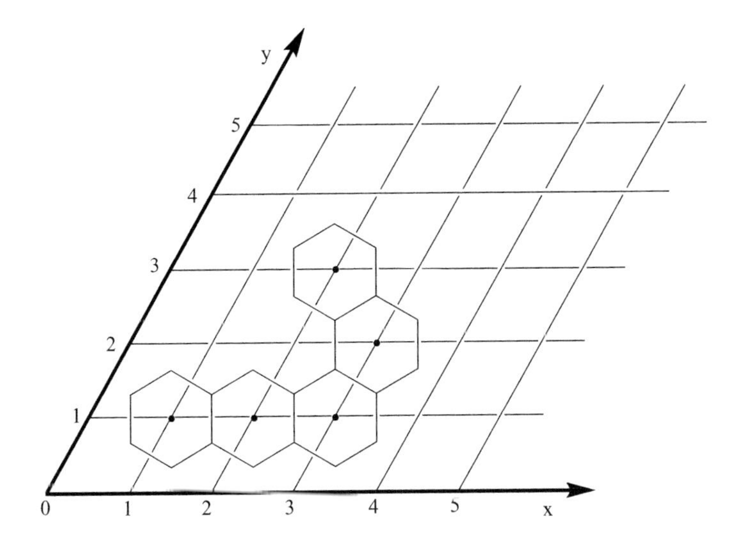

Fig. 9.1 The tilted Cartesian coordinate system with 60° angle between x and y axes and the tilted network in the xy-plane. A polyhex corresponding to the carbon skeleton of dibenzo[b,g] phenanthrene is placed onto this network in such a way that x,y intersections appear in the centre of each of its hexagons

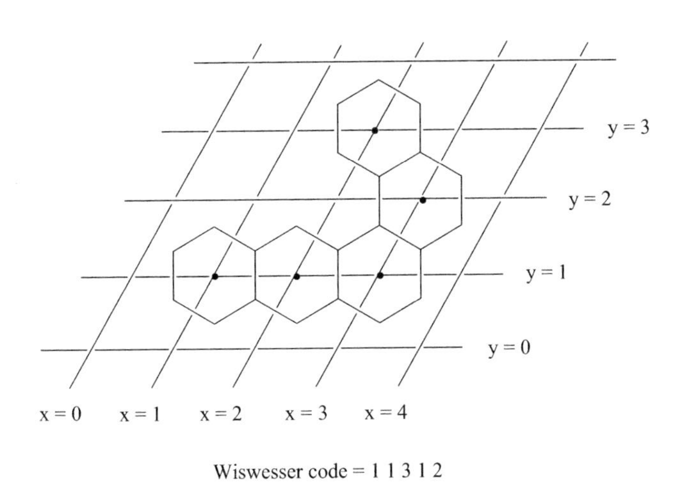

Fig. 9.2 An example for constructing the Wiswesser code for the polyhex depicting dibenzo[b,g] phenanthrene

is to give all the y-values for the same x-value before proceeding to the next x-value. In Fig. 9.2, an illustrative example the Wiswesser code for the dibenzo[b,g]phenanthrene graph is given. It should be noted that because of dots, Wiswesser and his co-workers called his codes the dot-plot numbers. They have also produced the computer program for setting up the Wiswesser codes.

The orientation of the polyhex should be such that the first entry is always unity or in some cases, e.g., perylenes, circulenes (or corona-condensed polyhexes) when this is not feasible, the smallest number available which leads to the smallest code of all possible Wiswesser codes for a given polyhex. This requirement eliminates from consideration all other orientations which would produce lexicographically higher codes. To illustrate this point, we give in Fig. 9.3 nine other orientations and the corresponding Wiswesser codes for the polyhex depicting dibenzo[*b*,*g*]phenanthrene. The Wiswesser code of dibenzo[*b*,*g*]phenanthrene, given in Fig. 9.2, is indeed smaller than any of the nine codes given in Fig. 9.3.

Since the convention for constructing the Wiswesser codes for polyhexes is understood, one can easily produce the Wiswesser code of any benzenoid by inscribing the appropriate number in each hexagon of a properly oriented polyhex. In Figs. 9.4 and 9.5, we give as examples Wiswesser codes for two structures in which the starting numeral is not unity: one representing perylenes, and the other circulenes.

A given polyhex can be recovered from the Wiswesser code by following steps presented in Fig. 9.6.

9.3 The Binary Boundary Code

Over the years several numerical codes have been developed for benzenoids, such as the boundary code (Trinajstić et al. 1983; von Knop et al. 1983), the DAST code (Müller et al. 1990), but the binary boundary code (Trinajstić et al. 1983; Herndon and Bruce 1987; Klein et al. 1988) appears to be the simplest to use.

The binary boundary code can be thought as a walk on the boundary of a polyhex that can be represented by a sequence of steps, each being labelled as 1 or 0 according the following convention. Each vertex on the polyhex boundary is labelled either 1 or 0, depending on whether its degree is 2 or 3. The degree of a vertex is the number of edges meeting at the vertex. For example, a single hexagon (monohex or hex) corresponding to the carbon skeleton of benzene is uniquely represented by the following sequence of numbers 111111. For two fused hexagons (bihex) corresponding to the carbon skeleton of naphthalene there are obviously five possible binary boundary codes, depending on the starting point of the walk on the boundary (see Fig. 9.7). The walk proceeds counter-clockwise.

Among the five possible binary boundary codes of the naphthalene-bihex, the maximum code is selected 1111011110. The binary boundary code is unique since two non-isomorphic polyhexes cannot produce the same code. In Fig. 9.8, we give the binary boundary code for the given orientation of the dibenzo[*b*,*g*] phenanthrene polyhex. A mirror-image orientation of the dibenzo[*b*,*g*]phenanthrene polyhex produces the maximum binary boundary code for this structure: 11110110110101111101000.

A way to recover a given polyhex from its binary boundary code is shown in Fig. 9.9.

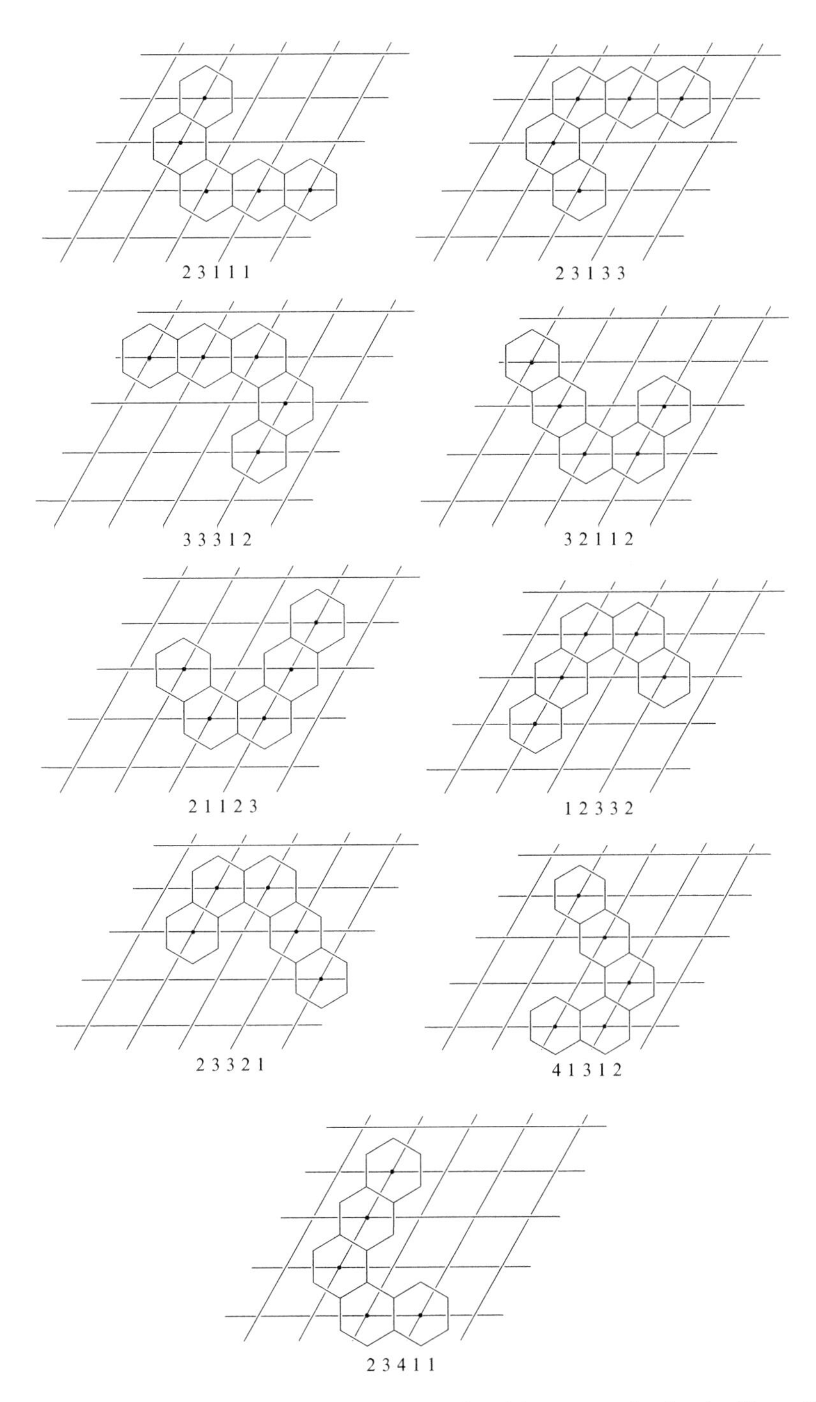

Fig. 9.3 Nine other orientations and the corresponding Wiswesser codes for the dibenzo[*b*,*g*] phenanthrene polyhex

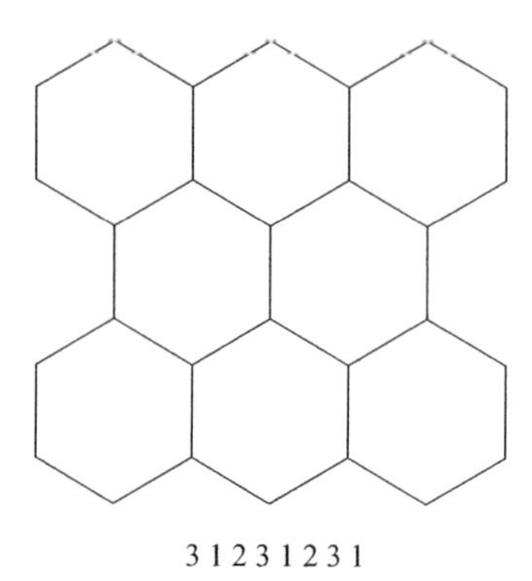

Fig. 9.4 Wiswesser code for dibenzo[*fg*,*op*]anthanthrene

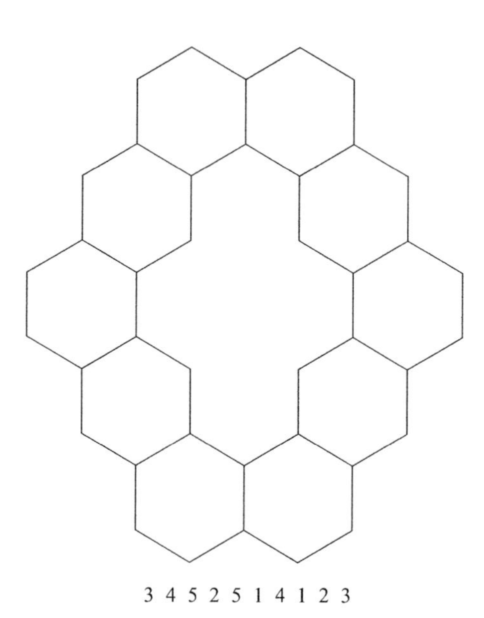

Fig. 9.5 Wiswesser code for [10] circulene

9.4 The Ordering of Benzenoids

We will order benzenoids using the concept of the lexicographic order (also known as dictionary order or alphabetic order). As an illustration, we lexicographically order benzenoids with five hexagons. In Fig. 9.10, we give lexicographically ordered Wiswesser codes for cata-condensed polyhexes representing carbon skeletons of 12 benzenoids with five six-membered rings $C_{22}H_{14}$. In Fig. 9.11, we give six lexicographically ordered peri-condensed polyhexes with five hexagons and one inner vertex. They model $C_{21}H_{13}$ benzenoids which are rather unstable systems. Finally, in Fig. 9.12, we lexicographically ordered three peri-condensed

a Wiswesser code

1 1 2 1 2 3 2 3 1

b Decoding Wiswesser code

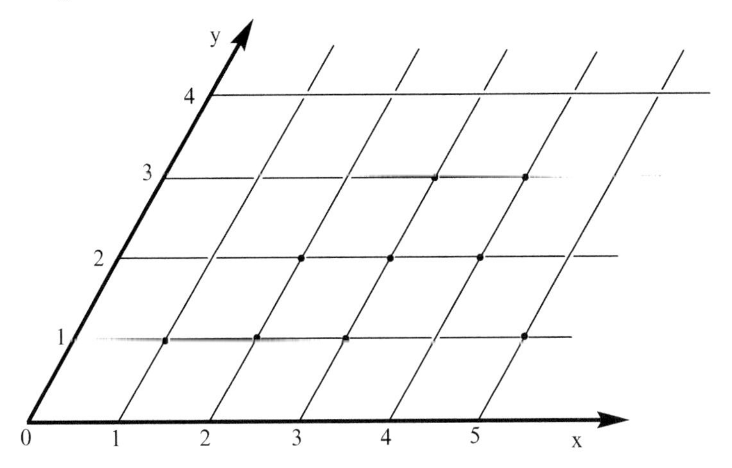

c Construction of the corresponding benzenoid

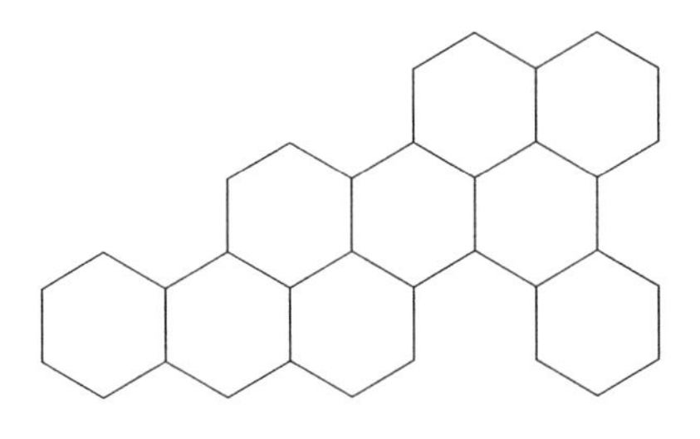

dibenzo[e,m]peropyrene

Fig. 9.6 Recovery of dibenzo[*e*,*m*]peropyrene from the corresponding (**a**) Wiswesser code 112123231, (**b**) Decoding Wiswesser code, (**c**) Construction of the corresponding benzenoid

polyhexes with five hexagons and two inner vertices. They model $C_{20}H_{12}$ benzenoids which are stable systems.

There are *in toto* 22 benzenoids with five hexagons. In Figs. 9.10–9.12, there are given 21 members of this set. In Fig. 9.13, we give for completeness the last member of the set, a polyhex representing unstable benzenoid $C_{19}H_{11}$. All benzenoids with up to nine hexagons are drawn in our book *Computer Generation of Certain Classes of Molecules* (von Knop et al. 1985).

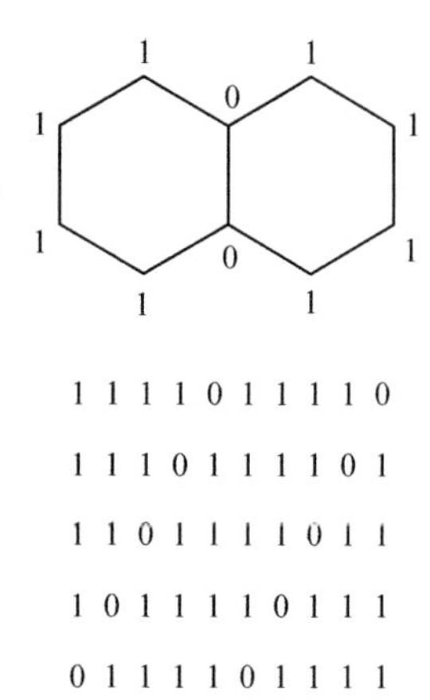

Fig. 9.7 A bihex representing the carbon skeleton of naphthalene and its five possible binary boundary codes

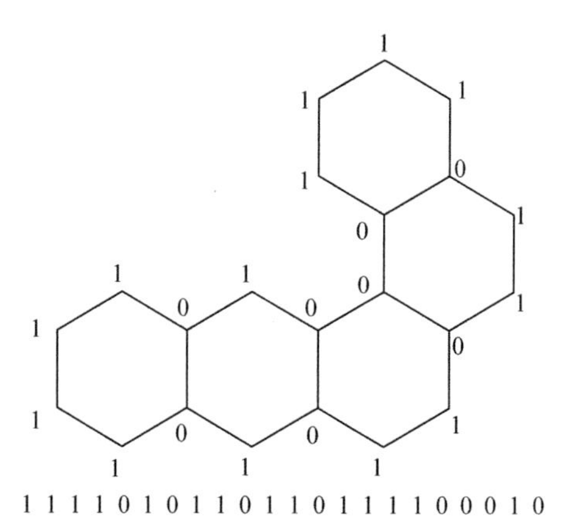

Fig. 9.8 The binary boundary code for a given orientation of the dibenzo[*b*, *g*]phenanthrene polyhex

9.5 The Coding and Ordering Kekulé Structures

The concept of *numeric* Kekulé structures will be used to order Kekulé structures. Randić introduced a novel description of Kekulé structures by replacing their standard representation by (what he called) algebraic representation (Randić 2004). Thus presented Kekulé structures Randić called the *algebraic* Kekulé structures, whilst their standard representation he called the *geometric* Kekulé structures. The algebraic representation of Kekulé structures appears to be useful for linear coding and lexicographic ordering Kekulé structures of benzenoids. We, however, replaced the term *algebraic* with *numeric* since the latter term better fits to this novel representation of Kekulé structures (Miličević et al. 2004).

The numeric Kekulé structures are generated in a rather simple way: Each unshared double bond in geometric Kekulé structures gets weight 2 for two π-electrons making up the double bond, but if the double bond is shared by two rings it gets weight 1. Then, the numbers assigned to each double bond in the ring are added up. Thus, every Kekulé structure can be linearly coded.

a Binary boundary code

1 1 1 1 0 1 0 1 1 0 0 0 1 1 1 1 0 0 1 1 1 0 1 1 0 0 1 1 0 0

b Construction of the polyhex boundary from the given binary boundary code

c Construction of the polyhex by filling in the inner space restricted by the boundary

d Recovered polyhex

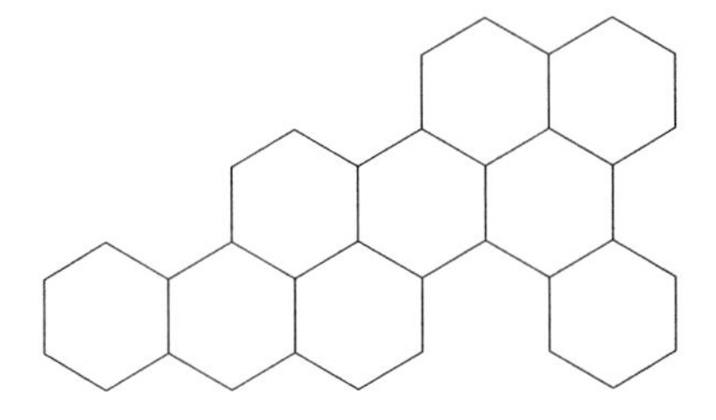

dibenzo[e,m]peropyrene

Fig. 9.9 Recovery of dibenzo[*e*,*m*]peropyrene from the corresponding binary boundary code (**a**) Binary boundary code 1 1 1 1 0 1 0 1 1 0 0 0 1 1 1 1 0 0 1 1 1 0 1 1 0 0 1 1 0 0, (**b**) Construction of the polyhex boundary from the given binary boundary code, (**c**) Construction of the polyhex by filling in the inner space restricted by the boundary, (**d**) Recovered polyhex

9.5.1 *Cata-Condensed Benzenoids*

We first consider the numeric Kekulé structures of cata-condensed benzenoids. As an example, we consider dibenzo[*b*,*g*]phenanthrene. In Fig. 9.14 we give geometric Kekulé structures of dibenzo[*b*,*g*]phenanthrene and in Fig. 9.15 their numeric counterparts. Kekulé structures of dibenzo[*b*,*g*]phenanthrene can be generated in

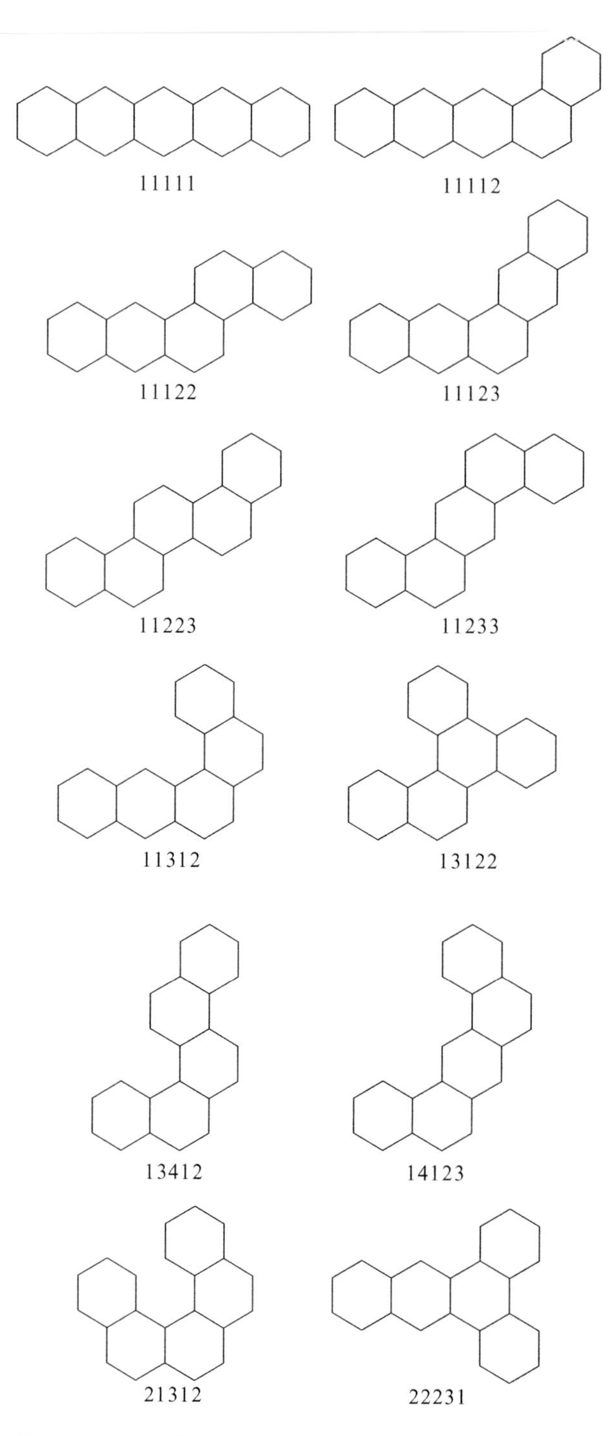

Fig. 9.10 The Wiswesser codes for cata-condesed polyhexes representing carbon skeletons of stable benzenoids with five six-membered rings $C_{22}H_{14}$

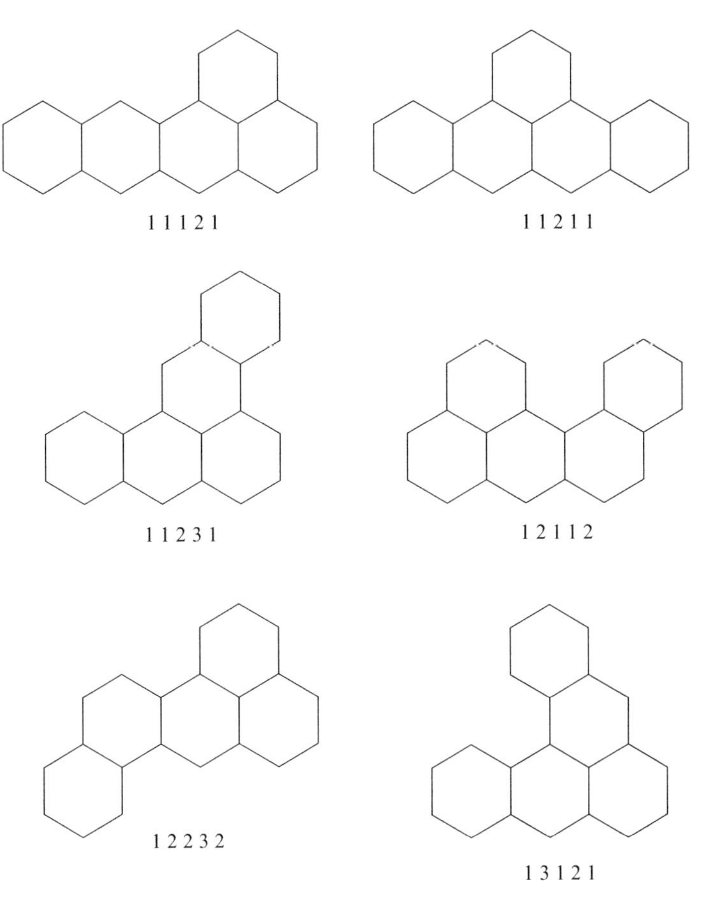

Fig. 9.11 The Wiswesser codes for peri-condensed polyhexes with a single inner vertex representing carbon skeletons of rather unstable benzenoids with five six-membered rings $C_{21}H_{13}$

a number of ways (Trinajstić 1983; Gutman and Cyvin 1989). Besides, the numbers of Kekulé structures for benzenoids with up to nine hexagons are given in our book *Computer Generation of Certain Classes of Molecules* (von Knop et al. 1985). We also devised computer programs for generating and enumerating the Kekulé structures (called in mathematical literature 1-factors) (Džonova-Jerman-Blažič and Trinajstić 1982; von Knop et al. 1984c).

From the numbers coding the numeric Kekulé structures of cata-condensed benzenoids, a given Kekulé structure can be easily recovered because the number of digits in the code determines the number of hexagons in the cata-condensed benzenoid, and the sum of digits gives the total π-electron count in the benzenoid.

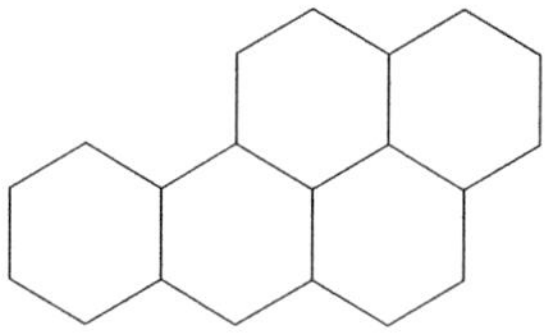

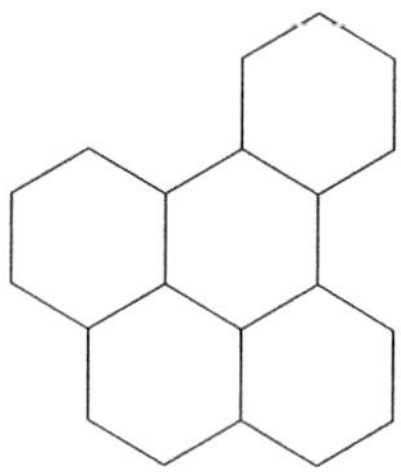

1 1 2 1 2

2 1 2 3 1

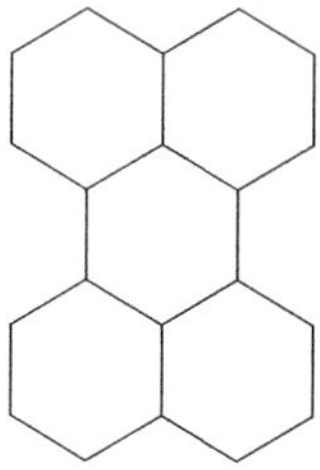

3 1 2 3 1

Fig. 9.12 The Wiswesser codes for peri-condensed polyhexes with five hexagons and two inner vertices representing carbon skeletons of stable benzenoids with five six-membered rings $C_{20}H_{12}$

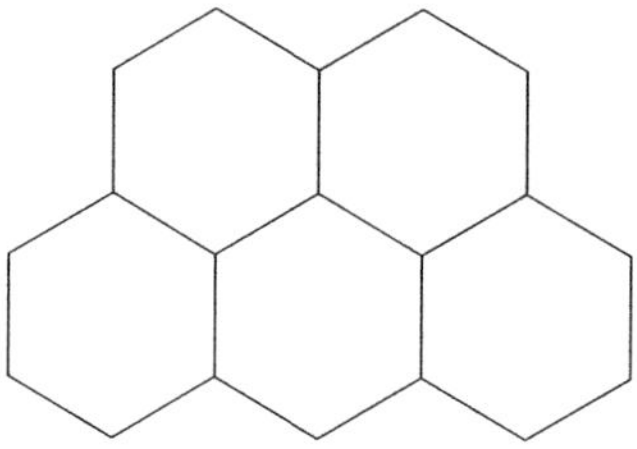

1 2 1 2 1

Fig. 9.13 The Wiswesser code of the last member of the set of polyhexes with five hexagons. This polyhex with three inner vertices represents the rather unstable benzenoid with five six-membered rings $C_{19}H_{11}$

For example, the code 44644 allows straightforward construction of one of the ten Kekulé structures of pentaphene (the fourth benzenoid in Fig. 9.10).

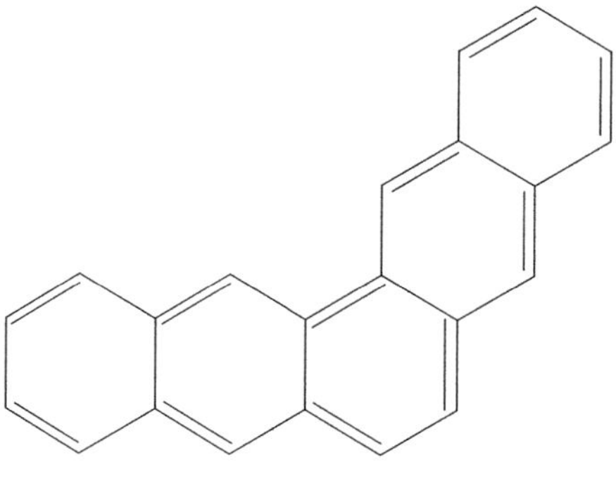

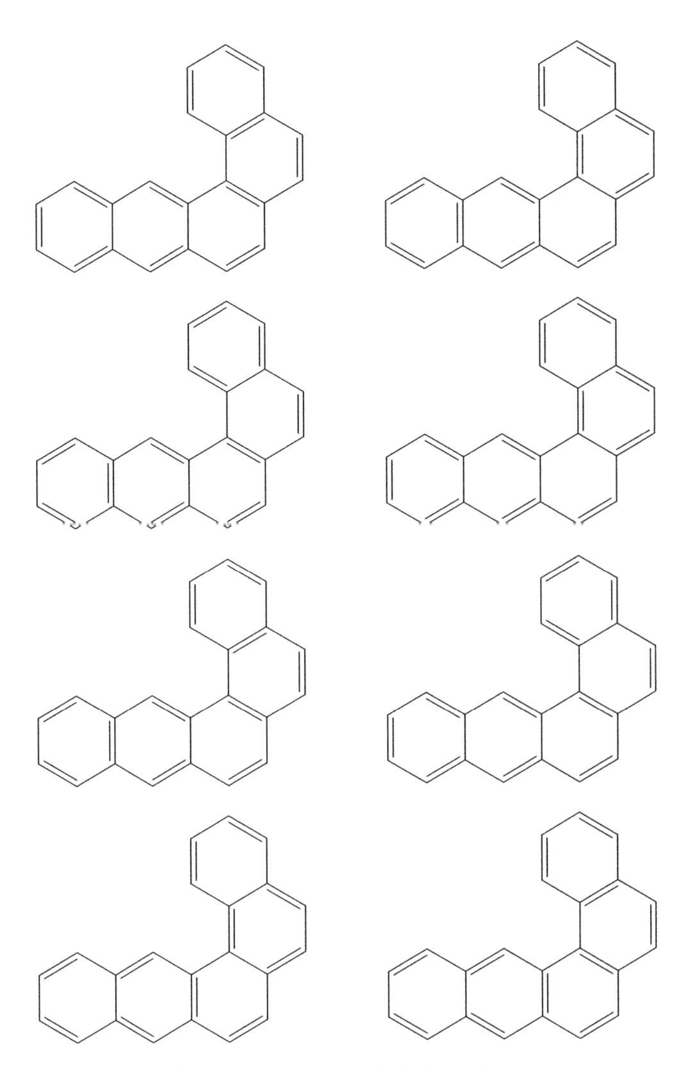

Fig. 9.14 Geometric Kekulé structures of dibenzo[*b*,*g*]phenanthrene

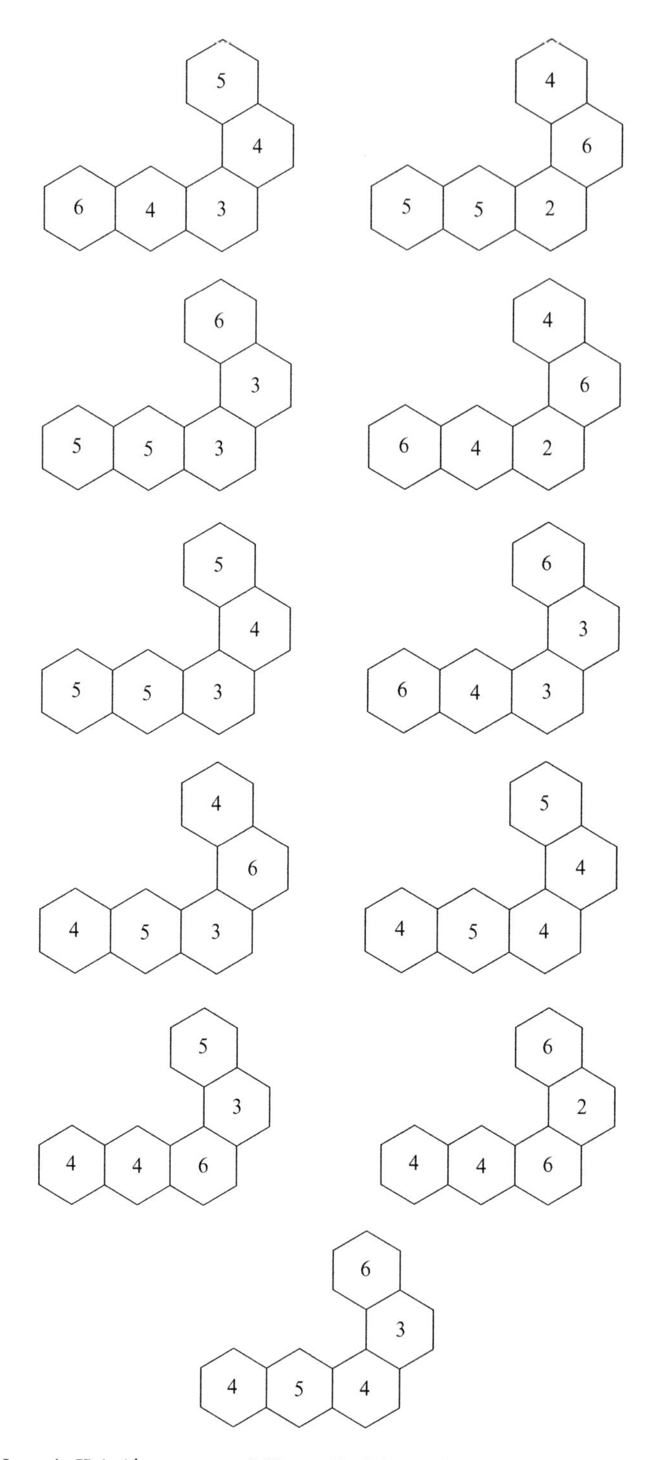

Fig. 9.15 Numeric Kekulé structures of dibenzo[*b*,*g*]phenanthrene

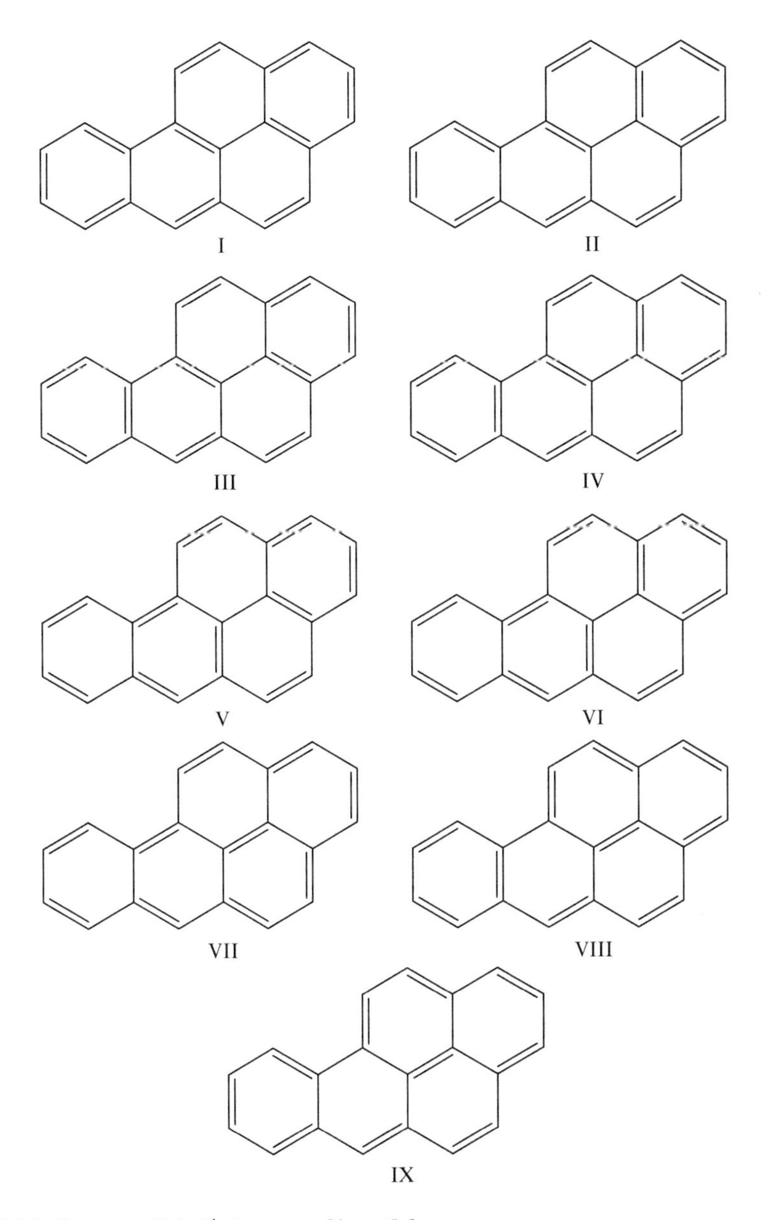

Fig. 9.16 Geometric Kekulé structures of benzo[*a*]pyrene

9.5.2 Peri-Condensed Benzenoids

At first, it seems that the generation of the numeric Kekulé structures for peri-condensed benzenoids is also straightforward. For example, this is shown in the case of benzo[*a*]pyrene. Its geometric and numeric Kekulé structures are given in Figs. 9.16 and 9.17.

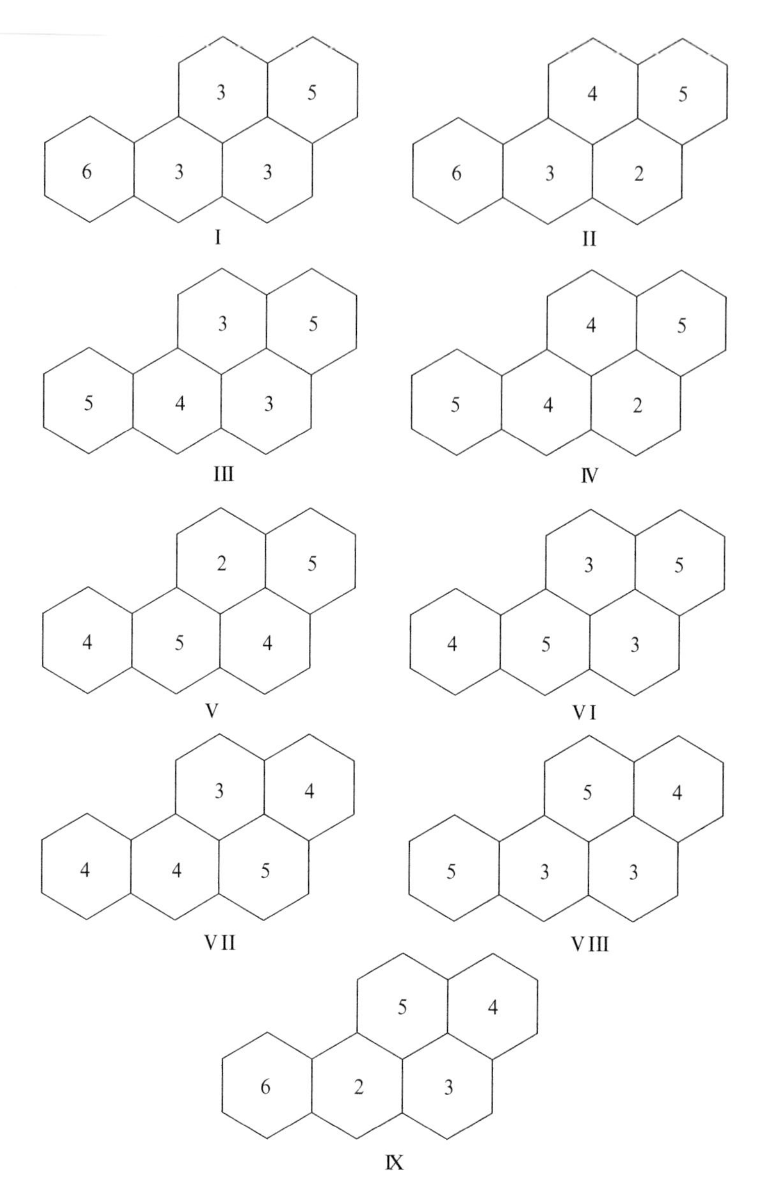

Fig. 9.17 Numeric Kekulé structures of benzo[*a*]pyrene

It should be noted that the numeric code is a linear code in which the codes for peri-condensed benzenoids are built in the same way as the Wiswesser codes. For example, the numeric code for the Kekulé structure VI (see Fig. 9.17) is 45335.

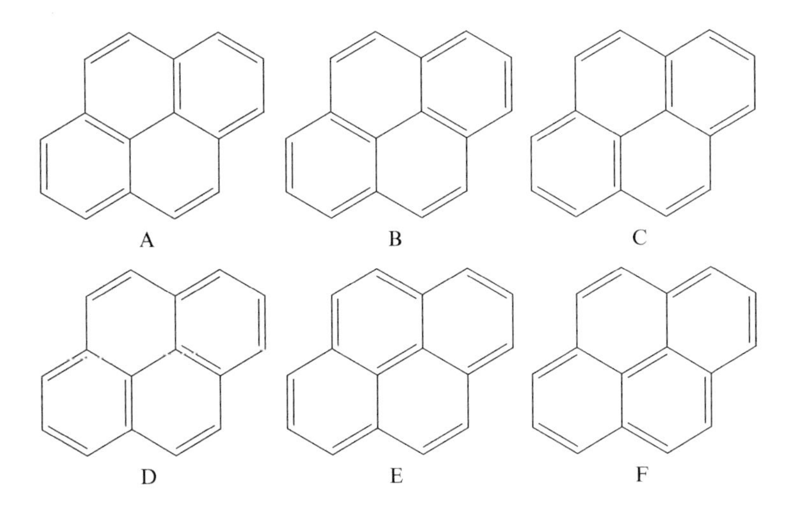

Fig. 9.18 Geometric Kekulé structures of pyrene

Below we order the numeric Kekulé structures of benzo[*a*]pyrene in reverse lexicographic order:

II	6 3 4 2 5
I	6 3 3 3 5
IX	6 2 5 3 4
IV	5 4 4 2 5
III	5 4 3 3 5
VIII	5 3 5 3 4
VI	4 5 3 3 5
V	4 5 2 4 5
VII	4 4 3 5 4

However, it appears that there exist peri-condensed benzenoids with distinct geometric Kekulé structures but with the identical numeric Kekulé structures (Gutman et al. 2004; Vukičević et al. 2004). The smallest example of such a case is pyrene. In Fig. 9.18, we give its six geometric Kekulé structures and in Fig. 9.19 its six numeric Kekulé structures. Note, structures denoted by B and C possess identical numeric codes. Therefore, there is no one-to-one correspondence between geometric and numeric Kekulé structures for some peri-condensed benzenoids. In such cases the recovery of geometric structures requires the amendment of the numeric codes.

One way to amend the identical numeric codes is to use additional codes. For example, one can use a perimeter code which consists of 1 for single bonds and 2 for double bonds on the boundary of a given Kekulé structure. We give in Fig. 9.20 two Kekulé structures of pyrene possessing identical numeric codes, but different perimeter codes. It should be noted that the benzenoid under the consideration should be oriented in such a way to have the smallest of all possible

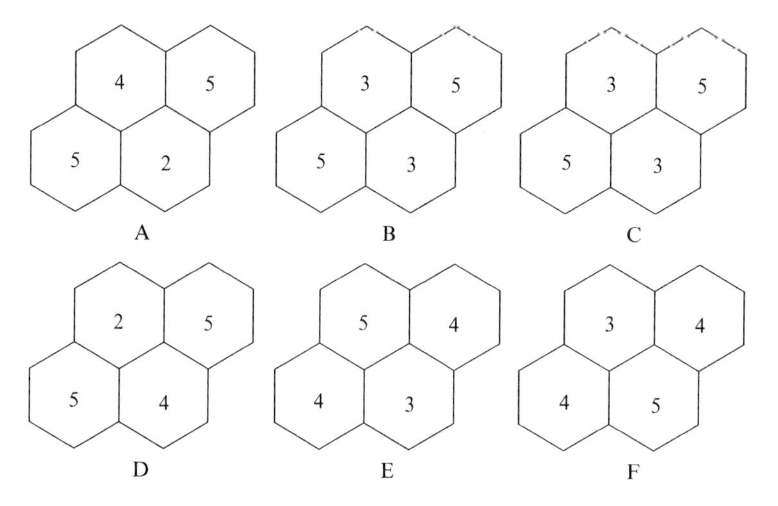

Fig. 9.19 Numeric Kekulé structures of pyrene

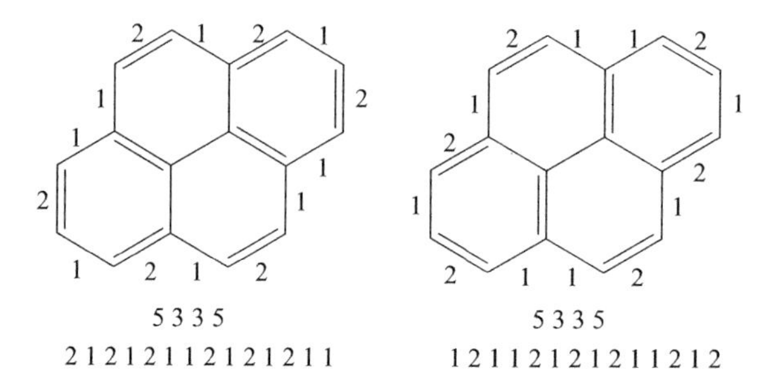

Fig. 9.20 Two Kekulé structures of pyrene with identical numeric 5335, but different perimeter codes

Wiswesser codes, and the starting point of the perimeter code should be at the vertical bond in the hexagon of a polyhex oriented in the Wiswesser manner, as discussed in the section 9.2. The perimeter code proceeds counter-clockwise.

In Fig. 9.21, we show how a pair of peri-condensed benzenoid with identical numeric codes can be recovered from their perimeter codes. As an illustrative example, we selected numeric code 5 5 2 2 5 5 which is identified to belong to a pair of Kekulé structures of dibenzo[*fg*,*op*]naphthacene.

It should be noted that dibenzo[*fg*,*op*]naphthacene has 20 Kekulé structures. Besides the pair of its Kekulé structures shown in Fig. 9.21, it possesses two more pairs of Kekulé structures with identical numeric codes (Vukičević et al. 2004).

a Wiswesser code

3 1 2 3 4 2

b Numeric code

5 5 2 2 5 5

c Perimeter codes

211121211212111212121121

121121211121211212121112

d Construction of the boundaries of benzenoids using the perimeter codes

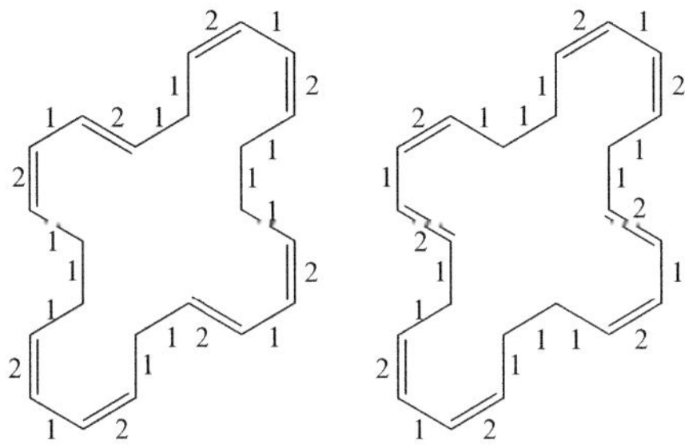

e Filling up the interiors of the boundaries

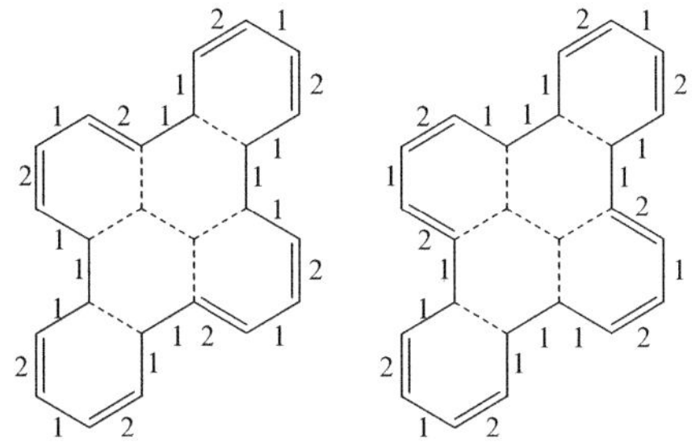

f Construction of the corresponding Kekulé structures

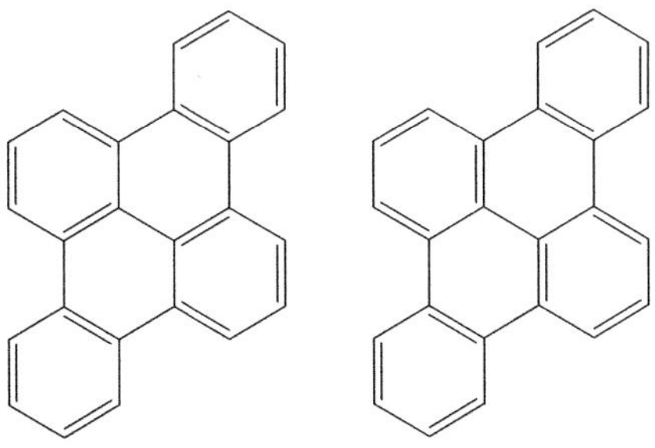

Fig. 9.21 The recovery of two Kekulé structures of dibenzo[*fg,op*]naphthacene with identical numeric codes 552255, from their perimeter codes: (**a**) Wiswesser code 3 1 2 3 4 2, (**b**) Numeric code 5 5 2 2 5 5, (**c**) Perimeter codes 211121211212111212121121, 121121211121211212121112, (**d**) Construction of the boundaries of benzenoids using the perimeter codes, (**e**) Filling up the interiors of the boundaries, (**f**) Construction of the corresponding Kekulé structures

9.6 Conclusion

Benzenoid hydrocarbons, or simply benzenoids, are attractive structures to study (Gutman and Cyvin 1988, 1989). In this report, we considered coding and ordering of benzenoids and their Kekulé structures. In this area of research seminal ideas had been introduced by Wiswesser, Balaban, Gutman, Klein, Randić, von Knop and their co-workers. We presented and illustrated the Wiswesser coding and the binary boundary coding systems. The Wiswesser coding system is rather simple and elegant. We used the Wiswesser codes to order cata- and peri-condensed benzenoids. The concept of numeric (algebraic called by its originator Randić) Kekulé structure is used for coding and ordering Kekulé structures of cata-condensed benzenoids. In the case of peri-condensed benzenoids this is not always possible, because the numeric code of their Kekulé structures is not discriminative enough, since there are Kekulé structures with identical numeric codes. For example, pyrene and benzenoids with pyrene as subunit exhibit this. In these cases an additional code is needed. It is suggested the perimeter code to be used. The concept discussed in the paper can be easily extended to a variety of conjugated systems such as azabenzenoids, [*N*] phenylenes, thio-perylenes, etc.

Acknowledgements This report was supported by the Croatian Ministry of Science, Education and Sports by grants No. 098-1770495-2919 and No. 022-1770495-2901.

References

Balaban AT, Brunvoll J, Cioslowski J et al (1987) Z Naturforsch 42a:863–870
Balaban AT, Furtula B, Gutman I, Kovačević R (2007) Polycycl Aromat Comp 27:51–63
Clar E (1941) Aromatische kohlenwasserstoffe – polycyclische systeme. Springer, Berlin
Dewar MJS, Trinajstić N (1970) Coll Czech Chem Commun 35:3137–3189
Dias JR (1987) Handbook of polycyclic hydrocarbons. Part A. Benzenoid hydrocarbons. Elsevier, Amsterdam
Džonova-Jerman-Blažič B, Trinajstić N (1982) Comput Chem 6:121–132
Faraday M (1825) Phil Trans Roy Soc (London) 115:440–466
Graovac A, Trinajstić N, Randić M (1980) Croat Chem Acta 53:571–579
Gutman I, Cyvin SJ (1988) Kekulé structures in benzenoid hydrocarbons. Springer, Berlin
Gutman I, Cyvin SJ (1989) Introduction to the theory of benzenoid hydrocarbons. Springer, Berlin
Gutman I, Trinajstić N (1976) Croat Chem Acta 48:297–299
Gutman I, Vukičević D, Graovac A, Randić M (2004) J Chem Inf Comput 44:296–299
Harary F, Palmer EM (1973) Graphical enumeration. Academic, New York
Henson RA, Windlinx KJ, Wiswesser WJ (1975) Comput Biomed Res 8:3–71
Herndon WC, Bruce AJ (1987) In: King RB, Rouvray DH, (eds) Graph theory and topology in chemistry. Elsevier Science Publishing, Amsterdam, pp 491–513
Klein DJ (2010) Acta Chim Slov 57:591–596
Klein DJ, Trinajstić N (1990) J Mol Struct (Theochem) 206:135–142
Klein DJ, Herndon WC, Randić M (1988) New J Chem 12:71–76
Lučić B, Trinajstić N, Zhou B (2009) Chem Phys Lett 475:146–148
Miličević A, Nikolić S, Trinajstić N (2004) J Chem Inf Comput Sci 44:415–421

Müller WR, Szymanski K, Knop JV et al (1990) J Comput Chem 11:223–235
Nikolić S, Miličević A, Trinajstić N (2006) Croat Chem Acta 79:155–159
Pompe M, Randić M, Balaban AT (2008) J Phys Chem 112:11769–11776
Randić M (2003) Chem Rev 103:3449–3605
Randić M (2004) J Chem Inf Comput Sci 44:365–372
Randić M, Balaban AT (2004) Polycycl Aromat Comp 24:173–193
Randić M, Nikolić S, Trinajstić N (1987) Gazz Chim Acta 117:69–73
Randić M, Plavšić D, Trinajstić N (1988) Gazz Chim Acta 118:441–446
Trinajstić N (1983) Chemical graph theory, vol II. CRC, Boca Raton
Trinajstić N (1990) J Math Chem 5:171–175
Trinajstić N (1992) Chemical graph theory, 2 revth edn. CRC, Boca Raton
Trinajstić N, Jeričević Ž, von Knop J et al (1983) Pure Appl Chem 55:379–390
Trinajstić N, Nikolić S, von Knop J et al (1991) Computational chemical graph theory. Simon & Schuster/Horwood, Chichester
von Knop J, Szymanski K, Jeričević Ž, Trinajstić N (1983) J Comput Chem 4:23–32
von Knop J, Szymanski K, Jeričević Ž, Trinajstić N (1984a) MATCH Commun Math Comput Chem 16:119–134
von Knop J, Szymanski K, Klasinc L, Trinajstić N (1984b) Comput Chem 8:107–115
von Knop J, Szymanski K, Trinajstić N, Křivka P (1984c) Comput Math Appl 10:369–382
von Knop J, Müller WR, Szymanski K, Trinajstić N (1985) Computer generation of certain classes of molecules. SKTH, Zagreb
von Knop J, Mûller WR, Szymanski K, Trinajstić N (1990) J Chem Inf Comput Sci 30:159–160
Vukičević D, Trinajstić N (2007) J Math Chem 42:575–583
Vukičević D, Randić M, Balaban AT (2004) J Math Chem 36:271–279
Wiswesser WJ (1954) A line-formula chemical notation. Crowell, New York
Živković TP, Trinajstić N (1987) Chem Phys Lett 136:141–144
Živković TP, Trinajstić N, Randić M (1981) Croat Chem Acta 54:309–320
Živković T, Randić M, Klein DJ et al (1995) J Comput Chem 16:517–526

Chapter 10
Prochirality and Pro-*RS*-Stereogenicity. Stereoisogram Approach Free from the Conventional "Prochirality" and "Prostereogenicity"

Shinsaku Fujita[1]

Abstract The stereoisogram approach concludes that chirality and *RS*-stereogenicity are independent concepts and that prochirality and pro-*RS*-stereogenicity are independent concepts. Accordingly, the conventional approach should be reorganized, where such confused terms as "chirality", "stereogenicity", "prochirality", and "prostereogenicity" should be replaced by newly-defined attributive terms based on the stereoisogram approach, e.g., *chirality* (as a purely geometric term), *RS-stereogenicity*, *prochirality* (as a purely geometric term), and *pro-RS-stereogenicity*. Such relational terms as "enantiotopic", "diastereotopic", and "stereoheterotopic" in the conventional approach should be replaced by newly-defined relational terms based on the stereoisogram approach, e.g., *enantiotopic* (as a purely geometric term), *RS-diastereotopic*, and *RS-stereoheterotopic*.

10.1 Introduction

The conventional term "prochirality" is ambiguous as pointed out by the IUPAC Recommendations 1996: "This term is used in different, sometimes contradictory ways; four are listed below." (IUPAC Organic Chemistry Division 1996). Although the IUPAC Recommendations 1996 has stated "Some misleading terms are included together with guidance on correct usage or acceptable alternatives" in its preface, these different, sometimes contradictory ways have been only enumerated without such guidance.

On the other hand, the term "prostereogenicity" or "prostereoisomerism", which was proposed as a generalization of "prochirality" (Hirschmann and Hanson 1971b; Eliel 1982), has been pointed out to be unnecessary but more appropriate than the term

[1] Shonan Institute of Chemoinformatics and Mathematical Chemistry, Kaneko 479-7, Ooimachi, Ashigara-Kami-Gun, Kanagawa-Ken, 258-0019, Japan
e-mail: shinsaku_fujita@nifty.com

M.V. Putz (ed.), *Carbon Bonding and Structures: Advances in Physics and Chemistry*, Carbon Materials: Chemistry and Physics 5, DOI 10.1007/978-94-007-1733-6_10,

"prochirality" to be abandoned (Helmchen 1996, pages 20 and 74). In contrast to this description, the glossary of a standard textbook (Eliel and Wilen 1994, page 1204) has adopted the terms "prochirality" and referred to "prostereoisomerism" as a more general term. The IUPAC Recommendations 1996 does not contain the definition of the term "prostereogenicity" or "prostereoisomerism" but adopts the ambiguous term "prochirality".

The confusion of these terms ("prochiral", "prostereogenic", and "prostereoisomerism") is so serious that most articles and textbooks other than those of biochemical fields tend to use "enantiotopic", "diastereotopic", and "stereoheterotopic" in place of these terms. However, such replacement conceals problematic situations to be settled, because the confusion reflects misleading situations concerned with "chiral", "stereogenic", and "stereoisomerism", which are the cruxes of stereochemistry.

After we clarify that the conventional approach has not been successful in settling the confusion of these terms, we will demonstrate the stereoisogram approach as a new terminology for settling the confusion.

10.2 Problem Setting — Aims and Scope

To begin with, the present section deals with surveying problematic usages of such terms as "chiral", "stereogenic", "prochiral", "prostereogenic", and so on in the conventional stereochemistry.

10.2.1 Problems of the Conventional "Chirality" and "Stereogenicity"

Since the Cahn-Ingold-Prelog (CIP) system for generating *RS*-stereodescriptors was originally proposed under the title "Specification of Molecular Chirality" (Cahn et al. 1966), such a confused recognition as "Chirality is a basis of the specification of configurations by the CIP system." has widely spread over organic stereochemistry. Although the CIP system has later changed its basis from chirality to stereogenicity (Prelog and Helmchen 1982; Helmchen 1996), the term "stereogenicity" itself has not been directly defined as found in a rulebook (IUPAC Chemical Nomenclature and Structure Representation Division 2004) and most textbooks on stereochemistry (Eliel and Wilen 1994; North 1998; Eliel et al. 2001), so that discrimination between chirality and stereogenicity has not been fully demonstrated. As a result, a related misleading remark such as "Each pair of *RS*-stereodescriptors is given to discriminate between a pair of enantiomers." is used even now, as found in the section title "P91.1 Enantiomers: the CIP priority system" of the IUPAC 2004 Provisional Recommendations (IUPAC Chemical Nomenclature and Structure Representation Division 2004), which presumes a direct linkage between the term *enantiomers* and the CIP system.

In particular, the term "chirality center" (Prelog and Helmchen 1982) has caused serious confusions, as found in a misleading or rather erroneous statement, "Stereogenic centers thus may be or may not be chiral (i.e., centers of chirality). Conversely, however, *all* chiral centers are stereogenic." (Eliel et al. 2001, page 33). For example, let us examine a molecule (**1**) with two achiral ligands (a) and (b) and two chiral ligands of the same kind (p), where the ligand a (or b) can be superposed on its mirror image in isolation and the ligand p cannot be superposed its mirror image in isolation. If the term *chiral* of "chiral center" indicates a geometric meaning, the carbon center (C) of the molecule (**1**) is chiral, because a reflection operated on **1** generates the carbon center of its enantiomer ($\overline{\mathbf{1}}$), where the ligand $\overline{\mathrm{p}}$ represents the mirror image of p in isolation. Note that the carbon center (C) of the molecule (**1**) is enantiomeric (a mirror image) to that of $\overline{\mathbf{1}}$. Because the two mirror-image carbons are not superposable, the carbon center of **1** (as well as that of $\overline{\mathbf{1}}$) is chiral, but "not stereogenic". Hence, the statement cited above (Eliel et al. 2001, page 33) is erroneous, so long as the term *chiral* of "chiral center" indicates a geometric meaning. From the viewpoint of the conventional stereochemistry coupled with the CIP system, on the other hand, the carbon center of **1** (as well as that of $\overline{\mathbf{1}}$) is "not chiral" and "not stereogenic". Thus, the statement (Eliel et al. 2001, page 33) would be justified only if the term *chiral* of a geometric meaning has suffered semantic transmutation (which has once been discussed in a review (Mislow 2002)) to give the term "chiral" of the CIP system.

As found in the preceding discussions, it is safe to say that the term "chirality" for explaining the CIP system (Cahn et al. 1966; Prelog and Helmchen 1982) no longer has the same meaning as the term *chirality* of an original geometric meaning. Note that the original term *chirality* is essential to discuss enantiomeric relationships in organic stereochemistry. It follows that the term "chirality" different from the original term *chirality* should be abandoned, after a theoretical framework is developed to rationalize the abandonment of the transmuted term "chirality". Moreover, the theoretical framework to be developed should explain the difference between chirality and stereogenicity.

10.2.2 Problems of the Conventional "Prochirality" and "Prostereogenicity"

The *pro-R*/*pro-S* system (Hanson 1966) (or the *Re*/*Si* system (Prelog and Helmchen 1972)) for specifying intramolecular environments has been widely adopted in organic stereochemistry. The term "prochirality" for supporting the *pro-R*/*pro-S* system (Hanson 1966) has originally coined to aim at indicating a geometric property, because the prefix *pro* designates a precursory stage of chirality. The original aim of the coinage has been a failure. For example, the central carbon atom of **1** is "prochiral" because the ligands (two p's) can be named by the *pro-R*/*pro-S* system. However, the carbon atom at issue is already chiral geometrically, as discussed in the preceding subsection.

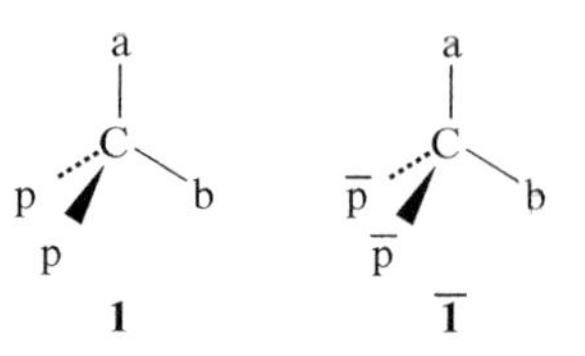

Fig. 10.1 Geometrically chiral but not stereogenic centers. The ligand $\overline{p}$ represents the mirror image of p in isolation and the ligand a (or b) is achiral when detached

Thereby, the same carbon of **1** is concluded (1) to be "prochiral" from the viewpoint of the *pro-R*/*pro-S* system, (2) not to be "chiral" from the viewpoint of the CIP system, and (3) to be chiral geometrically. Note that Item 1 can be deduced from Item 2 because not to be "chiral" is a prerequisite for being "prochiral". This means that the problem of the term "chiral" in the CIP system is immediately linked to the problem of the term "prochiral" of the *pro-R*/*pro-S* system. As a result, it is strange that Helmchen (1996, page 12) maintains the usage of the term "chirality center" in spite of the proposal of the term stereogenic units of type 1 (for chirality units) and type 2 (for pseudoasymmetric units) as being more appropriate. Moreover, it seems to be inconsistent to the usage of the term "chirality center" that he has stated (Helmchen 1996, page 20): "Terms containing *prochiral*, such as *prochiral unit* or *element*, *prochiral molecules* or *prochirality*, must be abandoned." The abandonment of "prochirality" for the *pro-R*/*pro-S* system should inevitably reach the abandonment of "chirality" for the CIP system because of the situation shown in Fig. 10.1.

In a parallel with the revision of the CIP system (Prelog and Helmchen 1982), the basis of the *pro-R*/*pro-S* system has been changed from "prochirality" into "prostereoisomerism" or "prostereogenicity" (Hirschmann and Hanson 1971b). Mislow and Siegel have recommended that the usage of "prochirality" with reference to prostereoisomerism should be altogether abandoned (Mislow and Siegel 1984) after their convincing discussions on geometric features of molecules. In spite of these situations, the term "prochiral" connected with the *pro-R*/*pro-S* system is still used in place of the term "prostereoisomerism" or "prostereogenicity" (IUPAC Organic Chemistry Division 1996). Moreover, discrimination between "prochirality" and prostereogenicity (or prostereoisomerism) has not been fully demonstrated, as symbolized by the title "Prostereoisomerism (Prochirality)" of a review (Eliel 1982). Although the review has expressed, "Correspondingly, the concept of prochirality must be generalized to one of prostereoisomerism" (Eliel 1982, page 4), is this a promising way to be followed?

10.2.3 Problems of the Conventional Dichotomy Between Enantiomers and "Diastereomers"

The conventional stereochemistry has heavily relied on the dichotomy between enantiomers and diastereomers, as found in reviews (Jonas 1988; Mislow 2002), textbooks on stereochemistry (Morris 2001; Mislow 1965; Eliel and Wilen 1994; North 1998; Eliel et al. 2001), glossaries (IUPAC Organic Chemistry Division 1996;

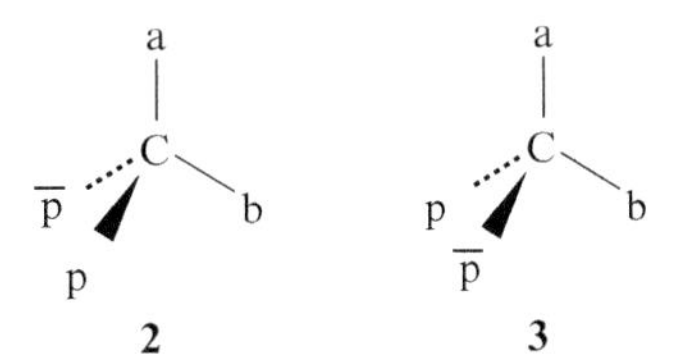

Fig. 10.2 Two achiral stereoisomers of meso-type (pseudo-asymmetric case), where the ligand $\overline{p}$ represents the mirror image of p in isolation

IUPAC Chemical Nomenclature and Structure Representation Division 2004), and several reports on a flow-chart approach (Jonas 1988; Black 1990; Mislow 1977; Vollhardt and Schore 2003). The dichotomy is typically expressed as: "Diastereomers (or diastereoisomers) are stereoisomers not related as mirror images." (Eliel and Wilen 1994, page 1196), "Any pair of stereoisomers which are not enantiomers of one another are called diastereomers." (North 1998, page 17), or other almost equivalent sentences.

Strictly speaking, these expressions involve enantiomeric relationships (based on reflection operations) and stereoisomeric relationships (based on projection operations for giving respective common graphs) as definitely specified categories. They do not directly define diastereomeric relationships, because stereoisomeric relationships minus enantiomeric relationships are generically called diastereomeric relationships. Note that a set of diastereomers does not provide a definite assembly, although a pair of enantiomers (or a set of stereoisomers) provides a definite assembly (an orbit, group-theoretically speaking).

Let us consider two achiral stereoisomers (**2** and **3**) shown in Fig. 10.2. They are also stereoisomeric to **1** and $\overline{\mathbf{1}}$ shown in Fig. 10.1. By the above definitions, the relationship between **1** and **2** (or **3**) is concluded to be diastereomeric, because the enantiomer $\overline{\mathbf{1}}$ is different to the achiral **2**. On the other hand, the relationship between the achiral **2** and **3** is also concluded to be diastereomeric, because their mirror images (**2** and **3** themselves) are different to each other. It should be noted, however, that the diastereomeric relationship between **1** and **2** is different from the diastereomeric relationship between **2** and **3** in whether there exist enantiomeric relationships or not.

From another point of view, an isomerization from **1** (chiral) to **2** (achiral) is different from an isomerization from **2** (achiral) to **3** (achiral), where both of the isomerizations correspond to the above-mentioned diastereomeric relationships of different types. Hence, we should say that there exist at least two diastereomeric relationships of different types. The conventional stereochemistry is silent about such diastereomeric relationships of different types. In other words, the dichotomy between enantiomers and diastereomers in the conventional stereochemistry is oversimplified.

10.2.4 Problems of the Conventional Dichotomy Between Enantiotopic Relationships and "Diastereotopic" Ones

The terms *enantiotopic* and "diastereotopic" have been first coined for describing geometric features (Mislow and Raban 1967). Note that a set (pair) of enantiotopic

a | C / X X b — 4 — e.g., C -1 of ethanol (X_2: enantiotopic)

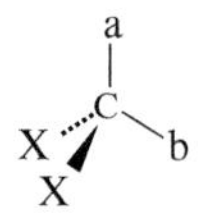

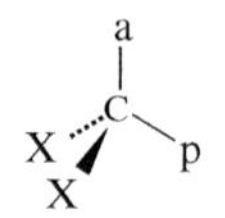

a | C / X X p — 5 — e.g., C-3 of a butan-2-ol (X_2: "diastereotopic")

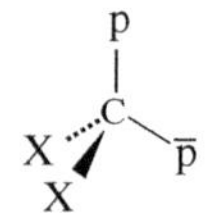

p | C / X X $\overline{p}$ — 6 — e.g., C-3 of a pentane-2,4-diol (X_2: "diastereotopic")

Fig. 10.3 "Prostereogenic centers" due to "stereoheterotopic" relationships of different types in the conventional terminology of stereochemistry, where the ligands a, b, and X represents achiral ligands in isolation and the ligand $\overline{p}$ represents a chiral ligand that is the mirror image of a chiral ligand p in isolation

ligands produces a definite assembly (i.e., an equivalence class), while a set (pair) of "diastereotopic" ligands is incapable of producing a definite assembly. Thus, the two terms for describing geometric features belong to different categories of concept. Later, the term "stereoheterotopic" has been defined as a combined concept of the conceptually different terms (enantiotopic and "diastereotopic") (Hirschmann and Hanson 1971a) in a parallel way to the dichotomy between enantiomers and "diastereomers" for classifying stereoisomers (stereoisomeric relationships). The term "stereoheterotopic" has been a key to test "prostereogenicity centers" (substituted for "prochiral centers"), where an atom of a molecule which becomes a "stereogenic center" (or "chiral center") by replacing one of the two "stereoheterotopic" ligands attached to it by a different ligand is said to be a "prostereogenicity center" (or a "prochirality center"), e.g., C-1 of ethanol (an enantiotopic case), C-3 of butan-2-ol (a diastereotopic case) (IUPAC Organic Chemistry Division 1996).

Figure 10.3 summarizes topicity relationships in the conventional terminology of stereochemistry. The two achiral ligands X's in **4** (the ligands a and b are also achiral in isolation) are enantiotopic to each other, while the two achiral ligands X's in **5** (a: achiral, p: chiral in isolation) as well as the two achiral ligands X's in **6** (p and $\overline{p}$: enantiomeric in isolation) are "diastereotopic" to each other. The difference between the enantiotopic X's in **4** and the "diastereotopic" X's in **5** (or **6**) stems from difference in their substitution products (enantiomers vs. "diastereomers").

The combination of the two terms (enantiotopic and "diastereotopic") into a single term "stereoheterotopic" has aimed at coining such a single term to give a basis to the *pro-R*/*pro-S* system (Hirschmann and Hanson 1971a). After this coinage, all the pairs of achiral ligands X's in the respective molecules listed in Fig. 10.3 (**3**, **4**, and **5**) can be apparently referred to as being "stereoheterotopic" so as to be named by the *pro-R*/*pro-S* system. This combination has brought about the dichotomy between enantiotopic relationships and "diastereotopic" relationships in a molecule ("stereoheterotopic") in parallel with the dichotomy between enantiomeric relationships and "diastereomeric" relationships among molecules ("stereoisomeric") (Mislow 1977).

However, is there a plausible rationalization to combine enantiotopic and "diastereotopic" into a single term "stereoheterotopic"?

One of such rationalizations belongs to a permutation category (concerned with permutation groups) such that two ligands X's of each molecule listed in Fig. 10.3 are exchangeable by a permutation between them without changing molecular properties. However, the term *enantiotopic* belongs to a geometric category (concerned with point groups), which is different from the permutation category. The preference of the term *enantiotopic* for **4** results in the neglect of the permutational exchangeability, which is common to **4**, **5**, and **6**. Hence, it is necessary to demonstrate whether the three terms (enantiotopic, "diastereotopic", and "stereoheterotopic") are properly related or not.

The discussion described in the preceding paragraph brings us back to an additional question on the dichotomy between enantiomeric relationships and "diastereomeric" relationships among molecules: Is there a plausible rationalization to combine enantiomeric (geometric category) and "diastereomeric" (permutation category) into a single term "stereoisomeric"?

10.2.5 Problems of the Transmuted Term "Enantiotopic"

The original term *enantiotopic* defined by Mislow and Raban (1967) has a geometric meaning. By applying this geometric definition strictly, the ligands p and $\overline{\mathrm{p}}$ in **2** (or **3** in Fig. 10.2 or **6** in Fig. 10.3) are concluded to be in an enantiotopic relationship. According to the conventional terminology described in Sect. 10.2.4, such enantiotopic ligands as p and $\overline{\mathrm{p}}$ should be named by the *pro-R*/*pro-S* system, although they do not require the differentiation by the *pro-R*/*pro-S* system. This inconsistency means that there is an enantiotopic pair beyond the scope of the *pro-R*/*pro-S* system, so that the criteria for rationalizing the *pro-R*/*pro-S* system, i.e., *enantiotopic*, "diastereotopic", and "stereoheterotopic", are proven to be futile.

To remedy this futileness, the geometric definition of the term *enantiotopic* has been changed into a transmuted one: "Enantiotopic ligands and faces. Homomorphic (q.v.) ligands in constitutionally equivalent locations that are symmetry plane (or center or alternating axis of symmetry) but not a (simple) symmetry axis." (Eliel and Wilen 1994, page 1198), where the modifier "homomorphic" has been added. This transmuted term can exclude such cases as a pair of p and $\overline{\mathrm{p}}$ (in **2**, **3**, or **6**) because p and $\overline{\mathrm{p}}$ are not homomorphic, where the term "homomorphic ligands" has been defined: "Ligands that are structurally (including configurationally) identical when detached." (Eliel and Wilen 1994, page 1200). By examining carefully, however, such exclusion by transmuting the geometrically defined term *enantiotopic* seems not to be rationalized, causing latent confusions. Note that p and $\overline{\mathrm{p}}$ (in **2** or **3** or **6**) can be differentiated by an appropriate chiral reagent to generate chiral products (i.e., *prochiral* geometrically). In other words, this excluded case exhibits the essentially same feature as the two X's of the molecule **4**. Hence, such exclusion demonstrates the limitation of the *pro-R*/*pro-S* system, even if it is unreasonably justified by the transmuted term "enantiotopic".

Moreover, the remedy by adding the modifier "homomorphic" is incomplete. Consider a case with p = C(H)(OH)COOH and $\overline{\mathrm{p}}$ = C(OH)(H)COOH in **2** (or **3** or **6**). Then, the ligand OH of p and the corresponding ligand OH of $\overline{\mathrm{p}}$ are in an enantiotopic relationship from a viewpoint of the original geometric term (Mislow and Raban 1967) as well as from a viewpoint of the transmuted term (Eliel and Wilen 1994, page 1198), because the ligand OH is homomorphic according to the definition described above (Eliel and Wilen 1994, page 1200). In other words, there again emerges an "enantiotopic" (also *enantiotopic*) case which indicates such inconsistency that there is an "enantiotopic" pair beyond the scope of the *pro-R*/*pro-S* system. To remedy this inconsistency, a further transmutation of the transmuted term "enantiotopic" would be necessary. Obviously, such multiple transmutation no longer indicates a proper way to be followed.

10.2.6 Aims and Scope

The problems of various types in the conventional stereochemistry are highly entangled as the result of the semantic transmutation described in the preceding subsections. It is impossible to solve the entangled problems if we maintain the conventional terminology of stereochemistry. Hence, the conventional terminology should be entirely reconsidered to restructure stereochemistry.

1. The term "prochirality" of the *pro-R*/*pro-S* system should be abandoned in the same way as the transmuted term "chirality" of the CIP system should be abandoned. In other words, the theoretical framework to be developed should define a term *prochirality* by starting from the term *chirality* of a purely geometric meaning (Fujita 1991b, 2007c).
2. The term "prostereogenicity" or "prostereoisomerism" of the *pro-R*/*pro-S* system and the term "stereogenicity" for giving *RS*-stereodescriptors of the CIP system should be abandoned. Instead, the newly-developed theoretical framework will define pro-*RS*-stereogenicity by starting from *RS*-stereogenicity, which has recently been developed (Fujita 2004a,c,d, 2005e).
3. The terms "enantiotopic", "diastereotopic", and "stereoheterotopic" should be abandoned to support the *pro-R*/*pro-S* system, just as the terms "enantiomeric", "diastereomeric", and "stereogenic" for supporting *RS*-stereodescriptors of the CIP system should be abandoned. In particular, the term *enantiotopic* will be used only to explain geometric aspects apart from the *pro-R*/*pro-S* system, while the newly-defined term *RS*-diastereotopic will be used to support the *pro-R*/*pro-S* system. On a similar line, the term *enantiomeric* will be used only to

explain geometric aspects apart from the CIP system, while the newly-defined term *RS*-diastereomeric will be used to support *RS*-stereodescriptors of the CIP system.
4. The use of the term "stereoheterotopic" should be entirely ceased to support the *pro-R*/*pro-S* system. Thereby, the dichotomy between enantiotopic and "diastereotopic" (for the term "stereoheterotopic") will be abandoned. The related terms such as "stereoheterotopism" or "stereoheterotopicity" (Hirschmann and Hanson 1971a; Hirschmann 1983) for supporting the *pro-R*/*pro-S* system will be also abandoned.
5. The transmuted term "enantiotopic" (Eliel and Wilen 1994, page 1198) should be abandoned. The term *enantiotopic* will be used in a purely geometric meaning.
6. In addition, the theoretical framework should explain the difference between the newly-defined prochirality (geometrically) and pro-*RS*-stereogenicity, just as it should explain the difference between chirality (geometrically) and *RS*-stereogenicity.

To prevent such semantic transmutation as described in the preceding subsections, we have developed a diagrammatic approach based on the group theory (point groups and permutation groups). The diagrammatic approach is here called *stereoisogram approach* because it stems from the concept of *stereoisograms* (Fujita 2004c,d, 2009b,d,e). The stereoisogram approach is a theoretical framework common to intermolecular and intramolecular features as well as to geometric and stereoisomeric features. Although the present discussions will omit the mathematical basis of the stereoisogram approach (Fujita 2004a,b, 2005d,e), it will be demonstrated elsewhere.

10.3 The Stereoisogram Approach

In the stereoisogram approach, reflection operations are differentiated distinctly from *RS*-permutations in whether ligand reflections are involved or not. This section is devoted to demonstrate that such differentiation generates three kinds of relationships for constructing stereoisograms. The crux of the stereoisogram approach is that chirality (related to reflections) and *RS*-stereogenicity (related to *RS*-permutations) are independent concepts.

10.3.1 *Reflections and* RS-*Permutations for Promolecules*

For the sake of simplicity, the present chapter is restricted to tetrahedral molecules, which are further simplified into tetrahedral promolecules substituted by a set of proligands. Promolecules have been defined as abstract molecules in which a set of proligands are placed on a given skeleton, where such proligands are defined as abstract ligands having either chirality or achirality (Fujita 1991b,a, 2000a).

This means that each proligand is achiral or chiral in isolation. When chiral, a pair of enantiomeric proligands in isolation is represents by a pair of an alphabet and an overlined alphabet, e.g., p and $\overline{\mathrm{p}}$ shown in Fig. 10.1. Differentiation between reflections and *RS*-permutations is essential to construct stereoisograms (Fujita 2004d,a,c). In addition, ligand reflections are created as operations of another type by operating reflections and *RS*-permutations successively.

Reflections A reflection ◉ operated on a promolecule generates the corresponding mirror image. When the original promolecule is not superposable on the mirror image, they are chiral and enantiomeric to each other. When the original promolecule is superposable on the mirror image, the promolecule is achiral. During such a reflection, each chiral proligand in the original promolecule is transformed into its mirror image in isolation and each achiral proligand is transformed into itself in isolation.

For example, a reflection operated on **1** generates its enantiomer $\overline{\mathbf{1}}$. On the other hand, a reflection operated on **2** (or **3**) generates **2** (or **3**) itself because of achirality.

It should be noted that two or more achiral ligands of the same kind can have local chiralities when they belong to an enantiospheric orbit in a molecule (Fujita 1991b, Chapter 10).

RS-Permutation An *RS*-permutation ○ corresponds to a reflection, where the chirality or achirality of each proligand (in isolation) is not changed. For example, an *RS*-permutation on the two proligands p's of **1** generates **1** itself. On the other hand, an *RS*-permutation on the proligands p and $\overline{\mathrm{p}}$ of **2** generates **3**.

Ligand reflections A ligand reflection (•) operated on a promolecule produces a promolecule which have the same skeleton as the original skeleton (not reflected) where all of the ligands are changed into their mirror images (in isolation).

10.3.2 Relationships of Three Types and Attributes of Three Types

The three types of operations bring about transformations of promolecules, where there emerge three types of relationships listed in Table 10.1 (Fujita 2004c). A reflection is related to an enantiomeric or self-enantiomeric relationship, by which a molecule is categorized to be chiral or achiral. An *RS*-permutation is related to an *RS*-diastereomeric or self-*RS*-diastereomeric relationship, by which a molecule is categorized to be *RS*-stereogenic or *RS*-astereogenic. A ligand reflection is related to a holantimeric or self-holantimeric relationship, by which a molecule is categorized to be scleral or ascleral.

Table 10.1 Three relationships and the corresponding attributes appearing in stereoisograms (Fujita 2004c)

Symbol	Relationship	Attribute
(Concerned with reflections ◉)		
←◉→	enantiomeric	chiral
=◉=	(self-enantiomeric)	achiral
(Concerned with *RS*-permutations ○)		
←○→	*RS*-diastereomeric	*RS*-stereogenic
=○=	(self-*RS*-diastereomeric)	*RS*-astereogenic
(Concerned with lig and reflections ●)		
←●→	holantimeric	scleral
=●=	(self-holantimeric)	ascleral

As summarized in Table 10.1, the newly-defined *RS*-diastereomeric (or holantimeric) relationship is a pairwise relationship, just as the enantiomeric relationship is a pairwise relationship. Thereby, a pair of attributes (properties of promolecules), i.e., *RS*-stereogenicity/*RS*-astereogenicity or sclerality/asclerality, is introduced in a similar way to a pair of chiralty/achirality.

Because reflections and *RS*-permutations are distinct operations as described above, chirality/achirality and *RS*-stereogenicity/*RS*-astereogenicity are distinct and independent concepts, as shown in Table 10.1. On the same line, enantiomeric relationships are distinct from and independent of *RS*-diastereomeric relationships, where the former may be superposed on the latter in special cases. In addition to the two categories, the existence of sclerality/asclerality provides us with a further theoretical framework, i.e., *RS*-stereoisomerism based on *RS*-stereoisomeric groups, which integrates chirality/achirality, *RS*-stereogenicity/*RS*-astereogenicity, and sclerality/asclerality (Fujita 2004c, 2005e).

In the conventional approach, on the other hand, the term "stereogenicity" and the term "stereoisomerism" have been used synonymously, as exemplified by the definition of "stereogenic unit" (IUPAC Organic Chemistry Division 1996) and by the phases of "stereogenic elements" (McCasland 1953; Prelog and Helmchen 1982) and "elements of stereoisomerism" (Hirschmann and Hanson 1971b). Such synonymous usage should be ceased so as to be harmonized with the present approach which distinguishes between *RS*-stereogenicity and *RS*-stereoisomerism, as discussed later in Sect. 10.3.5.3.

It should be emphasized that "diastereomeric" relationships of the conventional terminology are not pairwise relationships, whereas *RS*-diastereomeric (or holantimeric) relationships of the present terminology are pairwise relationships. This fact is a succinct piece of evidence for stating that the conventional dichotomy between enantiomers and "diastereomers" are oversimplified.

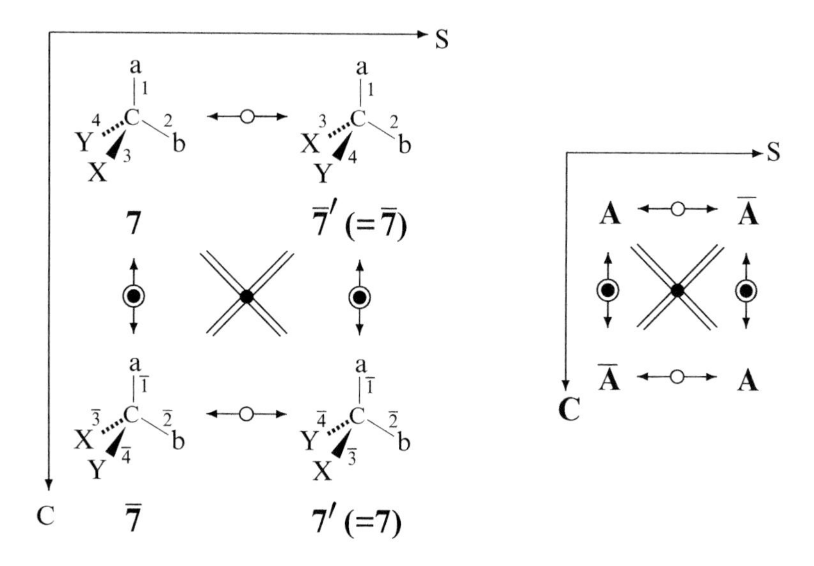

Fig. 10.4 Stereoisogram for characterizing promolecules of Type I, which is characterized by chiral, *RS*-stereogenic, ascleral attributes (stereoisogram index: $[-, -, a]$)

10.3.3 Construction of Stereoisograms

The construction of stereoisograms of Type I–V is illustrated by using representative examples so as to show how three relationships and the corresponding attributes (Table 10.1) participate in a stereoisogram.

10.3.3.1 Stereoisograms of Type I

Let us operate the three types of operations on a chiral molecule **7**, which have four achiral ligands (a, b, X, and Y) in isolation (Fig. 10.4).

A reflection operated on **7** is shown in the vertical direction of Fig. 10.4 (C-axis: chirality-axis), where its enantiomer ($\overline{\mathbf{7}}$) is generated to be combined to the original promolecule **7** by means of a vertical two-headed arrow modified by an encircled solid circle (Table 10.1). The locant of each position is represented by a number with an overbar, when the position accommodates a ligand with a changed chirality sense. Note that an achiral ligand (e.g., X on the position 3 of **7**) is changed into itself (e.g., X on the position $\overline{3}$ of $\overline{\mathbf{7}}$). This means that the mirror image of X (in isolation) is identical with the original X, i.e., $\overline{\mathrm{X}} = \mathrm{X}$.

An *RS*-permutation operated on **7** is shown in the horizontal direction of Fig. 10.4 (S-axis: *RS*-stereogenicity-axis), where its *RS*-diastereomer $\overline{\mathbf{7}}'$ (= $\overline{\mathbf{7}}$)

is generated so as to be combined to the original promolecule **7** by a horizontal two-headed arrow with a circle (Table 10.1). The locant of each position is represented by a number without an overbar, because the position accommodates a ligand with an unchanged chirality sense.

A ligand reflection operated on **7** is shown in the diagonal direction of Fig. 10.4, where its holantimer $\mathbf{7}'$ $(= \mathbf{7})$ is generated so as to be combined to the original promolecule **7** by a diagonal equality symbol with a solid circle (Table 10.1). Note that the original promolecule **7** is identical with its holantimer, i.e., $\mathbf{7}' = \mathbf{7}$, where they are self-holantimeric to each other. The locant of each position is represented by a number with an overbar, because the position accommodates a ligand with a changed (opposite) chirality sense.

After the generation of the quadruplet of the promolecules, i.e., $\mathbf{7}, \overline{\mathbf{7}}, \overline{\mathbf{7}}'$ $(= \overline{\mathbf{7}})$, and $\mathbf{7}'$ $(= \mathbf{7})$, the remaining relationships of respective pairs are specified, i.e., an *RS*-diastereomeric relationship between $\overline{\mathbf{7}}$ and $\mathbf{7}'$ $(= \mathbf{7})$ in the bottom horizontal direction, an enantiomeric relationship between $\overline{\mathbf{7}}'$ $(= \overline{\mathbf{7}})$ and $\mathbf{7}'$ $(= \mathbf{7})$ in the right-hand vertical direction, as well as a self-holantimeric relationship between $\overline{\mathbf{7}}$ and $\overline{\mathbf{7}}'$ $(= \overline{\mathbf{7}})$ in another diagonal direction. Thereby, we are able to construct the stereoisogram of Type I shown in Fig. 10.4, which is characterized by the presence of self-holantimeric relationships. The symbol $[-, -, a]$ for a Type I stereoisogram is called a stereoisogram index, which sequentially indicates chirality, *RS*-stereogenicity, and asclerality in accord with the existence (a) or non-existence ($-$) of the prefix 'a'.

It should be emphasized that reflection operations are conceptually independent of *RS*-permutation operations. The criteria collected in Table 10.1 indicate that a pair of chirality/achirality as attributes is conceptually independent of a pair of *RS*-stereogenicity/*RS*-astereogenicity and that enantiomeric relationships are independent of *RS*-diastereomeric relationships. Hence, the stereoisogram of Type I provides us with a new viewpoint to prevent the conventional dichotomy between enantiomers and "diastereomers":

1. In the present stereoisogram approach, the *RS*-diastereomer ($\overline{\mathbf{7}}' = \overline{\mathbf{7}}$) of **7** is conceptually different from the enantiomer ($\overline{\mathbf{7}}$) of **7**, even if they are identical with each other. The *RS*-diastereomeric relationship between **7** and $\overline{\mathbf{7}}'$ $(= \overline{\mathbf{7}})$ is recognized to be superposable on the enantiomeric relationship between **7** and $\overline{\mathbf{7}}$. This conceptual feature is emphasized by the recognition of the self-holantimeric relationship between **7** and $\mathbf{7}'$ $(= \mathbf{7})$ or between $\overline{\mathbf{7}}$ and $\overline{\mathbf{7}}'$ $(= \overline{\mathbf{7}})$.
2. In contrast, the enantiomeric relationship between **7** and $\overline{\mathbf{7}}$ is preferred in the conventional approach of stereochemistry. Once the relationship between **7** and $\overline{\mathbf{7}}$ has been recognized as being enantiomeric, the *RS*-diastereomeric relationship between **7** and $\overline{\mathbf{7}}'$ $(= \overline{\mathbf{7}})$ is neglected, i.e., recognized not to exist under the dichotomy between enantiomers and "diastereomers".

As a result, the conventional dichotomy between enantiomers and "diastereomers" is concluded to be misleading, because it neglects *RS*-diastereomeric relationships which are superposable on enantiomeric relationships (Fig. 10.4).

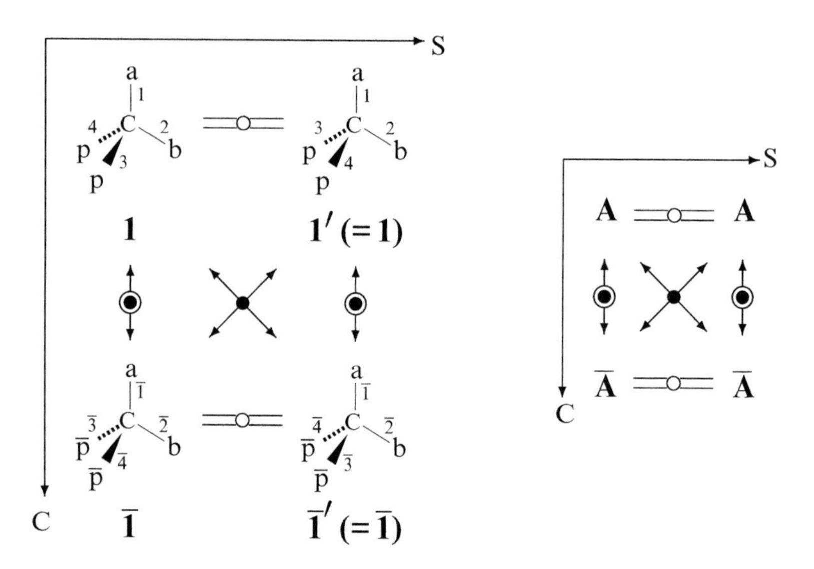

Fig. 10.5 Stereoisogram for characterizing promolecules (**1** etc.) of Type II, which is characterized by chiral, *RS*-astereogenic, scleral attributes (stereoisogram index: $[-, a, -]$)

10.3.3.2 Stereoisograms of Type II

Fig. 10.5 shows a stereoisogram concerned with **1** and $\overline{\mathbf{1}}$ shown in Fig. 10.1. They appear in the left-hand vertical direction, which shows the enantiomeric relationship between **1** and $\overline{\mathbf{1}}$. The stereoisogram (Fig. 10.5) is characterized by the presence of self-*RS*-diastereomeric relationships and referred to as belonging to Type II. The discussions described in Sect. 10.2.1 are more clearly demonstrated by means of the stereoisogram shown in Fig. 10.5. The stereoisogram index $[-, a, -]$ for a Type II stereoisogram sequentially indicates chirality, *RS*-astereogenicity, and sclerality.

10.3.3.3 Stereoisograms of Type III

A stereoisogram shown in Fig. 10.6 (a, b, and X are achiral in isolation, and p and $\overline{\mathrm{p}}$ represent a pair of enantiomeric ligands in isolation) is referred to as being Type III, where it is characterized by the feature that four components (**8**, $\overline{\mathbf{8}}$, **9**, and $\overline{\mathbf{9}}$) are different from one another. The stereoisogram index $[-, -, -]$ for a Type III stereoisogram sequentially indicates chirality, *RS*-stereogenicity, and sclerality.

10.3.3.4 Stereoisograms of Type IV

A stereoisogram shown in Fig. 10.7 is referred to as belonging to Type IV, where it is characterized by the feature that four components ($\mathbf{4} = \overline{\mathbf{4}} = \mathbf{4}' = \overline{\mathbf{4}}'$, cf. Fig. 10.3) represent the same promolecule. The stereoisogram index [*a*, *a*, *a*] for a Type IV stereoisogram sequentially indicates achirality, *RS*-astereogenicity, and asclerality.

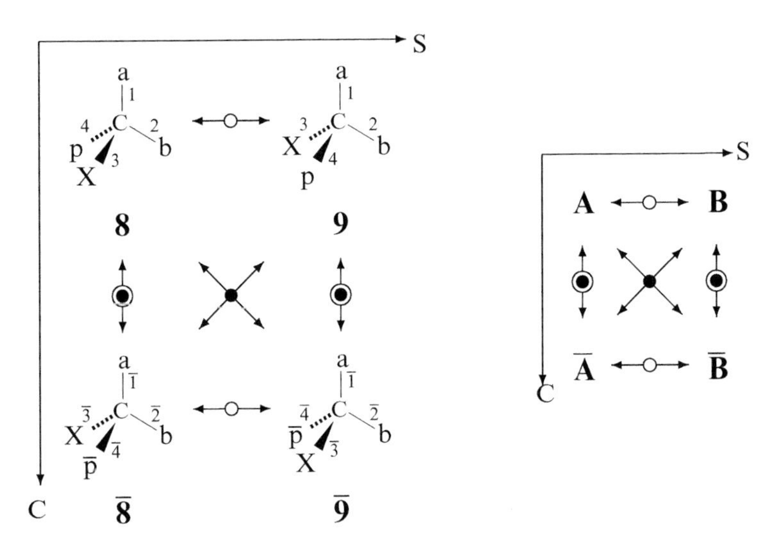

Fig. 10.6 Stereoisogram for characterizing promolecules of Type III, which is characterized by chiral, *RS*-stereogenic, scleral attributes (stereoisogram index: $[-,-,-]$)

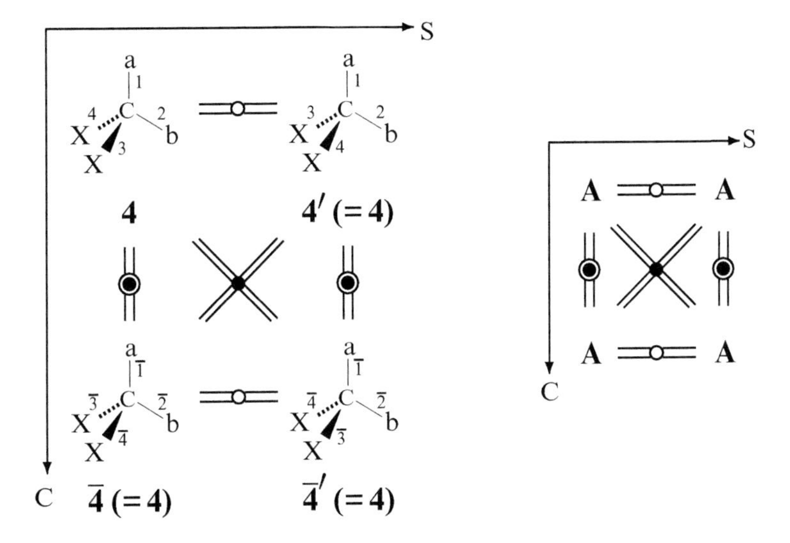

Fig. 10.7 Stereoisogram for characterizing promolecules of Type IV, which is characterized by achiral, *RS*-astereogenic, ascleral attributes (stereoisogram index: $[a, a, a]$)

10.3.3.5 Stereoisograms of Type V

A stereoisogram shown in Fig. 10.8 is concerned with **2** and **3** (Fig. 10.2). The stereoisogram is referred to as belonging to Type V, which is characterized by the presence of self-enantiomeric relationships. The stereoisogram index $[a,-,-]$ for a Type V stereoisogram sequentially indicates achirality, *RS*-stereogenicity, and sclerality.

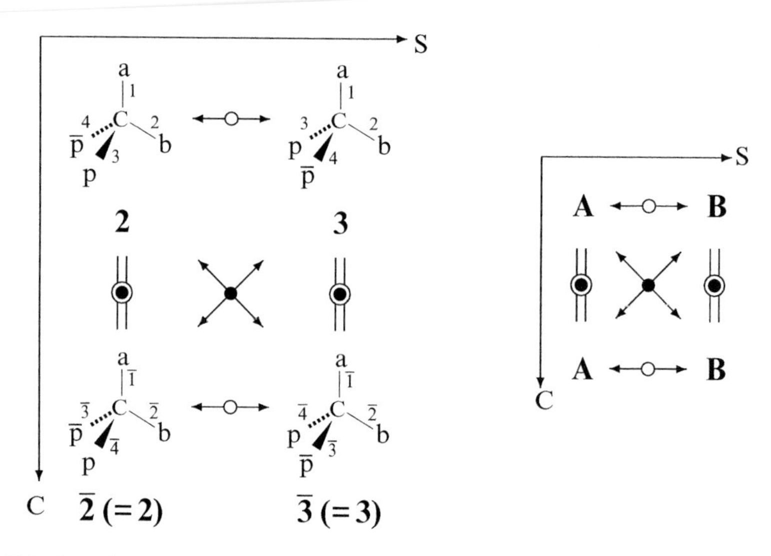

Fig. 10.8 Stereoisogram for characterizing promolecules of Type V, which is characterized by achiral, *RS*-stereogenic, scleral attributes (stereoisogram index: $[a, -, -]$)

10.3.4 Stereoisograms Categorized to Five Types

The quadruplet of promolecules in each stereoisogram is governed by an *RS*-stereoisomeric group defined by a point group and the corresponding permutation group (Fujita 2004b, 2005d). Such an *RS*-stereoisomeric group has been proven to have five types of subgroups, which directly indicates that there exist stereoisograms of five types (Fujita 2005e).

Each of the stereoisograms shown in the left-hand sides of Figs. 10.4–10.8 is abstractly represented by the corresponding abbreviated form shown in the right-hand side. They are collected in Fig. 10.9, which are characterized by virtue of three pairs of attributes, i.e., chirality/achirality, *RS*-stereogenicity/*RS*-astereogenicity, and sclerality/asclerality. For example, a stereoisogram of Type I is characterized by a set of attributes, i.e., chiral, *RS*-stereogenic, and ascleral, which is represent by a stereoisogram index $[-, -, a]$, where the italicized letter a stands for asclerality. The set of attributes corresponds to a set of enantiomeric, *RS*-diastereomeric, and self-holantimeric relationships, as found in Table 10.1.

The rows of Fig. 10.9 indicate chirality and achirality, where the upper row collects chiral stereoisograms (Types I–III) and the bottom row collects achiral stereoisograms (Type IV and V). On the other hand, the columns of Fig. 10.9 indicate *RS*-stereogenicity, where the right column collects *RS*-stereogenic stereoisograms (Types I, III, and V) and the left column collects *RS*-stereogenic stereoisograms (Types II and IV). Another category is possible, i.e., scleral stereoisograms (Types II, III, and V) and ascleral stereoisograms (Types I and IV), although Fig. 10.9 does not directly illustrate.

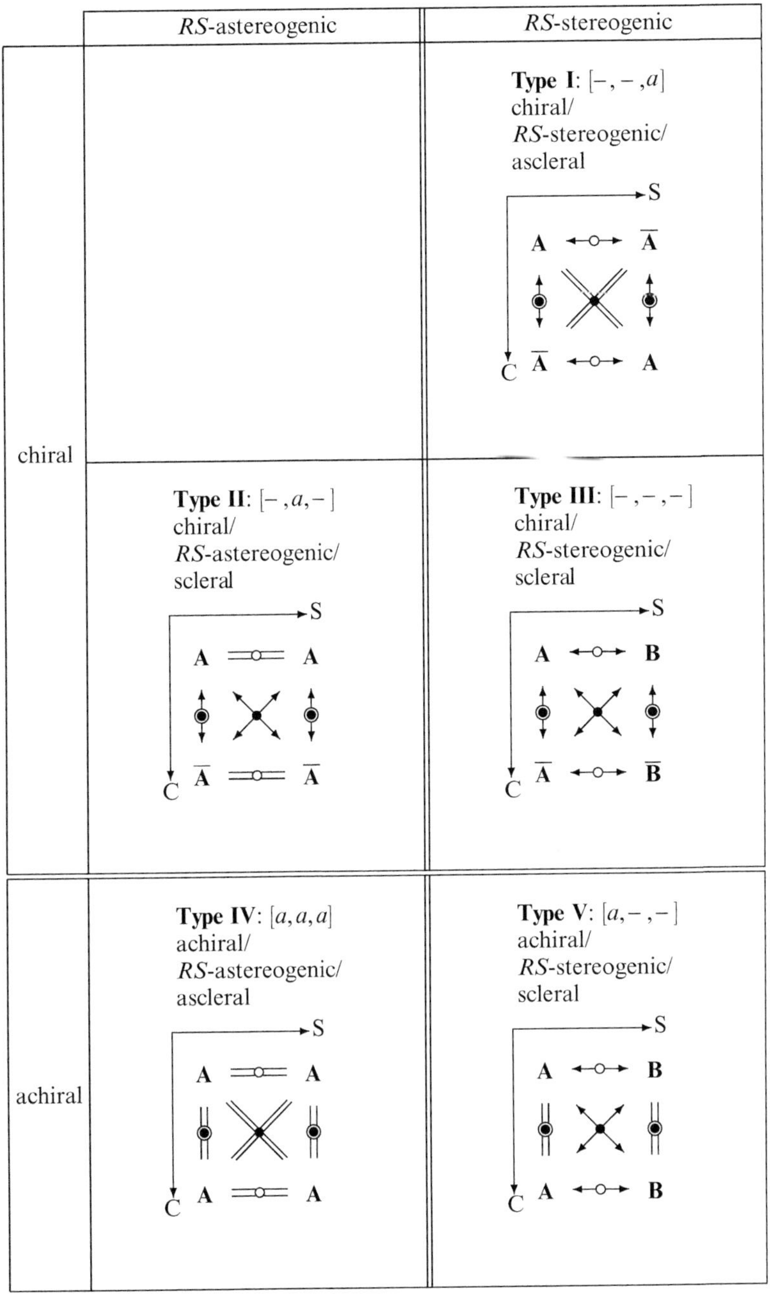

	RS-astereogenic	*RS*-stereogenic
chiral		**Type I**: $[-,-,a]$ chiral/ *RS*-stereogenic/ ascleral S A ↔ $\overline{\mathbf{A}}$ C $\overline{\mathbf{A}}$ ↔ A
	Type II: $[-,a,-]$ chiral/ *RS*-astereogenic/ scleral S A = A C $\overline{\mathbf{A}}$ = $\overline{\mathbf{A}}$	**Type III**: $[-,-,-]$ chiral/ *RS*-stereogenic/ scleral S A ↔ B C $\overline{\mathbf{A}}$ ↔ $\overline{\mathbf{B}}$
achiral	**Type IV**: $[a,a,a]$ achiral/ *RS*-astereogenic/ ascleral S A = A C A = A	**Type V**: $[a,-,-]$ achiral/ *RS*-stereogenic/ scleral S A ↔ B C A ↔ B

Fig. 10.9 Stereoisograms of five types (Fujita 2004c). The symbols **A** and $\overline{\mathbf{A}}$ (or **B** and $\overline{\mathbf{B}}$) represent a pair of enantiomers

The existence of stereoisograms of five types can be alternatively shown by a simple consideration on stereoisogram indices. Thus, the three pairs of attributes (chirality/achirality, *RS*-stereogenicity/*RS*-astereogenicity, and sclerality/asclerality) indicates 8 ($= 2^3$) combinations, i.e., [*a*, *a*, *a*] to [–, –, –]. Any two attributes (or relationships) selected from those having the prefix 'a' (or relationships having the prefix 'self'), i.e., achirality (self-enantiomeric relationship), *RS*-astereogenicity (self-*RS*-diastereomeric relationship), and asclerality (self-holantimeric relationship), are easily shown not to give an effective stereoisogram. This means that three cases selecting such two attributes should be omitted, i.e., [– , *a*, *a*], [*a*, – , *a*] and [*a*, *a*, –]. It follows that 5 ($= 8 - 3$) types are possible, as shown in Fig. 10.9.

10.3.5 RS-Stereodescriptors Assigned by RS-Stereogenicity

The stereoisogram approach and the conventional approach will be compared with respect to how the assignment of *RS*-stereodescriptors is rationalized.

10.3.5.1 Systematic Rationalization Due to the Stereoisogram Approach

RS-Stereodescriptors of the CIP system have pairwise nature, so that a pair of *RS*-stereodescriptors should be assigned to a pair of promolecules which are paired by a well-defined single criterion. The present approach has reached the conclusion that each pair of *RS*-stereodescriptors of the CIP system is assigned to a pair of *RS*-diastereomers, which is directly linked to *RS*-stereogenicity appearing in the horizontal direction (the S-axis: the *RS*-stereogenicity axis) of each stereoisogram, as shown in Fig. 10.9 (Fujita 2004a,c,d, 2005e, 2009b,e). Thus, a promolecule of Type I, III or V (the right-hand column of Fig. 10.9) can be named by using the CIP system, because they are commonly characterized by their *RS*-stereogenicity (corresponding to *RS*-diastereomeric relationships), as summarized in Table 10.2. The carbon centers are referred to as *RS*-stereogenic centers. The chirality of Type I, the chirality of Type III, and the achirality of Type V indicate that the chirality is by no means a common attribute for rationalizing *RS*-stereodescriptors of the CIP system. After specifying *RS*-stereodescriptors on the basis of

Table 10.2 Single criterion for giving *RS*-stereodescriptors of the CIP system in the stereoisogram approach

	Chirality	*RS*-stereogenicity
Type I	enantiomeric	*RS*-diastereomeric
Type III	enantiomeric	*RS*-diastereomeric
Type V	self-enantiomeric	*RS*-diastereomeric

(chirality-faithfulness)

RS-stereogenicity, the resulting *RS*-stereodescriptors are subsidiarily correlated to chirality by the newly-defined concept of *chirality-faithfulness* (Fujita 2009d).

Let us examine promolecules of Types I, III, and V to show a systematic rationalization provided by the stereoisogram approach (Table 10.2), where the concept of *chirality-faithfulness* (Fujita 2009d) assures a systematic but subsidiary linkage between chirality and *RS*-stereogenicity, which are independent concepts to each other.

1. The stereoisogram approach adopts *RS*-stereogenicity (or an *RS*-diastereomeric relationship) as a single criterion for rationalizing *RS*-stereodescriptors of the CIP system, as surrounded by a box in Table 10.2. For example, the Type I case of Fig. 10.4 exclusively takes account of the horizontal S-axis, where a pair of *RS*-stereodescriptors is given to a pair of *RS*-diastereomers, i.e., **7** and $\mathbf{7}'$ ($= \overline{\mathbf{7}}$). Because these promolecules have the common set of detached proligands a, b, X, and Y, they can be assigned to a pair of *RS*-stereodescriptors by applying such a common priority sequence as being a > b > X > Y (tentatively specified for the sake of explanation).

   The pair of *RS*-stereodescriptors assigned to the pair of *RS*-diastereomers are then reinterpreted to be assigned to the pair of enantiomers located at the vertical C-axis of Fig. 10.4, i.e., **7** and $\overline{\mathbf{7}}$. This reinterpretation is allowed by virtue of the concept of *chirality-faithfulness* (Fujita 2009d), because the priority sequence (a > b > X > Y) for the pair of *RS*-diastereomers **7** and $\mathbf{7}'$ ($= \overline{\mathbf{7}}$) is the same as the one for the pair of enantiomers **7** and $\overline{\mathbf{7}}$.
2. Each of Type III cases is characterized by a quadruplet of four different promolecules, where a pair of *RS*-stereodescriptors is also assigned to a pair of *RS*-diastereomers, as found in Table 10.2.

   a. For example, a pair of *RS*-diastereomers **8** and **9** (or $\overline{\mathbf{8}}$ and $\overline{\mathbf{9}}$) located at the horizontal S-axis of Fig. 10.6 is characterized by a pair of *RS*-stereodescriptors in terms of such a common priority sequence as a > b > X > p for **8**/**9** (or a > b > X > $\overline{\text{p}}$ for $\overline{\mathbf{8}}$/$\overline{\mathbf{9}}$), where the priority sequence is tentatively specified for the sake of explanation.

      The pair of *RS*-stereodescriptors assigned to the pair of *RS*-diastereomers (**8**/**9** or $\overline{\mathbf{8}}$/$\overline{\mathbf{9}}$) is then reinterpreted to be assigned to the pair of enantiomers located at the vertical C-axis of Fig. 10.4, i.e., **8**/$\overline{\mathbf{8}}$ (or **9**/$\overline{\mathbf{9}}$). In the process of such reinterpretation, the priority sequences a > b > X > p for **8** and a > b > X > $\overline{\text{p}}$ for $\overline{\mathbf{8}}$ (or a > b > X > $\overline{\text{p}}$ for **9** and a > b > X > p for $\overline{\mathbf{9}}$) are equalized by postulating the equality of p and $\overline{\text{p}}$, although they are different, strictly speaking. Such equalization of priority sequences allows us to reinterpret the assignments, which are referred to as being *chirality-faithful* (Fujita 2009d).

   b. Figure 10.10 illustrates a Type III case which is *chirality-unfaithful*. A pair of *RS*-diastereomers **10** and **11** (or $\overline{\mathbf{10}}$ and $\overline{\mathbf{11}}$) located at the horizontal S-axis of Fig. 10.10 is characterized by a pair of *RS*-stereodescriptors in terms of such a common priority sequence as a > p > $\overline{\text{p}}$ > q (or a > p > $\overline{\text{p}}$ > $\overline{\text{q}}$).

      The pair of *RS*-stereodescriptors assigned to the pair of *RS*-diastereomers cannot be reinterpreted to be assigned to the pair of enantiomers located at the

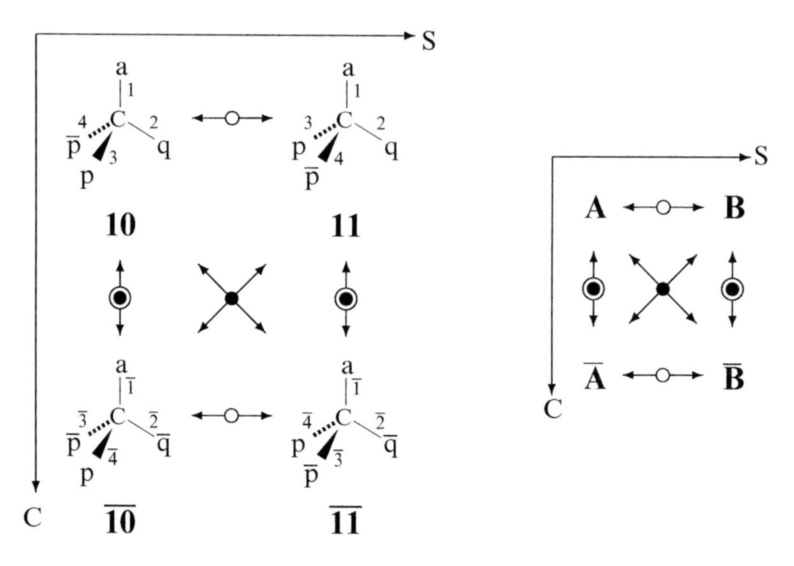

Fig. 10.10 Stereoisogram for characterizing promolecules of Type III, which is characterized by achiral, *RS*-stereogenic, scleral attributes (stereoisogram index: [−, −, −]), representing a chirality-unfaithful case

vertical C-axis of Fig. 10.10, i.e., **10** and $\overline{\mathbf{10}}$ (or **11** and $\overline{\mathbf{11}}$), because priority sequences a > p > $\overline{\text{p}}$ > q for **10** and a > p > $\overline{\text{p}}$ > $\overline{\text{q}}$ for $\overline{\mathbf{10}}$ (a > p > $\overline{\text{p}}$ > q for **11** and a > p > $\overline{\text{p}}$ > $\overline{\text{q}}$ for $\overline{\mathbf{11}}$) generate stereodescriptors of the same kind (not of pairwise nature). Hence, this case of Type III is referred to as being *chirality-unfaithful* (Fujita 2009d).

3. According to the stereoisogram approach, Type V cases (e.g., Fig. 10.2) are recognized to be named by the CIP system by virtue of *RS*-diastereomeric relationships on the same line as Types I and III. For example, a pair of *RS*-diastereomers **2** and **3** located at the horizontal S-axis of Fig. 10.8 is characterized by a pair of *RS*-stereodescriptors in terms of such a common priority sequence as a > b > p > $\overline{\text{p}}$ (tentatively specified for the sake of explanation).

   It should be emphasized that the pair of *RS*-stereodescriptors are given to the pair of *RS*-diastereomers (**2**/**3**), but not to a nonexistent pair of enantiomers (Table 10.2). Because each of Type V cases is concerned with achiral (self-enantiomeric) promolecules such as **2** and **3**, neither **2** nor **3** has a counterpart in an enantiomeric relationship. In other words, the lack of enantiomeric pairs does not allow any meaningful reinterpretation of the *RS*-stereodescriptors. This case is referred to as being *chirality-unfaithful* (Fujita 2009d).

As for Type III cases, it should be emphasized that enantiomeric relationships for Fig. 10.8 (a chirality-faithful case) and for Fig. 10.10 (a chirality-unfaithful case) represent situations of the same kind, whether they are chirality-faithful or not. This means that subdivision of Type III cases into chirality-faithful and chirality-unfaithful cases is concerned only with nomenclature due to the CIP system, but not with stereochemical discussions on reactivities. In other words, we are able to discuss their reactivities against chiral reagents equally, whether they are chirality-faithful or not in Type III cases.

10.3.5.2 Problematic Rationalization Due to the Conventional Approach

On the other hand, the conventional approach is bound by the dichotomy between enantiomeric relationships and diastereomeric ones, as found in Table 10.3.

Thus, geometric features (enantiomeric relationships, chirality) and permutational features (diastereomeric relationships, stereogenicity) are mixed up to give a seemingly single criterion, although the conventional approach has been unaware of this mixing-up. After that, such a misleading connotation as "chirality $\subset$ stereogenicity" seems to be postulated in the conventional approach (Helmchen 1996, page 12) and (Eliel and Wilen 1994, page 1208). After enantiomeric relationships are related to chirality and after diastereomeric relationships seem to be related to stereogenicity minus chirality, the former enantiomeric relationships are preferred over the latter diastereomeric relationships. Although Helmchen has discussed symmetry consistency of the CIP specification which mainly based on stereogenicity (Helmchen 1996, page 32), he has maintained the usage of the term "chirality center". The usage of the term "chirality center" in the CIP system (Prelog and Helmchen 1982; Helmchen 1996) is a result of preference of enantiomeric relationships over diastereomeric relationships.

To show such mixed-up situations more clearly, let us examine promolecules of Types I, III, and V (categorized by the present approach) in terms of the conventional approach:

1. To begin with, we should point out:

   > Discussions based on the conventional approach (e.g., *RS*-stereodescriptors) are incapable of arriving at the categorization of Types I–V.

   Hence the following discussions after categorizing into Types I, III, and V would be impossible so long as we continue in the conventional approach.
2. By virtue of the dichotomy between enantiomers and diastereomers in the conventional approach, Type I cases prefer enantiomeric relationships to rationalize *RS*-stereodescriptors of the CIP system, as found in Table 10.3. In other words, the conventional approach exclusively takes account of the vertical C-axis of Fig. 10.4 and results in the neglect of the horizontal S-axis, which causes systematic neglect of *RS*-diastereomeric relationships. Accordingly, *RS*-stereodescriptors are presumed to be given to an enantiomeric pair of promolecules **7** and $\overline{\mathbf{7}}$ by applying such as a priority sequence: a > b > X > Y. Fortunately, in the Type I cases, the priority sequence (a > b > X > Y) for the

Table 10.3 Entangled criteria to be abandoned in the conventional approach for giving *RS*-stereodescriptors of the CIP system. These criteria should be replaced by a single criterion shown in Table 10.2

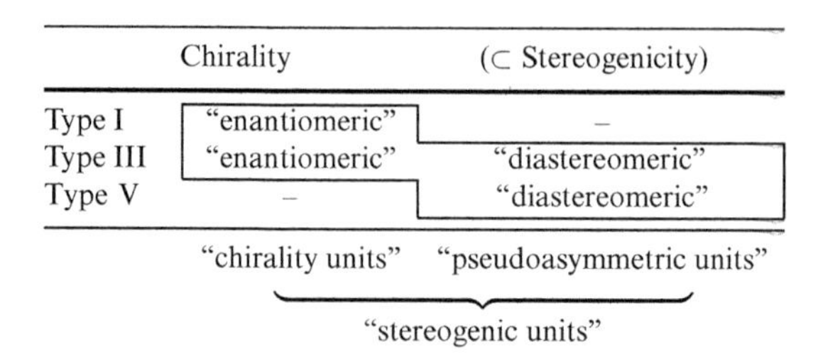

	Chirality	(⊂ Stereogenicity)
Type I	"enantiomeric"	–
Type III	"enantiomeric"	"diastereomeric"
Type V	–	"diastereomeric"
	"chirality units"	"pseudoasymmetric units"
	"stereogenic units"	

pair of enantiomers is the same as the one for the pair of *RS*-diastereomers **7** and $\overline{\mathbf{7}}'$ ($=\overline{\mathbf{7}}$) from a viewpoint of the present approach (along the horizontal S-axis of Fig. 10.4).

3. To make matters worse, the conventional approach has misinterpreted Type I cases as standards of chiral molecules, where enantiomeric relationships due to reflection operations have taken precedence over other relationships. The biased precedence influences the examination of Type II and III cases so that effects of the other relationships (e.g., diasteremeric relationships in the conventional terminology) have been underestimated.
4. In the conventional approach, most Type III cases also prefer enantiomeric relationships to rationalize *RS*-stereodescriptors of the CIP system (Table 10.3), where the Type III cases are unconsciously confused with Type I cases. For example, *RS*-stereodescriptors are presumed to be given to an enantiomeric pair of promolecules **8** and $\overline{\mathbf{8}}$ by applying different priority sequences, e.g., $\mathrm{a} > \mathrm{b} > \mathrm{X} > \mathrm{p}$ and $\mathrm{a} > \mathrm{b} > \mathrm{X} > \overline{\mathrm{p}}$ in the conventional approach. It should be noted that the conventional approach exclusively takes account of the vertical C-axis of Fig. 10.4 and neglects the horizontal S-axis. Fortunately in such Type III cases, the two priority sequences (e.g., $\mathrm{a} > \mathrm{b} > \mathrm{X} > \mathrm{p}$ and $\mathrm{a} > \mathrm{b} > \mathrm{X} > \overline{\mathrm{p}}$) are concordant, so as to be consistent to the apparent pairing of *RS*-stereodescriptors assigned to the enantiomeric pair of **8** and $\overline{\mathbf{8}}$. Strictly speaking, however, the detached set of a, b, X, and p is different from the detached set of a, b, X, and $\overline{\mathrm{p}}$, so that it is not logical to compare **8** and $\overline{\mathbf{8}}$ in order to assign such pairwise *RS*-stereodescriptors. In contrast, the horizontal S-axis of Fig. 10.4 shows that a common set of proligands a, b, X, and p can be detached from *RS*-diastereomers **8** and **9**.
5. Type V cases are named by the CIP system by virtue of diastereomeric relationships (Table 10.3), because Type V cases are concerned with achiral (self-enantiomeric) promolecules and lack enantiomeric relationships within the conventional terminology. Type V cases of the conventional approach are in the same situations as the present approach (e.g., Fig. 10.8), if the diastereomeric relationships are interpreted in terms of the *RS*-diastereomeric relationships of the present approach.

6. Such Type III cases as a promolecule with a, p, $\overline{\mathrm{p}}$ and q (a: achiral; p, $\overline{\mathrm{p}}$, q: chiral in isolation) are troublesome (Fig. 10.10), because two enantiomers of a pair give the same *r*-stereodescriptors (or *s*-stereodescriptors) as a result of reflection-invariance in terms of the CIP system (Prelog and Helmchen 1982). In such cases, the CIP system has been claimed to give precedence to stereogenicity over local symmetry, so as to demand a reflection-invariant descriptor (*r* or *s*) (Helmchen 1996, page 32). It should be noted, however, that the reflection-invariance is based on apparent reflection-invariance of priority sequences (e.g., $\mathrm{a} > \mathrm{p} > \overline{\mathrm{p}} > \mathrm{q}$ and $\mathrm{a} > \mathrm{p} > \overline{\mathrm{p}} > \overline{\mathrm{q}}$), where q and $\overline{\mathrm{q}}$ are tentatively equalized. Strictly speaking, the priority sequence $\mathrm{a} > \mathrm{p} > \overline{\mathrm{p}} > \mathrm{q}$ for one enantiomer is not equal to the priority sequence $\mathrm{a} > \mathrm{p} > \overline{\mathrm{p}} > \overline{\mathrm{q}}$ for the other enantiomer, so that they cannot be directly compared to judge their invariance. Obviously, the fact that p and $\overline{\mathrm{p}}$ are not equalized in the priority sequence is inconsistent with the fact that q and $\overline{\mathrm{q}}$ are equalized.

In terms of the CIP system (Prelog and Helmchen 1982; Helmchen 1996), Item 1 (Type I cases) and Item 3 (most of Type III cases) are categorized to be "chirality units" (stereogenic units of type 1), while Item 3 (Type V cases) and Item 4 (part of Type III cases) are categorized to be "pseudoasymmetric units" (stereogenic units of type 2). The categorization of the CIP system combined with the conventional dichotomy between enantiomers and diastereomers (in particular, the preference of enantiomers over diastereomers for Type I) results in an inconsistent theoretical framework shown in Table 10.3.

10.3.5.3 Comparison Between a Single Criterion and Entangled Criteria

Chirality and *RS*-Stereogenicityas Independent Concepts

The comparison between Tables 10.2 and 10.3 indicates that the present stereoisogram approach (Table 10.2) shows consistency, while the conventional approach (Table 10.3) lacks consistency from the present viewpoint.

1. In the stereoisogram approach, *RS*-stereogenic units modelled as promolecules are divided into Types I, III, and V (Fig. 10.9), which are all characterized by *RS*-stereogenicity as a single criterion, as shown in Table 10.2. Such *RS*-stereogenic units can be named by the CIP system, even if they are chiral (Type I and III) or achiral (Type V). The concept of *RS*-stereogenicity is concerned only with the capability of assigning *RS*-stereodescriptors, while the concept of chirality is concerned only with geometric aspects of stereochemistry. Although the two concepts are closely related, they are independent of each other, as discussed by using stereoisograms.
2. In the conventional approach, on the other hand, "stereogenic units" are divided into "chirality units" and "pseudoasymmetric units", which are characterized by entangled criteria shown in Table 10.3. In other words, the conventional approach involves a misleading basis that the concept of chirality is a subconcept

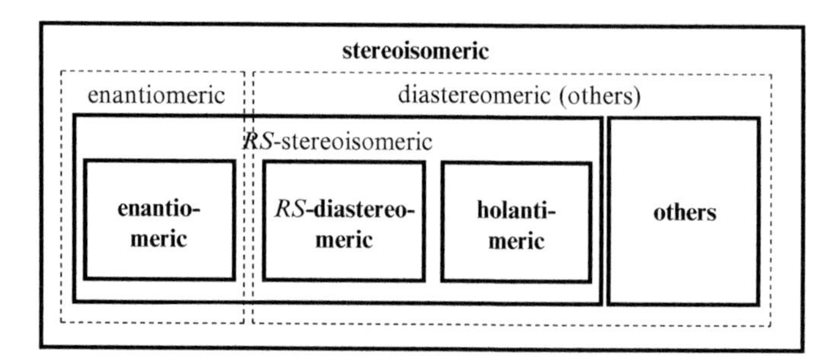

Fig. 10.11 Paradigm shift from the conventional terminology to the present terminology for stereoisomerism (Fujita 2009e). A *broken-lined box* represents a term of the conventional terminology, while a *solid-lined box* represents a term of the present terminology

of stereogenicity. In particular, Table 10.3 shows that the conventional approach has failed in looking for a consistent way by which the preference of chirality in the dichotomy between enantiomers and diastereomers is harmonized with the preference of stereogenicity in the CIP system.

Paradigm Shift Brought by the Stereoisogram Approach

The single criterion shown in Table 10.2 is based on the formulation of stereoisograms, which respectively contain enantiomeric, *RS*-diastereomeric, and holantimeric relationships. Each of these relationships is capable of producing a definite assembly. Moreover, these relationships are integrated into an *RS*-stereoisomeric relationship, which is also capable of producing a definite assembly (a quadruplet of *RS*-stereoisomers in a stereoisogram). Thereby, the dichotomy between *RS*-stereoisomers and others to cover stereoisomerism has been developed as shown in the solid-lined boxes in Fig. 10.11.

The newly-developed dichotomy has advantages over the conventional dichotomy between enantiomers and diastereomers (the broken-lined boxes in Fig. 10.11), because the latter causes the confusion due to the entangled criteria shown in Table 10.3. This means that a paradigm shift from the conventional terminology to the present terminology for stereoisomerism turns out to be inevitable (Fig. 10.11) (Fujita 2009e), where correspondence between terms for relationships and those for attributes are defined as follow: enantiomeric — chiral; *RS*-diastereomeric — *RS*-stereogenic; and *RS*-stereoisomeric — *RS*-stereoisomerism.

10.3.6 Itemized Enumeration of Stereoisograms

Tetrahedral promolecules have been enumerated in an itemized fashion due to point-group symmetry, where the USCI (unit-subduced-cycle-index) approach is applied to a tetrahedral skeleton with achiral and chiral ligands (Fujita 1991a), as described in a monograph (Fujita 1991b, Chapter 21). Tetrahedral promolecules have been also enumerated in an itemized fashion due to permutation-group symmetry (Fujita 2001). Each stereoisogram contains a quadruplet of tetrahedral promolecules, which is considered to be an entity to be counted just one, i.e., an *RS*-stereoisomer which can be regarded as an equivalence class under the *RS*-stereoisomeric group. Such quadruplets (*RS*-stereoisomers) have been enumerated in an itemized fashion (Types I–V) (Fujita 2009a). These results are summarized in Fig. 10.12 (Fujita 2004c, 2009b).

10.4 Prochirality and Pro-*RS*-stereogenicity as Independent Concepts

In parallel with the concepts of chirality and *RS*-stereogenicity for specifying molecules, the concepts of prochirality and pro-*RS*-stereogenicity are introduced to examine intramolecular aspects of stereochemistry. Three topic relationships correspond to three relationships appearing in a stereoisogram. Then, stereoisograms for testifying prochirality and/or pro-*RS*-stereogenicity are introduced by using typical examples of possible types.

10.4.1 Three Topic Relationships and Three Attributes

According to the stereoisogram approach, the concept of prochirality and enantiomeric relationships can be rationalized by examining the vertical C-axes of stereoisograms. Because prochirality has a purely geometric meaning just as chirality has a purely geometric meaning, several conversions of achiral promolecules (Types IV and V) into chiral promolecules (Types I, III, and V) are recognized to be linked to the concept of prochirality (cf. promolecules surrounded by a box in Fig. 10.12).

On the other hand, the concept of pro-*RS*-stereogenicity and *RS*-diastereotopic relationships can be rationalized by examining the horizontal S-axes of stereoisograms. Because pro-*RS*-stereogenicity for examining intramolecular environments has a purely *RS*-permutational meaning, just as *RS*-stereogenicity for examining intermolecular environments has a purely *RS*-permutational meaning, several conversions of *RS*-astereogenic promolecules (Types II and IV) into *RS*-stereogenic promolecules (Types I, III, and V) are recognized to be linked to the concept of pro-*RS*-stereogenicity (cf. promolecules surrounded by a box in Fig. 10.12).

To determine prochirality and pro-*RS*-stereogenicity, three types of criteria based on stereoisograms have been developed, i.e., the substitution criterion (Fujita 2009f), the membership criterion (Fujita 2006b, 2009f), and the symmetry

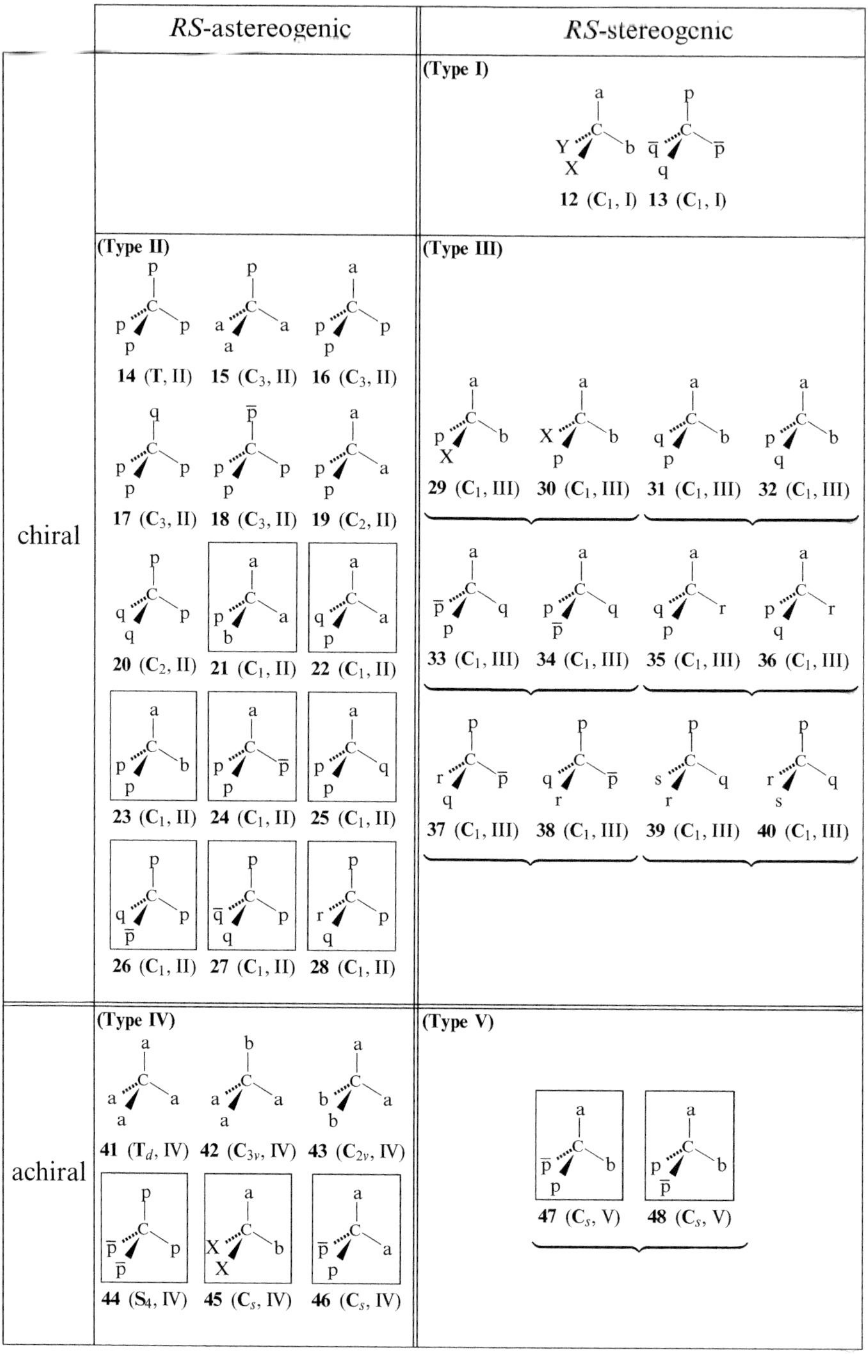

	RS-astereogenic	*RS*-stereogenic
chiral		**(Type I)** **12** (C_1, I) **13** (C_1, I)
chiral	**(Type II)** **14** (T, II) **15** (C_3, II) **16** (C_3, II) **17** (C_3, II) **18** (C_3, II) **19** (C_2, II) **20** (C_2, II) **21** (C_1, II) **22** (C_1, II) **23** (C_1, II) **24** (C_1, II) **25** (C_1, II) **26** (C_1, II) **27** (C_1, II) **28** (C_1, II)	**(Type III)** **29** (C_1, III) **30** (C_1, III) **31** (C_1, III) **32** (C_1, III) **33** (C_1, III) **34** (C_1, III) **35** (C_1, III) **36** (C_1, III) **37** (C_1, III) **38** (C_1, III) **39** (C_1, III) **40** (C_1, III)
achiral	**(Type IV)** **41** (T_d, IV) **42** (C_{3v}, IV) **43** (C_{2v}, IV) **44** (S_4, IV) **45** (C_s, IV) **46** (C_s, IV)	**(Type V)** **47** (C_s, V) **48** (C_s, V)

Fig. 10.12 Five *RS*-stereoisomeric types (Types I–V) for tetrahedral (pro)molecules (Fujita 2004c, 2009b). The symbols *a, b, X,* and *Y* represent atoms or achiral (pro)ligands. The symbols *p, q, r,* and *s* represents chiral (pro)ligands, while each symbol with an overbar represents the corresponding chiral (pro)ligand with the opposite chirality. Each promolecule surrounded by a box is a prochiral and/or pro-*RS*-stereogenic one

criterion (Fujita 2009c). They are capable of arriving at equivalent results. In the present chapter, the symmetry criterion is applied to candidates of showing prochirality or pro-*RS*-stereogenicity, which are surrounded by a box, as found in Fig. 10.12.

In accord with Table 10.1, three topic relationships and the corresponding attributes are collected in Table 10.2. Note that the term *holantitopic* does not appear in usual situations because such a process as Type I ($[-,-,a]$) to Type III ($[-,-,-]$) is usually regarded as a process of another *RS*-stereogenic center and because a promolecule $[\ *\ ,\ *\ ,\ -\ ]$ generated from Type IV ($[a,\ a,\ a]$) is restricted to $[-,a,-]$ or $[a,-,-]$, which is categorized preferably to enantiotopicity (for the first set of prefixes ($a/-$) in $[a,a,a] \to [-,a,-]$) or *RS*-diastereotopicity (for the second set of prefixes ($a/-$) in $[a,a,a] \to [a,-,-]$).

The symmetry criterion used in the present chapter adopts enantiotopic and/or *RS*-diastereotopic relationships after drawing stereoisograms for testifying prochirality and/or pro-*RS*-stereogenicity (Fujita 2009c).

10.4.2 Conversion of Type IV into Type I

As a promolecule of Type IV, let us examine **4** of Fig. 10.7 (**45** of Fig. 10.12) by using a stereoisogram for testifying prochirality or pro-*RS*-stereogenicity (Fujita 2009c). The two X's of **4** are differentiated by superscripts (e.g., X^{α} and X^{β}) so that all the promolecules contained in Fig. 10.7 are changed to give Fig. 10.13 ($[a,a,a] \to [-,-,a]$).

1. The geometric change is judged by the comparison of the vertical C-axis of Fig. 10.7 with the vertical C-axis of Fig. 10.13, which corresponds to the first set of prefixes ($a/-$) in $[a,a,a] \to [-,-,a]$. Note that the self-enantiomeric relationship ($\mathbf{4} = \overline{\mathbf{4}}$) in the original stereoisogram (Fig. 10.7) is changed into a hypothetically enantiomeric relationship ($\mathbf{4}^{\alpha}$ and $\overline{\mathbf{4}}^{\alpha}$) in the stereoisogram for testifying prochirality and/or pro-*RS*-stereogenicity (Fig. 10.13). Because $\mathbf{4}^{\alpha}$ is enantiomeric to $\overline{\mathbf{4}}^{\alpha}$ ($= \mathbf{4}^{\beta}$) along the vertical C-axis of Fig. 10.13, the two X's of **4**, which correspond to the proligands X^{α} and X^{β} appearing along the vertical C-axis of Fig. 10.13, are defined to be in an *enantiotopic* relationship.
   The term *prochiral* is used in accord with such an enantiotopic relationship judged by the examination of the vertical C-axis of Fig. 10.13. This usage is consistent to the process of converting the achiral promolecule **4** into a chiral promolecule which corresponds to $\mathbf{4}^{\alpha}$ or $\overline{\mathbf{4}}^{\alpha}$ ($= \mathbf{4}^{\beta}$) (e.g., **7** by replacing $X^{\alpha} \to X$ (unchanged) and $X^{\beta} \to Y$ or $\overline{\mathbf{7}}$ by replacing $X^{\alpha} \to Y$ and $X^{\beta} \to X$ (unchanged) in Fig. 10.4).
2. The *RS*-permutational change is judged by the comparison of the horizontal S-axis of Fig. 10.7 with the horizontal S-axis of Fig. 10.13, which corresponds to the second set of prefixes ($a/-$) in $[a,a,a] \to [-,-,a]$. Note that the self-*RS*-diastereomeric relationship ($\mathbf{4} = \mathbf{4}'$) in the original stereoisogram (Fig. 10.7)

Table 10.4 Three topic relationships and the corresponding attributes, which are characterized by stereoisograms for testifying prochirality and/or pro-*RS*-stereogenicity (Fujita 2009c)

Symbol	Relationship	Attribute
(Concerned with reflections ◉)		
←◉→	enantiotopic	prochiral
═◉═	(self-enantiotopic)	–
(Concerned with *RS*-permutations ○)		
←○→	*RS*-diastereotopic	pro-*RS*-stereogenic
═○═	(self-*RS*-diastereotopic)	–
(Concerned with ligand reflections •)		
←•→	(holantitopic)[a]	(proscleral)
═•═	(self-holantimeric)[a]	–

[a]The term *holantitopic* does not appear in usual situations.

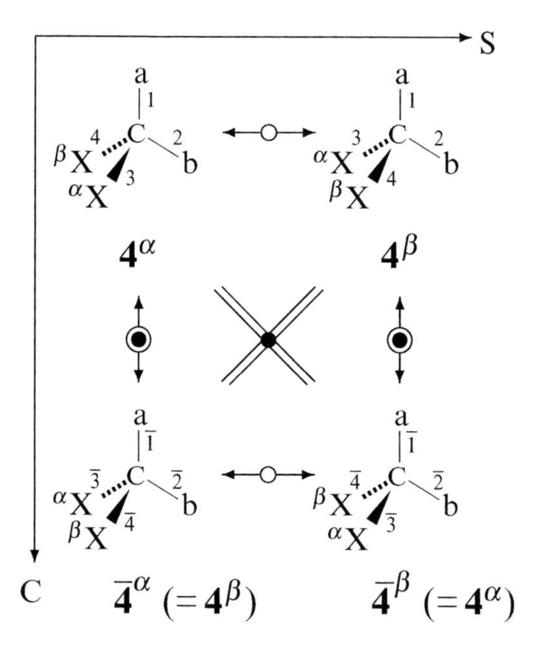

Fig. 10.13 Stereoisogram for Type I for testifying prochirality and pro-*RS*-stereogenicity, where it is characterized by tentative differentiation in a promolecule of Type IV (Fig 10.7), which generates a hypothetical promolecule of Type I

is changed into a hypothetically *RS*-diastereomeric relationship ($\mathbf{4}^{\alpha}$ and $\mathbf{4}^{\beta}$) in the stereoisogram for testifying prochirality and/or pro-*RS*-stereogenicity (Fig. 10.13). Because $\mathbf{4}^{\alpha}$ is *RS*-diastereomeric to $\mathbf{4}^{\beta}$ $(= \overline{\mathbf{4}}^{\alpha})$ along the horizontal S-axis of Fig. 10.13, the two X's of **4**, which correspond to the proligands X^{α} and X^{β} appearing along the horizontal S-axis of Fig. 10.13, are defined to be in an *RS*-*diastereotopic* relationship.

The term *pro-RS-stereogenic* is used in accord with such an *RS*-diastereotopic relationship judged by the examination of the horizontal S-axis of Fig. 10.13. This usage is consistent to the process of converting the *RS*-astereogenic promolecule **4** into an *RS*-stereogenic promolecule which corresponds to $\mathbf{4}^{\alpha}$ or $\mathbf{4}^{\beta}$ ($= \overline{\mathbf{4}}^{\alpha}$): e.g., **7** by replacing $X^{\alpha} \rightarrow X$ (unchanged) and $X^{\beta} \rightarrow Y$; or $\overline{\mathbf{7}}'$ by replacing $X^{\alpha} \rightarrow Y$ and $X^{\beta} \rightarrow X$ (unchanged) in Fig. 10.4.

It should be noted that the two X's can be named by the *pro-R*/*pro-S* system, where the capability of giving *pro-R*/*pro-S*-descriptors is ascribed to the pro-*RS*-stereogenicity. A pair of *pro-R*/*pro-S*-descriptors is assigned to such a pair of *RS*-diastereotopic ligands, as shown in Figs. 10.7 and 10.13.

Figure 10.13 along with Fig. 10.7 reveals an erroneous methodology in which the vertical C-axis and the horizontal S-axis are equalized in the conventional approach, even though they happen to become superposable in such Type I cases as described above. Thus the term "prochiral" of the *pro-R*/*pro-S* system, which should be replaced by the present term *pro-RS-stereogenicity*, has been erroneously linked to the vertical C-axis of Fig. 10.13. Figure 10.13 shows that the concept of prochirality is independent of pro-*RS*-stereogenicity. It follows that the expression in a review: "Correspondingly, the concept of prochirality must be generalized to one of prostereoisomerism" (Eliel 1982, page 4) causes a latent confusion, unless and until the concept of prostereoisomerism is properly defined, e.g., prostereoisomerism $\neq$ prostereogenicity.

10.4.3 Conversions of Type IV into Type V and into Type II

The *RS*-astereogenic nature of a Type IV promolecule indicates the capability of generating a promolecule of Type V (*RS*-stereogenic), while the achiral nature of a Type IV promolecule indicates the capability of generating a promolecule of Type II (chiral). This means that pro-*RS*-stereogenicity can coexist with prochirality in a single promolecule. Such coexistence is demonstrated clearly by the stereoisogram approach, showing that pro-*RS*-stereogenicity and prochirality are independent concepts.

10.4.3.1 Conversions of Type IV into Type V

The promolecule **46** of Fig. 10.12 belongs to Type IV, the stereoisogram of which is shown in Fig. 10.14a. The two a's of **46** are tentatively differentiated by superscripts (a^{α} and a^{β}) so as to give a stereoisogram of Type V for testifying prochirality and/or pro-*RS*-stereogenicity (Fig. 10.14b).

1. The geometric change is judged by the comparison of the vertical C-axis of Fig. 10.14a with the vertical C-axis of Fig. 10.14b, which corresponds to the first set of prefixes (a/a) in $[a, a, a] \rightarrow [a, -, -]$. Because $\mathbf{49}^{\alpha}$ and $\overline{\mathbf{49}}^{\alpha}$ are

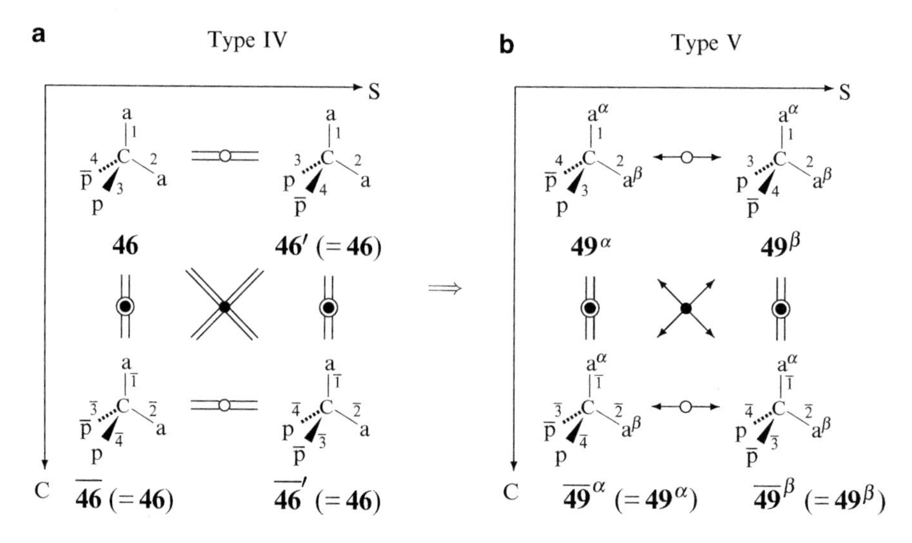

Fig. 10.14 Tentative differentiation of two a's in the original promolecule of Type IV (**a**), which generates a stereoisogram of Type V (**b**) for testifying prochirality and/or pro-*RS*-stereogenicity

identical (self-enantiomeric) to each other, the proligands corresponding to a^{α} and a^{β} are defined to be in a *self-enantiotopic* relationship. Hence, this case is not concerned with the concept of prochirality.

2. The *RS*-permutational change is judged by the comparison of the horizontal S-axis of Fig. 10.14a with the horizontal S-axis of Fig. 10.14b, which corresponds to the second set of prefixes $(a/-)$ in $[a, a, a] \rightarrow [a, -, -]$. Note that the self-*RS*-diastereomeric relationship (**46** $=$ **46**$'$) in the original stereoisogram (Fig. 10.14a) is changed into a hypothetically *RS*-diastereomeric relationship ($\mathbf{49}^{\alpha}$ and $\mathbf{49}^{\beta}$) in the stereoisogram for testifying prochirality and/or pro-*RS*-stereogenicity (Fig. 10.14b). Because $\mathbf{49}^{\alpha}$ is *RS*-diastereomeric to $\mathbf{49}^{\beta}$ along the horizontal S-axis of Fig. 10.14b, the two a's of **46**, which correspond to the tentatively differentiated proligands (a^{α} and a^{β}) appearing along the horizontal S-axis of Fig. 10.14b, are in an *RS-diastereotopic* relationship.

   The term *pro-RS-stereogenic* is used in accord with such an *RS*-diastereotopic relationship judged by the examination of the horizontal S-axis of Fig. 10.14b. This usage is consistent to the process of converting the *RS*-astereogenic promolecule **46** into an *RS*-stereogenic promolecule which corresponds to $\mathbf{49}^{\alpha}$ or $\mathbf{49}^{\beta}$, e.g., the conversion of **46** into **47** (or **48**).

It should be noted that the two a's can be named by the *pro-R*/*pro-S* system, where the capability of giving *pro-R*/*pro-S*-descriptors is ascribed to the pro-*RS*-stereogenicity. A pair of *pro-R*/*pro-S*-descriptors is assigned to such a pair of *RS*-diastereotopic ligands, as shown in Fig. 10.14a and b.

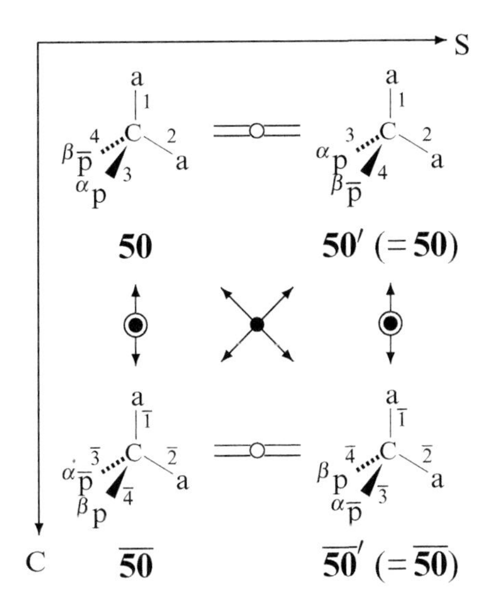

Fig. 10.15 Tentative differentiation in a promolecule of Type IV (Figure 10.14a), which generates a stereoisogram of Type II for testifying prochirality and/or pro-*RS*-stereogenicity

10.4.3.2 Conversions of Type IV into Type II

The promolecule **46** of Type IV (Fig. 10.14a) can be characterized in an alternative way shown in Fig. 10.15. The p and $\overline{\mathrm{p}}$ of **46** are tentatively differentiated by superscripts (p^{α} and $\overline{\mathrm{p}}^{\beta}$) so as to give a stereoisogram of Type II for testifying prochirality and/or pro-*RS*-stereogenicity (Fig. 10.15).

1. The geometric change is judged by the comparison of the vertical C-axis of Fig. 10.14a with the vertical C-axis of Fig. 10.15, which corresponds to the first set of prefixes $(a/-)$ in $[a,a,a] \rightarrow [-,a,-]$. Note that the self-enantiomeric relationship ($\mathbf{46} = \overline{\mathbf{46}}'$) in the original stereoisogram (Fig. 10.14a) is changed into a hypothetically enantiomeric relationship (**50** and $\overline{\mathbf{50}}$) in the stereoisogram for testifying prochirality and/or pro-*RS*-stereogenicity (Fig. 10.15). In accord with the enantiomeric relationship between **50** and $\overline{\mathbf{50}}$, the proligands p and $\overline{\mathrm{p}}$ corresponding to p^{α} and $\overline{\mathrm{p}}^{\beta}$ are defined to be in an enantiotopic relationship.

   If the ligand $\overline{\mathrm{p}}$ is converted into q in preference to p, the promolecule **46** of Type IV is converted into **22** of Type II. If the ligand p is converted into $\overline{\mathrm{q}}$ in preference to $\overline{\mathrm{p}}$, the promolecule **46** of Type IV is converted into $\overline{\mathbf{22}}$ of Type II. In accord with the enantiotopic relationship of p and $\overline{\mathrm{p}}$ in the starting promolecule **46**, the products **22** and $\overline{\mathbf{22}}$ are enantiomeric to each other.
2. The *RS*-permutational change is judged by the comparison of the horizontal S-axis of Fig. 10.14a with the horizontal S-axis of Fig. 10.15, which corresponds to the second set of prefixes (a/a) in $[a,a,a] \rightarrow [-,a,-]$.

Note that the self-*RS*-diastereomeric relationship ($\mathbf{46} = \mathbf{46'}$) in the original stereoisogram (Fig. 10.14a) is not changed so as to give a hypothetically self-*RS*-diastereomeric relationship ($\mathbf{50} = \mathbf{50'}$) in the stereoisogram for testifying prochirality and/or pro-*RS*-stereogenicity (Fig. 10.15). Hence, this case is not concerned with the concept of pro-*RS*-stereogenicity.

10.4.4 Conversions of Type V into Type III

The promolecules **2** and **3** in the stereoisogram shown in Fig. 10.8 (**47** and **48** in Fig. 10.12) are characterized to be prochiral in a purely geometric meaning. In terms of the stereoisogram approach, the p and $\overline{\mathrm{p}}$ of **2** (and **3**) are tentatively differentiated by superscripts (p^{α} and $\overline{\mathrm{p}}^{\beta}$) so as to give a stereoisogram of Type III for testifying prochirality and/or pro-*RS*-stereogenicity (Fig. 10.16).

1. The geometric change is judged by the comparison of the vertical C-axis of Fig. 10.8 with the vertical C-axis of Fig. 10.16, which corresponds to the first set of prefixes ($a/-$) in $[a, -, -] \rightarrow [-, -, -]$. Note that the self-enantiomeric relationship ($\mathbf{2} = \overline{\mathbf{2}}$ or $\mathbf{3} = \overline{\mathbf{3}}$) in the original stereoisogram (Fig. 10.8) is changed into a hypothetically enantiomeric relationship ($\mathbf{51}/\overline{\mathbf{51}}$ or $\mathbf{52}/\overline{\mathbf{52}}$) in the stereoisogram for testifying prochirality and/or pro-*RS*-stereogenicity (Fig. 10.16). In accord with the enantiomeric relationship between **51** and $\overline{\mathbf{51}}$ (or between **52** and $\overline{\mathbf{52}}$), the proligands p and $\overline{\mathrm{p}}$ of **2** (or **3**) which correspond to p^{α} and $\overline{\mathrm{p}}^{\beta}$ are defined to be in an enantiotopic relationship.
   If the ligand $\overline{\mathrm{p}}$ is converted into q in preference to p, the promolecule **2** of Type V is converted into **31** of Type III. If the ligand p is converted into $\overline{\mathrm{q}}$ in preference to $\overline{\mathrm{p}}$, the promolecule **2** of Type V is converted into $\overline{\mathbf{31}}$ of Type III. In accord with the enantiotopic relationship between p and $\overline{\mathrm{p}}$ in the starting promolecule **2**, the products **31** and $\overline{\mathbf{31}}$ are enantiomeric to each other. The conversion of $\mathbf{2} \rightarrow \mathbf{31}$ can be differentiated from the conversion of $\mathbf{2} \rightarrow \overline{\mathbf{31}}$ by means of an appropriate chiral reagent. Hence, the enantiotopic relationship between p and $\overline{\mathrm{p}}$ in **2** (or **3**) brings about the prochirality of **2** (or **3**).
2. The *RS*-permutational change is judged by the comparison of the horizontal S-axis of Fig. 10.8 with the horizontal S-axis of Fig. 10.16, which corresponds to the second set of prefixes ($-/-$) in $[a, -, -] \rightarrow [-, -, -]$. Note that the *RS*-diastereomeric relationship between **2** and **3** in the original stereoisogram (Fig. 10.8) is not changed so as to give a hypothetically *RS*-diastereomeric relationship between **51** and **52** in the stereoisogram for testifying prochirality and/or pro-*RS*-stereogenicity (Fig. 10.16). Hence, this case is not concerned with the concept of pro-*RS*-stereogenicity.

It should be noted that the enantiotopic relationship detected in Type V by the stereoisogram approach (in a purely geometric meaning) has been excluded from the territory of the *pro-R*/*pro-S* system by the transmuted term "enantiotopic" described in Sect. 10.2.5. As a result, the differentiation between the two

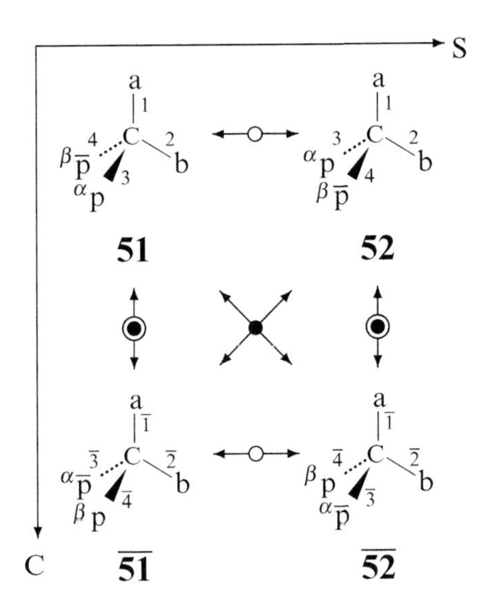

Fig. 10.16 Tentative differentiation in a promolecule of Type V (Fig. 10.8), which generates a stereoisogram of Type III for testifying prochirality and/or pro-*RS*-stereogenicity

conversions (**2** $\rightarrow$ **31** and **2** $\rightarrow$ $\overline{\mathbf{31}}$) is forced to be misplaced as an exceptional case, because this case has been regarded as a troublesome or undesirable thing beyond the *pro-R*/*pro-S* system. Obviously, such exclusion as inherent in the conventional approach has obstructed the proper view of geometric aspects of stereochemistry.

The fact that the *pro-R*/*pro-S* system uses such "enantiotopic" relationships as excluding a pair of p and $\overline{\text{p}}$ in **2** (or **3**) suffers from the same cause as the fact that the CIP system differentiates between p and $\overline{\text{p}}$ in specifying the pair of **2** and **3**. Even if a pair of *pro-R*/*pro-S*-descriptors is not given to a pair of such "different" proligands (p and $\overline{\text{p}}$ in **2** or **3**) in the conventional approach, this pair exhibits equivalence to be enantiotopic in terms of the stereoisogram approach (Fig. 10.16). Even if a pair of *RS*-stereodescriptors is given to a pair of **2** and **3** on the basis of such "different" proligands (p and $\overline{\text{p}}$), each promolecule of the pair is self-enantiomeric, where the pair of *RS*-stereodescriptors is concerned with an *RS*-diastereomeric relationship, not with an enantiomeric relationship, as detected by the stereoisogram approach (Fig. 10.8). In other words, the processes due to the *pro-R*/*pro-S* system and the CIP system are independent of geometric aspects of stereochemistry, so that the concept of chirality faithfulness should be consulted to assure correspondence (Fujita 2009d).

10.4.5 Conversions of Type II into Type III

The promolecules of Type II in the stereoisogram shown in Fig. 10.5 (**1** and $\overline{\mathbf{1}}$ corresponding to **23** in Fig. 10.12) are characterized to be pro-*RS*-stereogenic. In terms of the stereoisogram approach, the two p's of **1** are tentatively differentiated

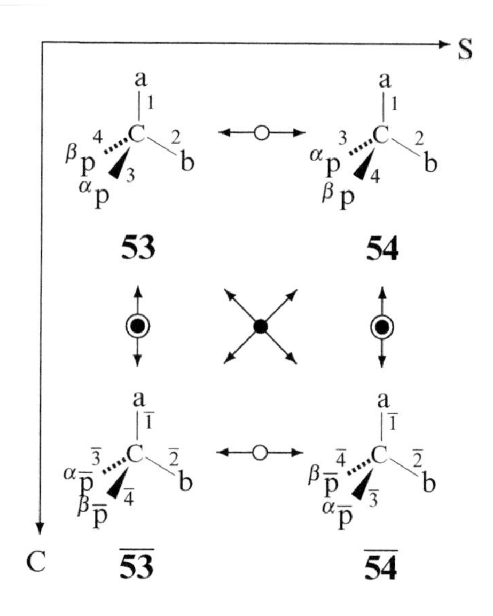

Fig. 10.17 Tentative differentiation in a promolecule of Type II (Fig. 10.5), which generates a stereoisogram of Type III for testifying prochirality and/or pro-*RS*-stereogenicity

by superscripts (p^{α} and p^{β}) so as to give a stereoisogram of Type III for testifying prochirality and/or pro-*RS*-stereogenicity (Fig. 10.17).

1. The geometric change is judged by the comparison of the vertical C-axis of Fig. 10.5 with the vertical C-axis of Fig. 10.17, which corresponds to the first set of prefixes ($-$ / $-$) in $[-, a, -] \rightarrow [-, -, -]$. Note that the enantiomeric relationship between **1** and $\overline{\mathbf{1}}$ in the original stereoisogram (Fig. 10.5) is not changed so as to give a hypothetically enantiomeric relationship between **53** and $\overline{\mathbf{53}}$ in the stereoisogram for testifying prochirality and/or pro-*RS*-stereogenicity (Fig. 10.17). Hence, this case is not concerned with the concept of prochirality.
2. The *RS*-permutational change is judged by the comparison of the horizontal S-axis of Fig. 10.5 with the horizontal S-axis of Fig. 10.17, which corresponds to the second set of prefixes ($a/-$) in $[-, a, -] \rightarrow [-, -, -]$. Note that the self-*RS*-diastereomeric relationship ($\mathbf{1} = \mathbf{1}'$ or $\overline{\mathbf{1}} = \overline{\mathbf{1}}'$) in the original stereoisogram (Fig. 10.5) is changed into a hypothetically *RS*-diastereomeric relationship (**53**/**54** or $\overline{\mathbf{53}}/\overline{\mathbf{54}}$) in the stereoisogram for testifying prochirality and/or pro-*RS*-stereogenicity (Fig. 10.17). In accord with the *RS*-diastereomeric relationship between **53** and **54** (or between $\overline{\mathbf{53}}$ and $\overline{\mathbf{54}}$), the two proligands p's of **1** (or $\overline{\mathbf{1}}$) which correspond to p^{α} and p^{β} are defined to be in an *RS*-diastereotopic relationship. Hence, this case is concerned with the concept of pro-*RS*-stereogenicity.

If one of the two ligands p's of **1** is predominantly converted into q, the promolecule **1** of Type II is converted into **31** of Type III. If the other ligand p is converted into q, the promolecule **1** of Type II is converted into **32** of Type III. In accord with the *RS*-diastereotopic relationship between the two p's in the starting promolecule **1**, the products **31** and **32** are *RS*-diastereomeric to each other.

The conversion of **1** → **31** can be differentiated from the conversion of **1** → **32**, automatically without chiral reagents. Hence, the *RS*-diastereotopic relationship between the two p's in **1** brings about the pro-*RS*-stereogenicity of **1**.

10.4.6 pro-R/pro-S-Descriptors Assigned by pro-RS-Stereogenicity

The stereoisogram approach and the conventional approach will be compared with respect to how the assignment of *pro-R*/*pro-S*-descriptors is rationalized.

10.4.6.1 Systematic Rationalization Due to the Stereoisogram Approach

The discussions described on the CIP system in Sect. 10.3.5.1 can be modified to discuss the *pro-R*/*pro-S* system. A pair of *pro-R*/*pro-S*-descriptors should be assigned to a pair of proligands which are paired by a well-defined single criterion. The present approach has reached the conclusion that each pair of *pro-R*/*pro-S*-descriptors is assigned to a pair of *RS*-diastereotopic proligands, which is directly linked to pro-*RS*-stereogenicity appearing in the horizontal direction (the *RS*-stereogenicity axis) of each stereoisogram for testifying prochirality and/or pro-*RS*-stereogenicity (Fujita 2006b, 2008b, 2009c,f).

Among promolecules of Type II or IV (the left-hand column of Fig. 10.9), there are pro-*RS*-stereogenic cases which can be named by the *pro-R*/*pro-S* system, because they are commonly characterized by their *RS*-astereogenicity. On the other hand, among promolecules of Type IV or V (the bottom of Fig. 10.9), there are cases which can be discussed in terms of prochirality, because they are commonly characterized by their achirality. Table 10.5 summarizes prochirality and pro-*RS*-stereogenicity, each of which is specified by a respective single criterion.

1. **Coincident appearance of prochirality and pro-*RS*-stereogenicity**: Fig. 10.13 shows a case of Type IV (into Type I) in which prochirality (achiral → chiral) coincides with pro-*RS*-stereogenicity (*RS*-astereogenic → *RS*-stereogenic). As summarized in Table 10.5, the prochirality is ascribed to the enantiotopic relationship between the two X's of **4**, while pro-*RS*-stereogenicity is ascribed to the *RS*-diastereotopic relationship between the same X's of **4**. The latter *RS*-diastereotopic relationship permits a pair of *pro-R*/*pro-S*-descriptors.
2. **Appearance of pro-*RS*-stereogenicity**: Fig. 10.14 shows a case of Type IV (into Type V), which is characterized in terms of pro-*RS*-stereogenicity (*RS*-astereogenic → *RS*-stereogenic). As summarized in Table 10.5, pro-*RS*-stereogenicity is ascribed to the *RS*-diastereotopic relationship between the two a's of **46**, which are named by a pair of *pro-R*/*pro-S*-descriptors.

   On the other hand, Fig. 10.5 along with Fig. 10.17 shows a case of Type II (into Type III), which is also characterized in terms of pro-*RS*-stereogenicity

Table 10.5 A single criterion for giving *pro-R*/*pro-S*-descriptors and another single criterion for discussing prochirality in the stereoisogram approach

Stereoisogram change			Prochirality	pro-*RS*-stereogenicity
(achiral → chiral & *RS*-astereogenic → *RS*-stereogenic)				
Type IV → I	$[\boldsymbol{a},\boldsymbol{a},a] \to [-,-,a]$	(Fig. 10.13)	enantiotopic	*RS*-diastereotopic
(*RS*-astereogenic → *RS*-stereogenic)				
Type IV → V	$[a,\boldsymbol{a},a] \to [a,-,-]$	(Fig. 10.14)	–	*RS*-diastereotopic
Type II → III	$[-,\boldsymbol{a},-] \to [-,-,-]$	(Fig. 10.17)	–	*RS*-diastereotopic
(achiral → chiral)				
Type IV → II	$[\boldsymbol{a},a,a] \to [-,a,-]$	(Fig. 10.15)	enantiotopic	–
Type V → III	$[\boldsymbol{a},-,-] \to [-,-,-]$	(Fig. 10.16)	enantiotopic	–

(chirality-faithfulness)

(*RS*-astereogenic → *RS*-stereogenic). As summarized in Table 10.5, pro-*RS*-stereogenicity is ascribed to the *RS*-diastereotopic relationship between the two p's of **1**, which are named by a pair of *pro-R*/*pro-S*-descriptors.

3. **Appearance of prochirality**: Fig. 10.14a along with Fig. 10.15 shows a case of Type IV (into Type II), which is characterized in terms of prochirality (achiral → chiral). As summarized in Table 10.5, the prochirality is ascribed to the enantiotopic relationship between p and $\overline{\text{p}}$ of **46** (Fig. 10.14a).

   Figure 10.8 along with Fig. 10.16 shows a case of Type V (into Type III), which is characterized in terms of prochirality (achiral → chiral). As summarized in Table 10.5, the prochirality is ascribed to the enantiotopic relationship between p and $\overline{\text{p}}$ of **2** (Fig. 10.16).

10.4.6.2 Problematic Rationalization Due to the Conventional Approach

The entangled situations in the discussion of the CIP system (described above in Sect. 10.3.5.2) become more complicated in discussing the *pro-R*/*pro-S* system, when we are restricted to the conventional approach. The following note should be added:

> Because of the lack of the categorization of Types I–V, discussions based on the conventional approach (e.g., *pro-R*/*pro-S*-descriptors) are incapable of arriving at categorized modes of conversions (e.g., Type IV → I in Fig. 10.13).

The problems of the conventional approach pointed out in Sect. 10.2.2, 10.2.4, and 10.2.5 can be more clearly demonstrated as summarized in Table 10.6 by adopting the categories of the stereoisogram approach (Table 10.5).

Table 10.6 Entangled criteria for giving *pro-R*/*pro-S*-descriptors and for discussing prochirality in the conventional approach. These criteria should be replaced by a set of independent and more succinct criteria shown in Table 10.5

	"Prostereogenicity" ("prochirality")
Type IV → I (Fig. 10.13)	"enantiotopic"
Type IV → V (Fig. 10.14)	"diastereotopic"
Type II → III (Fig. 10.17)	"diastereotopic"
Type IV → II (Fig. 10.15)	–
Type V → III (Fig. 10.16)	–

"stereoheterotopic"

By the stereoisogram approach, the case of Type IV → I (Fig. 10.13) is regarded as a coincident appearance of an enantiotopic case and an *RS*-diastereotopic case (Table 10.5). In contrast, by applying *pro-R*/*pro-S* system in the conventional approach, the case at issue is regarded as "enantiotopic" (Table 10.6), where geometric aspects (e.g. prochirality) of the term *enantiotopic* (Table 10.5) are neglected unconsciously after the coinage of the term "stereoheterotopic". This neglect has been forgotten in most discussions of the conventional approach, so that the transmuted term "enantiotopic" is used as if it still connotes the neglected geometric aspects (e.g., prochirality). In other words, *pro-R*/*pro-S*-descriptors to be restricted to nomenclature has been erroneously used to discuss geometric aspects of stereochemistry as a result of the confusion between the term *enantiotopic* of a geometric meaning and the transmuted term "enantiotopic". Note that such geometric aspects of stereochemistry can be discussed only by testifying difference between two ligands at issue and that the test for detecting the difference does not require *pro-R*/*pro-S*-descriptors.

The term "prochirality" of the *pro-R*/*pro-S* system, which is linked to the transmuted term "enantiotopic" (cf. Sect. 10.2.5) as well as to the term "diastereotopic", has different connotation from the term *prochirality* which is linked to the term *enantiotopic* of a geometric meaning. Such confused usage of the term "prochirality" seems to be convenient to cover the other cases listed in Table 10.6. To maintain the convenience by a semantic transmutation, the term "prochirality" has later been replaced by the term "prostereogenicity" to support the *pro-R*/*pro-S* system (cf. Sect. 10.2.2). Unfortunately, however, the resulting *pro-R*/*pro-S*-descriptors due to the transmuted term "enantiotopic" and the term "diastereotopic" (cf. Sect. 10.2.5) are still used to discuss prochirality of a geometric meaning, which has not been clearly differentiated from "prostereogenicity".

To conceal the inconsistency due to the treatment of the two terms of different categories on the same basis, the term "stereoheterotopic" has been inadequately coined by combining the transmuted term "enantiotopic" and the term "diastereotopic", as discuss in Sect. 10.2.4. Although the transmuted term "enantiotopic" neglects geometric aspects of the term *enantiotopic* as summarized in Table 10.6, it is used as if it still connotes the neglected geometric aspects (e.g., prochirality). The use of the term "stereoheterotopic" would reinforce such connotation unconsciously in comparison with the direct usage of the term "enantiotopic".

10.4.6.3 Comparison Between Independent Criteria and Entangled Criteria

The comparison between Tables 10.5 and 10.6 indicates that the present stereoisogram approach (Table 10.5) shows consistency, where prochirality and pro-*RS*-stereogenicity are respectively characterized by independent criteria. In contrast, the conventional approach (Table 10.6) lacks consistency from the present viewpoint, where "prochirality" and "prostereogenicity" are not adequately separated so as to cause neglect or confusion with respect to prochirality of a geometric meaning. The advantages of the stereoisogram approach over the conventional approach are summarized as follows:

1. In the stereoisogram approach, prochirality of a geometric meaning is characterized by means of a single criterion based on the term *enantiotopic* (Table 10.5), where such prochirality emerges in conversions of achiral promolecules into chiral ones, i.e., Type IV → I, Type IV → II, and Type V → III.
2. In the stereoisogram approach, pro-*RS*-stereogenicity to support the *pro-R*/*pro-S* system is characterized by means of a single criterion based on the term *RS-diastereotopic* (Table 10.5), where such pro-*RS*-stereogenicity emerges in conversions of *RS*-astereogenic promolecules into *RS*-stereogenic ones, i.e., Type IV → I, Type IV → V, and Type II → III.
3. The stereoisogram approach (Table 10.5) is capable of detecting a *coincident appearance* of an enantiotopic case and an *RS*-diastereotopic case (Figs. 10.7 and 10.13) as well as a *coexistence* of an enantiotopic case (e.g., Figs 10.14a and 10.15) and an *RS*-diastereotopic case (e.g., Figs. 10.14a and b) in a promolecule (**46**). The conventional approach (Table 10.6) is completely helpless to detect such interrelations between prochirality and pro-*RS*-stereogenicity.

Independent criteria shown in Table 10.5 are based on the formulation of stereoisograms for testifying prochirality and/or pro-*RS*-stereogenicity, where three kinds of intramolecular relationships are detected: enantiotopic, *RS*-diastereotopic, and holantitopic relationships. Each of these relationships is capable of producing a definite assembly. Moreover, these relationships are integrated into a more generic relationship, i.e., an *RS*-stereoheterotopic relationship, which is also capable of producing a definite assembly (a quadruplet of tentative *RS*-stereoisomers in a stereoisogram for testifying prochirality and/or pro-*RS*-stereogenicity). Thereby, the dichotomy between *RS*-stereoheterotopic relationships and others to cover stereoheterotopic relationships has been developed as shown in the solid-lined boxes in Fig. 10.18.

The newly-developed dichotomy shown in Fig. 10.18 corresponds to the above-mentioned dichotomy shown in Fig. 10.11. It has advantages over the conventional

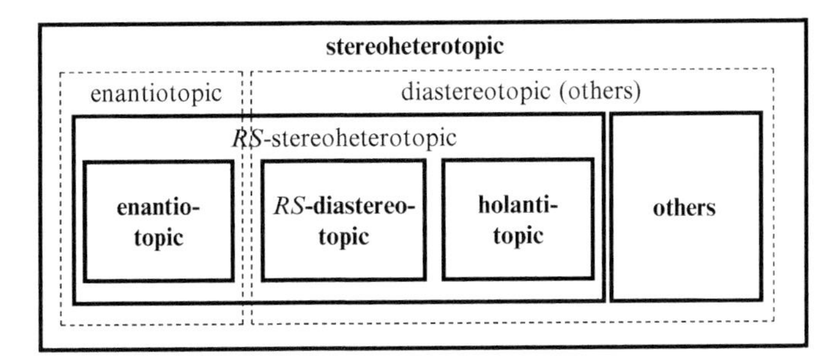

Fig. 10.18 Paradigm shift from the conventional terminology to the present terminology for topicities (Fujita 2009f). A *broken-lined box* represents a term of the conventional terminology, while a *solid-lined box* represents a term of the present terminology

dichotomy between enantiotopic and diastereotopic (the broken-lined boxes in Fig. 10.11), because the latter causes the confusion due to the entangled criteria shown in Table 10.6. This means that a paradigm shift from the conventional terminology to the present terminology for pro-*RS*-stereoisomerism turns out to be inevitable (Fig. 10.18) (Fujita 2009f), where correspondence between terms for relationships and those for attributes are defined as follow: enantiotopic — prochiral; *RS*-diastereotopic — pro-*RS*-stereogenic; and *RS*-stereoheterotopic — pro-*RS*-stereoisomerism.

10.5 Qualitative and Quantitative Aspects of Stereochemistry

As discussed above, relational terms such as *enantiomeric* and *RS-diastereomeric* are effective to demonstrate qualitative aspects of stereochemistry. In contrast, quantitative aspects of stereochemistry require more definite terms based on attributes of equivalence classes. This section is devoted to a brief survey of such quantitative aspects in comparison with such qualitative aspects.

10.5.1 Relationships vs. Attributes

The conventional approach mainly uses relational terms, e.g., enantiomeric and diastereomeric for representing intermolecular relationships between molecules as well as enantiotopic and diastereotopic for representing intramolecular relationships between ligands.

However, the relational term *self-enantiomeric* has not been available in the conventional approach so that an attributive term *achiral* is paired with the relational term *enantiomeric*. On the same line, the relational term *self-enantiotopic* has not been available in the conventional approach so that an attributive term *achirotopic*, which has been coined to be paired with an attributive term *chirotopic* (Mislow and Siegel 1984), is paired with the relational term *enantiotopic*.

Strictly speaking (cf. Table 10.1), the term *achiral* should be paired with the corresponding attributive term *chiral* and the term *achirotopic* should be paired with the corresponding attributive term *chirotopic*. On the same line the term *enantiomeric* should be paired with the corresponding relational term *self-enantiomeric* and the term *enantiotopic* should be paired with the corresponding relational term *self-enantiotopic*.

Pairing of a relational term (e.g., enantiomeric) with an attributive term (e.g., achiral) in the conventional approach overemphasizes "difference" between two mirror-image molecules of a pair, although the two molecules are equivalent to each other under reflection operations. Moreover, unconscious preference of such relational terms over such attributive terms has been spread in the conventional approach. For example, the dichotomy between enantiomers and diastereomers in the conventional approach does not explicitly involve the description on achiral molecules. In other words, the conventional approach underestimates the fact that a pair of enantiomeric molecules and an achiral molecule are equivalence classes to be treated in the same level.

10.5.2 Chirality and RS-Stereogenicity Based on Equivalence Classes

To prevent undesirable confusions, relational terms should be used in connection with equivalence classes. As for the term *enantiomeric*, a pair of enantiomers or an achiral molecule is an equivalence class to be connected with, as shown in Fig. 10.19a (Fujita 2009b).

By obeying the stereoisogram approach, the paradigm shift of relational terms shown in Fig. 10.11 results in the generation of equivalence classes under *RS*-stereoisomeric relationships (Fig. 10.19b) (Fujita 2009b), where the *RS*-stereoisomeric relationships are classified to enantiomeric, *RS*-diastereomeric, and holantimeric relationships appearing in stereoisograms (cf. Table 10.1). In terms of group theory, a quadruplet of promolecules contained in a stereoisogram is an equivalence class under the action of an *RS*-stereoisomeric group (Fujita 2004b, 2005d,e), which consists of reflections, *RS*-permutations, and ligand reflections (cf. Table 10.1).

Section 10.3, which has mainly based on such relational terms as enantiomeric and *RS*-diastereomeric, can be reformulated by attributive terms. The unit-subduced-cycle-index (USCI) approach developed by us (Fujita 1990, 1991b) is extended to be applied to *RS*-stereoisomeric groups. Because an *RS*-stereoisomeric group $\mathbf{G}$ contains a core group ($\mathbf{G}_C$) of all proper rotations, a point group ($\mathbf{G}_Y$) of all reflections, an *RS*-permutation group ($\mathbf{G}_X$) of all *RS*-permutations, and an inversion group ($\mathbf{G}_Z$) of all ligand reflections (Fujita 2004a), the respective types listed in Fig. 10.9 correspond to the following equivalence classes: Type I — $\mathbf{G}(/\mathbf{G}_Z)$; Type

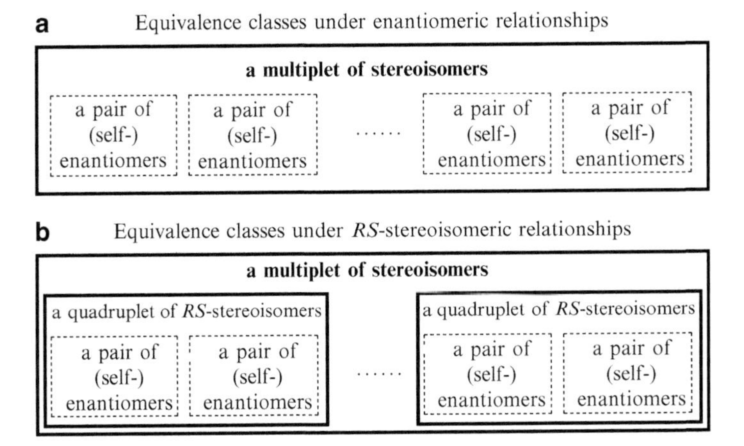

Fig. 10.19 A paradigm shift for equivalence classes. (**a**) Equivalence classes under enantiomeric relationship in the conventions of stereochemistry and (**b**) equivalence classes under *RS*-stereoisomeric relationships in the present approach (Fujita 2009b)

II — $\mathbf{G}(/\mathbf{G}_X)$; Type III — $\mathbf{G}(/\mathbf{G}_C)$; Type IV — $\mathbf{G}(/\mathbf{G})$; and Type V — $\mathbf{G}(/\mathbf{G}_Y)$. Note that each of the symbols represents a coset representation corresponding to an equivalence class (Fujita 1991b). These equivalence classes have been used in itemized enumeration of quadruplets of *RS*-stereoisomers under the action of *RS*-stereoisomeric group (Fujita 2009a).

10.5.3 Prochirality and Pro-*RS*-Stereogenicity Based on Equivalence Classes

Prochirality based on equivalence classes has been discussed in terms of *sphericities* (homospheric, enantiospheric, and hemispheric), which have been rather mathematically defined as attributive terms for specifying equivalence classes in molecules (Fujita 1990, 1991b). A non-mathematical discussion on prochirality has appeared on the basis of the term *sphericity* proposed as an attributive term (Fujita 2000b), although a preliminary definition of the term "prostereogenicity" has maintained the conventional methodology. A more sophisticated but non-mathematical discussion has appeared by emphasizing equivalence classes (Fujita 2002a). Advantages of *sphericity* as an attributive term over *topicity* as a relational term have been comprehensively discussed in reviews (Fujita 2002b,c).

After the proposal of pro-*RS*-stereogenicity, an integrated treatment of the pro-*RS*-stereogenicity with prochirality has been discussed (Fujita 2006b). Attributive terms *tropicities* (*RS*-homotropic, *RS*-enantiotropic, and *RS*-hemitropic) for discussing pro-*RS*-stereogenicity have been proposed in addition to sphericities (homospheric, enantiospheric, and hemispheric) as attributive terms for discussing prochirality (Fujita 2006b). The method using the tropicity terms is closely related to the membership criterion for detecting pro-*RS*-stereogenicity (Fujita 2006b, 2009f).

10.5.4 Multiple RS-Stereogenic Centers

The stereoisogram approach using promolecules and proligands is essentially concerned with local symmetry (local chirality and local *RS*-stereogenicity), although it is capable of specifying global symmetry. This means that we focus our attention on a single *RS*-stereogenic center during the construction of a stereoisogram. In order to treat molecules having multiple *RS*-stereogenic centers, more elaborate schemes should be developed. As such elaborate schemes, we have proposed *correlation diagrams of stereoisograms* (Fujita 2010a,b,c).

10.5.5 Combinatorial Enumerations

The USCI approach based on the sphericity concept of equivalence classes (Fujita 1991b, 2007c) has been applied to various enumerations. The USCI approach has been extended to treat *RS*-stereoisomers. Thus, *RS*-stereoisomers have been enumerated and itemized in terms of chirality, *RS*-stereogenicity, and sclerality (Fujita 2007b). Quadruplets of *RS*-stereoisomers are enumerated under the action of *RS*-stereoisomeric groups, where they are itemized into Types I–V (Fujita 2009a). As more elaborate models for supporting combinatorial enumerations in chemistry, we have proposed the concept of mandalas and related concepts (Fujita 2005c, 2006a,d, 2007c,h,i).

Recently, another approach (Fujita's proligand method) based on the sphericity concept of cycles (Fujita 2005a,b, 2006c, 2007f,g) has been developed and applied to enumeration of alkanes (Fujita 2007a,d,k, 2008a) and mono-substituted alkanes (Fujita 2007e,j). A review has discussed a long-standing interdisciplinary problem over 130 years on numbers of alkanes and monosubstituted alkanes (Fujita 2010d). The fates of asymmetry and pseudoasymmetry in the enumerations of mono-substituted alkanes (Fujita 2009h) and of alkanes (Fujita 2009g) have been discussed by using five types of promolecules on the basis of the stereoisogram approach.

10.6 Conclusion

The stereoisogram approach has reached the following conclusions:

1. Chirality and *RS*-stereogenicity are independent concepts.
2. Prochirality and pro-*RS*-stereogenicity are independent concepts.

Accordingly, the confused terminology of the conventional approach should be reorganized as follows:

1. The conventional "chirality" as a subconcept of the conventional "stereogenicity" has suffered transmutation apart from the geometric concept of chirality (cf. Item 1 in Sect. 10.2.6), so that it should be replaced by the *RS*-stereogenicity which is formulated to be independent of chirality in the stereoisogram approach.
2. The conventional "prochirality" as a subconcept of the conventional "prostereogenicity" is by no means the same as the geometric concept of prochirality based on the geometric concept of chirality. Hence, the conventional "prochirality" should be replaced by the pro-*RS*-stereogenicity which is formulated to be independent of prochirality in the stereoisogram approach (cf. Item 1 in Sect. 10.2.6).
3. The conventional terms "stereogenicity" and "prostereogenicity" should not be used for the purpose of giving *RS*-stereodescriptors of the CIP system and *pro-R*/*pro-S*-descriptors of the *pro-R*/*pro-S* system. They should be replaced by the terms on the stereoisogram approach, i.e., *RS*-stereogenicity and pro-*RS*-stereogenicity (cf. Item 2 in Sect. 10.2.6).
4. Such relational terms as "enantiotopic", "diastereotopic", and "stereoheterotopic" for supporting the *Pro-R*/*Pro-S* system in the conventional approach should be abandoned, as pointed out by Items 3–6 in Sect. 10.2.6.
5. *RS*-Stereodescriptors of the CIP system and *pro-R*/*pro-S*-descriptors of the *pro-R*/*pro-S* system should not be directly applied to the examination of geometric aspects of stereochemistry. They should be restricted to the nomenclature of configurations and to the description (not rationalization or not interpretation) of results of geometric examination, because the geometric examination is capable of such rationalization or interpretation without using *RS*-stereodescriptors or *pro-R*/*pro-S*-descriptors. When such results of geometric examination are described by using *RS*-stereodescriptors or *pro-R*/*pro-S*-descriptors, the concept of chirality-faithfulness (Fujita 2009d) should be taken into consideration.

Thereby, the conventional terminology for supporting *RS*-stereodescriptors of the CIP system and *pro-R*/*pro-S*-descriptors of the *pro-R*/*pro-S* system should be entirely replaced by the newly defined terms based on the stereoisogram approach.

References

Black KA (1990) J Chem Educ 67:141–142
Cahn RS, Ingold CK, Prelog V (1966) Angew Chem Int Ed Engl 5:385– 415
Eliel EL (1982) Top Curr Chem 105:1–76
Eliel EL, Wilen SH (1994) Stereochemistry of organic compounds. Wiley, New York
Eliel EL, Willen SH, Doyle MP (2001) Basic organic stereochemistry. Wiley-Interscience, New York
Fujita S (1990) J Am Chem Soc 112:3390–3397
Fujita S (1991a) Tetrahedron 47:31–46
Fujita S (1991b) Symmetry and Combinatorial enumeration in chemistry. Springer, Berlin/Heidelberg

Fujita S (2000a) J Chem Inf Comput Sci 40:426–437
Fujita S (2000b) Tetrahedron 56:735–740
Fujita S (2001) Bull Chem Soc Jpn 74:1585–1603
Fujita S (2002a) J Org Chem 67:6055–6063
Fujita S (2002b) Bull Chem Soc Jpn 75:1863–1883
Fujita S (2002c) Chem Rec 2:164–176
Fujita S (2004a) J Math Chem 35:265–287
Fujita S (2004b) MATCH Commun Math Comput Chem 52:3–18
Fujita S (2004c) Tetrahedron 60:11629–11638
Fujita S (2004d) J Org Chem 69:3158–3165
Fujita S (2005a) Theor Chem Acc 113:73–79
Fujita S (2005b) Theor Chem Acc 113:80–86
Fujita S (2005c) MATCH Commun Math Comput Chem 54:251–300
Fujita S (2005d) MATCH Commun Math Comput Chem 53:147–159
Fujita S (2005e) MATCH Commun Math Comput Chem 54:39–52
Fujita S (2006a) MATCH Commun Math Comput Chem 55:237–270
Fujita S (2006b) Tetrahedron 62:691–705
Fujita S (2006c) Theor Chem Acc 115:37–53
Fujita S (2006d) MATCH Commun Math Comput Chem 55:5–38
Fujita S (2007a) MATCH Commun Math Comput Chem 57:299–340
Fujita S (2007b) MATCH Commun Math Comput Chem 58:611–634
Fujita S (2007c) Diagrammatical approach to molecular symmetry and enumeration of stereoisomers. University of Kragujevac, Faculty of Science, Kragujevac
Fujita S (2007d) MATCH Commun Math Comput Chem 57:265–298
Fujita S (2007e) J Comput Chem Jpn 6:73–90
Fujita S (2007f) Theor Chem Acc 117:353–370
Fujita S (2007g) Theor Chem Acc 117:339–351
Fujita S (2007h) J Math. Chem 42:481–534
Fujita S (2007i) MATCH Commun Math Comput Chem 57:5–48
Fujita S (2007j) J Comput Chem Jpn 6:59–72
Fujita S (2007k) MATCH Commun Math Comput Chem 58:5–45
Fujita S (2008a) J Math Chem 43:141–201
Fujita S (2008b) Yuki Gosei Kagaku Kyokai-Shi/J Synth Org Chem Jpn 66:995–1004
Fujita S (2009a) MATCH Commun Math Comput Chem 61:71–115
Fujita S (2009b) Tetrahedron 65:1581–1592
Fujita S (2009c) J Comput Aided Chem 10:76–95
Fujita S (2009d) J Comput Aided Chem 10:16–29
Fujita S (2009e) MATCH Commun Math Comput Chem 61:11–38
Fujita S (2009f) MATCH Commun Math Comput Chem 61:39–70
Fujita S (2009g) MATCH Commun Math Comput Chem 62:65–104
Fujita S (2009h) MATCH Commun Math Comput Chem 62:23–64
Fujita S (2010a) MATCH Commun Math Comput Chem 63:3–24
Fujita S (2010b) J Math Chem 47:145–166
Fujita S (2010c) MATCH Commun Math Comput Chem 63:25–66
Fujita S (2010d) Bull Chem Soc Jpn 83:1–18
Hanson KR (1966) J Am Chem Soc 88:2731–2742
Helmchen G (1996) A. General aspects. 1. Nomenclature and vocabulary of organic stereochemistry. In: Helmchen G, Hoffmann RW, Mulzer J, Schaumann E (eds) Stereoselective Synthesis. Methods of Organic Chemistry (Houben-Weyl), Workbench Edition E21, vol 1, 4th edn. Georg Thieme, Stuttgart, New York, pp 1–74
Hirschmann H (1983) Trans N Y Acad Sci Ser II 41:61–69
Hirschmann H, Hanson KR (1971a) Eur J Biochem 22:301–309
Hirschmann H, Hanson KR (1971b) J Org Chem 36:3293–3306

IUPAC Chemical Nomenclature and Structure Representation Division (2004) Provisional Recommendations. Nomenclature of Organic Chemistry http://www.iupac.org/reports/provisional/abstract04/favre_ 310305.html. Accessed date June 17, 2011

IUPAC Organic Chemistry Division (1996) Pure Appl Chem 68:2193–2222

Jonas J (1988) Coll Czech Chem Commun 53:2676–2714

McCasland GE (1953) A New General System for the Naming of Stereoisomers. Chemical Astracts, Columbus

Mislow K (1965) Introduction to Stereochemistry. Benjamin, New York

Mislow K (1977) Bull Soc Chim Belg 86:595–601

Mislow K (2002) Chirality 14:126–134

Mislow K, Raban M (1967) Top Stereochem 1:1–38

Mislow K, Siegel J (1984) J Am Chem Soc 106:3319–3328

Morris DG (2001) Stereochemistry. Royal Soc Chem, Cambridge

North N (1998) Principles and Applications of Stereochemistry. Stanley Thornes, Cheltenham

Prelog V, Helmchen G (1972) Helv Chim Acta 55:2581–2598

Prelog V, Helmchen G (1982) Angew Chem Int Ed Eng 21:567–583

Vollhardt KPC, Schore NE (2003) Organic Chemistry. Structure and Function, 4th edn. Freeman, New York

Chapter 11
Diamond D_5, a Novel Class of Carbon Allotropes

Mircea V. Diudea[1], Csaba L. Nagy[1], and Aleksandar Ilić[2]

Abstract Design of hypothetical crystal networks, consisting of most pentagon rings and generically called diamond D_5, is presented. It is shown that the seed and repeat-units, as hydrogenated species, show good stability, compared with that of C_{60} fullerene, as calculated at DFT levels of theory. The topology of the network is described in terms of the net parameters and Omega polynomial.

11.1 Introduction

The nano-era, a period starting since 1985 with the discovery of C_{60}, is dominated by the carbon allotropes, studied for applications in nano-technology. Among the carbon structures, fullerenes (zero-dimensional), nanotubes (one dimensional), graphene (two dimensional), diamond and spongy nanostructures (three dimensional) were the most studied (Diudea 2005; Diudea and Nagy 2007). Inorganic compounds also attracted the attention of scientists. Recent articles in crystallography promoted the idea of topological description and classification of crystal structures (Blatov et al. 2004, 2009; Delgado-Friedrichs and O'Keeffe 2005).

Diamond D_6 (Fig. 11.1), the classical, beautiful and useful diamond has kept its leading interest among the carbon allotropes, even as the newer "nano" varieties. Along with electronic properties, the mechanical characteristics appear of great

[1] Faculty of Chemistry and Chemical Engineering, "Babes-Bolyai" University, Arany Janos Str. 11, 400028 Cluj, Romania
e-mail: diudea@gmail.com; nc35@chem.ubbcluj.ro

[2] Faculty of Sciences and Mathematics, University of Niš, Višegradska 33, 18000 Niš, Serbia
e-mail: aleksandari@gmail.com

M.V. Putz (ed.), *Carbon Bonding and Structures: Advances in Physics and Chemistry*, Carbon Materials: Chemistry and Physics 5, DOI 10.1007/978-94-007-1733-6_11,

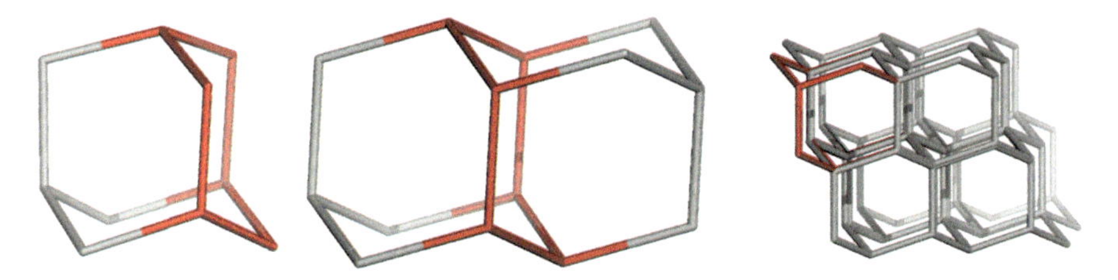

Fig. 11.1 Diamond, a triple periodic network: Ada(mantane) D_6_10_111 (*left*), Dia(mantane) D_6_14_211 (*central*) and Diamond D_6_52_222 net (*right*)

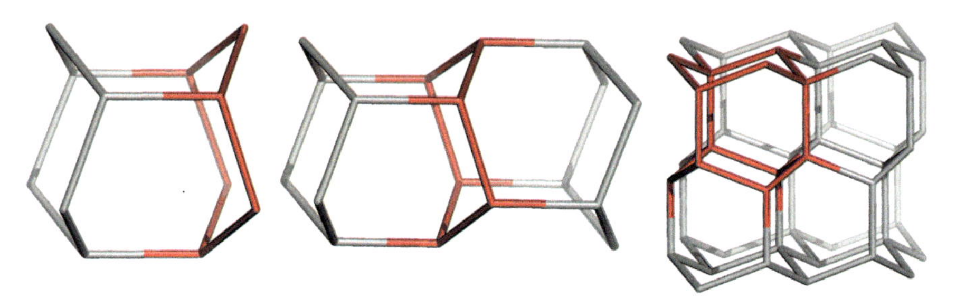

Fig. 11.2 Losdaleite, a double periodic network: L_6_12_111 (*left*), L_6_18_211 (*central*) and L_6_48_222 net (*right*)

importance, as the composites can overpass the resistance of steel or other metal alloys. A lot of efforts were done in the production and purification of "synthetic" diamonds, from detonation products (Decarli and Jamieson 1961; Aleksenskiĭ et al. 1997; Osawa 2007, 2008) and other synthetic ways (Khachatryan et al. 2008; Mochalin and Gogotsi 2009).

However, the diamond D_6 is not unique: out of the classical structure, showing all-hexagonal rings of sp^3 carbon atoms in a cubic network (space group $Fd3m$), there is Lonsdaleite (Frondel and Marvin 1967; He et al. 2002) a rare stone of pure carbon discovered at Meteor Crater, Arizona, in 1967 and also several hypothetical diamond-like networks (Sunada 2008; Diudea et al. 2010). The Lonsdaleite hexagonal network (space group $P6_3/mmc$) is illustrated in Fig. 11.2.

Dendrimers are hyper-branched nano-structures, made by a large number of (one or more types) substructures called monomers, synthetically joined within a rigorously tailored architecture (Tomalia et al. 1990; Newkome et al. 2001; Diudea and Katona 1999). They can be functionalized at terminal branches, thus finding a broad pallet of applications in chemistry, medicine, etc.

Multi-tori MT are structures of high genera (Diudea 2010a), consisting of more than one tubular ring. They are supposed to result by self-assembly of some repeat units (i.e., monomers) which can be designed by opening of cages/fullerenes or by appropriate map/net operations. Multi-tori, rather than dendrimers, appear in processes of self-assembling of some rigid monomers. Zeloites and spongy carbon, recently synthesized (Barborini et al. 2002; Benedek et al. 2003) also contain multi-tori.

Structures of high genera, like multi-tori, can be designed starting from the Platonic solids, by using appropriate map operations (Diudea 2010a). Such structures have before been modeled by Lenosky et al. (1992) and Terrones and Mackay (1997) etc.

11.2 Dendrimer Design and Stability

A tetrapodal monomer M_1(Fig. 11.3, *left*), designed by $Trs(P_4(T))$ sequence of map operations (Diudea and Ilić 2011) and consisting of all pentagonal faces, can self-arrange to a dendrimer M_5, at the first generation stage (Fig. 11.3, *right*).

The "growing process" is imagined occurring by identification of the trigonal faces of two opposite M_1 units; at the second generation, six pentagonal hyper-cycles are closed, as in molecule M_{17}, Fig. 11.4 (Diudea and Ilić 2011).

The process is drown as a "dendrimer growth", and is limited here at the fifth generation (Fig. 11.4), when a tetrahedral array results: 4S_MT = M_{57}.

Multi-tori herein considered can be viewed either as infinite (*i.e.*, open) structures or as closed cages; then, it is not trivial to count the number of simple tori (*i.e.*, the genus g) in such complex structures.

The Euler's formula [Euler (1758)]: $v - e + f = 2(1 - g)$, where v, e and f are the number of vertices/atoms, edges/bonds, and faces, respectively, is applicable only in case of single shell structures. In multi shell structures (Diudea and Nagy 2008), have modified the Euler formula as: $v - e + r - p(s - 1) = 2(1 - g)$, where r stands for the number of hard rings (*i.e.*, those rings which are nor the sum of some smaller rings), p is the number of smallest polyhedra filling the space of the considered structure while s is the number of shells. In case of an infinite structure, the external trigonal faces are not added to the total count of faces/rings. The calculated g-values are given in Fig. 11.4.

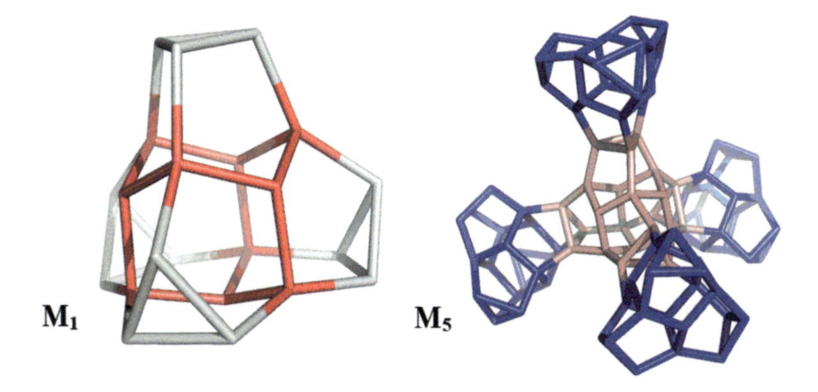

Fig. 11.3 Tetrapodal unit designed by $Trs(P_4(T))$ and the corresponding dendrimer, at first generation stage

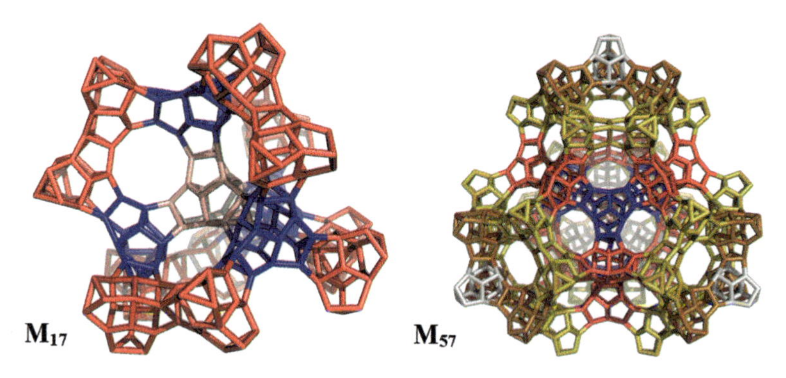

Fig. 11.4 Dendrimer at second (*left*) and fifth (*right*) generation stage; $M_{57} = 4S\_MT$; $v = 972$; $e = 1770$; $f_5 = 684$; $g = 58$ (infinite structure); adding $f_3 = 40$, then $g = 38$ (finite structure)

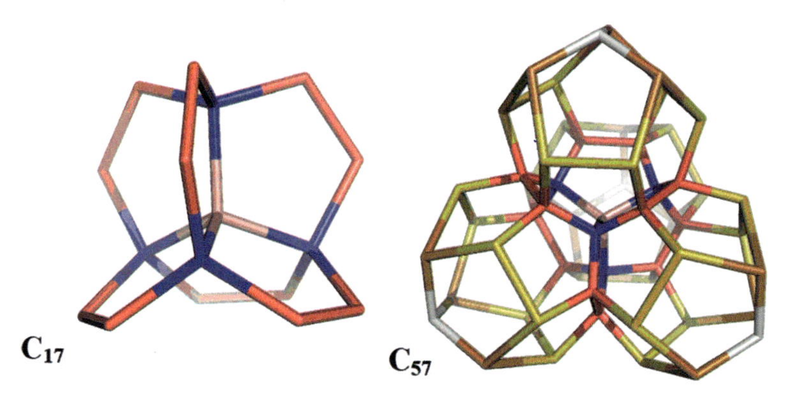

Fig. 11.5 Reduced graphs at 2nd (*left*) and 5th (*right*) generation stage; C_{57}: $v = 57$; $e = 94$; $r_5 = 42$; $g = 0.5$; $R(C_{57}, x) = 42x^5 + 82x^9 + 144x^{10}$

The number of tetrapodal monomers, added at each generation, up to the fifth one, realized as M_{57}, is: 1; 4; 12, 24, 12, 4. The connections in M_{57} are complex and to elucidate the large structures up to the fifth generation, the design of the corresponding reduced graphs (Fig. 11.5) was needed (Diudea 2010b, 2011).

The structure C_{17} (Fig. 11.5, *left*) we call the "seed" of all the hereafter structures. The structure C_{57} (Fig. 11.5, *right*) corresponds to the above M_{57} and is equivalent to 4 "condensed" dodecahedra, sharing a common point. By considering this common point as an internal shell s, the modified (Diudea and Nagy 2008) Euler formula will give (for $v = 57$; $e = 94$; $r = 42$; $p = 4$ and $s = 2$) a (non-integer) genus $g = 0.5$. The ring polynomial R(x) is also given, at the *bottom* of Fig. 11.5.

11.3 Diamond D_5 Networks

11.3.1 Spongy D_5

A monomer C_{81}, derived from C_{57} and consisting of four closed C_{20} units and four open units, and its mirror image (Fig. 11.6) was used by (Diudea and Nagy 2011a) to build the alternant network of spongy diamond SD_5 (Fig. 11.7). The nodes of diamond SD_5 network consist of C_{57} units and the network is triple periodic.

The number of atoms v, bonds e, and C_{57} monomers m, and the content in sp^3 carbon, given as a function of k – the number of monomers along the edge of a cubic (k,k,k) domain. At limit, in an infinitely large net, the content of sp^3 carbon approaches 77% (see Appendix 2, Table 11. 4). The density of the net varies around an average of d = 1.6 g/cm^3, in agreement with the "spongy" structure of D_5 net (Fig. 11.7).

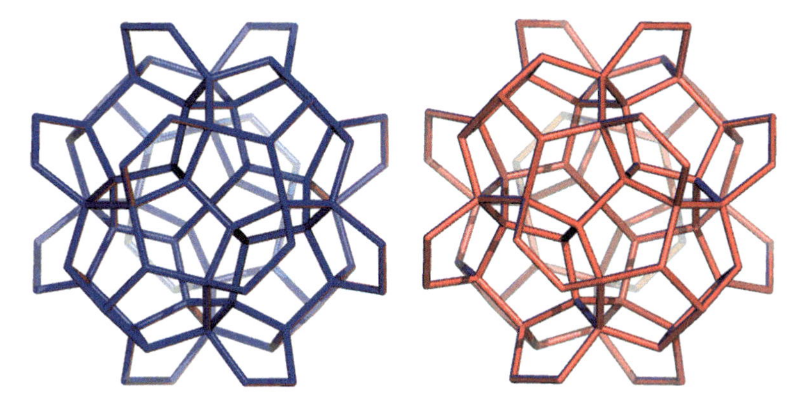

Fig. 11.6 Monomer C_{81} unit (*left-up*), and its mirror image-pair (*right-up*); the monomers as in the triple periodic network of diamond SD_5

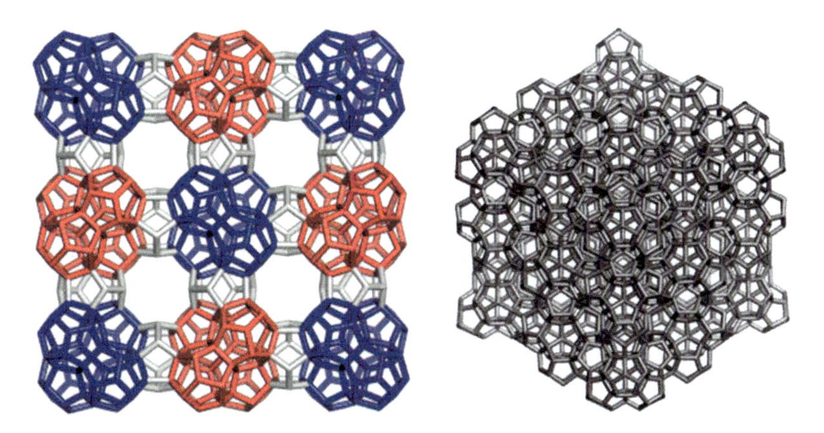

Fig. 11.7 SD_5 (C_{57}) triple periodic network: top view (*left*) and corner view (*right*)

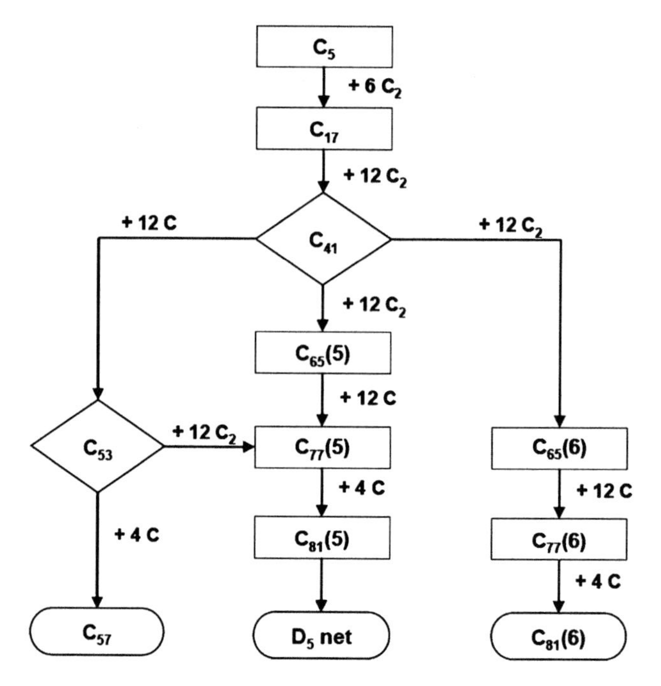

Fig. 11.8 Pathway to D_5

A possible pathway to D_5 was proposed by Diudea and Nagy (2011a) (Fig. 11.8): The main intermediate structures in this scheme are: C_{17} (the "seed" of D_5), C_{41}, C_{53} and $C_{81}(5)$ (see Fig. 11.8). The stability of these structures was evaluated as hydrogenated species. The C_{81} monomer has a C_{57} core and contains 12 flaps which represent half of the junctions between the SD_5 nodes.

To avoid the non-wished side products $C_{65}(6)$ to $C_{81}(6)$ (containing six-membered rings), the suggested way is through C_{53}. In a next step, one can reach either C_{57} as a final structure (which can, however, lead to some dense species of D_5) or go to $C_{81}(5)$, the monomer of SD_5 network.

Of course, the scientist will choose the most convenient route in an attempt to synthesize these structures.

11.3.2 Dense D_5

There is a chance to reach D_5 just from C_{17}, a centrohexaquinane (Paquette and Vazeux 1981; Kuck 2006) which can dimerize (Diudea and Nagy 2011b; Eaton 1979) to $2 \times C_{17} = C_{34}$ and this last condensing to $4 \times C_{17} = C_{51}$ (Fig. 11.9, *top row*).

A linear $4 \times C_{17} = C_{57}$ is also energetically possible (see Table 11.1). The angular tetramer $4 \times C_{17} = C_{51}$ will compose the six edges of a tetrahedron to form the corresponding Adamantane-like Ada_20_170, bearing six pentagonal

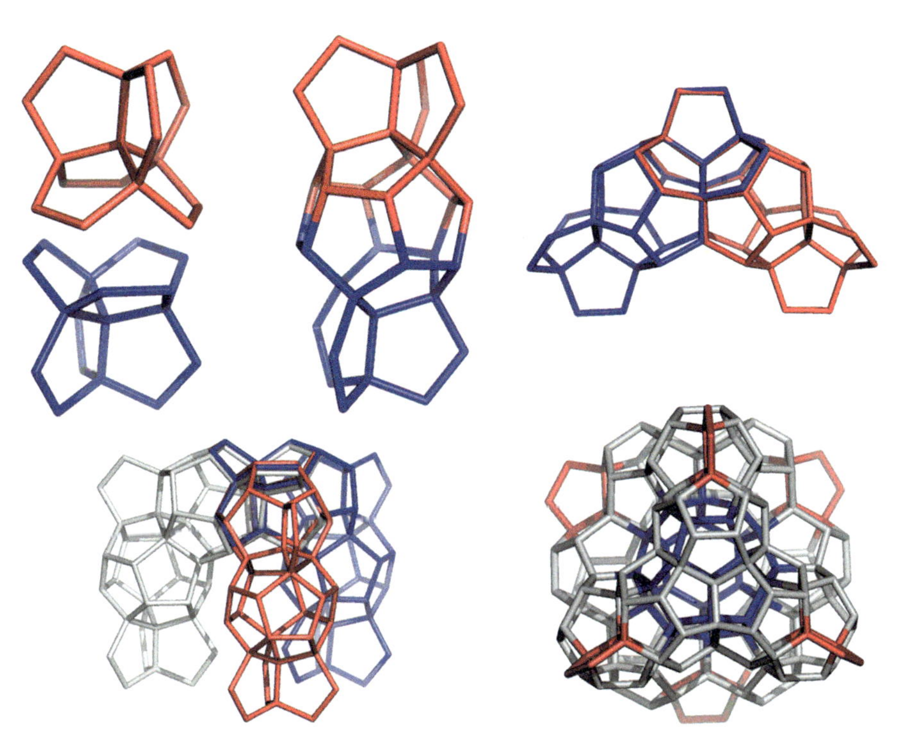

Fig. 11.9 Way to Ada_20: 2 × C_{17} = C_{34} (*top-left and central*), 4 × C_{17} = C_{51} (*top-right*), 3(4 × C_{17}) = C_{119} (*bottom-left*) and Ada_20_170 (*bottom-right*)

Table 11.1 Single point calculation results (HOMO-LUMO gap in eV and total energy E_{tot} in a.u.) at the B3LYP/6-31 G(d,p) levels of theory; C_{60} is taken as reference structure

		B3LYP		
Struct	Sym	E_{tot}	Gap	E_{tot}/N
$C_{17}H_{12}$	T_d	−650.66	6.04	−38.27
$C_{34}H_{12}$	C_{2h}	−1302.36	3.23	−38.31
$C_{51}H_{14}$-ang	C_{2v}	−1951.13	3.34	−38.26
$C_{57}H_{18}$-lin	D_{3h}	−2182.07	2.97	−38.28
$C_{158}H_{12}$	T_d	−6025.94	3.24	−38.139
$C_{170}H_{12}$	T_d	−6483.08	3.17	−38.136
$C_{20}H_{20}$	I_h	−774.21	8.00	−38.71
C_{20}	D_2	−761.44	1.94	−38.07
C_{60}	I_h	−2286.17	2.76	−38.10

wings (in red – Fig. 11.9, *right*, *bottom*), or without wings, as in Ada_20_158 (Fig. 11.10, *left*). Compare this with Adamantane (Fig. 11.1, *left*) in the structure of classical diamond D_6. In the above symbols, "20" refers to C_{20}, which is the main unit of the dense diamond D_5 (Figs. 11.10–11.12) while the last number counts the carbon atoms in the structures.

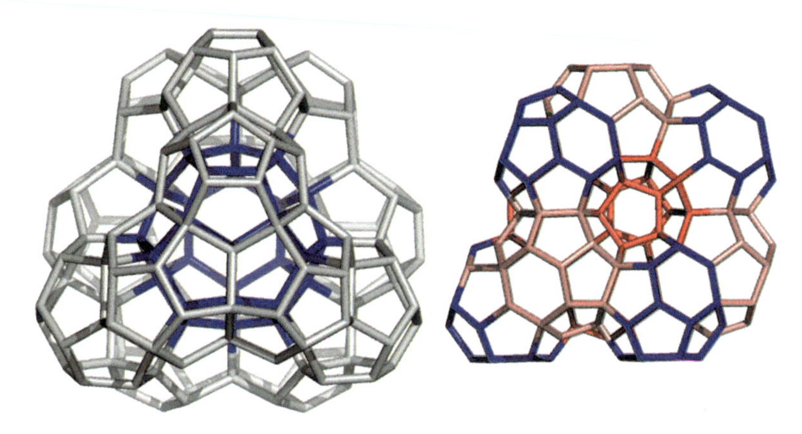

Fig. 11.10 Adamantane-like structures: Ada_20_170 (*left*) and Ada_28_213 (*right*)

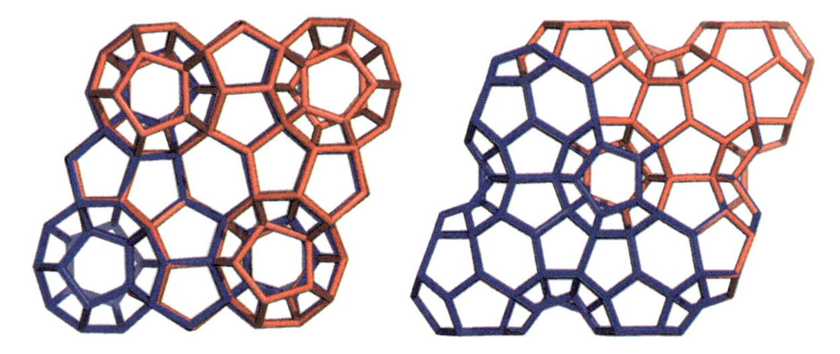

Fig. 11.11 Diamantane-like structures: Dia_20_226 net (*left*) and Dia_28_292 co-net (*right*)

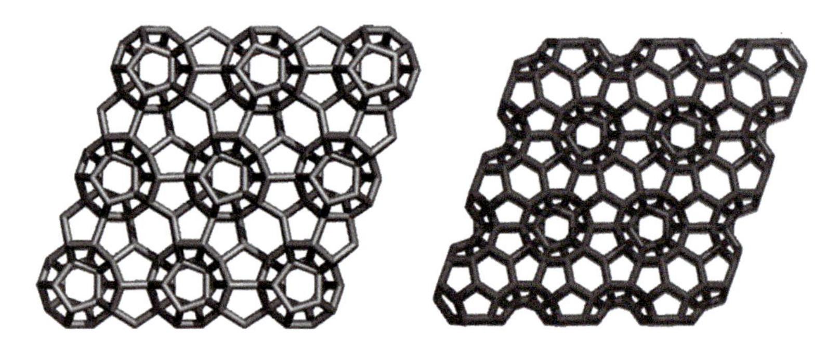

Fig. 11.12 Diamond D_5_20_860_333 net (*left*) and D_5_28_ 1022_333 co-net (*right*)

A diamantane-like unit is evidenced, as in Fig. 11.11 (see for comparison the diamantane, Fig. 11.1, *central*). Since any net has its co-net, the diamond D_5_20 net has the co-net D_5_28 (Fig. 11.12, *right*), of which corresponding units are illustrated in Figs 11.10 (*right*)–11.11 (*right*), respectively. In fact is one and the same *triple periodic* D_5 network, built up basically from C_{20} and having as hollows the fullerene C_{28} (Diudea and Nagy 2011b). The co-net D_5_28 cannot be reached from C_{28} alone since the hollows of such a net consist of C_{57} units (a C_{20}-based structure, see above) or higher tetrahedral arrays of C_{20} thus needing extra C atoms per Ada-like unit. The C_{28}-based hyperdiamond reported by Bylander and Kleinman (1993) consists of only C_{28} fullerenes joined by the tetrahedrally disposed neightbors by *only one covalent bond*.

Our D_5_20/28 hyperdiamond mainly consists of sp^3-bonded carbon atoms building Ada-like repeating units (including C_{28} as hollows), the cohesive bonds of these units being the same sp^3 covalent (and the same distributed) bonds as within the repeating units. This is in high contrast to the (Bylander and Kleinman 1993) hyperdiamond. The ratio C-sp^3/C-total trends to 1 in a large enough network. The topology of the D_5 networks is detailed in the last section.

Since the smallest C_{20} fullerene is the highest reactive one, it is expected to spontaneously stabilize in an sp^3-crystalline form, e.g., bcc-C_{20} (Chen et al. 2004; Ivanovskii 2008) or D_5_20/28 (Diudea and Nagy 2011b). A similar behavior is expected from the C_{20}-derivative $2 \times C_{17} = C_{34}$ (Fig. 11.9) to provide the Ada-like repeating units of the proposed D_5-diamond.

As the content of pentagons R[5] per total rings trend to 90% (see Table 11.5, entry 9) we called this, yet hypothetical carbon allotrope, the diamond D_5. Since the large hollows in the above spongy diamond are not counted, and the small rings are all pentagons, we also called it (S)D_5.

The presence of pentagons in diamond-like fullerides and particularly the ratio R[5]/R[6] seems to be important for the superconducting properties of such solid phases (Breda et al. 2000). In D_5 this ratio trends to 9 (see Table 11.5).

Energetic data, calculated at the DFT level in Table 11.1 – (Diudea and Nagy 2011a, b) show a good stability of the start and intermediate structures. Limited cubic domains of the D_5 networks have also been evaluated for stability, data proving a pertinent stability of the (yet) hypothetical D_5 diamond.

The calculated data show these structures as energetic minima, as supported by the simulated IR vibrations. All-together, these data reveal the proposed structures as pertinent candidates to the status of real molecules.

Density of the D_5 networks was calculated with the approximate (maximal) volume of a cubic domain. The values range from 1.5 (SD_5) to 2.8 (D_5).

11.4 Lonsdaleite L_5_28 Network

By analogy to D_5_20/28, a lonsdaleite-like net was built up (Fig. 11.13).

As a monomer, the hyper-hexagons L_5_28_134 (Fig. 11.13, *left* and *central*), in the chair conformation, of which nodes represent the C_{28} fullerene, was used. Its corresponding co-net L_5_20 was also designed. The lonsdaleite L_5_28/20 is a double periodic network, partially superimposed to the D_5_20/28 net (Diudea and Nagy 2011b).

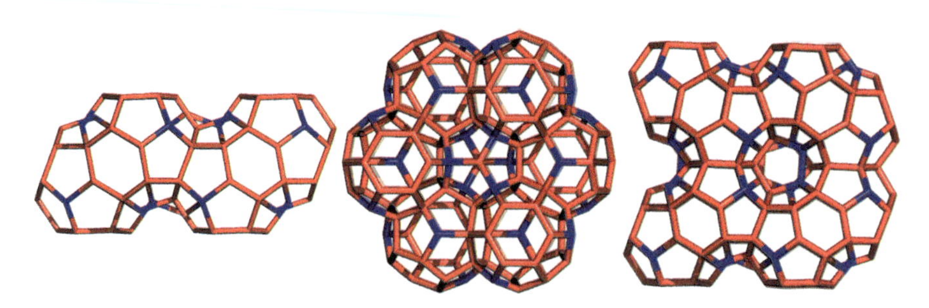

Fig. 11.13 Losdaleite, a double periodic network: L_5_28_134 (*left*), L_5_28_134 (*top view, central*) and L_5_28_250 (*side view, right*)

11.5 Omega Polynomial

In a connected graph *G*(*V*,*E*), with the vertex set *V*(*G*) and edge set *E*(*G*), two edges $e = uv$ and $f = xy$ of *G* are called *codistant e co f* if they obey the relation [John et al (2007)]:

$$d(v,x) = d(v,y) + 1 = d(u,x) + 1 = d(u,y) \tag{11.1}$$

which is reflexive, that is, *e co e* holds for any edge *e* of *G*, and symmetric, if *e co f* then *f co e*. In general, relation *co* is not transitive; if "*co*" is also transitive, thus it is an equivalence relation, then *G* is called a *co-graph* and the set of edges $C\times(e) := \{f \in E(G);\ f\ co\ e\}$ is called an *orthogonal cut oc* of *G*, *E*(*G*) being the union of disjoint orthogonal cuts: $E(G) = C_1 \cup C_2 \cup \ldots \cup C_k,\ \ C_i \cap C_j = \oslash,\ i \neq j$. Klavžar has shown (Klavžar 2008) that relation *co* is a theta Djoković-Winkler relation (1973–1984).

We say that edges *e* and *f* of a plane graph *G* are in relation *opposite, e op f,* if they are opposite edges of an inner face of *G*. Note that the relation *co* is defined in the whole graph while *op* is defined only in faces. Using the relation *op* we can partition the edge set of *G* into *opposite* edge *strips, ops*. An *ops* is a quasi-orthogonal cut *qoc*, since *ops* is not transitive.

Let *G* be a connected graph and $S_1, S_2, \ldots, S_k$ be the *ops* strips of *G*. Then the *ops* strips form a partition of *E(G)*. The length of *ops* is taken as maximum. It depends on the size of the maximum fold face/ring F_{max}/R_{max} considered, so that any result on Omega polynomial will have this specification.

Denote by *m*(*G*,*s*) the number of *ops* of length *s* and define the Omega polynomial as (Diudea 2006; Diudea et al. 2008; Diudea and Katona 2009):

$$\Omega(G,x) = \sum\nolimits_s m(G,s)\cdot x^s \tag{11.2}$$

Its first derivative (in $x = 1$) equals the number of edges in the graph:

$$\Omega'(G,1) = \sum\nolimits_s m(G,s)\cdot s = e = |E(G)| \tag{11.3}$$

On Omega polynomial, the Cluj-Ilmenau [John et al (2007)] index, $CI = CI(G)$, was defined:

$$CI(G) = \{[\Omega'(G,1)]^2 - [\Omega'(G,1) + \Omega''(G,1)]\} \tag{11.4}$$

This counting polynomial found utility in predicting stability of small fullerenes (Diudea 2010a) and in description of various polyhedral nanostructures.

11.5.1 Topology of Diamond D_6 and Lonsdaleite L_6 Nets

Topology of the classical diamond D_6 and Lonsdaleite L_6 are listed in Table 11.2. Along with Omega polynomial, formulas to calculate the number of atoms in a cuboid of dimensions (k,k,k) are given. In the above, k is the number of repeating units along the edge of such a cubic domain. One can see that the ratio $C(sp^3)/v(G)$ approaches the unity; this means that in a large enough net almost all atoms are tetra-connected, a basic condition for a structure to be diamondoid. Examples of calculus are given in Table 11.3.

11.5.2 Topology of Spongy Diamond SD_5

In describing the topology of the spongy diamond SD_5, we considered only the pentagons, the larger hollows being omitted. Thus, only the basic data of the net are presented in Table 11.4; the ratio $C(sp^3)/v(G)$ is here far from unity, because of many carbon atoms are exposed to exterior.

11.5.3 Topology of Dense Diamond D_5 and Lonsdaleite L_5

Topology of the dense diamond D_5 and lonsdaleite L_5 is presented in Tables 11.5–11.10: formulas to calculate Omega polynomial, number of atoms, number of rings and the limits (at infinity) for the ratio of sp^3 C atoms over total number of atoms and also the ratio R[5] over the total number of rings (Table 11.5). Numerical examples are given.

Table 11.2 Omega polynomial in Diamond D_6 and Lonsdaleite L_6 nets, function of the number of repeating units along the edge of a cubic (k,k,k) domain

	Network
A	Omega(D_6); R[6]
1	$\Omega(D_{6\_}k_{odd},x)=\left(\sum_{i-1}^{k}2x^{\frac{(i+1)(i+2)}{2}}\right)+\left(\sum_{i-1}^{(k-1)/2}2x^{\frac{(k+1)(k+2)}{2}+\frac{k\times k-1}{4}-i(i-1)}\right)+3kx^{(k+1)(k+1)}$
2	$\Omega(D_{6\_}k_{even},x)=\left(\sum_{i-1}^{k}2x^{\frac{(i+1)(i+2)}{2}}\right)+\left(\sum_{i-1}^{k/2}2x^{\frac{(k+1)(k+2)}{2}+\frac{k\times k}{4}-(i-1)(i-1)}\right)-x^{\frac{(k+1)(k+2)}{2}+\frac{k\times k}{4}}+3kx^{(k+1)(k+1)}$
3	$\Omega'(1)=e(G)=-1+6k+9k^2+4k^3$
4	$CI(G)=2-187k/10-k^2/4+305k^3/4+457k^4/4+1369k^5/20+16k^6$
5	$v(G)=6k+6k^2+2k^3$
6	$Atoms(sp^3)=-2+6k+2k^3$
7	$R[6]=3k^2+4k^3$
8	$\lim_{k\to\infty}\left[\frac{Atoms(sp^3)}{v(G)}=\frac{-2+6k+2k^3}{6k+6k^2+2k^3}\right]=1$
B	Omega(L_6); R[6]
1	$\Omega(L_6,x)=k\cdot x^{k(k+2)}+x^{(k+1)(3k^2+4k-1)}$
2	$\Omega'(1)=e(G)=-1+3k+9k^2+4k^3$
3	$CI(G)=k^2(k+2)(7k^3+15k^2+4k-2)$
4	$v(G)=2k(k+1)(k+2)=4k+6k^2+2k^3$
5	$Atoms(sp3)=2(k-1)\cdot k\cdot(k+1)=2k(k^2-1)$
6	$R[6]=-2k+3k^2+4k^3$
7	$\lim_{k\to\infty}\left[\frac{Atoms(sp^3)}{v(G)}=\frac{2k(k^2-1)}{4k+6k^2+2k^3}\right]=1$

Table 11.3 Examples, omega polynomial in diamond D_6 and lonsdaleite L_6 nets

k	Polynomial (Net)	Atoms	sp^3Atoms (%)	Bonds	CI(G)	R[6]
	Omega(D_6); R[6]					
1	$2x^3 + 3x^4$ (Diamantane)	14	–	18	258	7
2	$2x^3 + 2x^6 + 1x^7 + 6x^9$	52	26 (50.00)	79	5616	44
3	$2x^3 + 2x^6 + 2x^{10} + 2x^{12} + 9x^{16}$	126	70 (55.56)	206	39554	135
4	$2x^3 + 2x^6 + 2x^{10} + 2x^{15} + 2x^{18} + 1x^{19} + 12x^{25}$	248	150 (60.48)	423	169680	304
5	$2x^3 + 2x^6 + 2x^{10} + 2x^{15} + 2x^{21} + 2x^{25} + 2x^{27} + 15x^{36}$	430	278 (64.65)	754	544746	575
6	$2x^3 + 2x^6 + 2x^{10} + 2x^{15} + 2x^{21} + 2x^{28} + 2x^{33} + 2x^{36} + 1x^{37} + 18x^{49}$	684	466 (68.13)	1223	1443182	972
	Omega(L_6); R[6]					
1	$1x^3 + x^{12}$	12	–	15	72	5
2	$2x^8 + x^{57}$	48	12 (25.00)	73	1952	40
3	$3x^{15} + x^{152}$	120	48 (40.00)	197	15030	129
4	$4x^{24} + x^{315}$	240	120 (50.00)	411	67392	296
5	$5x^{35} + x^{564}$	420	240 (57.14)	739	221900	565
6	$6x^{48} + x^{917}$	672	420 (62.50)	1205	597312	960

Table 11.4 Topology of the spongy SD_5_20

	Formulas
	C_{57}-*based net*
1	$v(SD_5\_57) = k^2(69k - 12)$
2	$e(SD_5\_57) = 2k^2(65k - 18)$
3	$m(SD_5\_57) = k^3$; m = monomer; $k = 1,2,\ldots$
4	$Atoms(sp^3) = k^2(53k - 36)$
	$R[5] = 6k^2(11k - 4)$
5	$\lim_{k\to\infty} \frac{Atoms(sp^3)}{v(G)} = 53/69 \simeq 0.768116$
	C_{81}-*based net*
6	$v(SD_5\_81) = 3k^2[27 + 23(k - 1)] = 69k^3 + 12k^2$
7	$e(SD_5\_81) = 130k^3$
8	$m(SD_5) = k^3$; m = monomer
9	$Atoms(sp^3) = 53k^3 - 12k^2$
10	$\lim_{k\to\infty} \frac{Atoms(sp^3)}{v(G)} = 53/69 \simeq 0.768116$

Table 11.5 Omega polynomial in Diamond D_5_20 net function of k = no. ada_20 units along the edge of a cubic (k,k,k) domain

	Omega($D_5\_20^a$); R[6]; Formulas
1	$\Omega(D_5\_20b, x) = (4 - 54k + 36k^2 + 44k^3)\cdot x + (-3 + 18k - 27k^2 + 12k^3)\cdot x^2$
2	$\Omega'(1) = e(G) = -2 - 18k - 18k^2 + 68k^3$
3	$CI(G) = 12 + 54k + 468k^2 + 284k^3 - 2124k^4 - 2448k^5 + 4624k^6$
4	$v(D_5\_20b) = -2 - 12k + 34k^3 = v(Dia\_20a) + 20$
5	$Atoms(sp^3) = 2 - 36k^2 + 34k^3$
6	$R[5] = -6k - 18k^2 + 36k^3$
7	$R[6] = -1 + 6k - 9k^2 + 4k^3$
8	$\lim_{k\to\infty}\left[\frac{Atoms(sp^3)}{v(G)} = \frac{2 - 36k^2 + 34k^3}{-2 - 12k + 34k^3}\right] = 1$
9	$\lim_{k\to\infty}\frac{R[5]}{R(total\ no.)} = 9/10$; $\lim_{k\to\infty}\frac{R[5]}{R[6]} = 9$

[a]counting refers to lattices with complete tetrahedral ending

Table 11.6 Examples, omega polynomial in D_5_20 net

k	Omega($D_5\_20^a$); R[6]	Atoms	sp^3Atoms (%)	Bonds	CI	R[5]	R[6]
2	$392x^1 + 21x^2$	246	130 (52.85)	434	187880	204	7
3	$1354x^1 + 132x^2$	880	596 (67.73)	1618	2616042	792	44
4	$3180x^1 + 405x^2$	2126	1602 (75.35)	3990	15915300	1992	135
5	$6134x^1 + 912x^2$	4188	3352 (80.04)	7958	63319982	4020	304
6	$10480x^1 + 1725x^2$	7270	6050 (83.22)	13930	194027520	7092	575
7	$16482x^1 + 2916x^2$	11576	9900 (85.52)	22314	497886450	11424	972

[a]counting refers to lattices with complete tetrahedral ending

Table 11.7 Omega polynomial in D_5_28 co-net function of k = no. ada_20 units along the edge of a cubic (k,k,k) domain

	Omega (D_5_28[a]); R[6]
1	$\Omega(D_5\_28b, x) = (10 - 12k - 6k^2 + 44k^3) \cdot x + (9k^2 + 12k^3) \cdot x^2$
2	$\Omega'(1) = e(G) = 10 - 12k + 12k^2 + 68k^3$
3	$CI(G) = 12 + 54k + 468k^2 + 284k^3 - 2124k^4 - 2448k^5 + 4624k^6$
4	$v(D_5\_28b) = 4 - 6k + 18k^2 + 34k^3$
5	$Atoms(sp^3) = 8 - 6k - 30k^2 + 34k^3$
6	$R[5] = 6 - 18k^2 + 36k^3$
7	$R[6] = 5 + 6k - 9k^2 + 4k^3$
8	$\lim_{k\to\infty}\left[\frac{Atoms(sp^3)}{v(G)} = \frac{8 - 6k - 30k^2 + 34k^3}{4 - 6k + 18k^2 + 34k^3}\right] = 1$ The limit when k tends to infinity of sp^3 atoms to all atoms is 1. This means that for large k, almost all atoms are sp^3.

[a]counting refers to lattices with complete tetrahedral ending

Table 11.8 Examples, omega polynomial in D_5_28 co-net

k	Omega(D_5_28[a]), R[6]	Atoms	sp^3Atoms (%)	Bonds	CI	R[5]	R[6]
2	$314\ x^1 + 132\ x^2$	336	148 (44.05)	578	333242	222	44
3	$1108\ x^1 + 405\ x^2$	1066	638 (59.85)	1918	3675996	816	135
4	$2682\ x^1 + 912\ x^2$	2444	1680 (68.74)	4506	20297706	2022	304
5	$5300\ x^1 + 1725\ x^2$	4674	3478 (74.41)	8750	76550300	4056	575
6	$9226\ x^1 + 2916\ x^2$	7960	6236 (78.34)	15058	226722474	7134	972
7	$14724\ x^1 + 4557\ x^2$	12506	10158 (81.23)	23838	568217292	11472	1519

[a]counting refers to lattices with complete tetrahedral ending

Table 11.9 Omega polynomial in Lonsdaleite-like L_5_28 and L_5_20 nets function of k = no. repeating units along the edge of a cubic (k,k,k) domain

	Network
A	Omega (L_5_28); R[6]
1	$\Omega(L_5\_28, x) = 2k(-1 + 73k + 44k^2) \cdot x + 3k(4 + 21k + 8k^2) \cdot x^2$
2	$\Omega'(1) = e(G) = 2k(11 + 136k + 68k^2)$
3	$CI(G) = 2k(-23 + 43k + 5892k^2 + 39984k^3 + 36992k^4 + 9248k^5)$
4	$v(L_5\_28) = 2k(12 + 79k + 34k^2)$
5	$Atoms(sp^3) = 2k(-14 + 35k + 34k^2)$
6	$R[5] = 1 - 6k + 98k^2 + 72k^3$
7	$R[6] = k(4 + 21k + 8k^2)$
8	$\lim_{k\to\infty}\left[\frac{Atoms(sp^3)}{v(G)} = \frac{2k(-14 + 35k + 34k^2)}{2k(12 + 79k + 34k^2)}\right] = 1$
B	Omega (L_5_20); R[6]
1	$\Omega(L_5\_20, x) = (-2 - 69k + 36k^2 + 44k^3) \cdot x + 3(k-1)^2(4k-1) \cdot x^2$
2	$\Omega'(1) = e(G) = -8 - 33k - 18k^2 + 68k^3$
3	$CI(G) = 78 + 525k + 1449k^2 + 8k^3 - 4164k^4 - 2448k^5 + 4624k^6$
4	$v(L_5\_28) = -6 - 20k + 34k^3$
5	$Atoms(sp^3) = 2 - 6k - 36k^2 + 34k^3$
6	$R[5] = -2 - 14k - 18k^2 + 36k^3$
7	$R[6] = (k-1)^2(4k-1)$
8	$\lim_{k\to\infty}\left[\frac{Atoms(sp^3)}{v(G)} = \frac{2 - 6k - 36k^2 + 34k^3}{-6 - 20k + 34k^3}\right] = 1$

Table 11.10 Examples, omega polynomial in L_5_28 and L_5_20 nets

k	Polynomial (Net)	Atoms	sp^3Atoms (%)	Bonds	CI(G)	R[5]	R[6]
A	Omega(L_5_28); R[6]						
1	$232x + 99x^2$	250	110 (44.00)	430	184272	165	33
2	$1284x + 468x^2$	1224	768 (62.75)	2220	4925244	957	156
3	$3684x + 1251x^2$	3330	2382 (71.53)	6186	38257908	2809	417
4	$7960x + 2592x^2$	6976	5360 (76.83)	13144	172746408	6153	864
5	$14640x + 4635x^2$	12570	10110 (80.43)	23910	571654920	11421	1545
6	$24252x + 7524x^2$	20520	17040 (83.04)	39300	1544435652	19045	2508
B	Omega(L_5_28); R[6]						
2	$356x + 21x^2$	226	118 (52.21)	398	157964	186	7
3	$1303x + 132x^2$	852	578 (67.84)	1567	2453658	766	44
4	$3114x + 405x^2$	2090	1578 (75.50)	3924	15393042	1958	135
5	$6053x + 912x^2$	4144	3322 (80.16)	7877	62037428	3978	304
6	$10384x + 1725x^2$	7218	6014 (83.32)	13834	191362272	7042	575

11.6 Conclusions

A novel class of (hypothetical) carbon allotropes, consisting mostly of pentagon rings (going up to 90% in the total number of pentagon/hexagon rings), was here presented. The seed of these allotropes, C_{17} and the adamantane-like repeating-units, as hydrogenated species, show a good stability, comparable with that of C_{60} fullerene, as calculated at DFT levels of theory. The main representatives of these allotropes are the diamond D_5 and lonsdaleite L_5, in fact hyper-structures corresponding to the classical diamond D_6 and lonsdaleite L_6. The topology of the networks was described in terms of the net parameter *k* and Omega polynomial.

References

Aleksenskiĭ AE, Baĭdakova MV, Vul AY, Davydov VY, Pevtsova YA (1997) Phys Solid State 39(6):1007–1015

Barborini E, Piseri P, Milani P, Benedek G, Ducati C, Robertson J (2002) Appl Phys Lett 81(18):3359–3361

Benedek G, Vahedi-Tafreshi H, Barborini E, Piseri P, Milani P, Ducati C, Robertson J (2003) Diamond Relat Mater 12:768–773

Blatov VA, Carlucci L, Ciani G, Proserpio DM (2004) CrystEngComm 6:377–395

Blatov VA, O'Keeffe M, Proserpio DM (2009) CrystEngComm 12:44–48

Breda N, Broglia RA, Colò G, Onida G, Provasi D, Vigezzi E (2000) Phys Rev B Condens Matter 62(11.1):130–133

Bylander DM, Kleinman L (1993) Phys Rev B Condens Matter 47(16):10967–10969

Chen Z, Heine T, Jiao H, Hirsch A, Thiel W, Schleyer PVR (2004) Chem Eur J 10:963–970

Decarli PS, Jamieson JC (1961) Science 133(3467):1821–1822

Delgado-Friedrichs O, O'Keeffe M (2005) J Solid State Chem 178(8):2480–2485

Diudea MV (2005) Nanostructures, novel architecture. NOVA, New York

Diudea MV (2006) Carpath J Math 22:43–47

Diudea MV (2010a) Studia Univ Babes Bolyai Chemia 55:11–17
Diudea MV (2010b) Nanomolecules and nanostructures – polynomials and indices, vol 10, MCM. University of Kragujevac, Serbia
Diudea MV, Katona G (1999) In: Newkome GA (ed) Advances in Dendritic Macromolecules, JAI Press, Stamford 4:135–201
Diudea MV (2011) Int J Chem Model (accepted)
Diudea MV, Ilić A (2011) J Comput Theor Nanosci 8:736–739
Diudea MV, Nagy CL (2007) Periodic nanostructures. Springer Dordrecht, The Netherlands
Diudea MV, Nagy CL (2008) MATCH Commun Math Comput Chem 60:835–844
Diudea MV, Nagy CL (2011a) Diamond Relat Mater (submitted)
Diudea MV, Nagy CL (2011b) Diamond Relat Mater (submitted)
Diudea MV, Cigher S, John PE (2008) MATCH Commun Math Comput Chem 60:237–250
Diudea MV, Cigher S, Vizitiu AE, Florescu MS, John PE (2009) J Math Chem 45:316–329
Diudea MV, Bende A, Janežič D (2010) Fullerenes, Nanotubes, Carbon Nanostruct 18:236–243
Djoković Dž (1973) J Combin Theory Ser B 14:263–267
Eaton PE (1979) Tetrahedron 35(19):2189–2223
Euler L (1758) Novi Comment Acad Sci Imp Petrop 4:109–160
Frondel C, Marvin UB (1967) Nature 214:587–589
He H, Sekine T, Kobayashi T (2002) Appl Phys Lett 81:610–612
Ivanovskii AL (2008) Russ J Inorg Chem 53(8):1274–1282
John PE, Vizitiu AE, Cigher S, Diudea MV (2007) MATCH Commun Math Comput Chem 57:479–484
Khachatryan AK, Aloyan SG, May PW, Sargsyan R, Khachatryan VA, Baghdasaryan VS (2008) Diamond Relat Mater 17(6):931–936
Klavžar S (2008) MATCH Commun Math Comput Chem 59:217–222
Kuck D (2006) Pure Appl Chem 78:749–775
Lenosky T, Gonze X, Teter M, Elser V (1992) Nature 355:333–335
Mochalin VN, Gogotsi Y (2009) J Am Chem Soc 131(13):4594–4595
Newkome GR, Moorefield CN, Voegtle F (2001) Dendrimers and dendrons: concepts, syntheses, applications. Wiley-VCH, Weinheim
Osawa E (2007) Diamond Relat Mater 16(12):2018–2022
Osawa E (2008) Pure Appl Chem 80(7):1365–1379
Paquette LA, Vazeux M (1981) Tetrahedron Lett 22:291–294
Sunada T (2008) Notices Am Math Soc 55:208–215
Terrones H, Mackay AL (1997) Prog Cryst Growth Charact 34(1–4):25–36
Tomalia DA, Naylor AM, Goddard WAI (1990) Angew Chem Int Ed 29:138–175
Winkler PM (1984) Discrete Appl Math 7:221–225

Chapter 12
Empirical Study of Diameters of Fullerene Graphs

Tomislav Došlić[1]

Abstract We have computed the diameters of all fullerene graphs on $20 \leq n \leq 120$ vertices and of all fullerene graphs with isolated pentagons on $60 \leq n \leq 146$ vertices. The results are used to asses the quality of recently obtained linear upper bounds and square root-type lower bounds. It seems that the fullerenes with large diameters are exceedingly rare. Our results suggest that there is a linear upper bound on the diamater of the fullerenes with isolated pentagons and that the minimum diameter is achieved on an isomer with isolated pentagons for large enough number of vertices.

12.1 Introduction

One of the central themes of research in fullerene graphs has been a search for graph-theoretical invariants that could be used to predict fullerene stability. A number of invariants has been proposed, examined, and tested; so far, none of them were found completely satisfactory. The research effort resulted in a fairly thorough understanding of many aspects of fullerene graphs (Došlić 2002). However, some problems and invariants seem to have been neglected. Among them are such important quantities as the diameter of a fullerene graph. For example, no better upper and lower bounds on the diameter were available for fullerene graphs than for general planar cubic graphs. Further, no information on the distribution of actual values of diameters among all fullerenes on the given number of vertices seems to be available in the literature. Only very recently the logarithmic lower bound on the diameter of a fullerene graph has been improved to the one proportional to the square root of the number of vertices (Došlić et al. 2010). Similarly, the trivial upper bound has been upgraded to a non-trivial

[1] Faculty of Civil Engineering, University of Zagreb, Kačićeva 26, 10000 Zagreb, Croatia
e-mail: doslic@master.grad.hr

M.V. Putz (ed.), *Carbon Bonding and Structures: Advances in Physics and Chemistry*, Carbon Materials: Chemistry and Physics 5, DOI 10.1007/978-94-007-1733-6_12,

sharp linear upper bound (Andova et al. 2010). The main goal of this paper is to investigate the actual distributions of the diameters between the newly established theoretical bounds and to use the obtained information to asses the quality of the bounds.

12.2 Preliminaries

In this section we introduce the fullerene graphs and quote some auxiliary results. For graph-theoretic terms not defined here we refer the reader to any of standard monographs such as, e.g., (West 1996).

All graphs considered here are simple, finite, and connected. For a given graph G we denote its vertex set by $V(G)$ and its edge set by $E(G)$.

For two vertices u and v of $V(G)$ their **distance**$d(u, v)$ is defined as the length of any shortest path connecting u and v in G. For a given vertex u of $V(G)$ its **eccentricity** $\varepsilon(u)$ is the largest distance between u and any other vertex v of G. Hence, $\varepsilon(u) = \max_{v \in V(G)} d(u, v)$. The maximum eccentricity over all vertices of G is called the **diameter** of G and denoted by $D(G)$; the minimum eccentricity among the vertices of G is called the **radius** of G and denoted by $R(G)$. The set of all vertices of minimum eccentricity is called the **center** of G, while the set of all vertices of maximum eccentricity is called the **periphery** of G.

A **fullerene graph** is a planar, 3-regular and 3-connected graph 12 of whose faces are pentagons and any remaining faces are hexagons. If no two pentagons in a fullerene graph G share an edge, we say that G is an **isolated pentagon** (IP) fullerene.

It is well known that fullerene graphs on p vertices exist for all even $p \geq 24$ and for $p = 20$ (Grünbaum and Motzkin 1963). Similarly, IP fullerenes on p vertices exist for all even $p \geq 70$ and for $p = 60$ (Klein and Liu 1992). The smallest fullerene, $C_{20} : 1$, is the dodecahedron; the smallest IP fullerene is the buckminsterfullerene $C_{60} : 1812$. We refer the reader to reference Fowler and Manolopoulos (1995). for more background information on fullerene graphs.

The 3-connectedness of fullerene graphs results in an obvious upper bound $D(G) \leq n/3$. It is intuitively clear that linear upper bounds are actually achieved in long tubular fullerenes. It is also clear that the constant must depend on the circumference of the tubular part. Since the tubular fullerenes exist for all values of n for which there is a fullerene graph on n vertices, it follows that no sublinear upper bound is possible for general fullerenes. The above reasoning was formalized in a recent paper by Andova et al. (2010). Their main result is that the diameter of almost all fullerenes on n vertices is at most $n/6 + 2.5$.

Theorem A (Andova et al. 2010) The diameter of a fullerene graph G on n vertices cannot exceed $n/6 + 5/2$, unless G is a tubular fullerene on $10k$ vertices with two hemidodecahedral caps. In that case, the diameter is given by $n/5 - 1$ for $k \geq 5$, $n/5$ for $k = 3$ and 4, and $n/5 + 1$ for $k = 1$. □

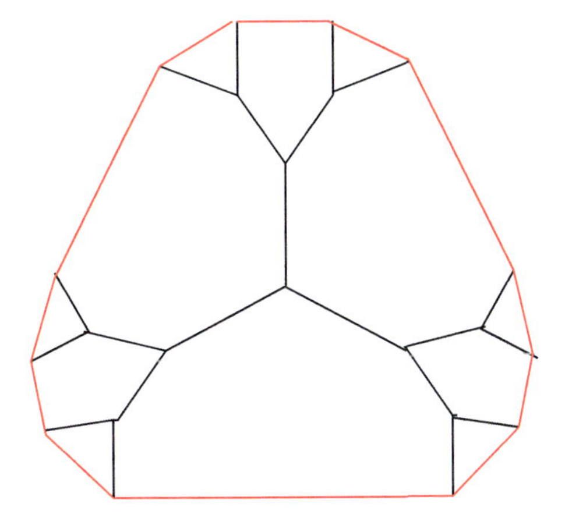

Fig. 12.1 A planar cubic graph with the diameter logarithmic in the number of vertices

As the number of fullerenes grows rather quickly with the number of vertices (proportional, roughly, to the ninth power of n), the number of exceptions becomes negligeably small even for moderate values of n.

The lower bound on $D(G)$ turns out to be much less obvious. A well known result on the degree-diameter problem states that the number of vertices in a planar graph with maximum degree 3 grows at most exponentially with the diameter.

Proposition B (Friedman and Pratt, preprint) Let G be a planar graph with maximum degree 3 and a given diameter $D(G)$. Then G has at most $2^{D(G)+1} - 1$ vertices. □

This results in a logarithmic lower bound on the diameter in terms of the number of vertices.

Corollary C Let G be a planar cubic graph on n vertices with the diameter $D(G)$. Then

$$D(G) \geq \lceil \log_2(n+1) - 1 \rceil. \qquad \square$$

Figure 12.1 shows an example of a planar cubic graph that actually achieves a lower bound on the diameter logarithmic in the number of vertices. The logarithmic character of the bound can be attributed to the presence of faces of large size. It would be reasonable to expect that better lower bounds exist for polyhedral graphs with bounded face size. Surprisingly, no such bounds seem to be available in the literature. That prompted some authors to try to formalize the above observation. The effort resulted in a paper Došlić et al. (2010), where it was found that the diameter of a fullerene graph must be at least proportional to the square root of the number of vertices.

Theorem D (Došlić et al. 2010) Let G be a fullerene graph on n vertices. Then its diameter $D(G)$ is at least $\Omega(\sqrt{n})$. □

The lower bound of Theorem D resulted also in a number of bounds on other invariants for fullerene graphs, such as. e.g., their independence number and bipartite edge frustration (Došlić et al. 2010).

12.3 Results

12.3.1 Upper Bound

As we have mentioned, the most obvious candidates for large diameters are tubular fullerenes. It can be shown that a long enough fullerene on $n = 10k$ vertices with two hemidodecahedral caps has the diameter of exactly $n/5 - 1 = 2k - 1$. By long enough we mean that a diametral path starting from the top of one of the caps reaches all vertices in the other cap in the same number of steps. In Fig. 12.2 the points represent the maximum diameters of all fullerenes on $20 \leq n \leq 120$ vertices. It can be seen that for $n \geq 76$ all of them lie on or below the line $y = x/5 - 1$. Hence, there are only finitely many exceptions to the upper bound $D(G_n) \leq n/5 - 1$, and the largest one has 74 vertices. Further, we can see that the upper bound $D(G_n) \leq n/6 + 5/2$ of reference Andova et al. (2010) is violated only for $n = 10k$, and our computations confirm that the unique offending fullerene (for a given number of vertices) is indeed the tubular isomer with two hemidodecahedral caps. The most interesting observation is that the upper bound $D(G_n) \leq n/6 + 5/2$ is not tight in the considered range. The results suggest that it could be improved by replacing $5/2$ by 2.

12.3.2 Lower Bound

Unlike the upper bound of reference Andova et al. (2010), which includes exact constants and is tight, the lower bound of reference Došlić et al. (2010) gives us only the order of magnitude, $D(G_n)$ is at least $\Omega(\sqrt{n})$. Hence we can expect somewhat less than perfect agreement between the bound and the actual values. Figure 12.3 shows the minimum diameters for all fullerenes on $20 \leq n \leq 120$ vertices; the curve $y = \sqrt{x}$ is added as a guide for the eye. It seems that the minimum diameter of a fullerene graph on n vertices grows somewhat faster than $\sqrt{n}$; in the considered range the lower envelope of the points shown in Fig. 12.3 is well approximated by the curve $y = x^{0.52}$. We have no plausible explanation for the value of 0. 52. An interesting observation is that the minimum diameter is almost always achieved on an IP fullerene, if such exists on a given number of vertices. The exceptions appear on 70, 82, 84 and 100 vertices.

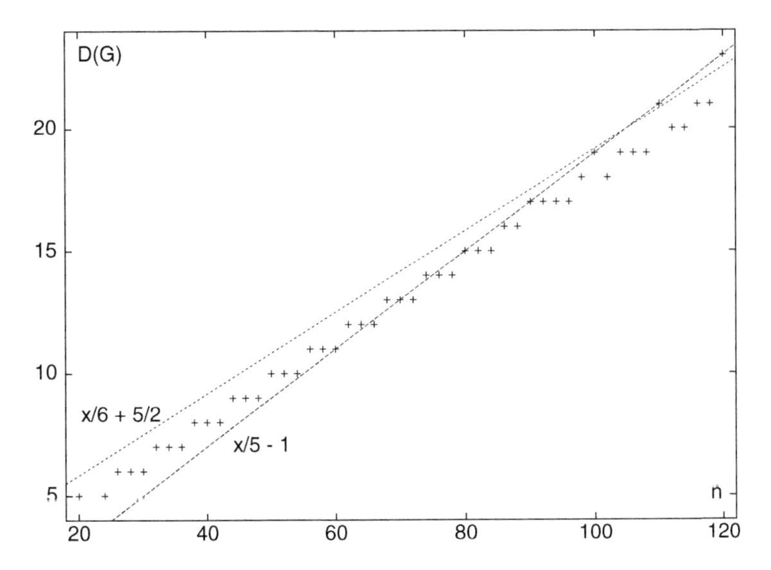

Fig. 12.2 The maximum diameter of a fullerene graph on n vertices

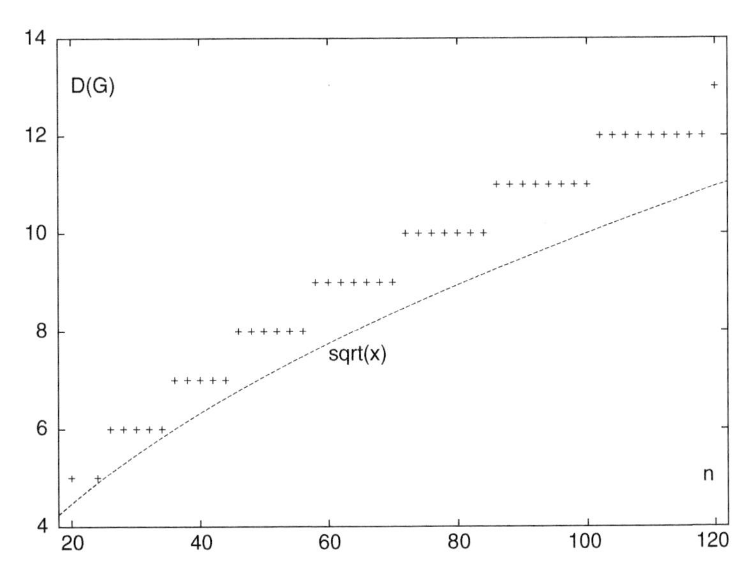

Fig. 12.3 The minimum diameter of a fullerene graph on n vertices

12.3.3 Average Diameter

In Fig. 12.4 we show the values of average diameters of fullerenes on $20 \leq n \leq 120$ vertices. The curve $y = x^{0.55}$ that seem to fit the points rather well is added as a guide to the eye. Again, we have no plausible explanation neither for the form

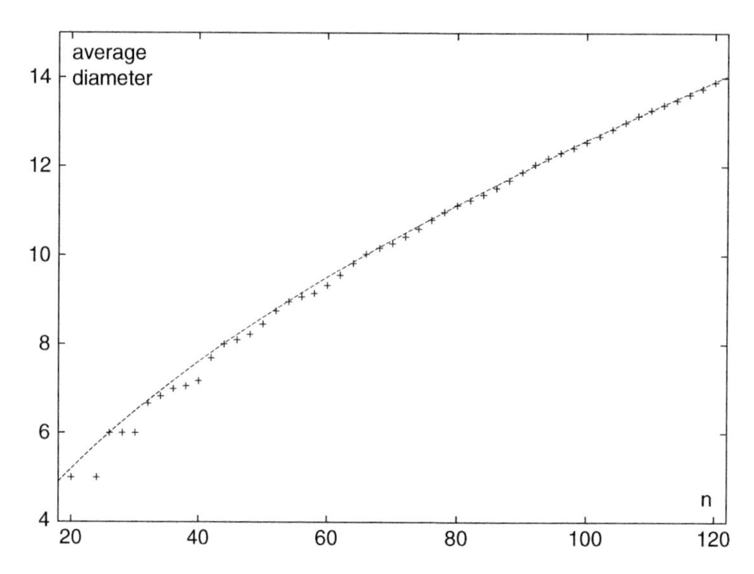

Fig. 12.4 The average diameter of a fullerene graph on n vertices. The curve $y = x^{0.55}$ is a guide for the eye

nor for the value of 0. 55. Evident from Fig. 12.4 is the fact that most fullerenes have their diameters close to the minimum possible value for the given number of vertices. The relationship between minimum, maximum, and average diameters of the fullerenes in the considered range is presented in Fig. 12.5.

12.3.4 Isolated Pentagon Isomers

The IP fullerenes tend to have smaller diameters than the general ones on the same number of vertices. This is evident from Fig. 12.6, where we show the extremal and the average diameters of IP fullerenes on $60 \leq n \leq 146$ vertices. It seems that the maximum diameters do not exceed $n/10 + 16/5$, the line shown in Fig. 12.6. It should not be too difficult to establish this bound by a reasoning similar to the one employed in Andova et al. (2010). The average diameter exhibits somewhat greater fluctuations than in general case, but it seems that it will settle down on a pattern similar to the general case. The minimum values seem to coincide with the minimum values for general fullerenes for $n > 100$; at least they coincide for $n \leq 120$. It would be interesting to explore if such behavior persists for larger values of n.

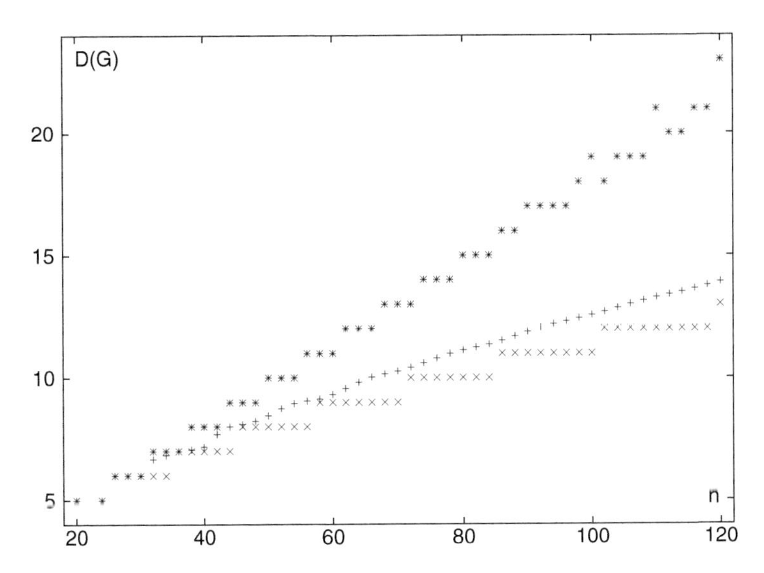

Fig. 12.5 The minimum ($\times$), the maximum ($*$), and the average (+) diameter of a fullerene graph on n vertices

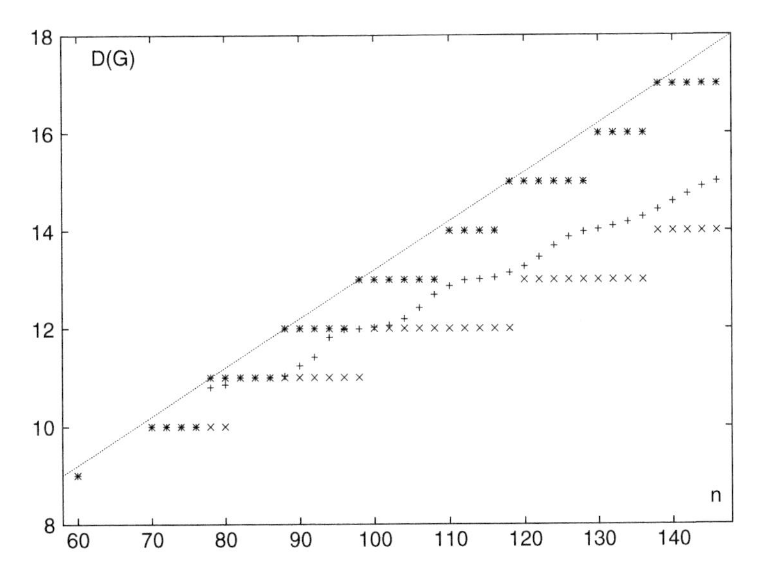

Fig. 12.6 The minimum ($\times$), the maximum ($*$), and the average (+) diameter of IP fullerene graphs on n vertices. The line $y = x/10 + 16/5$ is a guide for the eye

12.3.5 Distributions

It is clear from the previous subsections that the extremal values of the diameter must have a linear span. In this subsection we present a few examples of the distribution of the actual values within that span. The results on the average

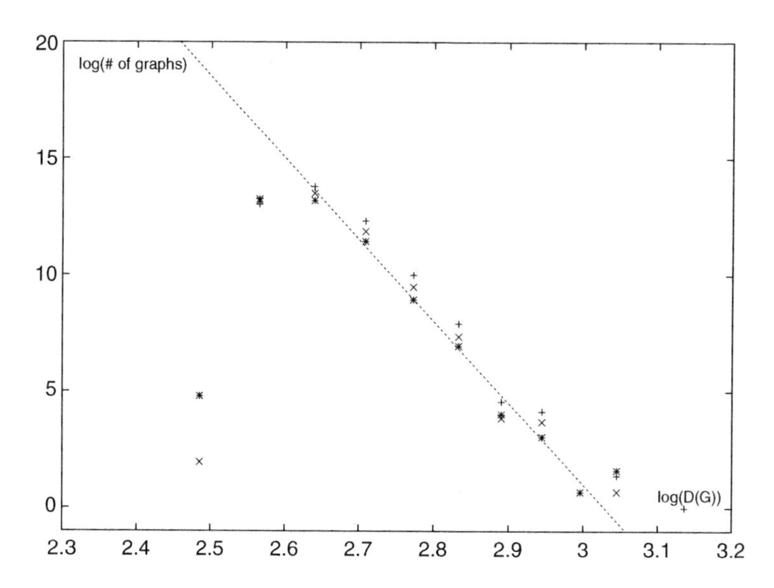

Fig. 12.7 Log-log plot of the distribution of diameters for fullerenes on 116 ($*$), 118 ($\times$) and 120 (+) vertices. The line $y = -35x + 106$ is a guide for the eye

diameter suggest that the distribution is skewed toward the lower values. Hence, the larger diameters must be rare. In Fig. 12.7 we show the log-log plot of distributions of diameters for fullerenes on 116, 118 and 120 vertices. (It is not very informative to look at the same plots in lin-lin scale, since the fine details at the upper end are completely dwarfed by the magnitude of the numbers on the lower end.) The line $y = -35x + 106$ is added as a guide for the eye. We see that in all three cases there is a power-law tail in the distributions. The significance of this power law remains unclear at the moment.

12.4 Conclusions

We have computed diameters of all fullerenes on $20 \leq n \leq 120$ vertices. We have also computed the diameters of all IP fullerenes on $60 \leq n \leq 146$ vertices. Our results suggest that the elongated fullerenes are exceedingly rare for large values of n. Most fullerenes seem to be of rather round shape and have the diameters closer to the lower end of the spectrum of possible values. The IP fullerenes tend to have significantly smaller diameters than the general ones on the same number of vertices. The upper bound on the diameter of IP fullerenes seem to be linear in the number of vertices, and our results suggest $n/10 + 16/5$ as the exact upper bound. The minimum values of diameter appear to grow somewhat faster than the lower bound of $C\sqrt{n}$ of reference Došlić et al. (2010). It seems that for large enough values of n the minimum values of diameter are achieved on IP isomers. An interesting open problem would be to verify (or falsify) that conjecture.

Acknowledgment Partial support of the Ministry of Science, Education and Sport of the Republic of Croatia (Grants No. 177–0000000-0884 and 037-0000000-2779) is gratefully acknowledged. The work was also partially supported by Croatian-Slovenian bilateral project BI-HR/09-10-015.

References

Andova A, Došlić T, Lužar B, Škrekovski R (2010) A linear upper bound on the diameter of fullerene graphs (submitted)
Došlić T (2002) J Math Chem 31:187
Došlić T, Krnc M, Lužar B (2010) New bounds on some invariants for fullerene graphs (submitted)
Fowler PW, Manolopoulos DE (1995) An Atlas of fullerenes. Clarendon Press, Oxford
Friedman E, Pratt RW (preprint) New bounds for largest planar graphs with fixed maximum degree and diameter
Grünbaum B, Motzkin TS (1963) Can J Math 15:744
Klein DJ, Liu X (1992) J Math Chem 11:199
West DB (1996) Introduction to graph theory. Prentice Hall, Upper Saddle River

Chapter 13
Hardness Equalization in the Formation Poly Atomic Carbon Compounds

Nazmul Islam[1] **and Dulal C. Ghosh**[1]

Abstract In this work we have basically launched a search whether there is a physical process of hardness equalization for molecules similar to the electronegativity equalization. We are tempted by the fact that the electronegativity equalization principle is widely accepted and theoretically justified and there is much communality in the basic philosophy of the origin and operational mechanism of the two fundamental descriptors– the electronegativity and the hardness of atoms. We have analyzed the origin and development in terms of the shell structure of atoms and molecules and classical theorems of electrostatics and put forward an alternative new definition of hardness. In the next venture, we have posited and logistically proved the occurrence of the physical process of hardness equalization principle at the event of molecule formation. Starting from our new definition of hardness and the new radial dependent formula of computing hardness of atoms and relying upon our newly introduced model of hardness equalization principle, we have derived an algorithm for the evaluation of the hardness of the hetero nuclear poly-atomic molecules. The algorithm is invoked to compute the hardness of as many as 22 poly atomic carbon containing molecules. In absence of any experimental benchmark, we have compared the computed hardness data of such molecules with the hardness data computed by an *ab- initio* quantum

[1] Department of Chemistry, University of Kalyani, Kalyani 741235, India
e-mail: nazmul.islam786@gmail.com; dcghosh1@rediffmail.com

M.V. Putz (ed.), *Carbon Bonding and Structures: Advances in Physics and Chemistry*, Carbon Materials: Chemistry and Physics 5, DOI 10.1007/978-94-007-1733-6_13,

chemical method. From comparative study we find that there is close correlation between the two sets hardness data one set evaluated through the algorithm suggested by us, and the other set evaluated through the *ab- initio* quantum chemical method.

Furthermore, we have performed a validity test of the suggested formula for evaluating molecular hardness by the application of the global hardness data of the carbon containing poly atomic molecules in the real field of hard-soft acid-base reactions. We have noticed that the hardness data of this calculation can draw the exact picture of the model of the chemical reaction surfaces. However, the hardness data evaluated through the ansatz and operational and approximate formula of Parr and Pearson poorly correlate the same reaction surfaces.

We, therefore, conclude that the present effort of deriving an algorithm for the evaluation of the hardness of the carbon containing poly atomic molecules is a successful venture. The detailed study would suggest that the paradigm of the hardness equalization principle may be another law of nature like the established electronegativity equalization principle.

13.1 Introduction

The hardness of atom is an important conceptual constructs of chemistry and physics. It has equal rank and status with other two very important conceptual constructs viz. atomic radius and the electronegativity. The importance of the hypothetical constructs self evident from the statement that, without the concept and operational significance of radius, hardness and electronegativity, chemistry and many aspects of condensed matter physics become chaotic and the long established unique order in chemico-physical world would be disturbed. The notion of hardness was first introduced by Mulliken (1952) when he pointed out that the 'Hard' and 'Soft' behavior of various atoms, molecules and ions can be conceived during acid-base chemical interaction. Soon after Mulliken's classification, the terms "hardness" and "softness" were in the glossary of conceptual chemistry and implicitly signified the resistance towards the deformability of atoms, molecules and ions under small perturbation. Thereafter, Pearson (1963) and Klopman (1964) tried to systematize and rationalize this intrinsic property of atoms and molecules. Pearson (1963) qualitatively classified molecules, atoms and ions in three classes, hard, soft and borderline- known as the HSAB principle and Klopman (1964) had drawn a link to Hard–Soft behavior with the HOMO-LUMO gap of the frontier orbital theory.

It is apparent that the hardness as conceived in chemistry fundamentally signifies the resistance towards the deformation or polarization of the electron cloud of the atoms, ions or molecules under small perturbation generated during the process of the event of chemical reaction. Thus, the general operational significance of the hard-soft behavior of a chemical species may be understood in the following statement. If the electron cloud is strongly held by the nucleus, the chemical species is 'hard' but if the electron cloud is loosely held by the nucleus, the system is 'soft' (Klopman 1964; Pearson 1963).

13.1.1 The Physical Hardness

The materials engineers have used for centuries the legend "hardness" to describe mechanical stability (Gilman 1997). In solid mechanics, hardness means the resistance to deformation, both elastic and plastic. The particular properties are the bulk modulus which measures the resistance to elastic volume changes, the shear modulus which measures the resistance to elastic shape changes, and dislocation mobility which determine rates of plastic shape changes. Thus, annealed brass in which dislocations are mobile is said to be soft, but in the same brass, the dislocation can be immobilized by cold-rolling until it is hard enough to be used for springs.

13.1.2 The HOMO-LUMO Gap

Happily the HOMO-LUMO gap is the one and the same unique concept applies for the rationalization of both chemical and physical hardness in chemistry and physics as well. The HOMO and the LUMO are the two frontier orbitals in the eigen value spectrum of atoms, molecules and the condensed matter. For molecules, the gap between the highest occupied MO and the lowest unoccupied MO in the bonding energy spectrum is called the HOMO-LUMO gap. For solids (crystals and glasses) the gap in the bonding spectrum is the difference between the bottom of the conduction energy band and the top of the valence energy band. The energy and symmetry type of the highest occupied molecular orbital (HOMO) and the lowest unoccupied molecular orbital (LUMO), labeled as frontier orbitals, play very important roles in deciding the orientation of interacting molecules with respect to each other and the feasibility of chemical reactions. The HOMO-LUMO gap was chosen by chemists in recent decades to describe atomic and molecular stability, to measure the electronegativity and hardness. The chemists also use the HOMO-LUMO gap as a definition of aromaticity (Zhou and Parr 1988, 1989) in recent times. In solids, the HOMO-LUMO gap is also used to evaluate the hardness, which describes its mechanical stability. The HOMO-LUMO gap also determines the optical hardness of transparent solids (Gilman 1997). Optical hardness is associated with the electronic polarizability which determines the refractive index and which is largely determined by the HOMO-LUMO gap. However, it seems that Pearson (1999) differs from engineers and material scientists in the conceptual structure of physical hardness.

13.1.3 Whether Electronegativity and Hardness Are Quantum Observable

Notwithstanding the erudite discussion of Putz (2009a) on the problem of observability of the electronegativity and chemical hardness, the hardness and the electronegativity are fundamentally hypothesis and conceptual structures and are not physical observables and, therefore, cannot be evaluated experimentally. As they are no observables, the possibility of their quantum mechanical evaluation is completely ruled out (Ghosh and Islam 2009, 2010a–e; Islam and Ghosh 2010, 2011a,b).

The chemical hardness, the atomic radius and the electronegativity are conundrums and objects of purely intellectual intuition and are not the things of real world. They occur in mind like the unicorn of mythical saga (Frenking and Krapp 2007). According Kant, these are noumenon – objects knowable by the mind or intellect but not by the senses. However, we are habituated in modeling chemistry in terms of these three components- chemical hardness, the atomic radius and the electronegativity. These quantities are qualitative concept per se and have to be evaluated qualitatively goaded by physical and chemical experience. Hence, before any algorithm of computing the radius, hardness and the electronegativity is developed, the reification of abstract concepts regarding the above descriptors into things of the real world is absolutely necessary. In order to assign some number to each of these abstract concepts like the atomic radius, the electronegativity and the hardness, it is required that these quantities should be reified (Ayers 2007) in terms of the physico-chemical behavior of such conundrums goaded by the quantum mechanical principles and then, some mathematical algorithm will be developed. Then the useful qualitative entities like hardness and electronegativity which were abstract semiotic representations can be considered as theoretical quantities of cognitive representations.

13.1.4 The Density Functional Underpinning of Hardness

There is a paradigm shift in the realm of conceptual chemistry due to the density functional theoretical (Hohenberg and Kohn 1964; Parr et al. 1978; Parr and Bartolotti 1982; Parr and Pearson 1983; Parr and Yang 1989; Parr and Chattaraj 1991) underpinning of Parr and co-workers (Parr et al. 1978; Parr and Bartolotti 1982; Parr and Pearson 1983; Parr and Yang 1989; Parr and Chattaraj 1991). The useful qualitative entities like hardness and electronegativity which were abstract semiotic representations are now considered as theoretical quantities of cognitive representations. Given the electron density function $\rho(r)$ in a chemical system and the energy functional $E(\rho)$, the chemical potential, μ of that system in equilibrium has been defined as the derivative of the energy with respect to the electron density at fixed molecular geometry.

The chemical potential (Gyftpoulous and Hatsopoulos 1968), μ, is given by

$$\mu = \left[\frac{\delta E(\rho)}{\delta \rho}\right]_v \tag{13.1}$$

where v is the external potential acting on an electron due to the presence of nucleus.

The differential definition more appropriate to atomic system is on the basis that for a system of N electrons with ground state energy E [N,v],

$$\mu = \left[\frac{\partial E}{\partial N}\right]_v \tag{13.2}$$

The absolute hardness (Parr and Pearson 1983) η, was defined as

$$\eta = \frac{1}{2}\left[\frac{\partial \mu}{\partial N}\right]_v = \frac{1}{2}\left[\frac{\partial^2 E}{\partial N^2}\right]_v \tag{13.3}$$

Although rigorous mathematical formulae were suggested but rigorous evaluation (Perdew 1988; Yang et al. 2000) of hardness η in terms of the Eq. 13.3 is difficult because the numerical method is required to be invoked to solve it. Sen and Vinayagam (1988) evaluated the hardness of 12 elements only in terms of Eq. 13.3 by numerical integration technique. Thus, the true density functional global hardness of the majority of atoms is still at large. However, calculus of finite difference approximation was invoked (Parr and Pearson 1983) to suggest an approximate and operational formula of hardness as under:

$$\eta = \frac{I - A}{2} \tag{13.4}$$

where '*I*' and '*A*' are the first ionization potential and electron affinity of the chemical species. Pearson (1986) proceeded further to evaluate 'I' and 'A' in terms of orbital energies of the highest occupied molecular orbital, HOMO and the lowest unoccupied molecular orbital, LUMO by connecting it with Hartree-Fock SCF theory and invoking Koopmans' theorem the hardness is reformulated as

$$\eta = \frac{-\varepsilon_{HOMO} + \varepsilon_{LUMO}}{2} \tag{13.5}$$

where $I = -\varepsilon_{HOMO}$, and $A = -\varepsilon_{LUMO}$.

Although, Perdew (1988), and Yang et al. (2000) have tried to formulate the E vs. N curve and justify the approximate formula of Parr and Pearson (1983), the incoherency between the definition and measurement of hardness is apparent (Reed 1997; Ghosh and Islam 2009).

Even the operational and approximate formulae of hardness in terms of the Eqs. 13.4 and 13.5 were not a respite because, I and A of all molecule are not known experimentally. Also accurate theoretical evaluation of I and A for every molecule is still at large and even an approximate calculation is very costly for sizable molecules. More ever, the evaluation of I and A through the Koopmans' theorem is very much approximate (Atkins 1991; Pillar 1968). Moreover, very accurate I and A does not improve the position because the ansatz η = (I-A)/2 is itself approximate.

However, the quest for the theoretical basis of the hard soft acid base behaviour has created such a surge of fundamental research in chemistry that it gave birth of a new branch of density functional based theoretical science known as 'Conceptual Density Functional Theory, CDFT' (Geerlings et al. 2003).

The conceptual density functional theory has added Maximum Hardness Principle, (MHP) (Pearson 1987) and Minimum Polarizability Principle, (MPP) (Chattaraj and Sengupta 1996) to the list of the fundamental laws of nature. The CDFT has been successfully exploited in elucidating and correlating mechanistic aspects viz. regio-selectivity, catalysis, aromaticity, intramolecualr rotation, inversion and isomerization reactions (Zhou and Parr 1989; Parr and Chattaraj 1991; Chattaraj et al. 1994; Pearson and Palke 1992; Pal et al. 1993; Chattaraj et al. 1995; Ayers and Parr 2000; Ghosh et al. 2000, 2002).

It is important to mention here some outstanding fundamental works of Putz and his coworkers (Putz et al. 2003, 2005; Putz 2006, 2007, 2008a,b, 2009a,b, 2010) on electronegativity and hardness and their usefulness for the theoretical prediction of several physicochemical properties-like the fundamentals of chemical bonding. The basic physico-chemical concepts of density functional theory are employed by Putz to highlight the energy role in chemical structure while its extended influence in electronic localization function helps in chemical bonding understanding. In this context the energy functionals accompanied by electronic localization functions may provide a comprehensive description of the global-local levels electronic structures in general and of chemical bonds in special. It is shown that the aromaticity of peripheral topological path may be well described by superior finite difference schemes of electronegativity and chemical hardness indices in certain calibrating conditions. They have also discussed at length the problem of observability to electronegativity and chemical hardness. Invoking a semi classical method, Putz further, have introduced the electronegativity of an element as the power by which the frontier electrons are attracted to the center of the atom being a stability measure of the atomic system as a whole. A new chemical hardness expression in terms of atomic radius is also given by Putz et al. (2003) A unified Mulliken valence with Parr ground-state electronegativity picture is presented by Putz which provides a useful analytical tool on which the absolute hardness as well ionization potential and electron affinity.

The hardness and shell structure of atoms and molecules must be inter connected. Parr and Zhaou (Parr and Yang 1989) discovered that the absolute hardness is a unifying concept for identifying shells and sub shells in nuclei, atoms, Molecules, and Metallic Clusters. In their consideration, the maxim of the maximum hardness principle is related to the close structure of shells and sub shells.

It, therefore, transpires that for the development of all algorithms for the computation of hardness of atoms, the electronic constitution of matter must be a necessary input.

Delving in to the fundamental nature of hardness and its logistic origin and development from the shell structure of the atom and physical process of screening, we have discovered a new electrostatic definition of the global hardness of atoms (Ghosh et al. 2002; Ghosh and Islam 2009, 2010a–e; Islam and Ghosh 2010, 2011a,b). It is, therefore, pertinent to mention that, following Feynman (Feynman et al. 1964) and relying upon the theorems of classical electrostatic, we (Ghosh and Islam 2009, 2010a–e; Islam and Ghosh 2010, 2011a,b) have suggested a new electrostatic definition and derived a new radial dependent formulae for computing atomic hardness as follows:

Classically, the energy E(N) of charging a conducting sphere of radius r with charge q is given by

$$E(N) = \frac{q^2}{2r} \text{ (in C.G.S. unit)} \tag{13.6}$$

$$E(N) = \frac{q^2}{4\pi\varepsilon_0(2r)} \text{ (in S.I. unit)} \tag{13.7}$$

In Eq. 13.7, E (N) is in ergs, q is in electrostatic unit and r is in cm. Now, for an atom, the change in energy associated with the change in q on removal of an electron (of charge, e), would be the ionization energy, I. Similarly, the energy evolved on addition of an electron with q would be the electron affinity, A. Hence,

$$I = E(N+1) - E(N) = \left[\frac{(q+e)^2}{2r} - \frac{q^2}{2r}\right] \tag{13.8}$$

and

$$A = E(N) - E(N-1) = \left[\frac{q^2}{2r} - \frac{(q-e)^2}{2r}\right] \tag{13.9}$$

Now, putting the values of I and A from above into the Eq. 13.4, we get:

$$\eta = \frac{1}{2}\left[\left[\frac{(q+e)^2}{2r} - \frac{q^2}{2r}\right] - \left[\frac{q^2}{2r} - \frac{(q-e)^2}{2r}\right]\right] \tag{13.10}$$

or,

$$\eta = \frac{e^2}{2r} \tag{13.11}$$

where e is the electronic charge in esu and r is the most probable radius of the atom in cm to calculate hardness in energy unit.

13.1.5 *Commonality in the Basis of the Physical Structures of Hardness and Electronegativity*

Let us contemplate over the subject that, as regards their origin and operational significance, the electronegativity and the hardness are akin and there is much commonality in their the physical structure and the philosophical basis. The hardness refers to the resistance of the electron cloud of the atomic and molecular systems under small perturbation of electrical field. An atom or molecule having least tendency of deformation are 'hard' and having large tendency of deformation are 'soft'. In other words, least polarizable means most hard and in such systems the electron clouds are tightly bound to the atoms or molecules. On the contrary, the most polarizable means least hard and in such systems the electron cloud is loosely bound to the atoms or molecules. Electronegativity though defined in many different ways, the most logical and rational definition of it is the electron holding power of the atoms or molecules. The more electronegative species hold electrons more tightly and the less electronegative species hold electrons less tightly.

Thus, if we compare the qualitative definitions of hardness and electronegativity stated above, the commonality of their conceptual structures and philosophical basis is self-evident. Thus the qualitative views of the origins of hardness and electronegativity nicely converge to the one and single basic principle that they originate from the same source –the electron attracting power of the screened nuclear charge.

There is also a group of scientists who believe that the electronegativity and hardness are linearly correlated and approximately proportional with each other (Ayers 2007; Putz 2007, 2008b, 2009b, 2010; Komorowski 1987; Pearson 1988).

13.2 Hardness Equalization Principle Analogous to the Electronegativity Equalization Principle

It is now accepted by the scientific community that the hardness is the cardinal index of stability and chemical reactivity and the concept was culminated with the introduction of the maximum hardness principle (Pearson 1987, 1993; Torrent-Sucarrat et al. 2001, 2002). After we have successfully posited above that the origin and the operational significance of the electronegativity and hardness are the same, we may conjecture that, in all probability, there is a hardness equalization process (Datta 1986; Yang et al. 1985; Ayers and Parr 2008a) similar to the physical process of electronegativity equalization (Sanderson 1951; Parr and Bartolotti 1982; Nalewajski 1985). The electronegativity equalization principle is now universally accepted and fundamentally justified (Datta 1986; Yang et al. 1985; Ayers and Parr 2008a,b; Islam and Ghosh 2010, 2011b).

To justify our hypothesis that- '*The Electronegativity and the Absolute Hardness are two different Appearances of the One and the Same Fundamental Property of Atoms and the Hardness Equalization Principle can be Equally Conceived like the Electronegativity Equalization Principle*' we have basically launched a search whether the molecular hardness, an important conceptual descriptor of chemistry and physics, can be evaluated in terms of the atomic hardness values.

Our perception on hardness equalization during the chemical event of the formation of hetero nuclear molecules is as follows:

With the physical process of charge transfer during chemical reaction leading to bond formation, the hardness kernel of atoms change and in the process, it would increase somewhere and decrease elsewhere ultimately the hardness values of the atomic fragments will equalize to some intermediate values common to all.

To justify the hypothesis and our perception on the hardness equalization principle, we, starting from the semi-empirical radial dependent formula of computing global hardness of atoms suggested by us (Ghosh and Islam 2009), have derived an algorithm of the molecular hardness assuming that the hardness equalization principle is operative and justifiably valid.

Now, let us visualize the hardness equalization on the process of the molecule formation from its constituent atoms. It is quite plausible that, while the physical process of charge transfer during the event of chemical reaction leading to the bond formation takes place, the hardness of the atoms change and in the process it would increase somewhere and decrease elsewhere so that the ultimate hardness values of all the participating atoms will equalize to some intermediate value common to all.

The amount and the direction of charge transfer mainly depend on the difference of electronegativity or hardness values of the atoms forming the molecule. However, for the derivation of the necessary formulae of the present work we arbitrarily assume that the central atom in the molecular cluster is of lower hardness and the other atoms are of higher hardness.

Now let us assume that during the formation of the polyatomic molecule, δ is the total amount of charge transferred from the central atom A to the n number of the ligands surrounding the central atom. Although the total amount of charge, (δ), is distributed among the ligands, the amount of charge received by the individual atom is governed by the difference of hardness of respective atom and the atom in the centre of the cluster.

Let, B, D,... n^{th} ligands have the charge δ_1, δ_2, ... δ_n respectively in the molecular cluster and then it is apparent that

$$\delta = \delta_1 + \delta_2 + ... + \delta_n \tag{13.12}$$

Now, after the charge transfer, the hardness of the central atom X in the poly–atomic molecule becomes, in terms of the Eq. 13.12 above

$$\eta_A^/ = K\frac{(e - \delta)^2}{2r_X^/} \tag{13.13}$$

and similarly the hardnesses of the ligands in the molecule become-

$$\eta_B^{/} = K\frac{(e+\delta_1)^2}{2r_B^{/}},$$

$$\eta_D^{/} = K\frac{(e+\delta_2)^2}{2r_D^{/}},$$

$$\eta_n^{/} = K\frac{(e+\delta_n)^2}{2r_n^{/}} \tag{13.14}$$

respectively, where $r_X^{/}$, $r_B^{/}$, $r_D^{/}$...... $r_n^{/}$ are the radii of atoms in the molecule and $\eta_X^{/}$, $\eta_B^{/}$, $\eta_D^{/}$, ... $\eta_n^{/}$ are the hardnesses of the atoms in the molecule.

Expanding the Eq. 13.13, $(e-\delta)^2$ and neglecting the δ^2 term we get the hardness of the central atom, X as–

$$\eta_X^{/} = K\frac{(e^2-2e\delta)}{2r_X^{/}} \tag{13.15}$$

and similarly expanding the Eq. 13.15 and neglecting the δ^2 terms, the formulae for hardness of atoms in the molecule look -

$$\eta_B^{/} = K\frac{(e^2+2e\delta_1)}{2r_B^{/}}, \eta_D^{/} = K\frac{(e^2+2e\delta_2)}{2r_D^{/}}, \ldots.., \eta_n^{/} = K\frac{(e^2+2e\delta_n)}{2r_n^{/}} \tag{13.16}$$

Now invoking hardness equalization principle after the formation of the molecule, the hardness of the individual constituents must be equalized i.e.

$$\eta = \eta_X^{/} = \eta_B^{/} = \eta_D^{/} \ldots.. = \eta_n^{/} \tag{13.17}$$

The Eq. 13.17 implies

$$\begin{aligned}\eta &= K\frac{(e^2+2e\delta)}{2r_X^{/}} = K\frac{(e^2+2e\delta_1)}{2r_B^{/}} = K\frac{(e^2+2e\delta_2)}{2r_D^{/}} = \ldots = K\frac{(e^2+2e\delta_n)}{2r_n^{/}}\\ &= K\frac{(e^2+2e\delta)+(e^2+2e\delta_1)+(e^2+2e\delta_2)+\ldots+(e^2+2e\delta_n)}{(2r_X^{/}+2r_B^{/}+2r_D^{/}+2r_n^{/})}\\ &= K\frac{(n+1)e^2}{2\sum_i r_i^{/}}\end{aligned} \tag{13.18}$$

Invoking a further approximation that atoms retain their identity in molecule (Murphy et al. 2000; Parr et al. 2005; Bader 1991), we can replace the r'term by the absolute radii r of the corresponding atom in the Eq. 13.18. Thus we obtain

$$\eta = K\frac{(n+1)e^2}{2\sum_{i} r_i} \tag{13.19}$$

The Eq. 13.19 computes hardness in esu (units) and the corresponding equation computing the hardness values in electron volt looks:

$$\eta = K\frac{7.2(n+1)}{2\sum_{i} r_i} \tag{13.20}$$

where the atomic radius, r_i must be expressed in Angstrom, K is the constant depending on the fundamental nature of hardness. However, we (Islam and Ghosh 2011a,b) have calculated standardized value of $K = 1$, 0.75 and 0.90 for single, double and triple bonded compounds respectively with the help of the Hyperchem 8.0 professional program (Hyperchem 8.0 2008).

In this essay we are applying our models of new definition of hardness and hardness equalization principle in computing the molecular hardness of some poly atomic carbon compounds.

The typical chemical formula of the carbon compounds chosen here is C–BDEF clusters, where C is the carbon atom at the centre of the clusters, B,D,E, and F are the ligands, attached to the C–centre. The procedure of the computation involves the sequences.

(i) we have computed the $\eta_{abinitio}$ of a series of selected molecules invoking the Eqn. (13.7). The geometry optimization of the corresponding molecules have been furnished using the 6–31 G* basis set of the Hyperchem 8.0 professional program (Hyperchem 8.0 2008). Such theoretically evaluated hardness of the compounds is labeled here as *ab initio* hardness.

   In this connection we must comment upon the accuracy of our theoretically evaluated hardness data. It transpires that calculation of I and A through the Koopmans' theorem is very much approximate (Atkins 1991; Pillar 1968). Moreover, very accurate I and A does not improve the position because the ansatz $\eta = (I\text{-}A)/2$ is itself approximate. We strongly do believe that electronegativity and hardness are conceptual structures and are not physical observables, therefore, these can neither be evaluated experimentally nor quantum mechanically. Therefore, we do not have any benchmark except the set of some SCF hardness data for the validity test of our computed hardness.

   Our main purpose of invoking some Hartree-Fock SCF formalism and Koopmans' theorem to calculate *ab initio* hardness is twofold. Firstly to evaluate the K in Eq. 13.20 and secondly to set up some benchmark for the validity test of our evaluated hardness data.

Table 13.1 Computed global hardness data (in eV) and the *ab initio* hardness data (or Parr and Pearson's hardness data) of 22 carbon containing poly atomic molecules

No	Molecule	η^a	η^b
1	CH_4	13.005	16.867
2	CH_3F	13.605	13.805
3	CH_3Cl	11.922	10.086
4	CH_3Br	11.278	9.3418
5	CH_3I	10.737	7.9889
6	CH_2F_2	14.264	13.318
7	CH_2Cl_2	11.005	9.3745
8	CH_2Br_2	9.9555	8.5343
9	CH_2I_2	9.1419	7.0322
10	CHF_3	14.989	13.573
11	$CHCl_3$	10.219	8.8028
12	$CHBr_3$	8.9107	7.9115
13	CHI_3	7.9597	7.9889
14	CF_4	15.792	14.211
15	CCl_4	9.5377	8.3107
16	CBr_4	8.0643	7.4183
17	CI_4	7.0482	5.6297
18	CF_3Cl	13.567	9.5182
19	CF_2Cl_2	11.893	8.7473
20	$CFCl_3$	10.586	8.4526
21	CH_2O	9.931	6.2
22	CO_2	10.24	9.500586

[a] Present calculation
[b] ab initio

(ii) The computed η_{abinitio} data of a series of selected molecules are divided by $\{7.2(n+1)/\sum_i r_i\}$ to get the value of K for individual molecular systems. The best mean K value for the single and double bonded carbon compounds are calculated as 1 and 0.75 respectively.

(iii) Thereafter using the K value and the Eq. 13.20, we have calculated the hardness of as many as 22 carbon containing poly atomic molecules.

In evaluating the hardness data of a polyatomic molecule we only need the absolute radii of the atoms. The absolute/or most probable radii of atoms are taken from our previous work (Ghosh et al. 2008). As a rationale of putting the absolute radii of the atoms in cases of multiply bonded molecules, we are to state that the K for such systems are adjusted accordingly (vide supra) and also the radii of atoms in multiply bonded molecules are at large.

We have presented the computed hardnesses data of 22 carbon containing molecular systems vis-à-vis their computed *ab initio* hardness data in Table 13.1. In Fig. 13.1 we have made a comparative study of the hardness values evaluated through the formula of this work vis-à-vis the corresponding *ab initio* hardness values of 22 carbon containing compounds.

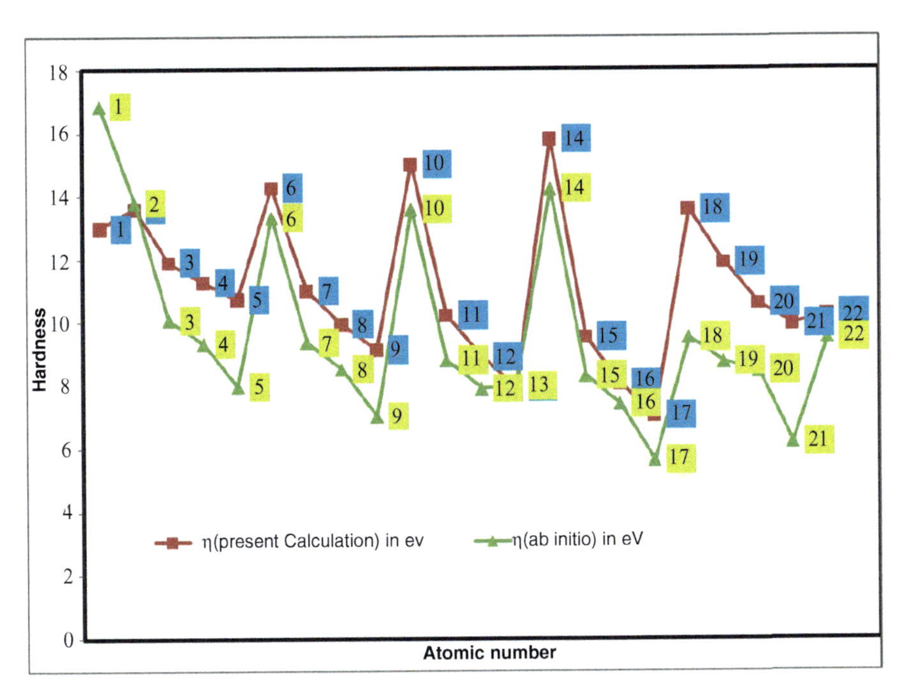

Fig. 13.1 Comparative study of the computed global hardness data and the *ab initio* hardness data of 22 carbon containing poly atomic molecules

13.3 Interaction Energy for Hard-Soft Acid-Base Exchange Reactions Involving Polyatomic Molecules

Let us consider the generalized acid-base exchange reaction as follows:

$$\text{Acid}_1 - \text{Base}_1 + \text{Acid}_2 - \text{Base}_2 = \text{Acid}_1 - \text{Base}_2 + \text{Acid}_2 - \text{Base}_1 \quad (13.21)$$

Now, the Hard-Soft Acid-Base (HSAB) principle suggests that if the reaction follows the sequence:

(i) hs + sh = hh + ss the reaction is spontaneous ($\Delta E < 0$)
but, if it follows the reverse sequence i.e.,
(ii) hh + ss = hs + sh the reaction is not spontaneous ($\Delta E > 0$).
where 'h' and 's' denotes harder part and softer part respectively.

However, the spontaneity of the reaction are dependent on other factors such as electronegativities, sizes, orbital symmetry and overlaps, steric repulsions etc. of the acids and bases involved in the reactions.

The reaction hardness may be defined as the difference between the sum of the hardness of the reactants and the products. The reaction hardness for a hard/soft exchange reaction can be computed by writing the exchange reactions, Eq. 13.21, in terms of fundamental acid–base reactions:

$$\text{Acid} + \text{Base} \rightarrow \text{Acid} - \text{Base} \tag{13.22}$$

We have already pointed out in a recent work that there are several models (Datta 1986; Ayers 2005; Chattaraj and Ayers 2005, 2007) for computing the reaction hardness for a hard/soft exchange reaction. But none of these models have included all the parameters that determine the reaction to proceed. In order to examine the HSAB rule, we also assumed, although it is admittedly impossible, that all the other factors mentioned above remain constant during the reaction.

Considering an exchange reaction between 'n' number of reactants and 'm' number of products, we can simply calculate the hardness difference, $\Delta\eta$, of the reaction as taking difference of the sum of the reactants hardness and that of the product hardness:

$$\Delta\eta = \sum_{i=1}^{n} (\eta_{\text{Product}})_i - \sum_{j=1}^{m} (\eta_{\text{Reactant}})_j \tag{13.23}$$

Gázquez (1997) provided a formula to estimate the interaction energy or reaction energy (ΔE_{Rxn}) as well as the bond energies (BE) using the chemical potentials and the hardnesses of the isolated species and the hardness of each of the interacting species at the equilibrium position as follows:

$$-\Delta E_{Rxn} \approx \left[\frac{1}{2}\frac{(\mu_{Acid} - \mu_{Base})^2}{(\eta_{Acid} + \eta_{Base})} + \frac{1}{2}N^2{}_{Acid-Base}\left[\eta_{(Acid-Base)} - \frac{(\eta_{Acid})(\eta_{Base})}{(\eta_{Acid} + \eta_{Base})}\right]\right] \tag{13.24}$$

and

$$BE = -\Delta E_{Rxn} \approx \left[\frac{1}{2}\frac{(\mu_{Acid} - \mu_{Base})^2}{(\eta_{Acid} + \eta_{Base})} + \frac{1}{2}N^2{}_{Acid-Base}\left[\eta_{(Acid-Base)} - \frac{(\eta_{Acid}\eta_{Base})}{(\eta_{Acid} + \eta_{Base})}\right]\right] \tag{13.25}$$

where μ's are chemical potentials, η's are the hardnesses and N's are the number of valence electrons of the species in the acid-base adduct.

Gázquez (1997) noted that, in general, the contribution from the first term on the right hand side of Eq. 13.25, which has been associated with the charge transfer process, is almost negligible in comparison with the second term. It implies that the change of energy associated with the second step, at constant chemical potential, represents the main contribution to the bond energy.

Now, let us consider the following type of reactions in which two old bonds are broken and two new bonds are formed.

$$\mathrm{A-B+C-D\rightarrow A-C+B-D} \tag{13.26}$$

where A,B,C and D are the atoms forming the molecules A-B,C-D,A-C and B-D.

This process may be divided into the following steps:

$$\mathrm{A-B\rightarrow A+B}$$

$$\mathrm{C-D\rightarrow C+D}$$

$$\mathrm{A-C\rightarrow A+C}$$

$$\mathrm{B+D\rightarrow B-D}$$

Now applying Eq. 13.25 to each one of these steps, the reaction energy is derived by Gázquez (1997) as follows:

$$\begin{aligned}\Delta E_{Rxn} = & -\frac{1}{2}\left(N^2{}_{AC}\eta_{AC}+N^2{}_{BD}\eta_{BD}-N^2{}_{AB}\eta_{AB}-N^2{}_{CD}\eta_{CD}\right)\\ & +\frac{1}{2}\left[N^2{}_{AC}\frac{\eta_A\eta_C}{(\eta_A+\eta_C)}+N^2{}_{BD}\frac{\eta B\eta_D}{(\eta_B+\eta_D)}-N^2{}_{AB}\frac{\eta_A\eta_B}{(\eta_A+\eta_B)}-N^2{}_{CD}\frac{\eta_C\eta_D}{(\eta_C+\eta_D)}\right]\\ & -\frac{1}{2}\left[\frac{\mu_A-\mu_C)^2}{(\eta_A+\eta_C)}+\frac{(\mu_B-\mu_D)^2}{(\eta_B+\eta_D)}-\frac{\mu_A-\mu_B)^2}{(\eta_A+\eta_B)}+\frac{(\mu_C-\mu_D)^2}{(\eta_C+\eta_D)}\right]\end{aligned} \tag{13.27}$$

Thereafter, Gázquez (1997) proceeded to arrive at the final formula by assuming that N = 1 in all cases and neglecting the second and third terms of Eq. 13.27

$$\Delta \mathrm{E}_{Rxn}\approx -\left[\frac{1}{2}\left\{(\eta_{AC}+\eta_{BD})-(\eta_{AB}+\eta_{CD})\right\}\right] \tag{13.28}$$

We have already applied this formula for the correlation of the reaction surface of a number of well known acid-base exchange reactions involving diatomic molecules using the global hardness data of the corresponding diatomic molecules computed through our algorithm.

Now, for a polyatomic molecule, the interaction energy can be written as

$$-\Delta E\approx\frac{1}{2}\left\{\sum_{i=1}^{n}(\eta_{\mathrm{Product}})_{\mathrm{i}}-\sum_{j=1}^{m}(\eta_{\mathrm{Reactant}})_{\mathrm{j}}\right\} \tag{13.29}$$

A look at the Eqs. 13.23 and 13.29 reveals that the hardness difference is directly related to the reaction energy of the acid/base exchange reactions. It may be represented by:

$$\Delta\eta \approx -2\Delta E_{Rxn} \tag{13.30}$$

or, more precisely,

$$\Delta\eta \propto -2\Delta E_{Rxn} \tag{13.31}$$

It follows immediately that if the sum of the hardness values of the products is greater than the sum of the hardness values of the reactants, then $\Delta E_{\text{Reac}} < 0$, and if the opposite case occurs, then $\Delta E_{\text{Reac}} > 0$.

These inequalities are in complete agreement with the experimental evidence that the exchange reactions almost always proceed in the direction that produces the hardest molecules (Datta 1992; Pearson 1993, 1995, 1999).

13.4 Results and Discussion

Chemical reactivity theory deals with the change of a molecule in response to the attack of different types of reagents. A parallel track of research, as a fruition of long time research, has focused on developing and using DFT descriptors to provide qualitative insights and to quantify intuitive chemical concepts.

Since, we cannot perform any validity test of our model because of the fact that there is neither any experimental nor any theoretical data to use as bench mark, we resort to semi-quantitative, qualitative and intuitive correlation. We have compared some selected set of hardness data of various carbon containing molecules computed through our suggested ansatz vis-à-vis a set of hardness data of such molecules computed by invoking an *ab initio* quantum chemical method mentioned above.

A look at the Table 13.1 and Fig. 13.1 reveals that both sets of the hardness data–the hardness data computed through the algorithm developed by us and their quantum mechanical counter parts show similar behavior in the nature of variation of the hardness data. Thus, it transpires from this comparative study that the suggested algorithm is useful theoretical model to evaluate the global hardness of CBDEF type carbon containing molecules.

In Table 13.2, we have studied, invoking the Eq. 13.29, the reaction surfaces of 8 well known chemical reactions in terms of the hardness data computed by us. Also we have evaluated the same reaction surface in terms of the hardness data of such chemical species in terms of the ansatz and operational and approximate formula of Parr and Pearson [9], Eq. 13.6. Such result is also presented in Table 13.2. We have chosen such carbon containing poly atomic molecules only whose enthalpy data are available.

Table 13.2 Verification of hard soft acid base rule entailing the maxim of the maximum hardness principle using the sets of hardness data of the present work and those computed through the ansatz and operational and approximate formula of Parr and Pearson

R_1		R_2		P_1		P_2	$\Delta\eta(\Sigma P\text{-}\Sigma R)^a$	$\Delta\eta(\Sigma P\text{-}\Sigma R)^b$	ΔH (kcal/mol)
$2\ CH_3F$			=	CH_4	+	CH_2F_2	0.05	2.575	−14
$2CF_2Cl_2$			=	CF_4	+	CCl_4	1.548	−4.1143	−16.3
$3CH_3F$			=	$2CH_4$	+	CHF_3	0.18	−10.975	−31.4
$4CHF_3$			=	CH_4	+	$3CF_4$	1.02	−23.214	−22.9
$4CH_3Cl$			=	$3CH_4$	+	CCl_4	0.888	−15.1663	−6
$4CH_3F$			=	$3CH_4$	+	CF_4	0.58	−24.142	−63
$2CH_2O$			=	CH_4	+	CO_2	3.388	6.7[c]	−56.5
CF_3H	+	CH_3F	=	CF_4	+	CH_4	0.2	3.7	−19

Here *R* is the reactant; *P* is the product
[a] Present work
[b] ab initio or Parr and Pearson
[c] Hardness data taken from (Pearson 1988)

A look at the Table 13.2 reveals that, for all cases, the hardness data computed through the present method predicts the proper reaction surfaces of all the eight double exchange reactions. It is worth mentioning that the reaction enthalpies of all the reaction can be beautifully correlated by the hardness difference of the corresponding reactants. However, the hardness data evaluated through the ansatz and operational and approximate formula of Parr and Pearson poorly correlate the same reaction surface. This proves unequivocally that the present method evaluates molecular hardness that can be successfully applied to study the chemical reactions and reaction surface.

13.5 Conclusion

In this work we have basically launched a search whether there is a physical process of hardness equalization for molecules similar to the electronegativity equalization. We are tempted by the fact that the electronegativity equalization principle is widely accepted and theoretically justified and there is much communality in the basic philosophy of the origin and operational mechanism of the two fundamental descriptors– the electronegativity and the hardness of atoms. We have analyzed the origin and development in terms of the shell structure of atoms and molecules and classical theorems of electrostatics and put forward an alternative new definition of hardness. In the next venture, we have posited and logistically proved the occurrence of the physical process of hardness equalization principle at the event of molecule formation. Starting from our new definition of hardness and the new radial dependent formula of computing hardness of atoms and relying upon our newly introduced model of hardness equalization principle, we have derived an algorithm for the evaluation of the hardness of the hetero nuclear poly-atomic molecules. The algorithm is invoked to compute the hardness of as many as 22 poly atomic carbon containing molecules. In absence of any experimental benchmark, we have compared the computed hardness data of such molecules with the hardness data computed by an *ab- initio* quantum chemical method. From comparative study we find that there is close correlation between the two sets hardness data one set evaluated through the algorithm suggested by us, and the other set evaluated through the *ab- initio* quantum chemical method.

Furthermore, we have performed a validity test of the suggested formula for evaluating molecular hardness by the application of the global hardness data of the carbon containing poly atomic molecules in the real field of hard-soft acid-base reactions. We have noticed that the hardness data of this calculation can draw the exact picture of the model of the chemical reaction surfaces. However, the hardness data evaluated through the ansatz and operational and approximate formula of Parr and Pearson poorly correlate the same reaction surfaces.

We, therefore, conclude that the present effort of deriving an algorithm for the evaluation of the hardness of the carbon containing poly atomic molecules is a successful venture. The detailed study would suggest that the paradigm of the hardness equalization principle may be another law of nature like the established electronegativity equalization principle.

References

Atkins PW (1991) QUANTA: a handbook of concepts, 2nd edn. Oxford University Press, Oxford
Ayers PW (2005) J Chem Phys 122:141102 (1–3)
Ayers PW (2007) Faraday Discuss 135:161–190
Ayers PW, Parr RG (2000) J Am Chem Soc 122:2010–2018
Ayers PW, Parr RG (2008a) J Chem Phys 128:184108 (1–8)
Ayers PW, Parr RG (2008b) J Chem Phys 129:054111 (1–7)
Bader RFW (1991) Chem Rev 91:893–928
Chattaraj PK, Ayers PW (2005) J Chem Phys 123:086101 (1–2)
Chattaraj PK, Sengupta S (1996) J Phys Chem 100:16126–16130
Chattaraj PK, Nath S, Sannigrahi AB (1994) J Phys Chem 98:9143–9145
Chattaraj PK, Liu GH, Parr RG (1995) Chem Phys Lett 237:171–176
Chattaraj PK, Ayers PW, Melin J (2007) Phys Chem Chem Phys 9:3853–3856
Datta D (1986) J Phys Chem 90:4216–4217
Datta D (1992) Inorg Chem 31:2797–2800
Feynman RP, Leighton RB, Sands M (1964) The Feynman lecture on physics, vol I. Addison-Wesley, Reading
Frenking G, Krapp A (2007) J Comput Chem 28:15–24
Gázquez JL (1997) J Phys Chem A 101:9464–9469
Geerlings P, Proft FD, Langenaeker W (2003) Chem Rev 103:1793–1874
Ghosh DC, Islam N (2009) Int J Quantum Chem 110:1206–1213
Ghosh DC, Islam N (2010a) Int J Quantum Chem. doi:10.1002/qua.22499 [Early View]
Ghosh DC, Islam N (2010b) Int J Quantum Chem. doi:10.1002/qua.22508 [Early View]
Ghosh DC, Islam N (2010c) Int J Quantum Chem. doi:10.1002/qua.22500 [Early View]
Ghosh DC, Islam N (2010d) Int J Quantum Chem. doi:10.1002/qua.22651 [Early View]
Ghosh DC, Islam N (2010e) Int J Quantum Chem. doi:10.1002/qua.22653 [Early View]

Ghosh DC, Jana J, Biswas R (2000) Int J Quantum Chem 80:1–26
Ghosh DC, Jana J, Bhattacharyya S (2002) Int J Quantum Chem 87:111–134
Ghosh DC, Biswas R, Chakraborty T, Islam N, Rajak SK (2008) J Mol Struct THEOCHEM 865:60–67
Gilman JJ (1997) Mat Res Innovat 1:71–76
Gyftpoulous EP, Hatsopoulos GN (1968) Proc Natl Acad Sci 60:786–793
Hohenberg P, Kohn W (1964) Phys Rev B 13:864–871
Hyperchem (2008) 8.0.6, Hypercube, Inc, Gainesville, FL 32608 (USA)
Islam N, Ghosh DC (2010) Eur J Chem 1:83–89
Islam N, Ghosh DC (2011a) Mol Phys 109:917–931
Islam N, Ghosh DC (2011b) Int J Quantum Chem. doi:10.1002/qua.22861 [Early View]
Klopman A (1964) J Am Chem Soc 86:4550–4557
Komorowski L (1987) Chem Phys 114:55–71
Mulliken RS (1952) J Am Chem Soc 74:811–822
Murphy LR, Meek TL, Allred AL, Allen LC (2000) J Phys Chem A 104:5867–5871
Nalewajski RF (1985) J Phys Chem 89:2831–2837
Pal S, Vaval N, Roy R (1993) J Phys Chem 97:4404–4406
Parr RG, Bartolotti LJ (1982) J Am Chem Soc 104:3801–3803
Parr RG, Chattaraj PK (1991) J Am Chem Soc 113:1854–1855
Parr RG, Pearson RG (1983) J Am Chem Soc 105:7512–7516
Parr RG, Yang W (1989) Density functional theory of atoms and molecules. Oxford University Press, New York
Parr RG, Donnely A, Levy M, Palke W (1978) J Chem Phys 68:3801–3807
Parr RG, Ayers PW, Nalewajski RF (2005) J Phys Chem A 109:3957–3959
Pearson RG (1963) J Am Chem Soc 85:3533–3539
Pearson RG (1986) Proc Natl Acad Sci 83:8440–8441
Pearson RG (1987) J Chem Educ 64:561–570
Pearson RG (1988) Inorg Chem 27:734–740
Pearson RG (1993) Acc Chem Res 26:250–255
Pearson RG (1995) Inorg Chim Acta 240:93–98
Pearson RG (1999) J Chem Educ 76:267–270
Pearson RG, Palke WE (1992) J Phys Chem 96:3283–3285
Perdew JP (1988) Phys Rev B 37:6175–6180
Pillar FL (1968) Elementary quantum chemistry. McGraw Hill, New York
Putz MV (2006) Int J Quantum Chem 106:361–389
Putz MV (2007) J Theor Comput Chem 6:33–47
Putz MV (2008a) Absolute and chemical electronegativity and hardness. Nova Science, New York
Putz MV (2008b) Int J Mol Sci 9:1050–1095
Putz MV (2009a) Int J Quantum Chem 109:733–738
Putz MV (2009b) J Mol Struct THEOCHEM 900:64–70
Putz MV (2010) MATCH Commun Math Comput Chem 64:391–418
Putz MV, Russo N, Sicilia E (2003) J Phys Chem A 107:5461–5465
Putz MV, Russo N, Sicilia E (2005) Theor Chim Acc 114:38–45
Reed JL (1997) J Phys Chem A 101:7396–7400
Sanderson RT (1951) Science 114:670–672
Sen KD, Vinayagam SC (1988) Chem Phys Let 144:178–179
Torrent-Sucarrat M, Luis JM, Duran M, Sola M (2001) J Am Chem Soc 123:7951–7952
Torrent-Sucarrat M, Luis JM, Duran M, Sola M (2002) J Chem Phys 117:10561–10570
Yang W, Lee C, Ghosh SK (1985) J Phys Chem 89:5412–5414
Yang WT, Zhang YK, Ayers PW (2000) Phys Rev Lett 84:5172–5175
Zhou Z, Parr RG (1988) Tetrahedron Lett 29:4843–4846
Zhou Z, Parr RG (1989) J Am Chem Soc 111:7371–7379

Chapter 14
Modeling of the Chemico-Physical Process of Protonation of Carbon Compounds

Sandip K. Rajak[1], Nazmul Islam[1], and Dulal C. Ghosh[1]

Abstract We have suggested a model for the evaluation of proton affinity of molecules in terms of so\me akin quantum mechanical descriptors that follow closely the physico-chemical process of protonation. Method relies upon the basic tenets of scientific modeling having four *akin* descriptors – the ionization energy (I), the global softness(S), the electronegativity (χ), and the global electrophilicity index (ω) as the components. These akin theoretical descriptors can be entailed in following and describing the alteration in geometrical parameters, the charge rearrangement and polarization in molecules as a result of protonation. The modeling has evolved an ansatz for the evaluation of gas phase proton affinity, PA, of molecules as PA $= C + C_1(-I) + C_2 S + C_3(1/\chi) + C_4(1/\omega)$, where C, C_1, C_2, C_3, and C_4 are the constants. The suggested ansatz is invoked to compute the protonation energy of as many as 88 carbon compounds of diverse physico-chemical nature viz, hydrocarbons, alcohols, carbonyls, carboxylic acids, esters, aliphatic amines, aromatic amines, pyridine derivatives and amino acids. A detailed comparative study of theoretically evaluated protonation energies of the above mentioned molecules vis-à-vis their corresponding experimental counterparts strongly suggest that the proposed modeling and the ansatz for computing the proton affinity of molecules are efficacious for studying the physico-chemical process of protonation and the hypothesis is scientifically acceptable 14.

14.1 Introduction

The protonation reactions or the physico-chemical process of protonation are ubiquitous in almost all the areas of chemistry and biochemistry (Stewart 1985; Carrol 1998; Zhao et al. 2004; Kennedy et al. 2003; Bouchoux 2007). The majority

[1] Department of Chemistry, University of Kalyani, Kalyani 741235, India
e-mail: sandip1ku@gmail.com; nazmul.islam786@gmail.com; dcghosh1@rediffmail.com

M.V. Putz (ed.), *Carbon Bonding and Structures: Advances in Physics and Chemistry*, Carbon Materials: Chemistry and Physics 5, DOI 10.1007/978-94-007-1733-6_14,

of chemical reaction occurs in acid medium. The chemical process of protonation or deprotonation is fundamental first step of many chemical rearrangements, and enzymatic reactions (Kennedy et al. 2003). The resulting protonated molecule is frequently a pivotal intermediate that guides the succeeding steps of the overall process. The knowledge of the intrinsic basicity and the site of protonation of a compound is central for the understanding of the mechanism of chemical reactions. The legend proton affinity is defined as the negative of the enthalpy change of a protonation reaction at the standard conditions. The gas-phase proton affinities are a quantitative measure of the intrinsic basicity of a molecule (Deakyne 2003). The study of thermochemistry of the proton transfer reaction in the gas phase is well-known experiment of acid-base reaction (Lias et al. 1984). Dynamics of proton transfer is also important for ionization processes in mass spectroscopy.

Basicity is defined (Bronsted 1923) as the tendency of a molecule, B, to accept a proton, H^+, in the following Base-Acid (Proton) adduct BH^+ formation reaction

$$B + H^+ = BH^+ - PA \tag{14.1}$$

where PA is the proton affinity of the base B.

During the physico-chemical process of protonation, electronic charge is soaked by the proton from the entire skeleton of the molecule. As a result, all the structural parameters i.e. bond lengths and bond angles, and other charge dependent physical properties like the polarizability and the dipole etc. are affected. A plethora of information has appeared on the study of this important chemico-physical process (Deakyne 2003; Lias et al. 1984).

Although, experimentally the proton affinity can be determined by several techniques like the measurement of the heats of formation (Bouchoux 2007) of the species involved in the adduct formation reaction, by mass spectroscopic measurement techniques (Lias et al. 1984; Bronsted 1923; Meot-Ner 1979) and by the measurement of the ionization thresholds. The ‘acid-base adducts’ are not always stable and/or does not exist in all cases and also it is well known (Dixon et al. 1987) that the experimental determination of the proton affinities of molecules is not easy always. For this reason, in recent years, much emphasis has been given to the possibility of the calculation of proton affinities through some quantum mechanical as well as density functional theoretical models (Curtiss et al. 1993; Del Bene 1983; Smith et al. 1995; Jursic 1999; Hammerum 1999).

It is now established (Hehre et al. 1986) that the *ab initio* quantum mechanical approaches are very successful in providing reliable values of proton affinities and gas phase basicities for small molecules. However, due to the reason of heavy computational expenses, application of *ab initio* methods for the estimation of proton affinities is still impractical for larger molecules (Labanowskiy et al. 2011). It is also recognized (Ozment et al. 1992) that the popular semi empirical methods such as AM1, MNDO and PM3 are not consistently reliable in calculating proton affinities. Although there are some attempts of modeling to compute protonation energy for specific groups of compounds (Deakyne 2003; Eckert-Maksic et al. 1995; Russo et al. 2000; Margabandu et al. 2010; Taft et al. 1978; Maksic et al. 2002; Perez et al. 2000), but fact remains that there is no report of universal model, mathematical or conceptualin nature, as a substitute for experimental or theoretical measurement of the energy of protonation.

Currently the conceptual density functional theory (Geerlings et al. 2003; Berkowitz et al. 1985; Ghosh et al. 1984; Hansch et al. 1979; Parr et al. 1978, 1983, 1984, 1989, 1999; Yang et al. 1985) of chemical reactivity have introduced, within the scope of sound theoretical framework, many descriptors like electronegativity, hardness, softness, fukui functions, electrophilicity index etc. in theoretical chemistry. Such descriptors have made serious inroad in science and opened a new paradigm of chemical thinking, modeling and computation (Geerlings et al. 2003; Berkowitz et al. 1985; Ghosh et al. 1984; Hansch et al. 1979; Parr et al. 1978, 1983, 1984, 1989, 1999; Yang et al. 1985).

In this work, we have developed a model for the evaluation of proton affinity in terms of some *akin* conceptual reactivity descriptors which can be conceptually linked and associated with the physico-chemical process of protonation. The *akin* descriptors are the ionization energy (I), the global softness(S), the electronegativity (χ), and the global electrophilicity index (ω).

14.2 The Physico-Chemical Process of Protonation

In the physico-chemical process of protonation, when a proton dynamically approaches towards a nucleophile from a long distance it is attracted by the electron cloud of the molecule. Thus a proton acting as an electrophile soaks the electron density from the entire skeleton of the nucleophile (Ghosh 1976). Ultimately, the electron cloud of the nucleophile is redistributed and remains under the influence of nucleus of the electrophile, the proton and it fixes at a site of lone pair of the molecule. However, if there is no lone pair in the structure of the molecule, the proton remaining weakly attached to the sphere of the charge cloud of the molecule. The polarizing power of the proton induces a physical process of structural and energetic changes in the molecule. This phenomenon is, in particular, at the origin of the site of protonation, has considerable effect on the strength and length of the bonds (Bouchoux 2007). The structural and energetic changes induced by the polarizing power of the proton are expected to be at its maximum at the gas phase of the molecule. Thus, the gas-phase basicity is certainly the ideal revelator of the structural and energetic characteristics of the molecular protonation process.

14.3 Physical Significances of the Ionization Energy and the Density Functional Theoretical Global Reactivity Descriptors – the Electronegativity, the Chemical Hardness, the Softness, and the Electrophilicity Index in the Context of Physico-Chemical Process of Protonation

The ionization energy is a fundamental descriptor of the chemical reactions and reactivity of atoms and molecules. High ionization energy indicates high stability and chemical inertness and small ionization energy indicates high reactivity of the atoms and molecules. The other fundamental important descriptors of chemical reactivity and property of atoms and molecule emanating from the density functional underpinning are the electronegativity, the hardness or softness and electrophilicity index (Parr et al. 1978, 1983, 1984, 1989, 1999; Yang et al. 1985; Pearson 1986). Mills et al. (1976) discovered a linear relationship between the proton affinity (PA) and additive inverse of the ionization potential (–I) of molecules.

It is apparent that the chemical hardness fundamentally signifies the resistance towards the deformation or polarization of the electron cloud of the atoms, ions or molecules under small perturbation of chemical reaction. The softness is simply the reciprocal of the hardness. Thus the general operational significance of the hard-soft chemical species may be understood in the following statement. If the electron cloud is strongly held by the nucleus, the chemical species is 'hard' but if the electron cloud is loosely held by the nucleus the system is 'soft' (Parr et al. 1983).

Electronegativity though defined in many different ways, the most logical and rational definition of it is the electron holding power of the atoms or molecules. Electronegativity is defined and measured as the power (force) with which the valence electron of an atom is held by its screened nuclear charge. The more electronegative elements hold electrons more tightly and the less electronegative elements hold less tightly (Ghosh et al. 2011b).

Worth mentioning here some outstanding fundamental works of Putz on electronegativity and hardness and their usefulness for the theoretical prediction of several physicochemical properties-like the fundamentals of chemical bonding, see Chaps. 1 and 13 of the present Volume and the related references.

In reference to nucleophilic-electrophilic, acid-base or donor-acceptor reaction, the electrophilicity index (Chaquin 2008; Noorizadeh 2007) of atoms and molecules seems to be an absolute and fundamental property of such chemical species because it signifies the energy lowering process on soaking electrons from the donors. This tendency of charge soaking and energy lowering must develop from the attraction between the soaked electron density and screened nuclear charge of the atoms and molecules. It, therefore, transpires that the conjoint action of the shell structure and the physical process of screening of nuclear charge of the atoms and molecules lead to the development of the new electrostatic property – the electrophilicity, electronegativity, hardness of atoms and molecules (Islam et al. 2010, 2011a, b; Ghosh et al. 2011a, c). It is also well known that the principal factor that controls ionization energy is the nuclear charge.

One may think that it is quite possible that logistically the electronic structure, especially the shell structure, and the physical process of screening of nuclear charge of atom are intimately linked to the origin of and development of the ionization energy, hardness, electronegativity and electrophilicity indices of atoms and molecules.

14.4 The Modeling of the Physico-Chemical Process of Protonation and Algorithm for Computing the Proton Affinity of the Molecules

We have posited above that the descriptors like the ionization process of atoms and molecules, the physical property like hardness, softness, the electronegativity and the electrophilicity have close relation with each other in their fundamental origin.

Hence, it is clear that fundamentally and operationally the physico-chemical process of protonation can be linked to the above akin descriptors – the ionization process, the hardness, softness, electronegativity and electrophilicity. Recently, we (Islam et al. 2010, 2011a, b; Ghosh et al. 2011a) have published good number of papers where we have discussed that the three descriptors, the electronegativity, the hardness and the electrophilicity index of atoms and molecules are fundamentally qualitative per se and operationally the same. All three represent the attraction of screened nuclei towards the electron pair/bond. Thus, we can safely and reasonably conclude that the proton affinity and the three descriptors have inverse relationship.

Thus, since the above four parameters have dimension of energy and can be linked to the process of charge rearrangement and polarization during the physico-chemical process of protonation, they can be components of a probabilistic scientific modeling of proton affinity.

The proton affinity or the ability of donating the lone pair of a Lewis base and the ability for the deformation of electron cloud of a species, the softness, and/or the tendency of the molecule to lose electron, the ionization potential, are fundamentally similar in physical appearance stemming from the attraction power of the nuclei of the atoms forming the molecule. Hence, the proton affinity and the softness and the ionization energy are directly proportional to each other.

Considering all the above mentioned fundamental nature of the physico-chemical process of protonation and its probable relationship with the quantum mechanical descriptors, we suggest an ansatz for the computation of the proton affinity in terms of these theoretical descriptors. The physico-chemical process and the energetic effect must entail the above stated four parameters. To derive an explicit relation to compute the proton affinity in terms of the above stated descriptors, we suggest explicit inter relationships between the protonation energy and the descriptors relying upon their response towards the protonation.

$$\mathrm{PA} \propto (-I) \tag{14.2}$$

$$\mathrm{PA} \propto \mathrm{S} \tag{14.3}$$

$$\mathrm{PA} \propto 1/\chi \tag{14.4}$$

$$\mathrm{PA} \propto 1/\omega \tag{14.5}$$

Combining the above four relations we get,

$$\mathrm{PA} = \mathrm{C} + \mathrm{C}_1(-\mathrm{I}) + \mathrm{C}_2\mathrm{S} + \mathrm{C}_3(1/\chi) + \mathrm{C}_4(1/\omega) \tag{14.6}$$

where PA is proton affinity, C, C_1, C_2, C_3, and C_4 are the constants I is ionization energy, S is global softness, χ is the electronegativity and ω is the global electrophilicity index of the molecule.

14.5 Mathematical Formulae of the Global Reactivity Descriptors Invoked in the Study

According to Koopmans' theorem the ionization potential (I) and the electron affinity (A) are computed as follows:

$$\mathrm{I} = -\varepsilon_{\mathrm{HOMO}} \tag{14.7}$$

$$\mathrm{A} = -\varepsilon_{\mathrm{LUMO}} \tag{14.8}$$

where $\varepsilon_{\mathrm{HOMO}}$ and $\varepsilon_{\mathrm{LUMO}}$ are the orbital energies of the highest occupied and the lowest unoccupied orbitals.

Parr et al. (1978, 1983) defined the chemical potential, μ, electronegativity, χ, and hardness, η, in the framework of density functional theory, DFT(Parr et al. 1989) as

$$\mu = (\partial \mathrm{E}/\partial \mathrm{N})_{\mathrm{v(r)}} = -\chi = (\mathrm{I} + \mathrm{A})/2 \tag{14.9}$$

$$\eta = \tfrac{1}{2}[\partial \mu/\partial \mathrm{N}]_{\mathrm{v(r)}} = \tfrac{1}{2}\left[\partial^2 \mathrm{E}/\partial \mathrm{N}^2\right]_{\mathrm{v(r)}} = \tfrac{1}{2}(\mathrm{I} - \mathrm{A}) \tag{14.10}$$

where E, N,v (r), I and A are the energy, the number of electrons, the external potential the ionization energy and the electron affinity of an atomic or molecular system respectively.

Table 14.1 Correlation coefficients and R^2 value for the set 1, set 2, set 3, set 4, set 5 and set 6

Sets	C	C_1	C_2	C_3	C_4	R^2
1	450	18.0	24100	−40539	8019	0.992
2	−113	−4.66	−1810	3561	−705	0.818
3	17.1	0.666	−0.1	−0.1	−0.15	0.995
4	−129	−7.94	147	167	−11.1	0.916
5	31.9	1.08	−31.4	−39.5	4.19	0.911
6	2.3	0.308	89.1	−44.1	4.06	0.88

Softness is a reactivity index and is defined as the reciprocal of hardness

$$S = (1/\eta) \tag{14.11}$$

Parr et al. (1999) defined electrophilicity index (ω) as

$$\omega = (\mu)^2/(2\eta) \tag{14.12}$$

In this study we have taken some hydrocarbons as Set 1, some alcohols, carbonyls, carboxylic acids and esters as Set 2, some aliphatic amines as Set 3, some aromatic amines as Set 4, some pyridine derivatives as Set 5 and some amino acids as Set 6 for which the experimental protonation (Lias et al. 1984; Hunter et al. 1998; National Institute of Standards and Technology; Wróblewski et al. 2007; Lias et al. 1988) energy are known. The PQS Mol 1.2-20-win software (PQSMol) has been used to calculate the global descriptors by using the *ab initio* Hartree-Fock SCF method with the 6–31 g basis set. The geometry optimization technique is adopted. The ionization energy, I, the electronegativity, χ, the global softness, S, and the global electrophilicity index, ω respectively of the molecules are computed by invoking the Koopmans' theorem and Eqs. 14.7, 14.9, 14.11 and 14.12.

A multi linear regression (Nantasenamat et al. 2009) is performed using Minitab15 (Minitab15) to compute the correlation coefficients C, C_1, C_2, C_3 and C_4 by plotting experimental PA along the abscissa and the values of the quantum mechanical descriptors along the ordinate. The computed correlation coefficients C, C_1, C_2, C_3 and C_4, for all the sets are tabulated in Table 14.1.

Thereafter, invoking the suggested ansatz, Eq. 14.6, and putting the quantum mechanical descriptors and the correlation coefficients in the Eq. 14.6, we have computed the PA's of six sets of carbon compounds. The comparative study of theoretically evaluated and experimentally determined PA's of the Set 1–Set 6 is performed in the Tables 14.2–14.7 respectively.

For better visualization of the comparative study, the results of the theoretically computed and experimentally determined proton affinities of the set 1–set 6 are depicted in the Figs. 14.1–14.6 respectively.

Table 14.2 Experimental and calculated PA (eV) for the set 1

Molecule	Experimental	Calculated
Methane	5.63294	6.01418
Ethane	6.17932	6.56332
Propane	6.48286	6.83883
Butane [a]	6.83237	7.07331
Isobutane	7.02491	7.34303
Pentane [a]	6.86533	7.13276
Hexane [a]	7.01407	7.37095

[a] P.A. calculated by Wróblewski et al. (2007)

Table 14.3 Experimental and calculated PA (eV) for the set 2

Molecule	Experimental	Caculated
Formaldehyde	7.38916	7.90889
Formic acid	7.68837	8.13938
Methanol	7.81846	8.57263
Ketene	8.55564	8.96532
Acetaldehyde	7.9659	8.34277
Ethanol	8.04829	8.57857
Acetic acid	8.12201	8.55023
Acetone	8.41254	8.90774
Propanol	8.15236	8.47112
Propionic acid	8.26077	8.50291
Methyl acetate	8.28679	8.61904
Butanol	8.17838	8.46674

Table 14.4 Experimental and calculated PA (eV) for the set 3

Molecule	Experimental	Calculated
NH_3	8.846181	8.860042
CH_3NH_2	9.284153	9.341572
$CH_3CH_2NH_2$	9.409908	9.399806
$(CH_3)_2CHNH_2$	9.47929	9.499455
$(CH_3)_2NH$	9.566017	9.583402
$(CH_3)_3CNH_2$	9.57469	9.596479
$(CH_3)_3N$	9.761153	9.794852

Table 14.5 Experimental and calculated PA (eV) for the set 4

Molecule	Experimental	Calculated
$3\text{-}H_3C_6H_4N(C_2H_5)_2$	9.925935	9.722904
$4\text{-}H_3C_6H_4N(C_2H_5)_2$	9.912926	9.706435
$C_6H_5N(C_3H_7)_2$	9.912926	9.673925
$C_6H_5N(CH_3)(C_2H_5)$	9.84788	9.522402
$C_6H_5NH(C_2H_5)$	9.618053	9.592654
$C_6H_5NHCH_3$	9.457608	9.44481
$C_6H_5CH_2NH_2$	9.401235	8.976198

(continued)

Table 14.5 (continued)

Molecule	Experimental	Calculated
2-(OH)$C_6H_4NH_2$	9.28849	9.197386
3-(OH)$C_6H_4NH_2$	9.28849	9.197251
4-$CH_3C_6H_4NH_2$	9.266808	9.06326
3-$CH_3C_6H_4NH_2$	9.253799	9.04584
3-$CH_3C_6H_4N(CH_3)_2$	9.253799	9.044886
1,2-$C_6H_4(NH_2)_2$	9.22778	9.031081
4-$ClC_6H_4NH_2$	9.045653	8.720894
3-$BrC_6H_4NH_2$	9.023971	8.683775
4-$FC_6H_4NH_2$	9.023971	8.763088
3-$CF_3C_6H_4NH_2$	8.854853	8.674228

Table 14.6 Experimental P.A (ev) and Calculated PA (ev) for the set-5

Molecule	Experimental	Calculated
pyridine	9.579025	9.70499435
3-Fluoropyridine	9.292825	9.45509615
4-Trifluoromethylpyridine	9.227780	9.24233169
2-Trifluoromethylpyridine	9.171407	9.27524221
4-cyanopyridine/4-pyridinecarbonitrile	9.119371	9.11862912
3-cyanopyridine/3-pyridinecarbonitrile	9.076007	9.16793132
4-methoxypyridine	9.869562	9.91634082
2-t-butylpyridine	9.860889	9.84251312
2,4-dimethylpyridine	9.856553	9.97185252
2-isopropylpyridine	9.852216	9.84129804
2-ethylpyridine	9.808853	9.75667056
2,3-dimethylpyridine	9.808853	10.0097702
3,4-dimethylpyridine	9.808853	9.88622832
2,5-dimethylpyridine	9.800180	9.89131007
pyridine-2-methoxymethyl	9.800180	9.71229849
4-tert-butylpyridine	9.795844	9.81876855
3,5-dimethylpyridine	9.778498	9.76297831
4-methylpyridine	9.765489	9.79359891
2-methylpyridine	9.756816	9.84350118
4-ethylpyridine	9.739471	9.7890431
3-methylpyridine	9.717789	9.73935956
3-ethylpyridine	9.709116	9.77988682
3-methoxypyridine	9.696107	9.72770612
4-vinylpyridine	9.678762	9.70246112
2-methoxypyridine	9.622389	9.76651527
2-(methylthio)-pyridine	9.626725	9.53136835
2-chloro-6-methylpyridine	9.496634	9.5342994
2-chloro-4-methylpyridine	9.479289	9.51421841
4-chloropyridine	9.444598	9.43047573

(continued)

Table 14.6 (continued)

Molecule	Experimental	Calculated
4-Fluoropyridine	9.392562	9.53212685
2-chloro-6-methoxypyridine	9.362207	9.45240344
3-bromopyridine	9.327516	9.45856933
2-chloropyridine	9.297162	9.45612427
3-chloropyridine	9.314507	9.44773003
2-bromopyridine	9.310171	9.46820189

Table 14.7 Experimental and calculated PA (eV) for set 6

Molecule	Experimental	Calculated
Glycine	9.175744	9.324845
Alanine	9.314508	9.33187
Cysteine	9.292826	9.386151
Serine	9.379553	9.445344
Tryptophan	9.774162	9.822751
Tyrosine	9.639735	9.639794
Methionine	9.600708	9.660442
Glutamic acid	9.388226	9.391748
(2 S, 3R) threonine	9.474953	9.431556
Aspartic acid	9.396899	9.340133

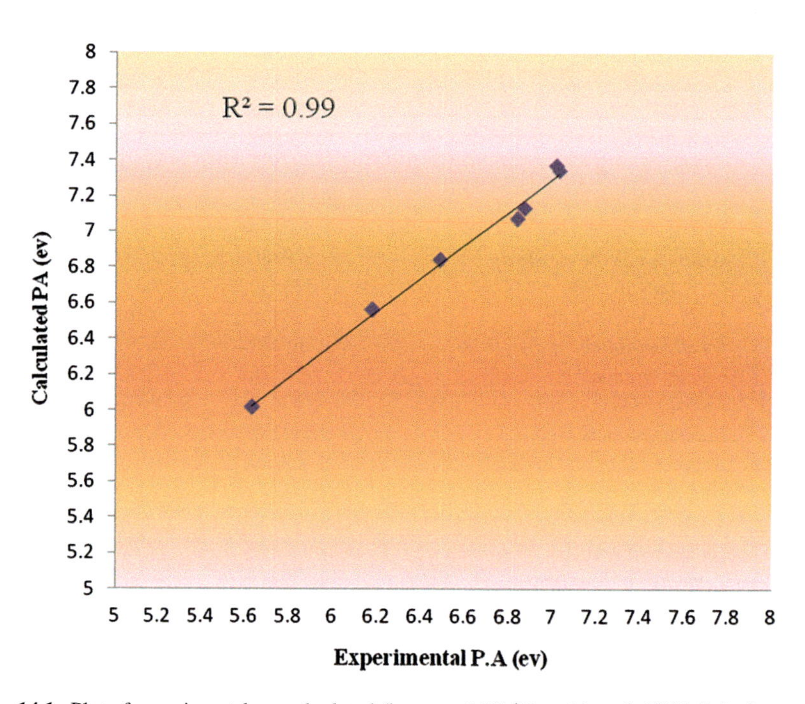

Fig. 14.1 Plot of experimental *vs.* calculated (by us and Wróblewski et al. 2007) P.A. for set 1

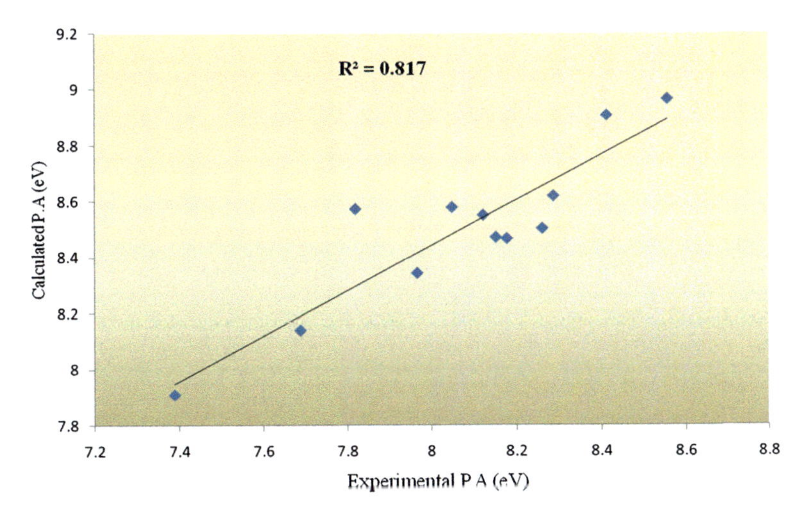

Fig. 14.2 Plot of calculated *vs.* experimental P.A. for set 2

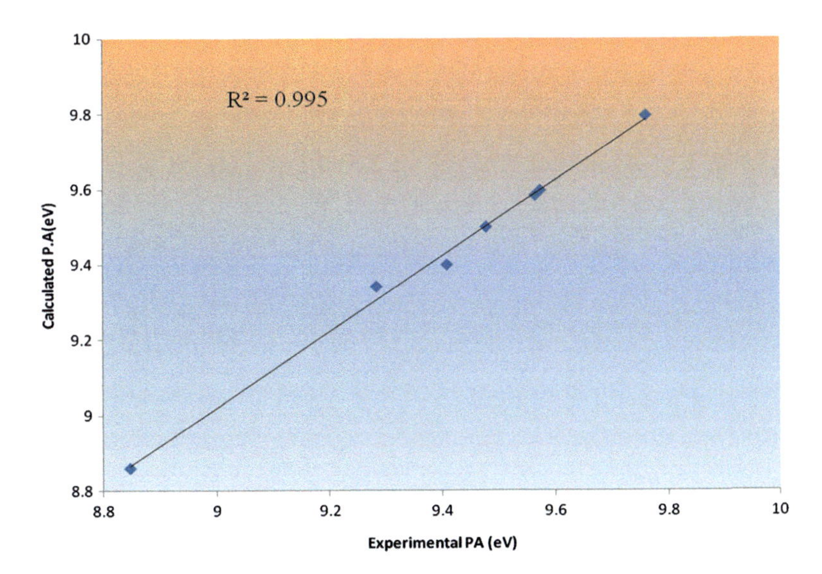

Fig. 14.3 Plot of calculated *vs.* experimental P.A. for set 3

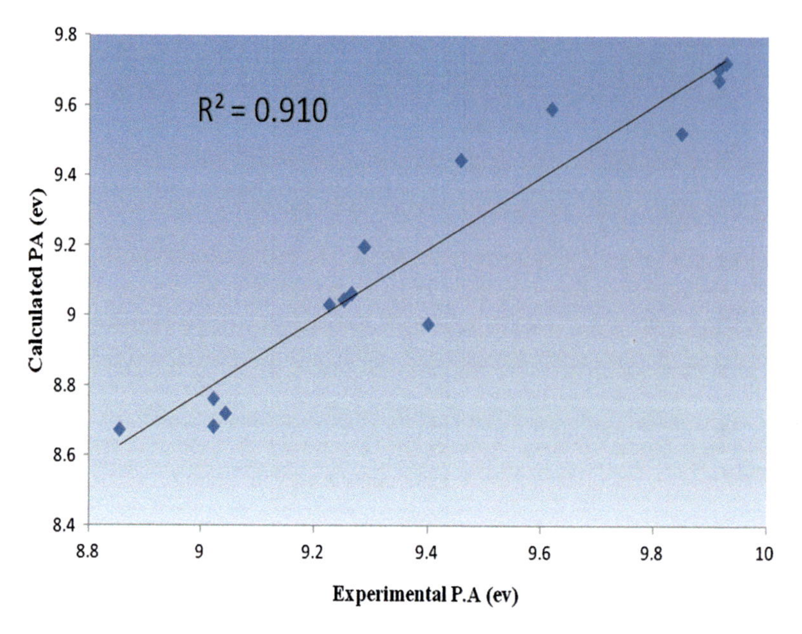

Fig. 14.4 Plot of calculated *vs.* experimental P.A. for set 4

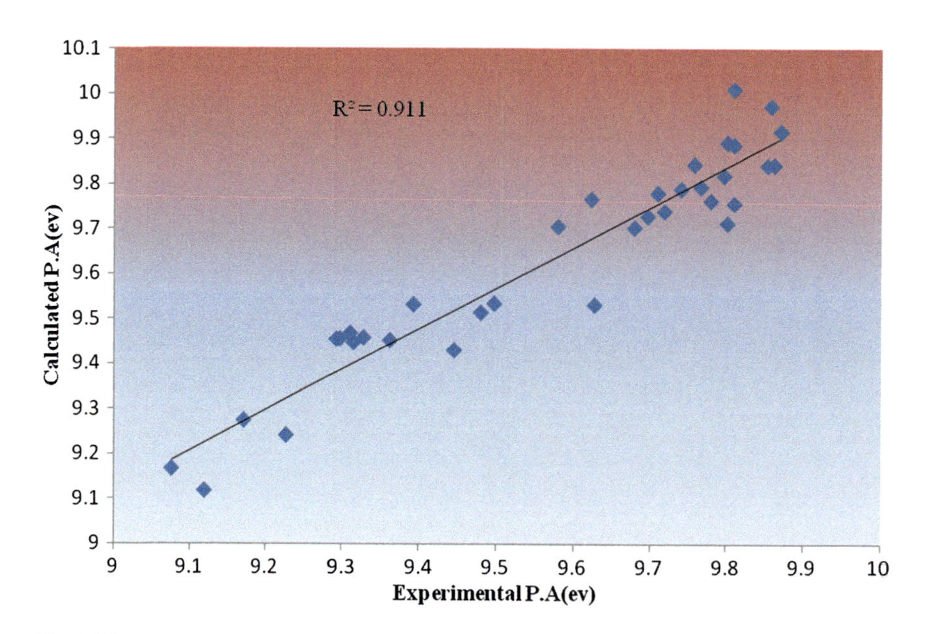

Fig. 14.5 Plot of calculated *vs.* experimental P.A. for set 5

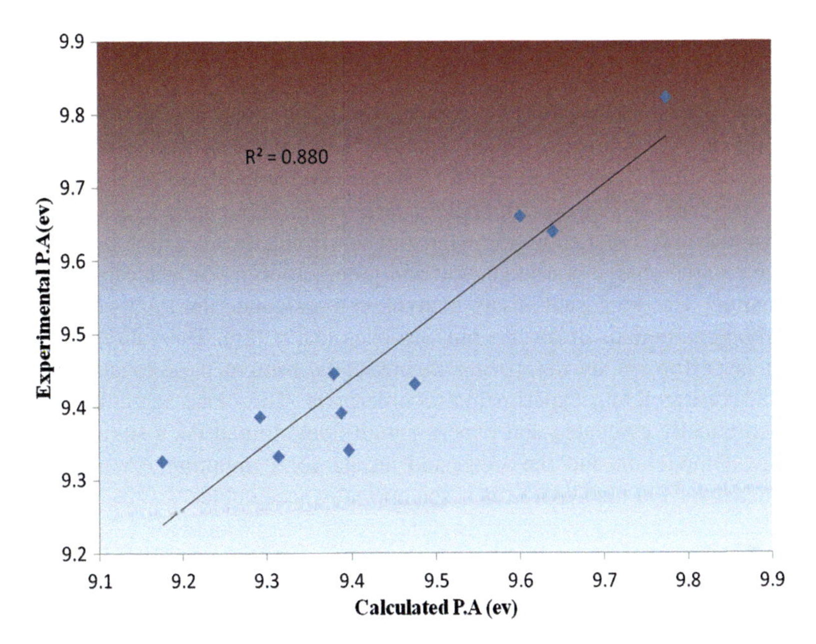

Fig. 14.6 Plot of calculated *vs.* experimental P.A. for set 6

14.6 Results and Discussion

A deep look on the Table 14.2 and Fig. 14.1 (for set 1), the Table 14.4 and Fig. 14.3 (for set 3) and Table 14.6 and Fig. 14.5 (for set 5) reveals that there are excellent correlation between the theoretically computed proton affinities of the seven hydrocarbons (set 1), seven aliphatic amines (set 3) and 40 pyridine derivatives (set 5) respectively. The R^2 value for correlation of set 1, set 3 and set 5 are 0.99, 0.995 and 0.911 respectively. A close look at the Figs. 14.1, 14.3 and 14.5 reveals that the two sets of PA's – experimental and theoretical of the three sets molecules viz the hydrocarbons(set 1), the aliphatic amines (set 3) and pyridine derivatives (set 5) are so close to each other that one curve just superimposes upon the other.

A look at the Table 14.3 and Fig. 14.2 (for set 2), Table 14.5 and Fig. 14.4 (for set 4) and Table 14.7 and Fig. 14.6 (for set 6) reveal that there is fairly a good correlation between the theoretically computed and experimentally determined proton affinities of as many as 12 compounds containing alcohols, carbonyls, carboxylic acids and esters (set 2), 17 aromatic amines (set 4) and 10 amino acids (set 6) respectively. The R^2 value for correlation of set 2, set 4 and set 6 are 0.817 and 0.91 and 0.88 respectively.

14.7 Conclusion

In this work, we have presented a scientific model for the evaluation of protonation energy of molecules in terms of four quantum theoretical descriptors – the ionization energy, the global softness, the electronegativity, and the global electrophilicity index. As a basis of scientific modeling, we have posited that these akin theoretical descriptors can be entailed in following and describing the alteration in geometrical parameters, the charge rearrangement and polarization in molecules as a result of protonation. The suggested ansatz is invoked to calculate the PA's of as many as 88 carbon compounds of diverse physico-chemical nature. The validity test of the model is performed by comparing theoretically computed protonation energies and the corresponding experimental counterparts. The close agreement between the theoretically evaluated and experimentally determined PA's suggests that the conceived modeling and the suggested ansatz for computing PA of molecules are efficacious and the hypothesis is scientifically acceptable.

References

Berkowitz M, Ghosh SK, Parr RG (1985) J Am Chem Soc 107:6811–6814
Bouchoux G (2007) Mass Spectrom Rev 26:775–835
Bronsted JN (1923) Recl Trav Chim Pays Bas 42:718–728
Carrol FA (1998) Perspectives on structure and mechanism in organic chemistry. Brooks Cole, New York
Chaquin P (2008) Chem Phys Lett 458:1439–1444
Curtiss LA, Raghavachari K, Pople PA (1993) J Chem Phys 98:1293–1298
Deakyne CA (2003) Int J Mass Spectrom 227:601–616
Del Bene JE (1983) J Phys Chem 87:367–371
Dixon DA, Lias SG (1987) In: Liebman JF, Greenberg A (eds) Molecular structure and energetics. vol. 2. Physical measurements. VCH, Deereld Beach
Eckert-Maksic M, Klessinger M, Maksic ZB (1995) Chem Phys Lett 232:472–478
Geerlings P, Proft FD, Langenaeker W (2003) Chem Rev 103:1793–1874
Ghosh DC (1976) A theoretical study of some selected molecules and their protonation by the application of CNDO method. Premchand Roychand Research Studentship Award, University of Calcutta
Ghosh DC, Islam N (2011a) Int J Quantum Chem. 111:1931–1941
Ghosh DC, Islam N (2011b) Int J Quantum Chem. 111:1961–1969
Ghosh DC, Islam N (2011c) Int J Quantum Chem 111:40–51
Ghosh SK, Berkowitz M, Parr RG (1984) Proc Natl Acad Sci 81:8028–8031
Hammerum S (1999) Chem Phys Lett 300:529–532
Hansch C, Leo A (1979) Substituent constants for correlation analysis in chemistry and biology. Wiley, New York
Hehre WJ, Radom L, PvR S, Pople JA (1986) Ab initio molecular orbital theory. Wiley, New York
Hunter EPL, Lias SG (1998) J Phys Chem Ref Data 27:413–656
Islam N, Ghosh DC (2010) Eur J Chem 1:83–89
Islam N, Ghosh DC (2011a) Mol Phys 109(6):917–931 (Accepted)
Islam N, Ghosh DC (2011b) Int J Quantum Chem. doi:10.1002/qua.22861 [Early View]
Jursic BS (1999) J Mol Struct THEOCHEM 487:193–203

Kennedy RA, Mayhew ChA, Thomas R, Watts P (2003) Int J Mass Spectrom 223:627–637
Labanowskiy JK, Hill RA, Heisterbergy DJ, Miller DD, Bender CF, Andzelm JW (2011) Proton affinities calculated by traditional ab initio approaches and by density functional methods. http://www.ccl.net/cca/documents/proton-affinity/affinities.pdf. Accessed 2011
Lias SG, Liebman JF, Levine RD (1984) J Phys Chem Ref Data 13:695–808
Lias SG, Bartmess JE, Liebman JF, Holmes JL, Levin RD, Mallard WG (1988) J Phys Chem Ref Data 17:1–861
Maksic ZB, Vianello R (2002) J Phys Chem A 106:419–430
Margabandu R, Subramani K (2010) Int J Chem Tech Res 2:1507–1513
Meot-Ner M (1979) J Am Chem Soc 101:2396–2403
Mills BE, Martin RL, Shirley DA (1976) J Am Chem Soc 98:2380–2385
MINITAB is a statistical software of STATSOFT Inc., USA
Nantasenamat C, Isarankura-Na-Ayudhya C, Naenna T, Prachayasittikul V (2009) EXCLI Journal 8:74–88
National Institute of Standards and Technology. http://webbook.nist.gov/chemistry/pa-ser.html
Noorizadeh S (2007) Chin J Chem 27:1439–1444
Ozment JL, Schmiedekamp AM (1992) Int J Quantum Chem 43:783–800
Parr RG, Pearson RG (1983) J Am Chem Soc 105:7512–7516
Parr RG, Yang W (1984) J Am Chem Soc 106:4049–4050
Parr RG, Yang W (1989) Density functional theory of atoms and molecules. Oxford University Press, New York
Parr RG, Donnely A, Levy M, Palke W (1978) J Chem Phys 68:3801–3807
Parr RG, Szentpaly LV, Liu S (1999) J Am Chem Soc 121:1922–1924
Pearson RG (1986) Proc Natl Acad Sci 83:8440–8441
Perez P, Toro-Labbe A (2000) J Phys Chem A 104:11993–11998
PQSMol 1.2-20-win, Parallel Quantum Solutions, LLC. http://www.pqs-chem.com
Russo N, Toscano M, Grand A, Mineva T (2000) J Phys Chem A 104:4017–4021
Smith BJ, Radom L (1995) J Phys Chem 99:6468–6471
Stewart R (1985) The proton: appellation to organic chemistry. Academic, New York
Taft RW, Taagepera M, Abboud JLM, Wolf JF, DeFrees DJ, Hehre WJ, Bartmess JE, McIver RT Jr (1978) J Am Chem Soc 100:7765–7767
Wróblewski T, Ziemczonek L, Alhasan AM, Karwasz GP (2007) Ab initio and density functional theory calculations of proton affinities for volatile organic compounds. Preprint. http://www.fizyka.umk.pl/~karwasz/publikacje/2007_Ab_initio_and_density_functional.pdf
Yang W, Parr RG (1985) Proc Natl Acad Sci 82:6723–6726
Zhao J, Zhang R (2004) Atmosph Environ 38:2177–2218

Chapter 15
Molecular Shape Descriptors: Applications to Structure-Activity Studies

Dan Ciubotariu[1], Vicentiu Vlaia[1], Ciprian Ciubotariu[2], Tudor Olariu[1], and Mihai Medeleanu[3]

Abstract Shape is a very important molecular feature for describing ligand molecules interacting with receptor, and other various complex chemical and biological processes. During the last years, there is a strong interest in developing several molecular shape descriptors. In this chapter we present our attempts for deriving van der Waals (vdW) shape molecular descriptors from two size molecular descriptors – the molecular vdW volume (V^W) and the molecular vdW surface (S^W). The ovality Θ_{iD}, i = 1,2,3 descriptors measure the deviation of the shape of a molecule from a spherical form in one-, two-, and three dimensions, respectively, whilst the {δ,G} molecular shape parameters take into consideration the cylindrical shape of a molecule or substituent. The application of these molecular descriptors in the study of the toxicity of a series of aliphatic amines on the protozoan ciliate *Tetrahymena pyriformis* (Θ_{iD}) and for the correlation of rate constants of bimolecular nucleophilic substitution, acid-catalyzed hydrolysis of esters and esterification reactions (δ) is also reported in this chapter. The MTD (minimal topologic difference) method allows a receptor site mapping in the frame of a series of bioactive compounds. The MVD (minimal volume difference) method is an improved variant of the MTD method, which takes into account the 3D extension of a molecule. These methods are described here in some detail. The results obtained in the

[1] Department of Organic Chemistry, Faculty of Pharmacy,
"Victor Babes" University of Medicine and Pharmacy, P-ta Eftimie Murgu No. 2,
300041, Timişoara, Romania
e-mail: dciubotariu@mail.dnttm.ro; vlaiav@gmail.com; rolariu@umft.ro

[2] Department of Computer Sciences, University "Politehnica",
P-ta Victoriei No. 2, 300006, Timişoara, Romania

[3] Department of Organic Chemistry, University "Politehnica",
P-ta Victoriei No. 2, 300006, Timişoara, Romania
e-mail: mihai.medeleanu@chim.upt.ro

M.V. Putz (ed.), *Carbon Bonding and Structures: Advances in Physics and Chemistry*,
Carbon Materials: Chemistry and Physics 5, DOI 10.1007/978-94-007-1733-6_15,

study of anti-carcinogenic activity of some retinoids and in the inhibition of carbonic-anhydrase (CA) by a series of sulfonamides are also presented in this chapter.

15.1 Introduction

Molecules may be viewed as three-dimensional objects, which fill some space. Consequently, when considering the space requirements of molecules, it is usual to associate with them a formal molecular body having a specific van der Waals volume, V^W, and a formal molecular surface characterized by its vdW aria, S^W (Gavezotti 1983; Connolly 1985). This surface is a formal vdW molecular boundary, in fact a closed surface that separates the vdW 3D space into two parts: the molecular body enclosed by the surface, which is composed by a collection of atomic spheres and is supposed to represent the entire molecule, and the 3D space that falls on the outside of the surface, hence on the outside of the molecule. These intuitive concepts of molecular vdW volume and molecular vdW surface are very useful for the interpretation of molecular size and shape properties within approximate models (Mezey 1993).

Isolated atoms show spherical symmetry, and it is obvious to consider a molecule as a collection of atomic spheres of some appropriate defined vdW radii. Because the vdW radii of atomic spheres used for the representations of molecular space are usually much too large for modeling molecules by simply placing the atomic hard vdW spheres side by side, commonly one generates various "fused sphere" models for molecules, that is the atomic vdW spheres are interpenetrated one with another. Positions of these spheres may be described by their Cartesian coordinates according to the 3D stereochemical bond pattern of a particular molecule. The envelope of the outer surface of the vdW atomic fused spheres may be regarded as a formal vdW molecular surface. This envelope embeds a formal vdW molecular volume.

Shape is a very important molecular feature for describing ligand molecules interacting with receptor, and other various complex chemical and biological processes. It has been long time recognized that determining molecular shape and variations in this property is essential to understanding the molecules involved in chemical reactions. Various studies have demonstrated that molecular shape has an important role in biological activity. Thus, in molecular recognition processes enzymes can differentiate between functional groups in a molecule by shape recognition. Natural products produced through biosynthetic pathways involve, also, shape recognition for selective oxidation. Chemical shape interaction has a key role in the smell, sight and taste, via specific receptors (Ballester and Richards 2007).

Irrespective of the type of definition used, the essence of shape is, therefore, very useful in describing a molecule or a part of a molecule (substituent). Hence the study of shape in molecular pharmacology has gained importance owing to its applicability in drug design *in silico* techniques widely employed to decrease the costs of drug discovery and development. These computational methods can enable rapid comparison between small molecules, or small molecules with protein receptor sites, mainly based on their shape and properties such as electrostatics (Kortagere et al. 2009).

During the last years, a variety of molecular shape descriptions have been devised. The shape is an interesting molecular feature due to its importance in the understanding, search, and prediction of molecular interaction and receptor inhibition, particularly in the field of drug design studies (Mezey 1993; Wilson et al. 2009). An early shape description was implemented in the MTD (minimal topological difference) algorithms, which describe ligand molecules interacting with a receptor with unknown structural topography (Simon et al. 1984). This was one of the first overlapping methods, and allows a description of the shape of receptor site by means of three types of vertices situated in the "cavity" (beneficial for the L – R interaction), wall (detrimental), and the "exterior" (irrelevant for affinity L – R). CoMFA (comparative molecular field analysis) algorithm (Cramer et al. 1988), which describes molecules by steric and electrostatic fields, was, probably, one of the most used grid-based methods of the molecular alignment and the calculations of molecular similarities. Other methods of huge interest for such areas as adsorption processes, chromatography, and various fundamental and applied physical-chemical properties include relevant works on topological indices (Wiener index, Randic index, and Balaban index), Weighted Holistic Invariant Molecular descriptors (WHIM) (Todeschini and Gramatica 1997), Shape Signatures (Zauhar et al. 2003), Path-Space Ratio (Edvinsson et al. 2003).

The development and application of molecular shape descriptors is an active area in computational chemistry and biology. The main goal of our work is to develop mathematical descriptors that can determine whether two molecules have comparable shapes. In this chapter we present a series of molecular shape descriptors developed on the basis of molecular vdW space. The molecules are treated in the "hard sphere" approximation, as a body composed from a collection of atomic fused spheres. Each sphere is centered in the corresponding nucleus and it is characterized by its Cartesian coordinates and by its vdW radius, r^W. These molecular vdW shape descriptors depend only on the internal structure of the molecule, being invariants to any translation and rotation movement. Consequently, they may inform us that two molecules have comparable shapes, but since they carry no information about the absolute orientation or position of the molecule, they are not useful for computing molecular superposition.

The essential step in any QSAR (quantitative structure – activity relationship, including in the term "activity" any biological, physical, chemical or toxicological properties) or virtual screening procedure consist in reducing the initial amount of information about the molecule to be investigated to a small set of molecular descriptors. It is very important to determine the essential and fundamental descriptors for the analyzed activity (Purvis 2008). Besides, these descriptors must have clear physical, chemical, or geometrical interpretation. Among the other basic reasons for using simple and concise size and shape descriptors is their ability to discard the molecular candidates before the thorough and expensive synthesis and activity studies have been undertaken (Zyrianov 2005).

In the last years, we developed some descriptors of molecular size and shape on the basis of standard or optimized geometry of molecules and the vdW space.

For this purpose we investigated the way in which the volume and surface of molecules or substituents, as well as some directions within the molecular vdW space – supposed to be homogeneous and isotropic, are responsible for the steric effects manifested during chemical organic reactions or in the course of interactions between ligand molecules and a given receptor or enzyme. Our attempts for deriving shape molecular descriptors of the vdW space of molecules and substituents were based on the molecular vdW volume and surface. With this end in view we developed a series of algorithms for computation of molecular vdW volume, V^W, and surface area, S^W (Ciubotariu et al. 1975, 1996, 2001a, b; Ciubotariu 1987) Once the parameters of molecular size (vdW volume, V^W, and surface, S^W) are defined this way, we introduced an additional group of molecular size descriptors that are the radii of spheres with the same values of V^W and S^W. The ovality Θ_{iD}, i = 1,2,3 molecular descriptors measure the deviation of the shape of a molecule from the spherical shape in one-, two-, and three dimensions, respectively, whilst the {δ,G} shape descriptors take into consideration the cylindrical shape of a molecule or substituent. These molecular shape descriptors are presented below, in this chapter, together with their application in the study of the toxicity of a series of aliphatic amines on the protozoan ciliate *Tetrahymena pyriformis* (Θ_{iD}) and for correlation the rate constants of the following chemical reactions (δ): bimolecular nucleophilic substitution, acid-catalyzed hydrolysis of esters and esterification reactions.

The molecular size and shape descriptors may indicate if a ligand molecule, L, does not fit the active site. Unfortunately, in many cases the receptor *site* cannot be described in simple terms of “large”, “width”, and “depth”. The MTD (minimal topologic difference) method allows a receptor site mapping in the frame of a series of bioactive compounds. The MVD (minimal volume difference) method is an improved variant of the MTD method, which takes into account the 3D extension of a molecule. These two methods are described here in some detail. The results obtained in the study of anti-carcinogenic activity of some retinoids and in the inhibition of carbonic-anhydrase (CA) by a series of sulfonamides are also presented in this chapter.

Although the shape is some characteristic of the molecule unrelated directly to its size, the common understanding of it still implies those metric relationships. The shape descriptors presented below should answer the following questions: how different is a molecule or substituent from a sphere or from a cylinder? It is more or less elongated? How different is the molecule when observed from different points of view?

15.2 Molecular Size

Let *S* be the *shape* for which we want to compute the *size*. From a mathematical point of view, we consider that *S* is a finite domain, not necessarily continuous, $S \subset R^n$. Let D be a well-known shape (like a sphere) which satisfies the following condition: $S \subset D \subset R^n$. One should be able to compute the size of D using a constant-time algorithm ($\tilde{O}(1)$), although this is not a strong requirement.

The most generic way to define S is to use a function $F : R^n \rightarrow \{0, 1\}$ that specifies whether a given point $\bar{x} \in R^n$ lies inside (1) or outside (0) our shape. Because $S \subset D \subset R^n$ we can safely define

$$F : D \rightarrow \{0, 1\} \quad F(\bar{x}) = \frac{0\ \bar{x} \in S}{1\ \bar{x} \notin S} \tag{15.1}$$

which describes S in a complete manner. The *size* of S will be

$$size(S) = \int_D F(\bar{x}) d\bar{x} \tag{15.2}$$

Because we are working in a discrete environment, having time and computer memory constraints, we can't use all possible points in D, but we will operate on a finite discrete subset of D, namely

$$M = \{\bar{x} | \bar{x} \in D\} \subset D \tag{15.3}$$

From these values one can select only those that fall inside S:

$$M' = M \cap S = \{\bar{x} \in M | \bar{x} \in \mathrm{S}\}$$

which, using $F(\bar{x})$, gives us the following expression:

$$M' = \{\bar{x} \in M | \mathrm{F}(\bar{x}) = 1\} \tag{15.4}$$

Because M and M' are discrete one can count their elements, and use these cardinalities (*card*) to calculate an approximate value for the size of the shape as follows,

$$\langle size(S) \rangle = \frac{card\ M'}{card\ \mathrm{M}} \cdot size(D) \tag{15.5}$$

Equation 15.5 shows that the precision of the integrator depends on *card M* and of the distribution of values within D. The error of the calculus cannot be easily determined (Vrânceanu et al. 1967).

15.2.1 Molecular van der Waals Volume – V^W

Volume and size are finite parameters of a molecule. The most common empirical references for the molecular volumes were the volumes estimated on the basis of vdW radii obtained from crystallographic studies (Bondi 1964). The distance where attractive forces between the unbounded atoms of a molecule or between atoms of different molecules are in equilibrium with repulsive forces is known as the van der Waals distance. One may define the van der Waals radius (r^w) as the half of the vdW distance. The van der Waals radii have long been considered as a measure of atomic size.

The space occupied by molecules can be conveniently described in the approximation of "hard spheres": each atom of the molecule M is represented by an isotropic sphere having the center in the equilibrium position (X_i, Y_i, Z_i) of the atomic nucleus i, and the radius equal with its van der Waals radius, r_i^w. A molecular van der Waals envelope, Γ, can be defined in the "hard-spheres" approximation as the external surface resulted from the interaction of all vdW spheres corresponding to the atoms of molecule M. The points (x, y, z) inside the envelope satisfy at least one of the following inequalities:

$$(X_i - x)^2 + (Y_i - y)^2 + (Z_i - z)^2 \leq (r_i^w)^2 \qquad i = \overline{1,m} \tag{15.6}$$

where m represents the number of atoms in M, and r_i^W is the Bondi vdW radius of atom i. Consequently, the total volume embedded by the envelope is the molecular vdW volume (V_M^w) of the molecule M.

Let it be a function F = f(x, y, z), continuous in a compact domain M. The integral:

$$I = \iiint_M f(x, y, z)\mathrm{dxdydz} \tag{15.7}$$

can be interpreted as a weight, f(x, y, z) being the repartition of density in space M. If we consider the density function$f(x, y, z) \equiv 1$, then the following integral:

$$V^w = \iiint_M dV; \qquad dV = \mathrm{dxdydz} \tag{15.8}$$

can be intuitively justified as a volume. This assumption is natural because the properties of molecular vdW space can be considered independent from the nature of the atoms, even in the case when domains of the vdW atomic spheres intersect.

To estimate the integral (15.8), the molecule is inserted into a bounding parallelepiped with the volume V_p. The random points are generated into the parallelepiped, which includes the domain M (Fig. 15.1). If n_t is the total number of generated points and n_s the number of points, which satisfy the inequalities in (15.6), then the van der Waals volume is:

$$V_M^w = \frac{n_s}{n_t} \cdot V_p \tag{15.9}$$

One of the main problems is to find out the volume of the intersection of the spheres, in order to avoid multiple computation of the same volume. One of the

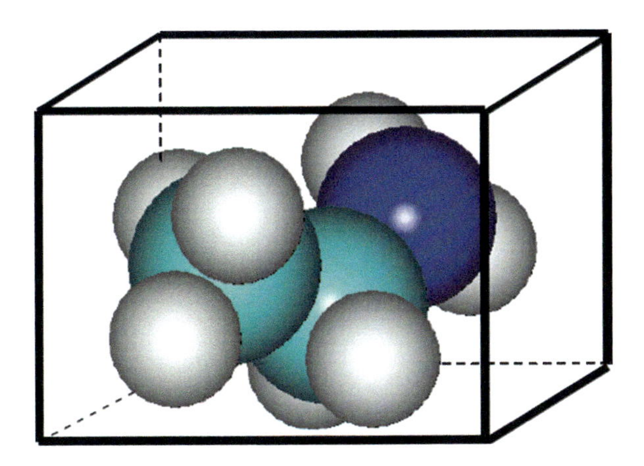

Fig. 15.1 Ethylamine molecule inserted into a bounding parallelepiped

approaches is based on the Monte Carlo methods (Ciubotariu et al. 1975, 1990, 1993; Ciubotariu 1987; Niculescu-Duvaz et al. 1991; Muresan et al. 1994). The accuracy ε of the estimate (15.9) for a given maximum probability is inversely proportional to the square root of the number of trials, or

$$\varepsilon = \frac{1}{2 \cdot \sqrt{\delta \cdot N}} \tag{15.10}$$

This circumstance causes the relatively slow convergence of the Monte Carlo methods. For example, in order to reduce the error of the result 10-fold, the number of trials must be increased 100-fold. If the accuracy of the estimate ε and the guarantee probability 1-δ are given, then from formula (15.10) one derives the necessary number of trials (Demidovich and Maron 1987):

$$N = \frac{1}{4\varepsilon\delta^2} \tag{15.11}$$

Taking into consideration the precision and the accuracy of chemical and biological experiments, for $\varepsilon = 0.05$ and $\delta = 0.01$, the number of necessary points is N = 10,000. This makes the Monte Carlo method not difficult to apply, due to the performances of nowadays computers. In order to increase the accuracy of the method the calculus must be repeated at least 10–20 times for each volume. The final result, the mean value of these computed volumes, is validated by statistical method.

15.2.2 Molecular van der Waals Surface – S^W

The van der Waals volume of the envelope Γ, defined in the previous paragraph, can be a measure of the molecules' size. Obviously, this envelope is a surface, and

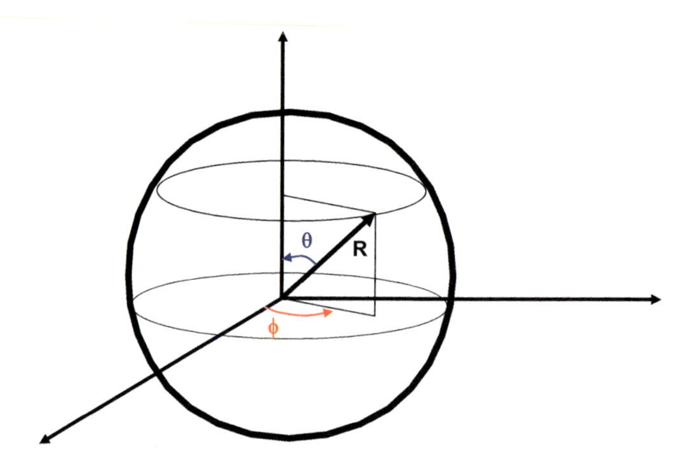

Fig. 15.2 The parametric representation of a sphere in 3D space

there were methods developed to compute the area of this surface (Pearlmann 1983; Meyer 1985; Gogonea et al. 1991). Some of them are based on a Monte Carlo method (Ciubotariu 1987), others on an analytical algorithm (Gogonea 1996). The computed surfaces were especially used to characterize the shape and the similarity of the molecules and their graphical representation (Cohen 1979; Gogonea 1996).

The molecules are treated within the hard sphere approximation (see previous section). Points (x, y, z) residing on vdW envelope Γ of a molecule M satisfy one of the equations (15.6) where (X_i, Y_i, Z_i) are the Cartesian coordinates of the *m* atoms belonging to molecule M, and r_i^W is the vdW radius of atom *i*.

Equation 15.6 was used together with the numerical integration technique to obtain accurate values of S^W. One determines the finite, countable sets ξ_i (whose elements $\xi_j(\xi_{jx}, \xi_{jy}, \xi_{jz})$ are points in the Euclidian 3D-space):

$$\xi_i = \{\xi_{ji} | P_1, P_2\}, i = 1, 2, ..., m$$

which satisfy the following properties:

(P_1): ξ_j (for all j) lie on the envelope of vdW spheres described by Eq. 15.6:

$$\xi_{jx} = R \cos\varphi \sin\theta, \quad \xi_{jy} = R \sin\varphi \sin\theta, \quad \xi_{jz} = R \cos\theta \tag{15.12}$$

where Eqs. 15.12 represent the parametric equations of sphere with $R = r^W$, and θ, φ defined in Fig. 15.2 ($\theta \in [-\pi/2, \pi/2]$ and $\varphi \in [0, 2\pi]$).

(P_2): ξ_j (for all j) are uniformly distributed random points in space. Next, one determines the sets ζ_i^M,

$$\zeta_i^M = \{\zeta_{ji} | P_3, P_4\}$$

whose elements $\zeta_{ji}(\zeta_{jix}, \zeta_{jiy}, \zeta_{jiz})$ satisfy the following properties:

$(P_3) : \zeta_i^M \subset \xi_i$

(P_4) : ζ_{ji} (for all j) satisfy at least one of Eqs. 15.6 with the supplementary conditions:

$$(X_k - \zeta_{jix})^2 + (Y_k - \zeta_{jiy})^2 + (Z_k - \zeta_{jiz})^2 > (r_k^W)^2 \tag{15.13}$$

for all k ≠ i and i, k = 1,m.

As S^W may be regarded as a non-negative continuous function in the closed bounded domain defined by the vdW envelope of each atomic sphere i, i = 1,m, the S^W value is estimated by

$$S^W \sum_{i=1}^{m} (card\zeta_i^M | card\xi_i) \cdot s_i^W \tag{15.14}$$

where s_i^W represents the surface of the corresponding atomic sphere i,

$$s_i^W = 4\pi (r_i^W)^2,$$

and *card* denotes the cardinal number.

ξ_i sets are constructed using either Monte Carlo (Demidovich and Maron 1987) or deterministic algorithms (Gogonea et al. 1981, 1991; Ciubotariu 1987). Within the first procedure $\xi_{ji} \in \xi_i$ is given by relations (15.12) in which

$$\theta = -\pi/2 + \pi \cdot \lambda_n, \quad \varphi = 2\pi \cdot \lambda_{n+1} \tag{15.15}$$

and $\{\lambda_n\}$ are uniformly distributed random sequences in the unit interval, generated using standard algorithms (Ciubotariu et al. 2006a, b, c). The deterministic algorithm simulates the stochastic approach (15.15), and consists in dividing the surface of each atomic sphere into $card\xi_i$ subspace (areas) whose centers $(\xi_{jx}, \xi_{jy}, \xi_{jz})$ define $\xi_{ji} \in \xi_i$.

The accuracy of the estimate (15.14), for given values of s_i^W, is proportional to$(card\xi_i)^{-1/2}$; this leads to relatively slow convergence of the algorithm and imposes the use of ξ_i sets with high cardinality.

The number of points generated on each sphere can be calculated as follows,

$$n^{(k)} = \frac{4\pi (r_i^W)^2}{g^2}; k = 1, m \tag{15.16}$$

where $n^{(k)}$ denotes $card\xi_i$ from (15.15) and g is the distance between grid points lying on vdW surface; each point contribution to the area of vdW surface is g^2.

Relation (15.16) establishes a relation between the grid (g) generated on each vdW sphere and the number of random points $\left(\xi_{jix}, \xi_{jiy}, \xi_{jiz}\right)$ which has to be generated to guarantee the uniformity pf the sphere network.

15.3 Cylindrical Descriptors of Molecular Shape

Although in the process of quick recognition of major trends and dominant common features, a somewhat imprecise shape perception based on intuitive, subjective shape concepts are useful, the needs for clearly defined shape measures and for precise and accurately shape evaluation are evident if details of shape are important and if it is not well understood which shape characteristics are potentially responsible for a given molecular property (Mezey 1993).

The measures of linear elongation of a molecule along the three orthogonal directions as described in the previous section may give some representation of a molecule shape, and, also, the size of that molecule.

Another approach is to use an expected value of radius averaged in a certain way. Let r_0 be a center of a molecule, also defined in certain way. Let r_i be a vector from r_0 to point i on the surface of a molecule. If r_0 is a center of mass supposing an uniform distribution of mass within the boundaries of the molecule and vectors r_i are being chosen randomly and with probability density uniformly distributed along all directions in the space, expectation of $|r|$ is defined probabilistically as average length of r_i's determined in the infinite number of experiments. This may be written as the following limit (Zyrianov 2005):

$$r = \lim_{n \to \infty} \frac{1}{n} \sum_{i} |r_i| \tag{15.17}$$

The expectation of a molecular radius calculated by averaging the radius-vector within all the volume of the molecule is:

$$r_V = \frac{1}{V} \iiint_{V} r \cdot dV \tag{15.18}$$

The development of cylindrical descriptors of molecular shape (δ, G) has been made only for the substituents, and on the basis of the following assumptions (Ciubotariu 1987):

(i) The substituents are treated according to hard sphere model, as a collection of atomic fused vdW spheres. Consequently, a relatively impenetrable vdW envelope Γ may be uniquely defined by the outer surface of the intersected vdW spheres.
(ii) The vdW molecular space embedded by Γ is supposed homogeneous, but not sterically isotropic.

Table 15.1 Values of (δ,G) vdW descriptors for some substituents, X

X	δ_i	G_i	δ_e	G_e	X	δ	G
Et	1.420	4.876	1.625	4.276	Me	1.244	3.400
n-Pr	1.628	5.799	2.187	4.418	FCH_2	1.437	3.474
i-Pr	2.332	4.123	2.748	3.506	$ClCH_2$	1.473	4.387
n-Bu	1.656	7.409	2.121	5.622	$BrCH_2$	1.527	4.688
i-Bu	2.786	4.416	2.234	5.453	ICH_2	1.598	5.059
s-Bu	3.133	3.918	2.689	4.585	F_2CH	1.701	3.379
t-Bu	3.605	3.399	3.593	3.399	Cl_2CH	2.453	3.523
n-Am	1.811	8.318	2.640	5.492	Br_2CH	2.776	3.609
s-$BuCH_2$	3.267	4.589	2.570	5.772	I_2CH	3.265	3.712
i-$BuCH_2$	2.627	5.667	2.113	6.703	F_3C	1.889	3.400
Et_2CH	3.617	4.059	3.798	3.835	Cl_3C	3.177	3.400
c-$HxCH_2$	2.553	7.279	2.687	6.928	Br_3C	3.794	3.400
$PhCH_2$	2.334	6.727	2.311	6.771	I_3C	4.668	3.400
$Ph(CH_2)_2$	2.106	8.787	2.353	7.868	H_2N	1.055	3.083
Ph(Me)CH	2.600	7.088	2.631	7.252	O_2N	1.604	3.190
Ph(Et)CH	2.991	6.985	3.088	6.653	HO	0.898	3.017
MeO	1.209	4.558	1.303	4.148	CN	0.991	4.410
EtO	1.455	5.608	1.900	4.160	COOH	1.409	4.184

Subscript *i* refers to intercalated conformer in its anti conformation, and *e* refers to eclipsed conformer

(iii) The substituent shows a rotational symmetry round the axis linking the molecular gravity center and the contact point of the substituent with the parent structure at which it is attached.

Taking into consideration the above suppositions, we can consider the steric anisotropy of molecular vdW space along the two directions, i.e. (1) the direction of rotation axis (G), and (2) a direction perpendicular to it (δ). G measures the dimension of the substituent on the direction of rotation axis. For estimating the value of δ, we supplementary supposed that

(iv) the vdW space of a substituent is relatively compressible to a cylinder having a volume equal to the vdW volume of that substituent, V^W.

Center of gravity corresponds to the center of vdW space of a molecule and also to the mass center. The supposed uniform distribution of mass avoids a direct dependence of the calculated shape descriptors on the chemical composition of a molecule.

In Table 15.1 are summarized the values of (δ, G)-descriptors for a series of substituents.

The development of these molecular descriptors have been based on the physical model of transition states in acid-catalyzed esterification reaction of carboxylic acids and alcohols and acid hydrolysis of esters – standard reactions used by Taft for the development of E_S's empirical steric parameter in the frame of LFER (linear free energy relationship). The physical meaning of the (δ, G) shape descriptors is depicted in Fig. 15.3.

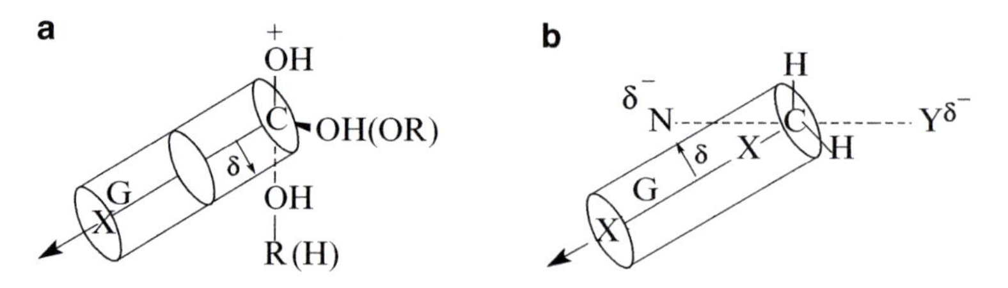

Fig. 15.3 Physical significance of (δ,G) parameters for the transition states in reactions: (**a**) acid-catalyzed hydrolysis of esters and esterification of carboxylic acids with alcohols XCOOH + ROH; (**b**) nucleophilic bimolecular substitution, S_N2

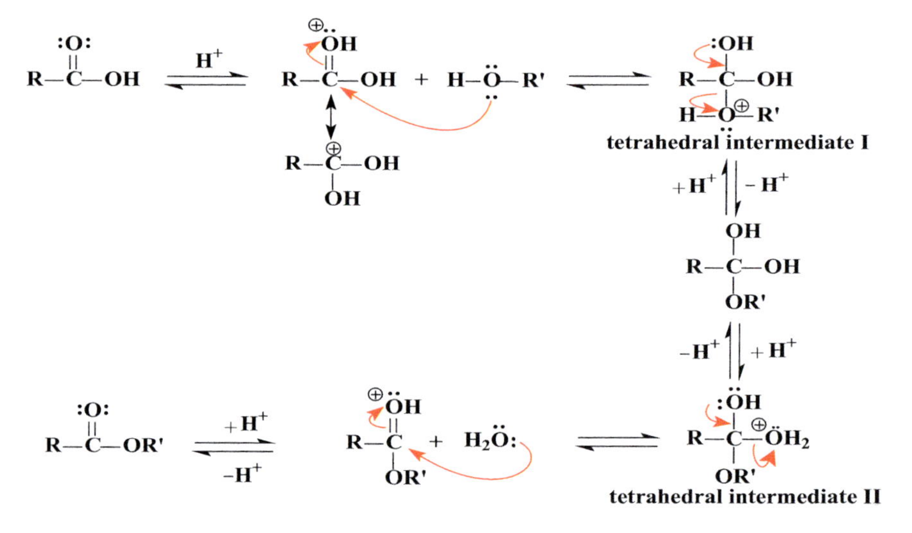

Fig. 15.4 Mechanism for acid-catalyzed ester hydrolysis and carboxylic acid hydrolysis

One may observe from this figure that δ-value measure a radius of a cylinder having a volume equivalent to the vdW volume of a given substituent, while G represents the generating line of this cylinder, which is the length of the substituent upon the rotational axis.

The δ-parameter was used with good results in structure-activity correlations for modeling the acid-catalyzed carboxylic acid esterification and ester hydrolysis, and bimolecular substitution reactions (Chiriac et al. 1996).

The mechanism for the acid-catalyzed reaction of a carboxylic acid and an alcohol to form an ester is the exact reverse of the mechanism for the acid-catalyzed hydrolysis of an ester to form a carboxylic acid and an alcohol – see Fig. 15.4. H^+ increases the rate of formation of tetrahedral intermediates I and II (Fig. 15.4) by protonating the carbonyl oxygen. Protonated carbonyl groups are more susceptible than non-protonated carbonyl groups to nucleophilic attack, because positively charged oxygen is more electron withdrawing than uncharged oxygen. But, H^+ increases also the rate

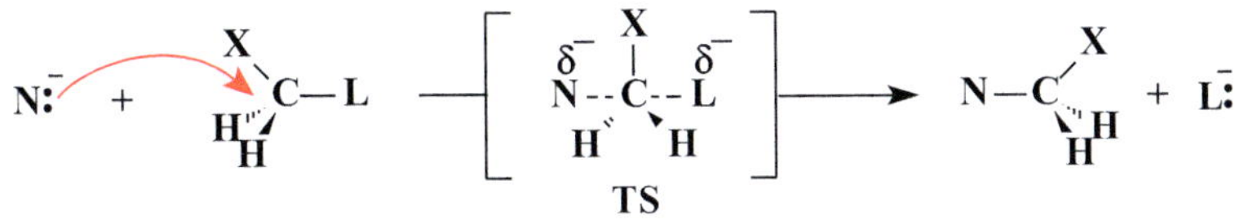

Fig. 15.5 Mechanism of bimolecular nucleophilic substitution reaction, S_N2

of collapse of a tetrahedral intermediates I and II (Fig. 15.4) by decreasing the basicity of the leaving group (ROH and H_2O), thereby making it easier to eliminate.

The general form of a nucleophilic substitution reaction is as follows,

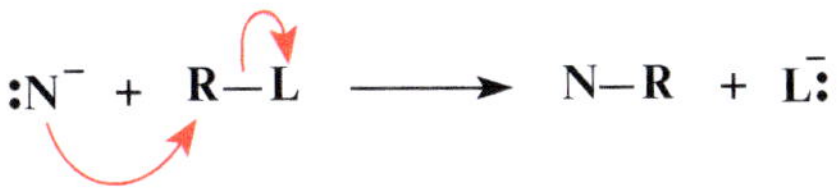

$$:N^- + R{-}L \longrightarrow N{-}R + L:^-$$

where N is a nucleophilic agent, L is a leaving group and R represents a substituent. The rate of a bimolecular nucleophilic substitution reaction, S_N2, depends on the concentrations of both reagents: the nucleophile, N, and the substrate, R–L.

The mechanism proposed by Hughes and Ingold for an S_N2 reaction has one step. The nucleophile attacks the carbon bearing the leaving group and displaces the leaving group. Because the nucleophile hits the carbon on the side opposite to the side bonded to the leaving group, the carbon is said to undergo *back side attack*. Thus, the nucleophile displaces the leaving group in a single step. The mechanism shows the substrate R–L and the nucleophile coming together in the transition state (TS) of the one-step reactions – see the Fig. 15.5.

We used the δ structural descriptor for correlation the rate of the ester hydrolysis, esterification, and S_N2 reactions by means of linear equations as follows:

$$\log k = \alpha + \beta \cdot \delta \tag{15.19}$$

where k represents the rate constant and α and β are parameters of the linear model (15.19) obtained by least square method. The values of rate constants k are given in Table 15.2 and the results obtained by statistical regression are summarized in Table 15.3.

The values of δ-parameter reported in Table 15.1 have been obtained for ground state geometry of alkyl substituents. The δ-parameters quantify very well the steric effects and their physical meaning is very clear (see Fig. 15.3 and Tables 15.1 and 15.3). The δ-values remain approximately constant when the number of Carbon atoms of the substituents (in extended conformations) is increasing, but the δ-values are increasing with the branching degree of the radical alkyl, R.

On the other hand, G-parameter, representing the length of the substituent in the direction of the rotational axis, could be very useful in QSAR studies to detect the steric requirements of the (eventually) hydrophobic pockets of the receptor site. Therefore, the quantification of steric interactions by a set of only two parameters

Table 15.2 Rate constants (k) for (a) esterification and acid-catalyzed hydrolysis of esters (Charton 1975a), and (b) bimolecular nucleophilic substitution reactions (Charton 1975b)

Set	(a) values of k for esterification and acid-catalyzed hydrolysis of esters
1	k, XCO_2H + MeOH in MeOH at 15°C, catalyzed by HCl: H = 1124.0; Me = 104.00; CCl_3 = 0.969
2	k, XCO_2H + EtOH in EtOH at 14.5°C, catalyzed by HCl: Me = 3.661; t-Bu = 0.0909; CCl_3 = 0.0371; CBr_3 = 0.0135
3	10^4k, XCO_2H + EtOH in EtOH at 25°C: Me:148; t-Bu:2.19; CCl_3:1.14
4	10^3k, XCO_2H + MeOH in MeOH at 20°C, catalyzed by HCl: H = 632; Me = 44; t-Bu = 1.4
5	10^3k, XCO_2H + MeOH in MeOH at 30°C, catalyzed by HCl: H = 1100; Me = 81.4; t-Bu = 2.68
6	10^3k, XCO_2H + MeOH in MeOH at 40°C, catalyzed by HCl: H = 1730; Me = 132; t-Bu = 4.93
7	10^5k, Et_2OCX + H_2O in 70% MeAc-H_2O at 24.8°C, catalyzed by HCl: Me = 4.470; Et = 3.700; Pr = 1.960; i-Pr = 1.460; Bu = 1.790; i-Bu = 0.572; t-Bu = 0.128; $BuCH_2$ = 1.770; $PhCH_2$ = 1.580
8	10^5k, Et_2OCX + H_2O in 70% MeAc-H_2O at 35.0°C, catalyzed by HCl: Me = 10.9; Et = 9.24; Pr = 4.83; i-Pr = 3.43; Bu = 4.45; i-Bu = 1.46; t-Bu = 0.36; $BuCH_2$ = 4.3; $PhCH_2$ = 3.84
9	10^5k, Et_2OCX + H_2O in 70% MeAc-H_2O at 20°C, catalyzed by HCl: Me = 24.7; Et = 20.7; Pr = 10.8; i-Pr = 7.46; Bu = 10.2; i-Bu = 3.3; t-Bu = 1.1; $BuCH_2$ = 9.76; $PhCH_2$ = 8.82
10	10^5k, Et_2OCX + H_2O in 70% MeAc-H_2O at 25°C, catalyzed by HCl: Me = 4.55; Et = 3.77; Pr = 1.86; Bu = 1.82; $PhCH_2$ = 1.47; $PhCH_2CH_2$ = 1.28; PhCHMe = 0.277; PhCHEt = 0.11; c-$C_6H_{11}CH_2$ = 0.481
11	10^2k, XCO_2H + MeOH in MeOH at 50°C, catalyzed by HCl: Me = 21.9; Et = 19.3; Pr = 10.3; i-Pr = 7.27; Bu = 10.1; i-Bu = 2.48; s-Bu = 2.19; t-Bu = 0.858; $BuCH_2$ = 10.2; s-$BuCH_2$ = 2.4;; $PhCH_2$ = 9.44; $PhCH_2CH_2$ = 9.2; PhCHEt = 1.05; c-$C_6H_{11}CH_2$ = 2.79; Et_2CH = 0.253
12	10^2k, XCO_2H + MeOH in MeOH at 25°C, catalyzed by HCl: Me = 5.93; Et = 5.73; i-Pr = 1.95; t-Bu = 0.194; $PhCH_2$ = 2.62
13	10^6k, XCO_2Et + H_2O in 60%EtOH-H_2O v/v at 60.04°C: Me = 785; CH_2Cl = 619; HCl_2 = 270.4; CCl_3 = 7.61
14	10^3k, XCO_2-2-$C_{10}H_8$ + MeOH in MeOH at 25°C, catalyzed by HCl: Me = 378; Et = 238; Pr = 140; i-Pr = 54.9; t-Bu = 2.71; $BuCH_2$ = 116
15	10^3k, XCO_2-2-$C_{10}H_8$ + EtOH in EtOH at 25°C, catalyzed by HCl: Me = 91.2; Et = 50.7; i-Pr = 8.58
16	10^3k, XCO_2-2-$C_{10}H_8$ + PrOH in PrOH at 25°C, catalyzed by HCl: Me = 87.6; Et = 45.8; Pr = 24; i-Pr = 7.3; t-Bu = 0.21; $BuCH_2$ = 23.3
17	10^3k, XCO_2-2-$C_{10}H_8$ + PrOH in PrOH at 40°C, catalyzed by HCl: Me = 233; Pr = 69.2; t-Bu = 0.725; $BuCH_2$ = 64.4
18	10^3k, XCO_2-2-$C_{10}H_8$ + i-PrOH in i-PrOH at 25°C, catalyzed by HCl: Me = 5.43; Et = 2.62; Pr = 1.36; i-Pr = 0.3; t-Bu = 0.00595; $BuCH_2$ = 1.24
19	10^3k, XCO_2-2-$C_{10}H_8$ + i-PrOH in i-PrOH at 40°C, catalyzed by HCl: Me = 15.1; Et = 7.77; Pr = 3.8; i-Pr = 0.849; t-Bu = 0.026; $BuCH_2$ = 3.55
20	10^3k, $XCO_2C_6H_4NO_2$ + H_2O in H_2O at 30°C: Me = 7.25; Et = 7.45; Pr = 4.33; i-Pr = 4.33; i-Bu = 2.48; t-Bu = 0.884

(continued)

Table 15.2 (continued)

Set	(b) values of k for bimolecular nucleophilic substitution reaction
1	10^5k_2, X–CH_2–Br + LiCl in MeAc at 25°C: H = 600; Me = 9.88; Et = 6.45; i-Pr = 1.53; t-Bu = 0.00026
2	10^5k_2, X–CH_2–Br + LiBr in MeAc at 25°C: H = 13; Me = 170; Et = 110; i-Pr = 5.7; t-Bu = 0.00026
3	10^5k_2, X–CH_2–Br + LiI in MeAc at 25°C: H = 25; Me = 166; Et = 137; i-Pr = 6; t-Bu = 0.002
4	10^5k_2, X–CH_2–I + LiCl in MeAc at 25°C: H = 468; Me = 42; Et = 24.6; i-Pr = 1.62; t-Bu = 0.00058
5	10^3k_2, X–CH_2–Br + NaOEt in EtOH at 55°C: H = 34.4; Me = 1.95; Et = 0.547; Pr = 0.396; i-Pr = 0.058; Bu = 0.357; t-Bu = 0.00000826
6	10^3k_2, X–CH_2–Br + NaOEt in EtOH at 95°C: H = 965; Me = 64.7; i-Pr = 2.62; t-Bu = 0.000649
7	k, X–CH_2–I + sodium eugenoxid in EtOH at 68°C: Me = 2.5; Et = 1.04; Pr = 0.94; i-Pr = 0.34; i-PrCH_2 = 0.51; Bu = 0.87
8	10^4k_2, X–CH_2–Br + $Na_2S_2O_3$ in 50%EtOH-H_2O at 12.5°C: H = 285; Me = 2.43; Et = 1.266; i-Pr = 0.0459
9	10^5k_2, X–CH_2–I + Bu_3P in MeAc at 34.97°C: H = 26; Me = 154; Et = 63.7; Pr = 56.8; i-Pr = 4.94
10	10^4k_2, X–CH_2–Br + NaOMe in MeOH at 80°C: H = 814; Me = 90.6; Et = 33.5; Pr = 33.4; i-Pr = 6.75
11	10^4k_2, X–CH_2–Br + NaOPh in MeOH at 80°C: H = 284; Me = 62.1; Et = 28.9; Pr = 30.8; i-Pr = 5.12
12	10^6k_2, X–CH_2–Br + $PhNMe_2$ in MeOH at 80°C: H = 27.8; Me = 178; Et = 86.7; Pr = 90.3; i-Pr = 3.1
13	10^6k_2, X–CH_2–Br + pyridine in MeOH at 80°C: H = 27.1; Me = 152; Et = 84.2; i-Pr = 2.54
14	10^5k_1, X–CH_2–OTs + AcOH in AcOH at 70°C: H = 0.0516; Me = 0.0442; Et = 0.0355; i-Pr = 0.0179; t-Bu = 0.00417

(δ and G), which are easily calculable, may be very useful for structure – activity studies, especially when the ligand – receptor steric interaction is important for the expressed biological activity.

Analysis of the δ-parameter for alkyl substituents in reactivity correlation data suggests its possible utility in quantitative treatment of steric effects in organic reactions (see, for example, the results in Table 15.3).

15.4 Ovality Descriptors of Molecular Shape

Assuming that a molecule can be characterized by two spheres, corresponding to the vdW volume, V^w, and to the vdW surface, S^w, respectively, and taking into account the fact that for a given volume the spherical shape presents the minimum surface, we extended the ovality measure reported in the literature (Bodor et al. 1989; Todeschini and Consonni 2000) to three ovality measures, Θ_{iD}, i = 1,2,3 (Ciubotariu et al. 2006b, c). This hypothesis is based on the known conformational

Table 15.3 Results of correlation log k vs. δ: (a) esterification and acid-catalyzed hydrolysis of esters; (b) bimolecular nucleophilic substitution

Set	r	a	b	s	F	r_{adj}^2	n
(a)							
1	−0.9497	4.1421	−1.3115	0.6903	9.2017	0.8040	3
2	−0.9477	1.5964	−0.8600	0.4147	17.6454	0.8473	4
3	−0.9568	3.1902	−0.8725	0.4714	10.8409	0.8311	3
4	−0.9074	3.2997	−0.8777	0.7911	4.6606	0.6467	3
5	−0.9089	3.5390	−0.8657	0.7727	4.7529	0.6523	3
6	−0.9059	3.7106	−0.8399	0.7641	4.5746	0.6412	3
7	−0.9617	1.3707	−0.5898	0.1355	86.1028	0.9141	9
8	−0.9660	1.7233	−0.5685	0.1226	97.6640	0.9236	9
9	−0.9696	2.0026	−0.5264	0.1070	109.9444	0.9316	9
10	−0.9449	1.7434	−0.8356	0.1865	58.3248	0.8775	9
11	−0.9317	2.2153	−0.6540	0.2075	85.5756	0.8580	15
12	−0.9723	1.6786	−0.6278	0.1637	51.9691	0.9272	5
13	−0.9167	4.2430	−0.9551	0.4569	10.5278	0.7605	4
14	−0.9952	3.6569	−0.8807	0.0838	416.0111	0.9881	6
15	−0.9960	3.0508	−0.9115	0.0676	123.8801	0.9840	3
16	−0.9954	3.2454	−1.0744	0.1005	430.8433	0.9885	6
17	−0.9980	3.6432	−1.0492	0.0843	511.2897	0.9942	4
18	−0.9969	2.2227	−1.2213	0.0941	634.0398	0.9922	6
19	−0.9977	2.5500	−1.1426	0.0758	855.6431	0.9942	6
20	−0.9598	1.3487	−0.3646	0.1090	46.7464	0.9015	6
(b)							
1	−0.9433	4.4524	−2.1508	0.8982	24.2326	0.8531	5
2	−0.9396	4.7611	−2.1680	0.9371	22.6190	0.8439	5
3	−0.9558	4.4573	−1.8827	0.6875	31.6962	0.8847	5
4	−0.9800	4.7937	−2.1744	0.5245	72.6240	0.9471	5
5	−0.9702	3.6015	−2.3550	0.5498	80.2115	0.9296	7
6	−0.9798	5.3504	−2.3118	0.6555	48.0261	0.9400	4
7	−0.8388	0.7013	−0.4194	0.1799	9.4960	0.6295	6
8	−0.8179	4.1342	−2.4100	1.1016	1.1016	0.5035	4
9	−0.8075	3.1109	−0.9841	0.3826	5.6203	0.5360	5
10	−0.8355	3.9188	−1.3865	0.4852	6.9382	0.5975	5
11	−0.8982	3.5031	−1.2283	0.3199	12.5224	0.7423	5
12	−0.8064	3.5108	−1.2110	0.4726	5.5771	0.5336	5
13	−0.8778	3.4982	−1.2988	0.4605	6.7181	0.6559	4
14	−0.9958	−0.7940	−0.4335	0.0471	357.6054	0.9889	5

The statistical indicators in table are: r – correlation coefficient; s – standard deviation; F – Fisher test; r^2_{adj} – explained variance; n – number of points in data set

flexibility of the molecules and on the fact that the molecules are relatively compressible (Ciubotariu 1987). Therefore, one may suppose that a molecule should be compressed from the greatest sphere (S_G), corresponding to the vdW surface area of Γ, equal to S^w, to the smallest sphere (S_S), concordant to the vdW volume embedded by Γ, equal to V^w (Ciubotariu et al. 2006a, b, c; Vlaia et al. 2009; Vlaia 2010).

The vdW radius, r_S^w, and the vdW volume, V_S^w, of the molecular S_G sphere are calculated as follows,

$$r_S^w = [S^w/4\pi]^{1/2} \quad (15.20)$$

$$V_S^w = 4\pi(r_S^w)^3/3 \quad (15.21)$$

The molecular S_G sphere can be compressed to a molecular S_S sphere, which has a volume equal to the molecular van der Waals volume, V^w. The vdW radius, r_V^w, and the vdW surface area S_V^w of this molecular S_S sphere are calculated with the following relations:

$$r_V^w = [3V^w/4\pi]^{1/3} \quad (15.22)$$

$$S_V^w = 4\pi(r_V^w)^2 \quad (15.23)$$

In this way, the molecular S_G and S_S spheres are characterized by the following two triplets:

$$\{S_G\} : \left(r_S^w, S^w, V_S^w\right) \quad (15.24)$$

$$\{S_S\} : \left(r_V^w, S_V^w, V^w\right) \quad (15.25)$$

The molecular vdW ovality measures, Θ_{iD}, i = 1,2,3 can be easily defined from the triplets (15.24) and (15.25), as the ratio between the corresponding values of vdW radius, surface area, and volume of S_G sphere and S_S sphere, as follows,

$$\Theta_{1D} = \frac{r_S^w}{r_V^w} \quad (15.26)$$

$$\Theta_{2D} = \frac{S^W}{S_V^W} \quad (15.27)$$

$$\Theta_{3D} = \frac{V_S^W}{V^W} \quad (15.28)$$

$V^w$and S^w were calculated with computer package IRS (investigation of receptor space; http://irs.cheepe.homedns.org/) (Ciubotariu et al. 2006a). The geometry of molecules was optimized with MM + and AM1 algorithms from the HyperChem software package (HyperChem version 7.0, HyperCube Inc.).

The original ovality index, O, was introduced (Bodor et al. 1989) as a measure of deviation of a molecule from the spherical shape. It was calculated from the ratio between the actual molecular surface area, S^w, and the minimum surface area,

S_V, corresponding to the actual molecular vdW volume, V^w, (Todeschini and Consonni 2000; Bodor et al. 1998) as follows:

$$O = \frac{S^w}{S_V} = \frac{S^w}{4\pi(r^w)^2} = \frac{S^w}{4\pi \cdot \left(\frac{3 \cdot V^w}{4\pi}\right)^{2/3}} \quad (15.29)$$

In relation (15.29) r^w represents the vdW radius of the given molecule calculated from its actual vdW volume. The ovality index is equal to 1 for spherical top molecules characterized by a more or less spherical shape, and increases with increasing the elongation of the molecule, with oval or elongated shape.

In fact, the reciprocal of the ovality index, $\Psi = O^{-1}$, was introduced (Wadell 1935) before, in 1935, as sphericity index, to measure how spherical (or round) an object is. The sphericity, Ψ, is the ratio of the surface area of a sphere (with the same volume as the given object) to the surface area of the object.

One may observe that the relations (15.27) and (15.29) are the same. Consequently, the two-dimensional (2D) ovality molecular vdW descriptor, Θ_{2D}, is identical with the ovality index, O. The relations (15.30) and (15.32) extend the index O so that one can also measure the deviation of a molecule from the spherical shape on one- (1D) and on three-dimensions (3D) of the vdW space.

Ovality descriptors presented here have been applied to study the toxic activity of various compounds including alcohols, esters, carboxylic acids, amines (Vlaia et al. 2009; Vlaia 2010). Here we present a toxicological QSAR study realized on a series of 28 amines.

One knows that aliphatic amines belong to a class of compounds which are largely synthesized because they show various biological activities. They are also used as intermediaries in pharmaceutical and chemical industry for obtaining very useful chemicals such as the colorants and drugs. Therefore, their toxicities on the humans or environment have been investigated. These studies reveal a narcotic toxicity of amines, greater than that of non-polar narcotics, because at physiological pH they lie predominantly in the protonated ionic form, which presents a greater affinity for cellular membrane (Vlaia 2010).

The measured toxicity is represented by the molar concentration that produces a 50% growth inhibition (IGC_{50}) of the protozoan ciliate *Tetrahymena pyriformis*. The biological activities, $A = \log(1/IGC_{50})$, together with the values of ovality descriptors Θ_{iD} (i = 1,2,3) are systematized in Table 15.4.

The linear models A *vs.* Θ_{iD} (i = 1,2,3) are the following:

$$\hat{A} = -7,1927(\pm 0,5505) + 4,2275(\pm 0,3376) \cdot \Theta_{3D}$$
$$n = 28; s = 0,347; R = 0.926; r^2_{adj} = 0.852; F = 157 \quad (15.30)$$

$$\hat{A} = -10,7662(\pm 0,8626) + 7,5594(\pm 0,6239) \cdot \Theta_{2D}$$
$$n = 28; s = 0,357; R = 0.922; r^2_{adj} = 0.844; F = 147 \quad (15.31)$$

Table 15.4 Biological activity, A_i, and ovality descriptors, Θ_{iD} (i = 1,2,3) for a series of 28 amines

No.	Amines	$A = \log(1/IGC_{50})$	Θ_{1D}	Θ_{2D}	Θ_{3D}
1	1,1-Dimethylpropargilamine	−0.9104	1.150	1.323	1.522
2	1,2-Dimethylpropylamine	−0.7095	1.161	1.348	1.564
3	1,3-Diaminopropane	−0.7045	1.150	1.324	1.523
4	1-Dimethylamino-2-propine	−1.1451	1.147	1.316	1.507
5[a]	1-Ethylpropylamine	−0.8129	1.171	1.371	1.599
6[a]	1-Methylbutylamine	−0.6846	1.171	1.371	1.607
7	2-Methylbutylamine	−0.4774	1.168	1.363	1.586
8[a]	Amylamine	−0.4848	1.180	1.393	1.642
9	Butylamine	−0.5735	1.156	1.336	1.543
10[a]	Decylamine	2.0555	1.276	1.627	2.071
11	Heptylamine	0.2109	1.222	1.496	1.825
12	Hexylamine	−0.2197	1.202	1.444	1.736
13	Isoamylamine	−0.5774	1.169	1.366	1.597
14	Isobutylamine	−0.2616	1.146	1.314	1.503
15	Isopropylamine	−0.8635	1.127	1.271	1.430
16	n,n-Diethylmethylamine	−0.7559	1.165	1.358	1.579
17[a]	n,n-Dimethylethylamine	−0.9083	1.145	1.311	1.493
18	n-Methylbutylamine	−0.6784	1.180	1.393	1.644
19	n-Methylpropargylamine	−0.9818	1.131	1.279	1.448
20	n-Methylpropylamine	−0.8087	1.156	1.336	1.546
21	Nonylamine	1.7011	1.259	1.586	1.994
22	Octylamine	0.3509	1.241	1.540	1.913
23[a]	Propargylamine	−0.8260	1.100	1.210	1.326
24	Propylamine	−0.7075	1.129	1.275	1.442
25	Secbutylamine	−0.6708	1.145	1.313	1.496
26	Tert-amylamine	−0.6978	1.157	1.339	1.543
27[a]	Tert-butylamine	−0.8973	1.150	1.322	1.519
28	Undecylamine	2.3279	1.291	1.668	2.144

[a]Compounds randomly selected for test set

$$\hat{A} = -21,4710(\pm 1,8372) + 18,0081(\pm 1,5651) \cdot \Theta_{1D}$$
$$n = 28; s = 0,373; R = 0.914; r_{adj}^2 = 0.830; F = 132 \quad (15.32)$$

One observe that the quality of the above linear models slightly decreases in the order $\Theta_{3D} > \Theta_{2D} > \Theta_{1D}$. Similar results were obtained when testing the stability of the models (15.30) – (15.32) – see the values of the corresponding statistics in Table 15.5.

The obtained results in Table 15.5 are fairly good, the prediction coefficients being greater than 0.500 (q^2 values). The predictive ability of these models decreases also as their fitting power.

The external validation of the models (15.30)–(15.32) was made by (75%/25%) randomization of the series under study in two sets, namely the training set and the test set. In this way, the training set was composed of 21 amines randomly selected,

Table 15.5 Cross validation results of models (15.30)–(15.32)

Model	Descriptors	q^2_{LOO}	q^2_{BOOT}	SDEP	SDEC	r^2_{Y-s}	q^2_{Y-s}
15.30	Θ_{3D}	0.817	0.776	0.379	0.334	−0.025	−0.167
15.31	Θ_{2D}	0.806	0.758	0.390	0.344	−0.031	−0.194
15.32	Θ_{1D}	0.787	0.739	0.409	0.359	−0.016	−0.182

Table 15.6 Cross validation results of models (15.33)–(15.35)

QSTR model	Descriptors	q^2_{LOO}	q^2_{BOOT}	q^2_{ext}	SDEP	SDEC	r^2_{Y-s}	q^2_{Y-s}
15.33	Θ_{3D}	0.805	0.742	0.775	0.374	0.309	0.042	−0.162
15.34	Θ_{2D}	0.797	0.730	0.753	0.381	0.314	0.015	−0.222
15.35	Θ_{1D}	0.780	0.714	0.727	0.397	0.327	−0.012	−0.191

and the remainder 7 amines make up the test set. The toxicological QSAR models are as follows:

$$\hat{A} = -7,1566(\pm 0,6178) + 4,1994(\pm 0,3781) \cdot \Theta_{3D}$$
$$n = 21; s = 0,325; R = 0.931; r^2_{adj} = 0.860; F = 123 \qquad (15.33)$$

$$\hat{A} = -10,7273(\pm 0,9570) + 7,5245(\pm 0,6913) \cdot \Theta_{2D}$$
$$n = 21; s = 0,331; R = 0.928; r^2_{adj} = 0.855; F = 119 \qquad (15.34)$$

$$\hat{A} = -21,4933(\pm 2,0340) + 18,0191(\pm 1,7315) \cdot \Theta_{1D}$$
$$n = 21; s = 0,344; R = 0.922; r^2_{adj} = 0.843; F = 108 \qquad (15.35)$$

Statistical indicators obtained by the internal validation of linear models (15.33)–(15.35) corresponding to the training set are given in Table 15.6.

The results summarized in Tables 15.5 and 15.6 prove that the statistical models (15.30)–(15.32) well predict the biological activity A, i.e. the toxicity of amines. The best molecular shape descriptor is Θ_{3D}.

The conditions proposed (Golbraikh and Tropsha 2002) for verifying the external validation are the following:

- $q^2 > 0,5$
- $r^2 > 0,6$
- $\dfrac{r^2 - r_0^2}{r^2} < 0,1$ and $0,85 \leq k \leq 1,15$
- $\dfrac{r^2 - r'^2_0}{r^2} < 0,1$ and $0,85 \leq k' \leq 1,15$

As a matter of fact these conditions suppose the line correlating the experimental and predicted activities is going by the origin of the coordinate system, and the slope (k, k') must be close to 45°. The predictor variables have to explain at least 60% of the variance of experimental data; this is required to assure the stability of the proposed model.

Table 15.7 External validation of models (15.30)–(15.32)#

QSTR model	Descriptors	r^2	$(r^2-r_0^2)/r^2$	$(r^2-r'^2_0)/r^2$	k	k′
15.30	Θ_{3D}	0.842	0.0029	0.0320	1.0051	0.8541
15.31	Θ_{2D}	0.827	0.0021	0.0151	1.0015	0.8449
15.32	Θ_{1D}	0.807	0.0015	0.0161	0.9887	0.8388

see above the conditions for external validation (Golbraikh and Tropsha 2002)

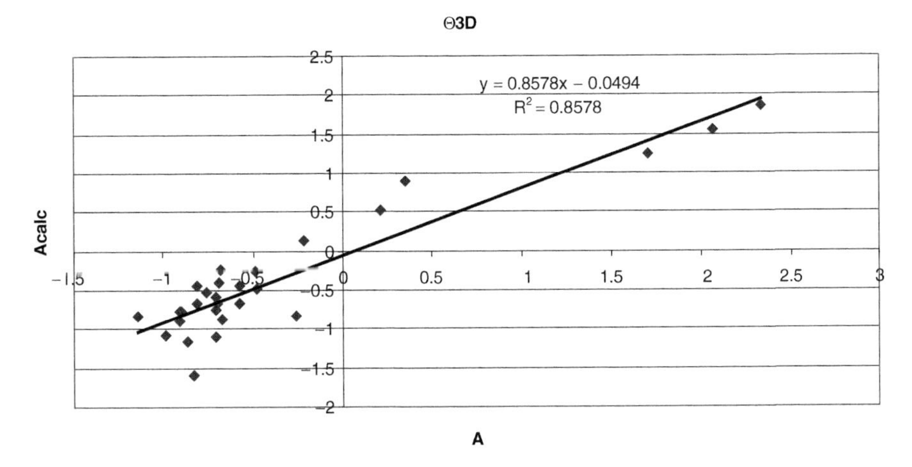

Fig. 15.6 The linear model between predicted (*Acalc*) and experimental (*A*) toxicity of amines: Acalc vs. A (Eq. 15.30)

The values in Table 15.7 prove the achievement of the conditions imposed for the external validation of models (15.30)–(15.32). The linear relation between predicted and experimental toxicity for tridimensional ovality descriptor Θ_{3D} is displayed in Fig. 15.6.

This study suggests that the ovality descriptors codify a quantity of molecular shape information which can be used in structure – activity studies. The more or less spherical or elongated shape of amine molecules quantified by these ovality descriptors can be an important structural requirement for a good interaction between the toxic molecule and its target.

15.5 Comparison of Molecular Shape by Means of MTD Method

The importance of molecular stereochemistry for biological activity was recognized 100 years ago, by Emil Fisher, with his "lock and key" theory for enzymatic reactions (Motoc 1983). The steric fit of drug molecule (L-ligand) with its target (usually, receptor, R) depends upon both the shape of the biological receptor and the shape of ligand molecule. MTD (minimal topological difference) method takes into

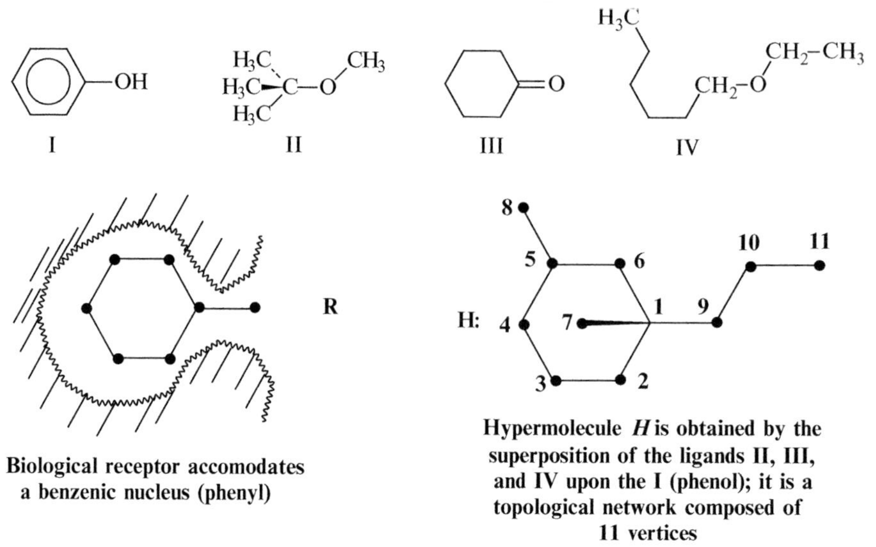

Fig. 15.7 Construction of hypermolecule **H**; an illustrative example

consideration the shape of the both L and R, and, also, experimentally measured biological activities A_i – in fact, the unique observables of the given biological response, resulting from specific action of L.

The MTD approach is based on the notion of hypermolecule, **H**, which is obtained by approximately atom by atom superposition of the all molecules, M_i, of the data base, neglecting the hydrogen atoms and seeking maximal superposition. **H** is used as topological framework for describing each molecule i involved in the building of **H**. The hypermolecule **H** can be considered as a topological network whose vertices correspond to atoms and edges to chemical bonds. The fundamentals of the MTD method is that the affinity of ligand molecule to a receptor is a linearly decreasing function of the steric misfit of the drug molecule M_i and the receptor cavity (space of interaction situs) (Fig 15.7).

During the optimization process, the vertices of **H** are assigned to the receptor cavity, to the receptor walls or to the exterior of cavity that is the so-called steric irrelevant zone. The degree of steric misfit for the molecule M_i, i.e. the value of MTD_i, is defined as the sum of the number of cavity vertices in **H** unoccupied by M_i and the number of wall vertices in **H** occupied by M_i. The whole number of **H** (meaning cavity, wall and irrelevant zone) obtained by the optimization procedure makes up a hypothetical steric receptor map. Really, the optimization procedure is a mapping of a receptor space by means of regression analysis (Simon et al. 1978,1984; Ciubotariu et al. 2001a, b).

Information for the description of the receptor are given as a set of experimental biological activities, A_i, corresponding to molecules M_i, $i = 1,N$. It is supposed that the biological activities are measured in the same conditions for all the M_i compounds, with the same accuracy and precision.

The topography of the receptor map – hypermolecule **H,** is described by the vector Λ:

$$\Lambda = \{\lambda_j\},\ j = 1,m$$

where

$$\lambda_j = \begin{cases} -1 & \text{for the interior of the cavity (beneficial vertices),} \\ 0 & \text{for the steric irrelevant zone (irrelevant vertices),} \\ +1 & \text{for the cavity walls (detrimental vertices).} \end{cases}$$

H is used as topological framework for describing each molecule i by the vector $\mathbf{X} = \{x_{ijk}\}$, $i = 1,N$ molecules in the data set, $j = 1,m$ vertices in **H** and $k = 1,C_i$, where C_i is the number of conformations of molecules M_i. The entry x_{ijk} is taken to be 1 and 0, as the δ Kronecker symbol:

$$x_{ijk} = \begin{cases} 1 & \text{if the molecule } M_i \text{ in conformation k contains a non - hydrogen atom in the vertex j of } \mathbf{H}; \\ 0 & \text{if the molecule } M_i \text{ does not occupy the vertex j.} \end{cases}$$

The computational procedure implies the evaluation of the shape descriptor MTD_i for each molecule M_i in comparison with the receptor map **Λ** by means of the following relation:

$$MTD_{ik} = \left(s + \sum_{j=1}^{m} \lambda_j \cdot x_{ijk}\right); MTD_i = \min_k (MTD_{ik}) \qquad (15.36)$$

where m stands for the whole vertex number in **H**, and s is the number of cavity vertices in **H**.

The correlation coefficient is used as optimization criterion to estimate the quality of the model (15.35).

The MTD procedure consists in the following steps:

(s1) The start receptor map (denoted S^0) is derived from **H** using a statistical analysis based on interquartile range.

(s2) The shape of the receptor site is changed vertex by vertex, with no restrictions as follows:

(a) if $\lambda_j = 0$, then $v_j = -1$ and $v_j = +1$;
(b) if $\lambda_j = +1$, then $v_j = -1$ and $v_j = 0$;
(c) if $\lambda_j = -1$, then $v_j = 0$ and $v_j = +1$.

(In the connectivity variant of MTD method is imposed that cavity, wall and irrelevant vertices must form connected graphs, respectively).

(s3) One estimates the quality of the new model **Λ** of the receptor shape for each of the above mentioned cases (a), (b) and (c);

(s4) One determines the maximal value of r (r_{max}) from a number of 2·m correlation coefficients and one retains the corresponding map of the receptor S = Λ (r_{max}).

(s5) The resulted S map is considered as initial maps and the steps (s2) – (s4) are resumed.

(s6) The procedure is stopped when the correlation coefficient obtained at step n, r_n, is less than the correlation coefficient obtained at step n + 1, r_{n-1}. The resulted S_n map is termed optimal and is denoted by $S^* = \left\{\lambda_j^*\right\}; j = 1, m$. The best shape of receptor is done by the vector $\Lambda = \left\{\lambda_j^*\right\}; j = 1, m$.

To assess the predictive ability of a QSAR in the frame of MTD method the cross-validation technique is used, in which one supposes that one or more of the known experimental values are in fact "unknown". The analysis is repeated, excluding the temporarily "unknown" compounds. The resulting equations are used to predict the experimental measurements for the omitted compound(s), and the resulting individual squared errors of prediction are accumulated. The cross-validation cycle is repeated, leaving one out (LOO) or more (LMO) different compound(s), until each compound has been excluded and predicted exactly once. The result of cross-validation is the predictive discrepancy sum of squares, sometimes called PRESS (Predictive REsidual Sum of Squares):

$$PRESS = \sum_{i=1}^{N} \left(A_i - \hat{A}_i\right)^2$$

The PRESS is a measure of the predictive power of the obtained QSAR model.

In this QSAR study with the aid of the MTD method we apply the leave half out (LHO) model, that is the half of the data is used to generate a correlation (simple linear) model and the other half to validate it, and the PRESS value is transformed in a dimensionless term, called cross-validation coefficient, r_{CV}^2:

$$r_{CV}^2 = 1 - \frac{PRESS}{\sum \left(A_i - \overline{A}\right)^2} \tag{15.37}$$

The cross-validation coefficient is the complement to the fraction of unexplained variance over the total variance.

The MTD method was applied to study the biological activity of a series of retinoic acids. One knows that all *trans* retinoic acids (see Fig. 15.8) and its derivatives show a large variety of biological effects, such as the inhibition of tumor cell growth and the induction of cell differentiation (Sporn et al. 1994). The adjustment of the gene's activity seems to be the main process, although the varied biological effects suggest complex action mechanisms involving more retinoid receptors (Mangelsdorf et al. 1994).

Although the role of retinoids in carcinogenesis is not completely known yet, these compounds have been tested as inhibitors of developing some cancer forms. Thus, it has been shown that they can block the phenotype expression of tumor

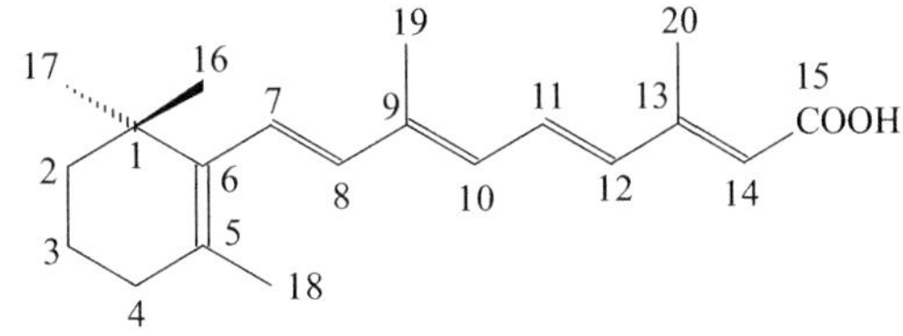

Fig. 15.8 General structure of trans-retinoic acids (RAs); (the chain is numbered in agreement with the IUPAC rules)

cells, inhibiting their growth and inducing differentiation processes (Sherman 1986). Here we present the results of a QSAR study upon a series of 38 compounds (retinoids and their analogues) that inhibit the keratinization of hamster's trachea (*ICTH*) and induce the decarboxylation of ornitine by ornitine decarboxylase (*IODC*) – see Table 15.6. The biological activities *ICTH* and *IODC* were taken from literature (Dawson et al. 1981, 1984, 1990). A_{ICTH} represents the logarithm of reciprocal molar concentration (ED_{50}) of the retinoid ligand L_i which produces an inhibition of 50% in the tracheal cell keratinization from the organ culture (Dawson et al. 1981); A_{IODC} represents the logarithm of reciprocal molar concentration (mol/L) of the ligand L_i which produces an inhibition of 50% in the ODC induction of the epidermis. The values are listed in Table 15.8.

The construction of the hypermolecule **H** was realized by overlapping the ligands L_i, i = 2,38 in Table 15.8 on the most active compound Ret_1 (see Table 15.8 and Fig. 15.9). Ret_1 is also codified as Ro 13–7410 and known as TTNPB. For this superposition, the molecular structures of all retinoid ligands, Ret_i, i = 1,38, were optimized initially with the MM + method (the force field Allinger), then by means of MO semiempirical AM1 calculation.

Previous studies (Niculescu–Duvăz et al. 1990) of standard molecular geometry, for active conformation based on the Dreiding models suggested a distance of 11–14 Å between the atom C_{15} of carboxyl group and the atom C_6 of the retinoid cycle (see Fig. 15.1), and an angle of 90–107° between the atoms C_5-C_6-C_{15}. As a result, the modifications of the median segment (polyenic chain) correspond to the hypothesis that retinoids link to the receptors through means of two situes: one placed within the area of retinoid cycle and the other one within the area of carboxyl group.

The results obtained with the aid of AM1 method confirm this hypothesis and sustain a topological construction of the basic network of hypermolecule $\hat{\mathbf{H}}$. Thus, the following C_8-C_9-C_{10}-$C_{11}= -179, 99°$; C_8-C_7-C_6-$C_1 = -179, 47°$; C_7-C_6-C_5-C_4 $= -179,99$ are the dihedral angles for the optimized conformation of Ret_1 (TTNPB). It is observed from the above values of dihedral angles that the carbon atoms numbered 1, 4–11 are practically on the same plane, while the atoms 2 and 3 are outside the plane, being placed with 14, 65° under the plane of the rest of the atoms which compose the molecule. The structures of all the 38 retinoid ligands (Ret_1 – Ret_{38}) were optimized. The geometry of the molecule was considered optimal when the gradient RMS was under 0,1 kcal/(Å mol).

In Table 15.9 we have given several geometric parameters useful for the construction of the hypermolecule. It can be seen in the table that the distance between the carbon atoms C_{15} (carboxyl) and the atom C_6 from the retinoid cycle which links the polyenic chain (see Fig. 15.8) is around 10 Å, the minimum value being of 8,18 Å

Table 15.8 Biological activities (A_{ICTH} and A_{IODC}), structural parameters ($\mathbf{X}_{ij}$), and the optimized MTD values (M*) used in the QSAR analysis on the retinoid ligands

No.	Retinoid	j, x_{ij} = 1 (X)	A_{ICTH}	MTD*	A_{IODC}	MTD*
1.	Ret_1	1–5, 10–15, 18, 19, 25, 26	11.7	9	10.5	11
2.	Ret_2	1–5, 10–15, 18, 19, 25, 26, 33	11.5	10	8.7	12
3.	Ret_3	1–5, 10–14, 18, 19, 23, 25, 26	11.2	10	10.5	11
4.	Ret_4	1–4, 12–16, 18, 19	11.0	10	10.4	11
5.	Ret_5	1–5, 10–14, 18, 19, 23–26	11.0	9	10.2	10
6.	Ret_6	1–5, 12–15, 18, 19, 25, 26	10.5	11	9.0	12
7.	Ret_7	1–5, 10–15, 18, 19, 25, 26, 33, 34	10.5	10	8.0	13
8	Ret_8	1–5, 11–15, 18, 19, 25, 26	10.4	10	10.3	11
9.	Ret_9	1–5, 12–15, 18, 19, 25, 26	10.3	11	9.3	12
10.	Ret_{10}	1–5, 10–15, 18, 19, 23, 25, 26, 33	10.1	10	7.2	12
11.	Ret_{11}	1–4, 12–15, 18, 19, 26	10.0	11	8.4	12
12.	Ret_{12}	1–4, 12–15, 18, 19, 25, 26	10.0	11	7.7	13
13.	Ret_{13}	1–5, 10–14, 18, 19, 25, 26	10.0	10	8.5	11
14.	Ret_{14}	1–4, 12–16, 18, 19, 30–34	9.7	11	7.3	14
15.	Ret_{15}	1–4, 12–15, 18, 19, 25, 26	9.7	11	8.4	12
16.	Ret_{16}	1–4, 12–15, 18, 19, 23, 25, 26	9.7	11	8.7	12
17.	Ret_{17}	1–5, 10–16, 18, 19, 25, 26, 33	9.7	10	8.5	12
18.	Ret_{18}	1–4, 12–15, 18, 19, 25, 26, 40	9.7	11	8.5	12
19.	Ret_{19}	1–5, 12–15, 18, 19, 25, 26	9.7	11	8.6	12
20.	Ret_{20}	1–4, 12–15, 18, 19, 25, 26, 38	9.5	11	9.4	11
21.	Ret_{21}	1–4, 12–15, 18, 19, 25, 26	9.5	11	8.6	12
22.	Ret_{22}	1–5, 10–12, 15, 18, 19, 25, 26, 33	9.5	11	8.1	13
23.	Ret_{23}	1–5, 12–14, 18, 19, 23, 25, 26	9.4	12	9.2	12
24.	Ret_{24}	1–5, 10–15, 18, 19, 25, 26, 33, 36	9.4	11	8.8	12
25.	Ret_{25}	1–4, 12–16, 19, 27–29, 39	9.4	11	8.1	13
26.	Ret_{26}	2, 4, 7–9, 12–15, 18, 19, 25, 26	9.4	11	7.8	13
27.	Ret_{27}	1–4, 12–15, 18, 19, 25, 26	9.2	11	8.2	12
28.	Ret_{28}	1, 3–5, 11–15, 18, 19, 25, 26	9.2	11	9.4	11
29.	Ret_{29}	5, 12–15, 17–19, 25, 26, 42, 43	9.0	12	8.5	13
30.	Ret_{30}	1–5, 12–15, 18, 19, 25, 26, 33	9.0	12	7.6	13
31.	Ret_{31}	1–5, 12–15, 18, 19, 25, 26, 33	8.7	12	7.5	13
32.	Ret_{32}	5, 12–14, 17–19, 25, 26, 42, 43	8.5	13	7.8	13
33.	Ret_{33}	1–5, 10–12, 15, 18, 19, 25, 26, 33, 36	8.5	12	8.2	13
34.	Ret_{34}	1–4, 12–16, 18, 20–22, 26, 35, 36	8.3	13	7.8	13
35.	Ret_{35}	5, 12–14, 17–19, 25, 26, 41–43	8.0	13	7.8	13
36.	Ret_{36}	1, 2, 5, 6, 8, 12–15, 18, 19, 25, 26, 33	8.0	13	7.8	13
37.	Ret_{37}	1–4, 12–16, 19, 20–22, 28	7.5	14	7.8	13
38.	Ret_{38}	1–4, 15, 18, 19, 25, 26, 33, 34, 36, 37	7.5	14	7.8	13

$A_{ICTH} = -\lg ED_{50}$; ED_{50} is the molar concentration of the retinoic ligand L_i (mol/L) that gives a 50% inhibition of keratinization of tracheal cells in culture (Ciubotariu et al. 2001b)
$A_{IODC} = -\lg ED_{50}$; ED_{50} is the molar concentration (mol/L) of L_i that gives a 50% inhibition of epidermal ODC induction (Ciubotariu et al. 2001b)

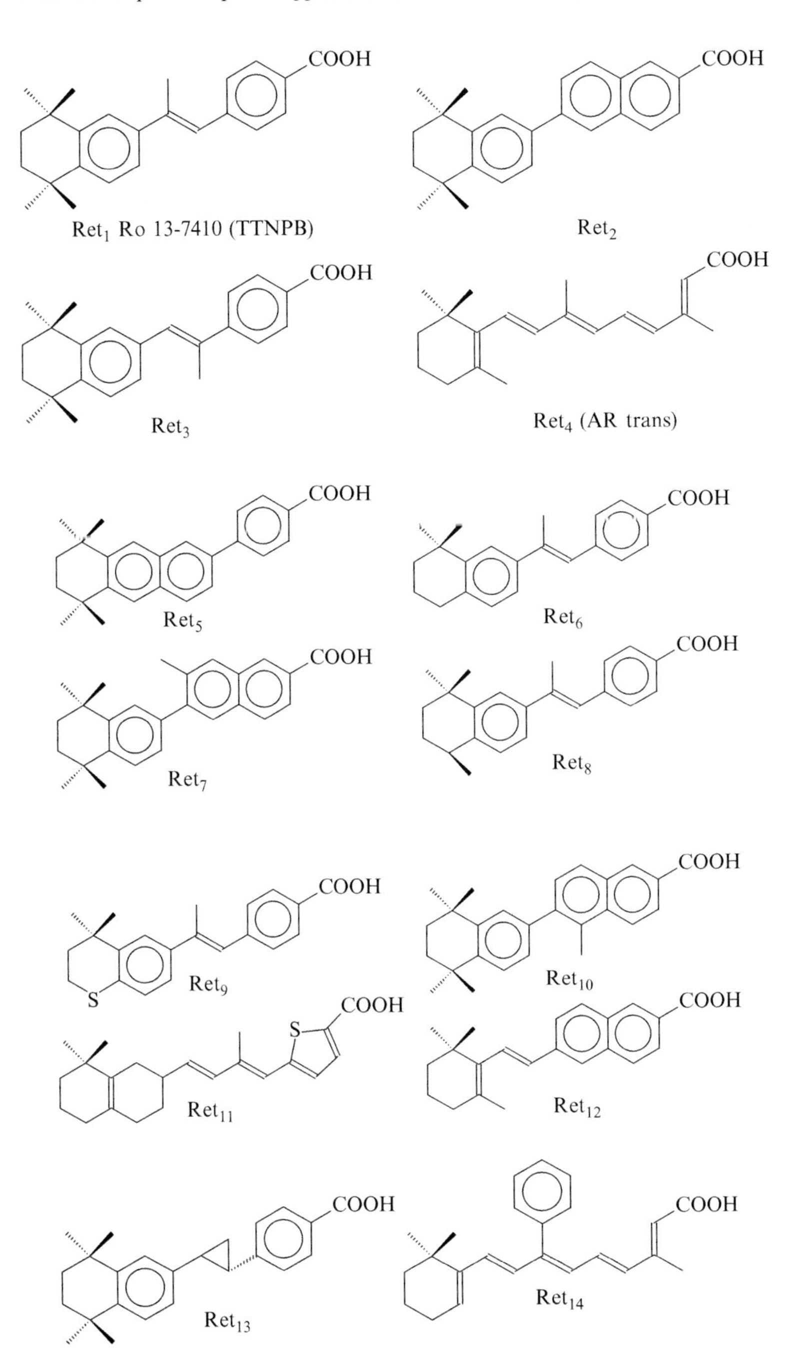

Fig. 15.9 The structures of retinoid ligands whose biological activities are presented in Table 15.8

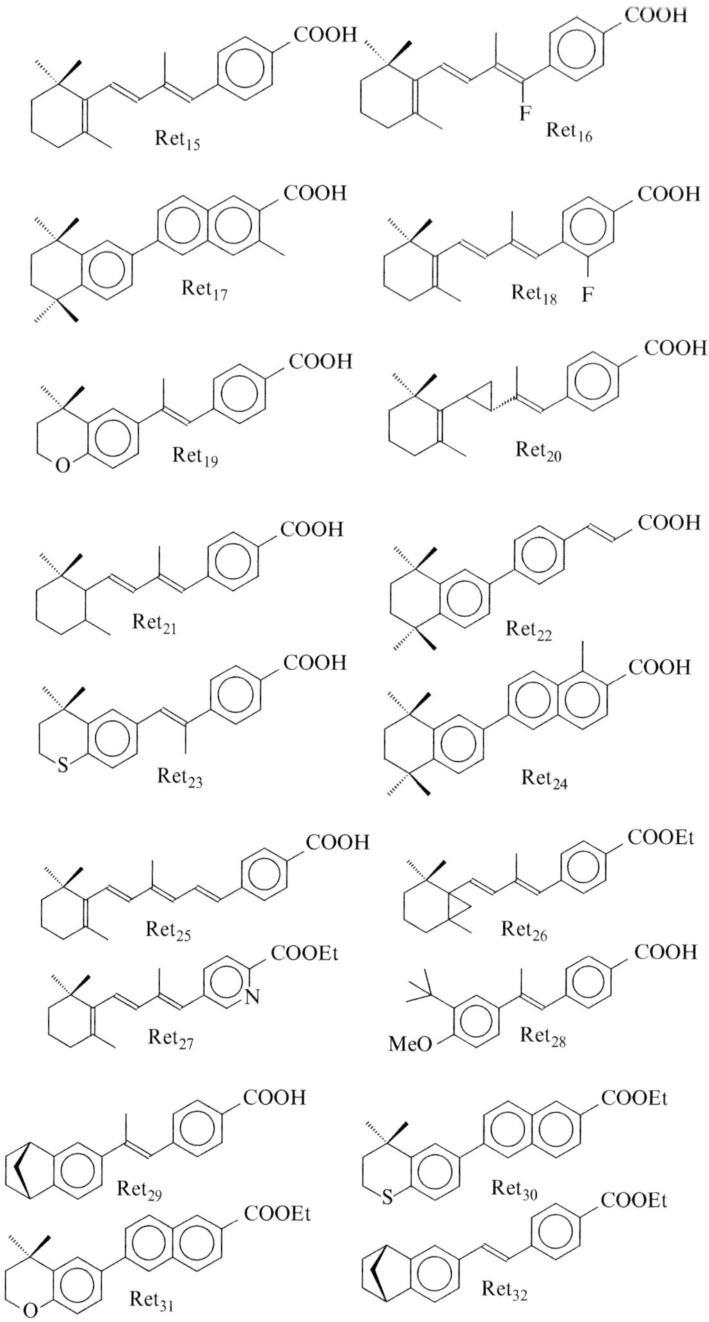

Fig. 15.9 (continued)

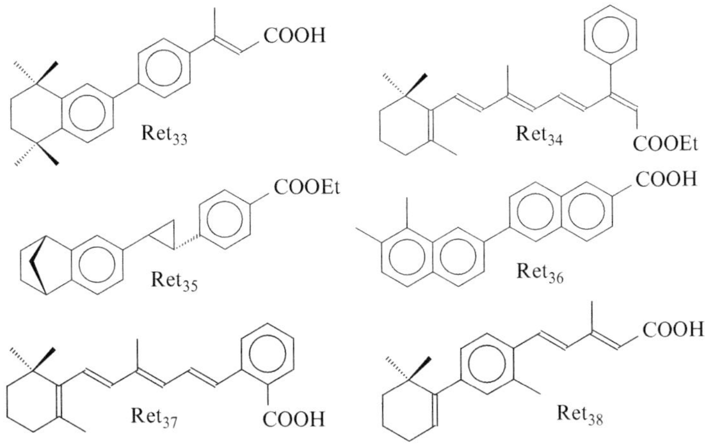

Fig. 15.9 (continued)

for Ret_{28} and the maximum one being of 12,83 Å for Ret_{25}. Actually, only 5 of the 38 retinoid derivates have the C_{15}-C_6 distance smaller than 10 Å-Ret_{11} (9,71 Å), Ret_{28} (8,18 Å), Ret_{32}(9,79 Å), Ret_{35} (9,79 Å), Ret_{37}(9,74 Å) but the values are very close to 10,00 Å (>9,70 Å), only one having the distance larger than 10,75 Å (corresponding to Ret_5). The polyenic chain modulates the biological activity, the carboxyl group disposing of a bigger conformational variability.

These values allowed us to advance the hypothesis that the isomeric process E/Z in the polienic conjugated chains is much faster than the one which takes place in the phenomena of reversion of keratinization. Therefore, in the construction of the hypermolecule we have followed a maximum overlapping on the network formed by the (flat) structure of molecule R_0 13–7410 (TTNPB), in accordance to a previous study executed upon another series of retinoids (Ciubotariu et al. 2001b).

The overlapping procedure used for constructing the hypermolecule was much more simplified by the fact that TTNPB presents some conformational restrictions; it can be considered as having a practically flat structure (see Ret_1 and the optimized image in Fig. 15.10).

For the construction of the hypermolecule, the cyclic portion of retinoids was analyzed, the distance between the pairs of the correspondent atoms in the 6 structures being closely observed (see Fig. 15.11).

As a result of the overlapping process B – F over A and F over D and E it has been found that of all the corresponding vertices, only the ones numbered 5 (from F), 7/8 from (E) and the cyclopropanic vertex linked to the atoms 2 and 3 (from D) are to be found at a distance larger than 0.5 Å, but that in all cases smaller than 0.8 Å (The previous studies indicated a maximum distance of 0.5 Å between the vertices so that the atoms could occupy the same vertex) (Ciubotariu et al. 2001b).

Table 15.9 The selected geometrical parameters correspondent to the optimal structures of minimal energy, calculated with MM + and AM1 [a]

Nr.	L_i	d(6–15)	α	φ	Nr.	L_i	d(6–15)	α	φ
1.	Ret_1	10.38	104.7	−170.7	23.	Ret_{20}	10.26	84.3	83.8
2.	Ret_2	10.31	106.6	178.4	24.	Ret_{21}	10.07	123.3	118.3
3.	Ret_3	10.16	113.4	83.3	25.	Ret_{22}	10.34	108.3	172.9
4.	Ret_4	10.41	98.5	−170.6	26.	Ret_{23}	10.16	113.2	82.7
5.	Ret_5	10.75	105.7	−168.7	27.	Ret_{24}	10.35	105.7	−177.8
6.	Ret_6	10.01	121.7	160.5	28.	Ret_{25}	12.83	134.2	−43.3
7.	Ret_7	10.24	106.7	165.2	29.	Ret_{26}	10.12	119.7	149.0
8.	Ret_8	10.01	122.1	166.1	30.	Ret_{27}	10.46	107.1	−141.1
9.	Ret_9	10.01	121.4	159.4	31.	Ret_{28}	8.18	142.4	0.0
10.	Ret_{10}	10.23	109.6	148.0	32.	Ret_{29}	10.33	104.6	171.4
11.	Ret_{11}	9.71	116.4	150.3	33.	Ret_{30}	10.26	106.9	−166.0
12.	Ret_{12}	10.32	127.2	−41.00	34.	Ret_{31}	10.31	105.5	−178.1
13.	Ret_{13}	10.37	92.9	−80.1	35.	Ret_{32}	9.97	110.1	−59.6
14.	Ret_{14}	10.37	96.2	−140.5	36.	Ret_{33}	10.00	120.6	175.8
15.	Ret_{15}	10.12	110.2	−67.9	37.	Ret_{34}	10.61	126.6	−66.7
16.	Ret_{16}	10.21	113.7	−128.2	38.	Ret_{35}	9.97	110.1	−59.6
17.	Ret_{17}	10.25	107.6	166.6	39.	Ret_{36}	10.32	105.3	180.0
18.	Ret_{18}	10.14	109.7	−65.7	40.	Ret_{37}	9.74	151.6	152.8
19.	Ret_{19}	10.01	120.6	160.6	41.	Ret_{38}	10.12	118.7	148.5

[a] d(6–15) represents the distance between the carbon atoms C_6–C_{15}; α represents the angle between the carbon atoms C_5–C_6–C_{15}; φ is the torsion angle C_5–C_6–C_{14}–C_{15}

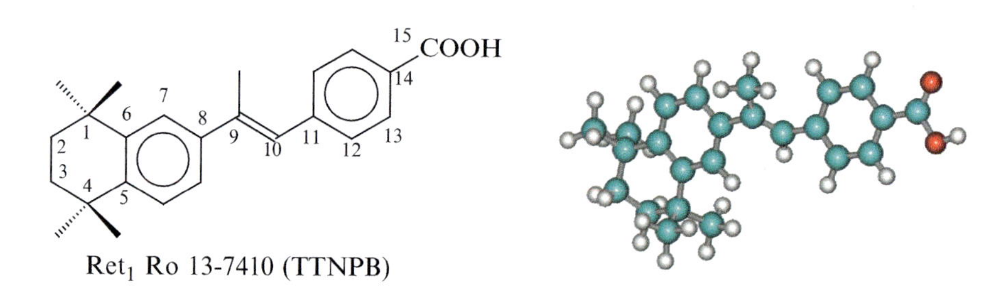

Fig. 15.10 The structure of the most active compound (*TTNPB*) and its optimized geometry with MM + and AM1

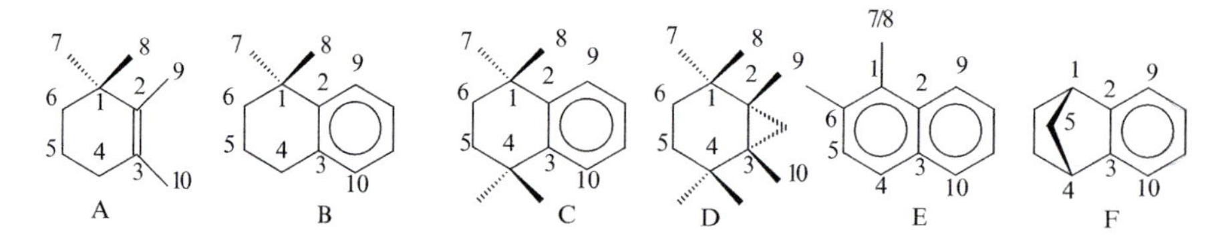

Fig. 15.11 Retinoic cycles (common to the ligands L from Table 15.9) used for overlapping and in the numbering of atoms, which could be considered equivalents

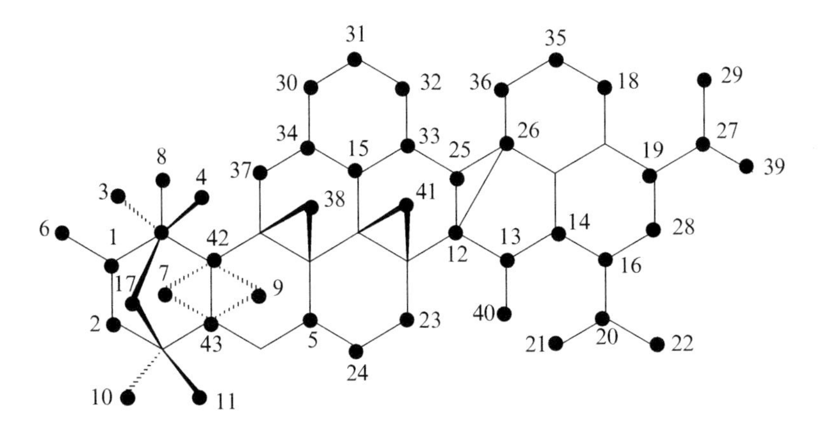

Fig. 15.12 The hypermolecule **H** constructed by the overlapping of the 38 retinoid ligands from Fig. 15.9

The hypermolecule **H** for the series of 38 retinoids considering only one conformation and expecting a maximum overlap, as resulted from the above discussions, is presented in Fig. 15.12. The common vertices of all compounds in **H** were not numbered, because they bring either constant or null contribution to the biological activity of each molecule. In fact, the hypermolecule **H** is a topological network with some 3D elements, composed of 43 vertices. It describes in an indirect manner the topological and stereo-chemical structural aspects of the retinoid receptors in interaction with their ligands presented in Fig. 15.9 (Ciubotariu et al. 2001b).

The optimized maps of receptors in which take place keratinization of hamster's trachea receptor, S^{*}_{ICTH}, and decarboxylation of the ornitine, S^{*}_{IODC}, developed on the basis of ligands in Table 15.8 and Fig. 15.12, are shown in Figs. 15.13 and 15.14, respectively.

The corresponding correlation equations are the following:

$$\hat{A}_{ICTH} = 18,246(\pm 1,254) - 0,775(\pm 0,112) MTD^{*}$$
$$n = 38;\ r = 0{,}920;\ s = 0{,}403;\ F = 96{,}8;\ r^2_{cv} = 0,800 \qquad (15.38)$$

$$\hat{A}_{IODC} = 19,125(\pm 2,450) - 0,868(\pm 0,201) MTD^{*}$$
$$n = 38;\ r = 0{,}826;\ s = 0,\ 520;\ F = 37{,}5;\ r^2_{cv} = 0,682 \qquad (15.39)$$

Equations 15.38 and 15.39 were obtained using simple linear regression *A* vs. MTD*; the values of *A* and MTD* are from Table 15.8. In these equations r represents correlation coefficient, s is standard deviation, F is the Fisher statistic indices and r^2_{CV} (cross-validation coefficient) measures the predictive powers of the linear equations. One may observe that we have obtained fairly good statistical results, especially for ICTH interactions.

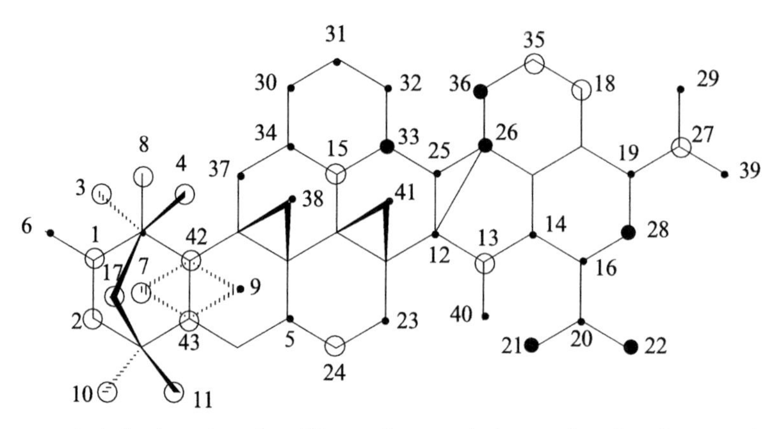

- – steric irrelevant vertices (disposed outer the intercation site of receptor)
- ○ – beneficial vertices (disposed within the interaction site - cavity - of receptor)
- ● – detrimental vertices (located within the wall of receptor)

$$S^*_{ICTH} \begin{cases} j(\lambda_i = -1): 1-4, 7, 8, 10, 11, 13, 15, 17, 18, 24, 27, 35, 42, 43 \\ j(\lambda_i = 0): 5, 6, 9, 12, 14, 16, 19, 20, 23, 25, 29, 30\,32, 34, 37, 41 \\ j(\lambda_i = +1): 21, 22, 26, 28, 33, 36 \end{cases}$$

Fig. 15.13 Optimized map of the ICTH receptor, *S*ICTH*

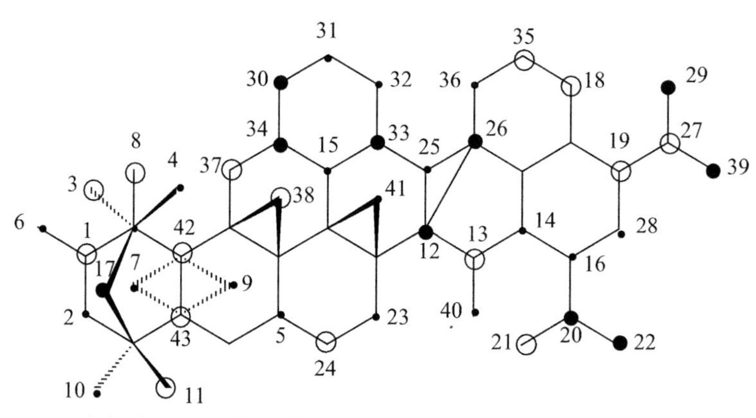

- – steric irrelevant vertices (disposed outer the intercation site of receptor)
- ○ – beneficial vertices (disposed within the interaction site - cavity - of receptor)
- ● – detrimental vertices (located within the wall of receptor)

Fig. 15.14 Optimized map of the IODC receptor

The real predictive power of our QSAR results was tested by a cross-validation like procedure (LHO). The molecules, previously arranged in the decreasing order of activity for each biological assay, were separated in a EVEN (i = 1,3,5, . . .) and ODD (i = 2,4,6, . . .) subseries. Afterwards, the receptor maps were again optimized and the corresponding correlation equation for each test subseries was in turn used to calculate activities for the other (predictive) series. Correlation between experimental and calculated activities for entire series yields the cross-validation coefficient, r^2_{CV}, reported above in Eqs. 15.38 and 15.39. One can remark that these cross-validation coefficients are greater than 0.500. Consequently, we can conclude that the predictive ability of QSAR models (15.38) and (15.39) is reliable for predicting other retinoid molecules which present or not an inhibitory activity of *ICTH* or *IODC* type. We remind that MTD_i value is a measure of steric misfit of molecule L_i in relation with the receptor map, and is equal to the number of unoccupied cavity vertices plus occupied wall vertices – see relations (1), with $k = 1$ (one single conformation for each L_i).

Comparison of *ICTH* and *IODC* results, i.e. the optimized maps S^*_{ICTH} and S^*_{IODC}, indicates a bulkier region in the iononic cycle zone of *ICTH* receptor, accommodating the greater cycles, than for the *IODC* receptor. In this last case, most steric constraints in the polyenic chain and terminal –COOH region are required (see Fig. 15.14). The binding site for the iononic cycle presents for the *ICTH*-assay receptor a more spatially extended attractive region in comparison with the *IODC*-assay receptor, where the steric strain is less severe. At the level of the second binding site, for the *IODC*-assay, there is an about equal increase in number of both cavity and wall vertices (comparatively with *ICTH*-assay receptor) and the global effect is, approximately, the same for both receptors. Therefore, we can assume that, at this level, the retinoids are more sterically constrained by the *IODC*-assay receptor than by the *ICTH*-assay receptor.

According to the optimized receptor maps and correlation results obtained for both types of biological activity data, some aspects related to general structural requirements of retinoid receptor and, also, to resemblance and differences between these types of receptors, can be summarized as follows:

1. Both receptors for retinoids possess at least two binding sites. One site accommodates the iononic ring and the other should partly bind the polyenic chain and the negatively charged carboxyl group. At these requirements, one must add a predominant general hydrophobic character and the possibility to form hydrogen bonds with hydrophilic groups in different positions of the retinoid molecules (if available).
2. The binding site of each receptor presents some differences from one to another, and these can be resumed like that:
3. The binding site for the iononic cycle presents for the *ICTH*-assay receptor a more spatially extended attractive region in comparison with the *IODC*-assay receptor, where the steric strain is less severe.
4. The retinoids are more sterically constrained by the *IODC*-assay receptor than by the *ICTH*-assay receptor.

15.6 Comparison of Molecular Shape by Means of MVD Method

The Monte Carlo version of minimal steric difference (denoted as MCD) improves the computation of non-overlapping volumes in the standard-ligand superposition, "translating" thus the topological MSD parameter into the (3D) metric context (Moţoc et al. 1975; Ciubotariu et al. 1983). In order to calculate the MCD, the molecules are described by the Cartesian coordinates and vdW radii of their atoms. The atomic coordinates implicitly specify the way one achieves the superposition: all molecules of the series are represented in the same Cartesian coordinate system. The mathematical method used in the MCD-technique for computation of non-overlapping volumes is the Monte Carlo method (Demidovich and Maron 1987).

MCD is a 3D-measure of steric misfit between the most active compound and the others within a given series of ligands under study. It translates the topological similarity/dissimilarity MSD parameter, which is an extended Hamming distance, from 2D space into a 3D space (Ciubotariu et al. 1990).

MVD (molecular volume difference) method presented here was based on MCD method and on MTD multi-conformational method. MVD allows a logical construction of a start map of receptor, S^0, and assures the differentiation of the vertices in hypermolecule **H**, translating all the ligands into their vdW space.

The multi-conformational MTD method was described above – see Sect. 15.6. The optimization procedure is beginning by choosing a start receptor map $S^0 = \{\lambda_j^0\}$, $j = 1,m$, where λ may be -1 (beneficial-cavity vertices), +1 (detrimental-wall vertices), or 0 (irrelevant-vertices situated in outer space of the receptor site) – see step s1 in Sect. 15.6. To eliminate the (possible) ambiguities which may appear in selection of a start map we introduced (Muresan et al. 1994) a procedure based on the statistical concept of the inter-quartile range. (Lee and Lee 1982) With this end in view, one arranges the experimental values A_i (the bioactivities) in descending order and one chooses the ligands corresponding to one-quarter and three-quarters of the ligand set. Thus, the atoms of all compounds from one-quarter are assumed to have a beneficial contribution to the interaction L – R because the corresponding activities are the greatest. Therefore, the vertices in hypermolecule **H** occupied by these compounds will be denoted by $\lambda_j = -1$. On the contrary, the vertices corresponding to the ligands in three-quarters can be assumed to have a detrimental contribution to the interaction L – R, i.e. $\lambda_j = +1$. The others in set are considered of no importance for L – R interaction, i.e. $\lambda_j = 0$ (their vertices are located in the exterior of the interaction site).

This procedure was based on the assumption that the experimental bioactivities, in fact the unique observable data in QSAR studies, contain all the information about L – R interaction. Therefore, the vertices of **H** occupied by the most active compounds must have a beneficial contribution because the corresponding atoms from L interact with those of R, thus increasing the stability of the complex L – R. On the contrary, the common atoms from last quarter of bioactive molecules L interact with the walls of the receptor R, having a detrimental contribution to experimentally measured biological activity. The interaction between L and R is decreasing. For a normal distribution the inter-quartile range represents 1.35 standard deviations (Lee and Lee 1982). This domain corresponds to the molecules for which common vertices are irrelevant ($\lambda_j = 0$).

The steric misfit of molecule is measured by MTD_i, calculated with relation (15.36). If a molecule M_i has several low energy conformations, it will adopt the one which best fits to the receptor site. This is the conformation k of L_i with the lowest MTD_{ik} value.

In some cases the MTD_{ik} values are degenerated because of topological nature of the descriptor **X**, whose values x_{ijk} are only 0 or 1. This degeneracy will affect the statistical quality of the QSAR's and the reliability of inferences based on the MTD procedure. For example, the MTD method in its original form cannot discriminate between different non-hydrogen atoms or groups, such as O, C, N, S, OH, NH_2, CH_2, CH_3, etc.

In order to eliminate this kind of problem we introduced a supplementary vdW descriptor, $\mathbf{V} = \{v_{ijk}\}$, whose entry, v_{ijk}, represents the vdW volume of vertex j for each molecule M_i in conformation k. The v_{ijk} values were calculated with the aid of Monte Carlo techniques, with our in home computer package IRS (http://irs.cheepe.homedns.org/). In this context, the MTD value for molecule M_i in conformation k, denoted by MVD_{ik} was calculated with the following relation:

$$MVD_{ik} = s_v + \sum_{j=1}^{m} \varepsilon_j \lambda_{ijk} x_{ijk} \tag{15.40}$$

Because we supposed that any ligand molecule will adopt the conformation that best fit the receptor, the MVD_i is

$$MVD_i = \min_k MVD_{ik} \tag{15.41}$$

The hypermolecule **H** is characterized, in addition to the ternary column vector$\Lambda = \{\lambda_j\}; j = 1, m$, by a vdW descriptor $\mathbf{W} = \{w_j\}$ whose elements are the highest volumes of vertices within the receptor site. In relation (15.40) s_v is the total volume of vertices (atoms and/or atomic groups) situated within the interaction site (cavity) of the receptor ($\lambda_j = -1$).

The calculated biological activity $\hat{A}_i$is given by the following relation:

$$\hat{A}_i = a_0 + a_1 \cdot P_{i1} + a_2 \cdot P_{i2} + \cdots - b \cdot MVD_i \tag{15.42}$$

where a_0, a_1, a_2,..., b are regression coefficients and P_{ij}, j = 1,2,...,M, stand for various molecular structural descriptors describing other effects (for example, electronic, hydrophobic, etc.).

The optimization procedure is the same to the algorithm described above for the MTD method. In order to avoid the appearance of different cavity clusters in **H** during the optimization procedure, in the MVD method we introduced a

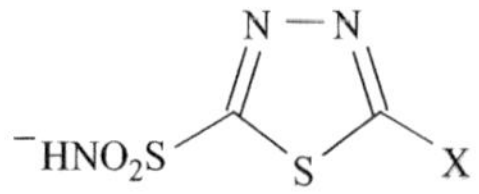

Fig. 15.15 2-Substituted-1, 3,4-thiadiazole-5-sulfonamide derivatives (L)

Table 15.10 Enzymatic inhibitor indices $\log II_{50}$, structural data, and MTD and MVD values for a series of twelve 2-substituted-1, 3, 4-thiadiazole-5-sulfonamide derivatives

No.	X	A[a]	$j(x_{ij} = 1)$[b]	MTD[c]	MVD[d]
1.	$NHSO_2C_6H_4$-4-$NHCOCH_3$	3.38	1–5,7–15	4	77.11
2.	$NHSO_2C_6H_4$-Cl	3.29	1–5,7–12	4	78.47
3.	$NHSO_2C_6H_5$	3.16	1–5,7–11	5	99.11
4.	$NHCOC_6H_5$	2.95	1–5,8–11	5	102.96
5.	$N(CH_3)COCH_3$	2.66	1–3,5,6	6	139.44
6.	NHC_6H_5	2.62	1–5,16,17	6	138.25
7.	$NHCOCH_3$	2.52	1–3,5	6	141.78
8.	NHCHO	2.28	1,2,5	7	166.24
9.	$NHCH_3$	1.96	1,2	8	177.07
10.	H	1.90	–	8	185.67
11.	NH_2	1.64	1	9	205.95
12.	$NHCOCH_2NH_2$	1.02	1–5	–	–

[a] Enzymatic inhibition indices
[b] j vertices of hypermolecule **H** occupied by molecule i (see also Fig. 15.16)
[c] Used in Eq. 15.43
[d] Used in Eq. 15.44

connectivity test for cavity vertices, characterized by$\lambda_j = -1$. As a result, the physical meaning of the optimized receptor map is more easily made because the vertices within the cavity can be assimilated with atoms connected each other.

The MVD method has been applied in a QSAR study of twelve 2-substituted-1, 3,4-thiadiazole-5-sulfonamide derivatives (L) – see Fig. 15.15 and Table 15.10 – together with the MTD method. (Mureşan et al. 1993; Vlaia et al. 2005). These compounds are known to bind specifically to the zinc ion of carbonic anhydrase (CA), a metalloenzyme that catalysis the interconversion of bicarbonate and carbon dioxide. These compounds are, in general, more potent inhibitors of CA than aromatic sulfonamides. Primary sulfonamides are known to bind specifically to the zinc ion using a deprotonated form of the NH_2 group (Menziani and De Benedetti 1992).

The biological activity $A = \log II_{50}$ was measured on bovine erythrocyte B. The inhibition constants, $\log II_{50}$, normalized with respect to p-amino-benzensulfonamide, represent on a logarithmic scale how many times a given heterocyclic sulfonamide is more active than the reference compound, namely p-amino-benzensulfonamide. The experimental data are from Menziani and De Benedetti (1992).

The hypermolecule, **H**, depicted in Fig. 15.16 was obtained by superposition of ligands L_i, $i = 2{,}12$ in Table 15.10 upon the L_1 – see Table 15.10 and Fig. 15.15.

The calculated vdW volumes of the atoms and atomic groups are summarized in Table 15.11. The standard geometrical parameters – atomic distances, dihedral angles

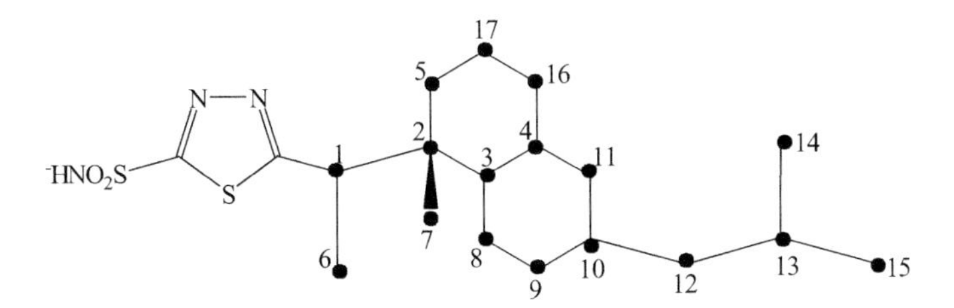

Fig. 15.16 Hypermolecule **H** obtained by superposition of ligands L_i, i = 2,12 on ligand L_1

Table 15.11 Van der Waals volumes of atomic groups used in QSAR study

Atomic groups	Volume [Å^3]	Atomic groups	Volumes [Å^3]
C	20.58	O	14.71
CH	22.39	N	15.60
CH_2	24.52	NH	17.94
CH_3	26.54	NH_2	20.28
S	24.43	Cl	22.45

and Bondi's vdW radii – have been used for computation of Cartesian coordinates and vdW volumes of atomic groups from Table 15.11. For non-hydrogen atoms as C, O, N, S the volumes have been calculated by the well-known relation: $V = 4/3\pi r^3$.

The receptor starting map S^0 (denoted as standard) has the vertices j = 4,8-15 situated within cavity of receptor map, and no detrimental (wall) vertices were chosen.

$$S^0 = \begin{cases} j(\varepsilon_j = -1): & 4, 8-15 \\ j(\varepsilon_j = 0): & 1-3, 5-7, 16, 17 \\ j(\varepsilon_j = +1): & - \end{cases}$$

The best receptor maps of receptor site achieved after optimization using MTD, S^*_{MTD}, and MVD, S^*_{MVD}, are described below together with the corresponding regression Eqs. 15.43 and 15.44, respectively.

$$S^*_{MTD} = \begin{cases} j(\varepsilon_j = -1): & 2-5, 10-13 \\ j(\varepsilon_j = 0): & 6, 7, 15, 17 \\ j(\varepsilon_j = +1): & 1, 8, 9, 14, 16 \end{cases}$$

$$\hat{A}_{MTD} = 4.745(\pm 0.213) - 0.350(\pm 0.033) \cdot MTD$$
$$n = 11,\ \mathrm{r} = 0.992,\ r^2_{CV} = 0.962, s = 0.078,\ F = 249.8,\ r^2 = 0.984 \quad (15.43)$$

$$S^*_{MVD} = \begin{cases} j(\varepsilon_j = -1): & 2-5, 10-13 \\ j(\varepsilon_j = 0): & 6-8, 15, 17 \\ j(\varepsilon_j = +1): & 1, 9, 14, 16 \end{cases}$$

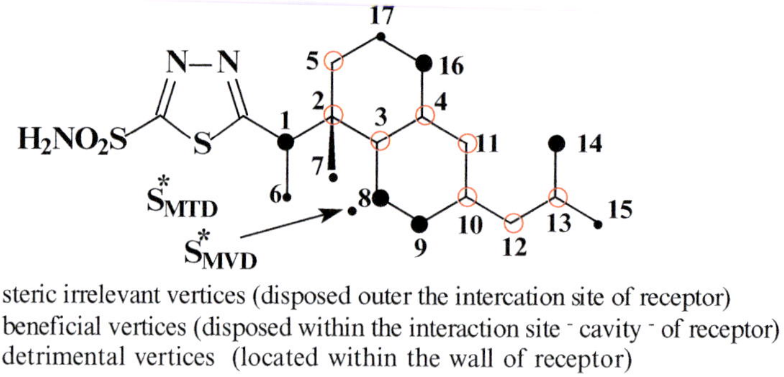

• – steric irrelevant vertices (disposed outer the intercation site of receptor)
○ – beneficial vertices (disposed within the interaction site - cavity - of receptor)
● – detrimental vertices (located within the wall of receptor)

Fig. 15.17 The optimised maps of CA receptor S^*_{MTD} and S^*_{MVD}, corresponding to the MTD and MVD method, respectively

$$\hat{A}_{MVD} = 4.406(\pm 0.175) - 0.013(\pm 0.001) \cdot MVD$$
$$n = 11,\ r = 0.993,\ r^2_{CV} = 0.945,\ s = 0.075, F = 271.6,\ r^2 = 0.985 \quad (15.44)$$

The statistical quality of the QSAR models (15.43) and (15.44) is evaluated by means of r (correlation coefficient), s (standard deviation from the regression line), F (Fisher test), and r^2_{CV}(cross-validation coefficient). The statistic r^2 represents the explained variance, and the numbers in brackets give the 95% confidence limits. In these equations we used only 11 data form data set because compound 12 from Table 15.10 is an outlier. The possible explanation of this exception may be connected with the peculiar nature of the substituent $NHCOCH_2NH_2$, which contain a supplementary amino (NH_2) group.

The QSAR model for the all data in Table 15.10 is as follows,

$$\hat{A}_{MVD} = 5.114(\pm 0.847) - 0.015(\pm 0.005) \cdot MVD$$
$$n = 12,\ r = 0.916,\ r^2_{CV} = 0.696,\ s = 0.303,\ F = 23.4,\ r^2 = 0.839 \quad (15.45)$$

One may observe that the statistical quality of the model (15.45) is considerably reduced. This is probably due to the stereochemical characteristics of the ligand L_{12}, in which the $NHCOCH_2NH_2$ side chain is constrained into a zone disposed along the hydrophobic wall of the active site. Therefore, one appears an unfavorable interaction that could explain the depression of the biological activity of L_1.

The optimized receptor map S^*_{MTD} is close toS^*_{MVD}. They are depicted in Fig. 15.17.

The QSAR study presented here is in accordance with some X-ray results on the sulfonamide-CA complex (Menziani and De Benedetti 1992). Our findings reveal some requirements imposed to the shape and size of the ligand's side chain, which are important for inhibition of CA enzyme. The linear models (15.43)–(15.45) are consistent with the fact that the interaction between side-chain, X, and part of the CA enzyme, namely Phe 131, seems to be responsible for the modulation of the inhibitory activity in this congeneric series – see Fig. 15.17. The specific interaction become stronger with the ability of the substituents to reduce the charge separation on the atoms of the thiadiazole ring, predisposing both the heterocyclic moiety and side-chain X to van der Waals interactions. In addition to the known mechanism of CA inhibition by heterocyclic sulfonamides we can make the following supplementary remark: the interaction with the protein side-chains contribute significantly to the stability of the complex L-CA and can be quantitatively modeled by means of van der Waals molecular characteristics. The results reported above indicate a reliable and significant predictive power of the multiconformational MVD method.

Obviously, the biological activities experimentally observed and measured encode various molecular structural features that can be bringing out to light. To gain this purpose it is necessary hard work for testing various molecular descriptors and QSAR techniques.

15.7 Conclusions

The molecular shape descriptors presented here are typically of a global nature. They offer ample evidence that shape as a volume or as a surface is an exciting and useful concept when applied to drug discovery. They provide a reliable scaffold for virtual screening and lead optimization by "decoration" with chemical intuition, but also have their unadorned use as in ligand fitting, prediction, or active site description.

The purpose of a molecular descriptor in quantitative structure-activity relationship (QSAR) applications is to provide a measure of a particular feature of the molecular structure of the compounds being studied. The goal is simply to measure the feature in question as accurately and unambiguously as possible.

In this chapter we presented two structural measures of molecular shape that can be used as predictor variables in MLR (multiple linear regression) analysis of structure-activity studies – cylindrical (δ,G) and ovality (Θ_{iD}, i = 1,2,3) molecular descriptors – and two inexpensive overlapping methods useful for quick receptor mapping – MTD (minimal topological difference) and MVD (minimal volume difference). A subsequent statistical analysis of QSAR models developed with these shape molecular descriptors explained well the variance in the observed reactivity data (δ descriptor of cylindrical shape) and biological activity of retinoids (MTD) and sulfonamides (MVD).

These last two methods build up the dissimilarity measures of topological (MTD) and vdW (MVD) shape of ligand molecules in the process of interaction with its biological targets. At the same time, the shape of the ligands is compared with a hypothetical shape of the enzyme's (or receptor's) site. The optimization procedure tries to guess the best map of receptor. This one, denoted by S^*, represents in fact a complementary shape of the receptor site. A detailed analysis of the S^* in connection with the structure of the ligands can lead to supplementary information on the shape of the receptor-ligand complex. This way, one can identify some important sites on the drug molecule that are necessary for a favorable (or unfavorable) interaction. Even if it is difficult to guess a new molecular structure with greater bioactivity, it is certain that one can eliminate some drug candidates whose synthesis will rise up the costs of drug design.

References

Ballester PJ, Richards WG (2007) J Comput Chem 28:1711–1723

Bodor N, Gabany Z, Wong CK (1989) J Am Chem Soc 111:3783–3786

Bodor N, Buchwald P, Huang M-J (1998) SAR QSAR Environ Res 8:41–92

Bondi AJ (1964) J Phys Chem 68:441–451

Charton M (1975a) J Am Chem Soc 97:1552–1556

Charton M (1975b) J Am Chem Soc 97:3694–3697

Chiriac A, Ciubotariu D, Simon Z (1996) Quantitative Structure – Activity Relationship (QSAR) The MTD method. Mirton, Timisoara

Ciubotariu D (1987) Structure-reactivity relationships in the class of carbonic derivatives, Ph.D. thesis, Polytechnical Institute of Bucharest, Romania

Ciubotariu D, Holban Ş, Moţoc I (1975) Fac St Nat Ser Chim 3:1–8

Ciubotariu D, Holban S, Mihalas I, Simon Z, Chiriac A (1983) Fac St Nat Ser Chim 3:1–19

Ciubotariu D, Deretey E, Medeleanu M, Gogonea V, Iorga I (1990) Chem Bull PIT 35:83–98

Ciubotariu D, Gogonea V, Iorga I, Deretey E, Medeleanu M, Mureşan S, Bologa C (1993) Chem Bull Tech Timişoara 38:63–75

Ciubotariu D, Mureşan S, Gogonea V, Medeleanu M, Dragoş D (1996) In: Chiriac A, Ciubotariu D, Simon Z (eds) Quantitative Structure-Activity Relationship (QSAR). The MTD method. Mirton, Timişoara, pp 40–138

Ciubotariu D, Gogonea V, Medeleanu M (2001a) In: Diudea MV (ed) QSPR/QSAR studies by molecular descriptors. NOVA Science, Huntington, pp 281–362

Ciubotariu D, Grozav A, Gogonea V, Ciubotariu C, Medeleanu M, Dragos D, Pasere M, Simon Z (2001b) Match 44:65–92

Ciubotariu C, Medeleanu M, Ciubotariu D (2006a) Chem Bull "Politehnica" Univ Timisoara 51:13–16

Ciubotariu D, Vlaia V, Olariu T, Ciubotariu C, Medeleanu M, Dragos D, Ursica L (2006b) Generalized topological descriptors based on reciprocal distance matrix for development of toxicological QSARs, 12th International workshop quantitative structure-activity relationship environmental toxicology, 8–12 May, Lyon, p 89

Ciubotariu D, Vlaia V, Olariu T, Ciubotariu C, Medeleanu M, Ursica L, Dragos D (2006c) Molecular van der Waals descriptors for quantitative treatment of toxicological effects. 12th International workshop quantitative structure-activity relationship environmental toxicology, 8–12 May, Lyon, p 90

Cohen C (1979) Beyond the 2-D chemical structure. In: Computer assisted drug design, Olson EC, Christoffersen RE (eds), ACS Symposium Series, vol 112, Washington DC, p 371–382

Connolly ML (1985) J Am Chem Soc 107:1118

Cramer RD, Patterson DE, Bunce JD (1988) J Am Chem Soc 110(18):5959–5967

Dawson MJ, Hobbs PD, Chan RL, Chao W, Fung VA (1981) J Med Chem 24:583–592

Dawson MJ, Hobbs PD, Derdzinski K, Chan RL, Gruber J, Chao W, Smith S, Thies RW, Schiff LJ (1984) J Med Chem 27:1516–1531

Dawson MI, Chao W, Hobbs PD, Delair T (1990) In: Dawson MI, Okamura WH (eds) Chemistry and biology of synthetic retinoids. CRC Press, Boca Raton, pp 385–466

Demidovich BP, Maron IA (1987) Computational mathematics. MIR Publishers, Moscow, pp 649–656

Edvinsson T, Arteca GA, Elvingson C (2003) J Chem Inf Comput Sci 43:126–133

Gavezotti A (1983) J Am Chem Soc 105(16):5220–5225

Gogonea V (1996) An approach to solvent effect modelling by the combined scaled-particle theory and dielectric continuum-medium method, Ph.D. thesis, Toyohashi University of Technology, Japan
Gogonea V, Motsenigos A, Ciubotariu D, Deretey E, Chiriac A, Simon Z (1981) Molecular mechanics and molecular shape descriptors. 2. Comparative treatment of four methods for molecular surface calculation, Preprint, University of Timisoara, Ser. Chimie. 1–12
Gogonea V, Ciubotariu D, Deretey E, Popescu M, Iorga I, Medeleanu M (1991) Rev Roum Chim 36:465–471
Golbraikh A, Tropsha A (2002) J Comput Aided Mol Des 16:357–369
HyperChem version 7.0 (2002), HyperCube Inc., Gainesville
Kortagere S, Krasowski MD, Ekins S (2009) Trends Pharmacol Sci 30:138–147
Lee DJ, Lee TD (1982) Statistics and computer methods in BASIC. Van Nostrand Reinhold, New York, p 15
Mangelsdorf DJ, Umesono K, Evans RM (1994) In: Sporn MB, Roberts AB, Goodman DS (eds) The retinoids, biology, chemistry, and medicine. Raven, New York, pp 319–349
Menziani MC, De Benedetti PG (1992) Struct Chem 3:215–219
Meyer AY (1985) J Chem Soc Perkin Trans II:1161–1169
Mezey PG (1993) Shape in chemistry. An introduction to molecular shape and topology. VCH Publishers, New York
Motoc I (1983) In: Charton M, Motoc I (eds) Steric effects in drug design. Springer, Berlin, pp 93–106
Moţoc I, Holban S, Ciubotariu D, Simon Z (1975) Fac St Nat Ser Chim 16:1–8
Muresan S, Bologa C, Medeleanu M, Gogonea V, Dragos D, Ciubotariu D (1994) Chem Bull Tech Univ Timişoara 39:23–33
Mureşan S, Ciubotariu D, Simon Z (1993) unpublished data
Niculescu-Duvaz I, Ciubotariu D, Simon Z, Voiculetz N (1991) In: Voiculetz N, Balaban AT, Niculescu-Duvăz I, Simon Z (eds) Modeling of cancer genesis and prevention. CRC Press, Boca Raton, pp 157–214
Niculescu–Duvăz I, Simon Z, Voiculetz N (1990) In: Dawson MI, & Okamura WH (eds) Chemistry and biology of synthetic retinoids. CRC Press, Boca Raton, pp 575–606
Pearlmann RS (1983) SAREA: van der Waals (and accessible) surface area of molecules, program QCPE No. 413
Purvis GD (2008) J Comput Aided Mol Des 22:461–468
Sherman MI (1986) In: Dawson P, Nakamura M (eds) Retinoids and cell differentiation. CRC Press, Boca Raton, chapter 8
Simon Z, Bădilescu I, Chiriac A, Ciubotariu D, Szabadai Z (1978) In: Francke R, Oehme P (eds) Quantitative structure-activity analysis. Akademie Verlag, Berlin, pp 161–167
Simon Z, Chiriac A, Holban S, Ciubotariu D, Mihalas GI (1984) Minimum steric difference. The MTD method for QSAR studies. Research Studies Press, Letchworth/New York
Sporn MB, Roberts AB, si Goodman DS (1994) The retinoids. biology, chemistry, and medicine. Raven Press, New York, pp 319–349
Todeschini R, Consonni V (2000) Handbook of molecular descriptors. Wiley-VCH, New York
Todeschini R, Gramatica P (1997) Quant Struct Act Relat 16:113–119
Vlaia V (2010) Contributions to the toxicity study of substances by means of quantitative structure – activity relationships, PhD thesis, Romanian Academy
Vlaia V, Olariu T, Urşica L, Ciubotariu C, Medeleanu M, Ciubotariu D (2005) Farmacia 6:89–101
Vlaia V, Olariu T, Ciubotariu C, Medeleanu M, Vlaia L, Ciubotariu D (2009) Revista de Chimie (Bucureşti) 60(12):1357–1361
Vrânceanu Gh, Hangan Th, Teleman C (1967) Geometrie elementară din punct de vedere modern. Ed. Tehnică, Bucureşti, pp 56–78
Wadell H (1935) J Geol 43:250–280
Wilson JA, Bender A, Kaya T, Clemons P (2009) J Chem Inf Model 49:2231–2241
Zauhar RJ, Moyna G, Tian LF, Li ZJ, Welsh WJ (2003) J Med Chem 46:5674–5690
Zyrianov Y (2005) J Chem Inf Model 45:657–672

Chapter 16
Recent Advances in Bioresponsive Nanomaterials

Cecilia Savii[1] and Ana-Maria Putz[1]

Abstract The development of effective nanodelivery systems capable of carrying a drug specifically and safely to a desired site of action can be considered as one of the most challenging tasks of pharmaceutical formulation. Stimuli-responsive materials, as called intelligent or smart, could be the key of the best developments in the area. Controlled Drug Delivery Systems (CDDSs) have almost three decades since they were first investigated. Their formulation, at the beginning, was almost exclusively based on biomaterials ((biomaterials or (bio)compatible polymers)) as carriers and targeting functionalities. Despite intense scientific work, focused in the field of delivery systems, the number of systems that have undergone clinical trials and then reached the market has been not as expected. Starting from early 1990 ordered mesoporous materials and hybrids were first synthesized and in 2000 they started to be used as part of CDDS, complementary inducing major improvements. The evolutions in the field from microtechnology toward nanotechnology caused tremendous improvements and developing of nanomedicine. Because of the huge literature of scientific papers on CDD systems, this chapter attempts to limit sensing material's examples to some CDDSs encompassing (meso) porous and magnetic silica based nanomaterials as drug vehicle, trying also to present their complex functionalities, as a part of an intelligent material concept. By studying this extremely huge and fascinating domain, we tried to hear "the voice" of inorganic materials, actually, a very tiny part of it. The studies were done in order to better understand the basics related to both drug and CDDSs interaction and behavior within biosystems in order to find the required essentials for inorganic components to fit and even improve such a complex systems.

[1] Laboratory of Inorganic Chemistry, Institute of Chemistry Timişoara
of Romanian Academy, Ave. Mihai Viteazul, No. 24, Timişoara, RO 300223, Romania
e-mail: ceciliasavii@yahoo.com; putzanamaria@yahoo.com

M.V. Putz (ed.), *Carbon Bonding and Structures: Advances in Physics and Chemistry*,
Carbon Materials: Chemistry and Physics 5, DOI 10.1007/978-94-007-1733-6_16,

16.1 Introduction

By environmental stimuli-responsive materials, generally, it can be understood as such objects that can change their properties, in response to external environment stimuli (Kono and Takagishi 1996). The response to external stimuli can be observed, qualitatively evaluate or even quantitatively measured. In this context, the meaning of "environmental" was not literarily understood as the well known equivalent term for natural environment. Environmental stimuli, in this concern, encompass any signal which can alter the status of the sensitive-responsive object. The stimuli could originate in specificity of chemical nature of compounds or changes of temperature, pH, ionic strength, redox conditions, solubility, light or ultrasonic, magnetic and electric field etc. The stimuli can be generated by or inside of many environmental places, such as in nature, in laboratory, industrial or office buildings, in live bodies as plant, animal, and human. Stimuli could be considered the signals which can inform about possible changes in the conditions which can allow, promote, slow or forbid certain actions. When we are talking about the stimuli generated in live bodies, the materials which can sense such signals could have been named bio-sensing materials. Bio-responsive materials could be considered such materials or devices which can sense and react to bio-stimuli, in such cases modifying their functionalities or their behaviour. Resuming, it can be observed these materials taking an action as a response to stimuli interaction. This kind of behaviour is leading us to the concept as named "intelligent" or "smart materials" such as memory shape metallic alloys, ceramics or ceramic composites, polymers or polymer composite as well as (magneto- or electro-) reologic fluids, when it is about of the changes at macroscale. Actually, all cited materials can accomplish one or more functions in a "smart material" device. As it was defined by Haddad in a very simple and accessible way, the "intelligent material" concept imply three major components, "sensor", "microprocessor" and "actuator", each of them having specific functions (Iyer and Haddad 1994).

"Smart materials" are modelled upon biological systems with sensors acting as a nervous system, actuators acting as muscles and microprocessor controllers acting as a brain. These concepts are currently being applied to advanced composite materials where sensors and actuators can be embedded during fabrication (Davidson 2002).

A sensing mechanism referrers to perceive the external stimuli (e.g., skin which senses thermal gradients, an eye that senses optical signals, etc.), termed "sensor" function.

A communication network by which the sensed signal would be transmitted to a decision-making mechanism (e.g., the nervous system in humans and animals), referred to as "memory" function.

A decision-making mechanism which has the capability of reasoning (e.g., the brain) is designated as "processor" function.

An actuating mechanism could be inherent in the parent material or externally coupled with it (e.g., stiffening of muscles in humans and animals to resist any strain due to external loading), is referred to as "actuator" function (Iyer and Haddad 1994).

According to recent discoveries in biology and chemistry of materials, it can be clearly seen that the component mechanisms of an intelligent material concept, could have been much smaller entities as nanoparticles or even molecules (Yang et al. 2005).

External stimuli could cause changes in the stimuli-responsive material properties, as particle dimensions, structure, and interactions, and lead to their rearrangement or changes in their aggregation state. Motornov et coworkers compare the stimuli-responsive small (micro or nano scale) particle structure, in its complexity, to a living cell, where many particle's compartments interact, exchange chemicals, receive energy, perform mechanical work, alter the chemical and physical properties, all of these in response to environmental stimuli.

An increase in complexity of the particle and the material structures and immersion of this particle in a complex environment will result in a broad variety of functional systems with important applications (Motornov et al. 2010). One of the application for which the suitability of such a stimuli responsive or intelligent materials is of a first evidence, namely **C**ontrolled **D**rug **D**elivery **S**ystem, CDDS. Because many drugs act as protagonists or antagonists to different chemicals in the body, a delivery system that can respond to the concentrations of certain molecules in the body or other stimuli is invaluable (Peppas 2004). For this purpose, intelligent therapeutics or smart drug delivery calls for the design of the newest generation of sensitive materials based on molecular recognition. Biomimetic polymeric networks can be prepared by designing interactions between the building blocks of biocompatible networks and the desired specific ligands and by stabilizing these interactions by a three-dimensional structure. These structures are at the same time flexible enough to allow for diffusion of solvent and ligand into and out of the networks.

Synthetic networks that can be designed to recognize and bind biologically significant molecules are of great importance and influence a number of emerging technologies. These artificial materials can be used as unique systems or can be incorporated into existing drug delivery technologies that can aid in the removal or delivery of biomolecules and restore the natural profiles of compounds in the body (Peppas 2004).

Inorganic nanomerials, including silica mesoporous matrix and superparamagnetic iron oxide silica nanocomposite have advantages as small size, huge total specific surface area, large inner volume, stable membrane, and tunable permeability, they have wide applicability in submicro- to- micrometer entrapping of drugs, enzymes, DNA, and other active macromolecules (Guo et al. 2005; Yang et al. 2005), and have found many applications in various fields, such as drug and gene delivery (Liu et al. 2005; Yang et al. 2005) biosensoring (Chinnayelka and McShane 2005; Yang et al. 2005), electrocatalysis (Ren et al. 2004; Yang et al. 2005) and enzyme immobilization (Taqieddin and Amiji 2004; Yang et al. 2005). We obtained iron oxide silica nanocomposite, of different concentrations and iron oxide polymorphes, alpha, gama and epsilon, having para, superpara and ferromagnetic behaviour (Savii et al. 2002; Popovici et al. 2004a, b, c, 2005, 2006; Lancok et al. 2008).

Especially in controlled drug delivery systems, that materials are considered to be suited and potentially useful, because at least some of the essential requirements of a controlled drug delivery system, which is supposed to assure an improved drug treatment (outcome) through rate- and time-programmed and site-specific drug delivery (Breimer 1999; Yang et al. 2005) could be met in inorganic materials too.

Because of the huge literature of scientific papers on CDD systems, including excellent books providing the both basics and new development (Lu 2004; Wang et al. 2005; Gupta and Kompella 2006; Swarbrick 2009; Pathak and Thassu 2009) and relevant findings in the field, this chapter attempts to limit sensing material's examples to some CDDSs encompassing (meso)porous and magnetic silica based nanomaterials as drug vehicle, trying also to present their more complex functionalities, as a part of an intelligent material. By studying this extremely fascinating domain, we tried to hear "the voice" of inorganic materials (Eisenberg 2011) actually, of a very small part of them. Even so, it was impossible to know of and include all such relevant aspects of the problem we assumed and to properly credit all the key people who helped to bring essential issues in the field. The authors apologize in advance for all omissions and express the receptivity and gratitude for all observations and suggestions.

16.2 CDDS Background

Hoffman (2008) who has been personally involved in the origins and evolution of this field for the past 38 years, describes the evolution of the "controlled drug delivery" (CDD) field from its origins in the 1960s, starting with macroscopic "controlled" drug delivery (DD) devices and implants or ingestible capsules (1970–1980) (Peppas and Colombo 1997; Langer and Folkman 1976) continuing in 1980s and 1990s with microscopic degradable polymer depot DDSs (Kent et al. 1987) and reaching at the currently nanoscopic era of targeted nano-carriers (Iwai et al. 1984) in a sense bringing to life Ehrlich's imagined concept of the "Magic Bullet" (Ehrlich 1954).

Considerable work there has been carried out in preparing materials and finding new uses for nanoscale structures based on biomaterials, such as carriers for controlled and targeted drug delivery, micropatterned devices, systems for biological recognition as indicated by Langer and Peppas (Langer and Peppas 2003; Peppas 2004).

Drug delivery system mainly is consisting of two components, the active part, the drug itself and the drug carrier. CDDS or stimuli responsive drug delivery systems has the third part, the sensitive or sensitive-reactive functionality and it is required to act following some strict rules in order to improve the therapeutic efficacy. They have to precisely spatial and temporal delivery of therapeutic agents to the target site. More than that, the controlled drug delivery systems have to maintain the concentration of drugs in the precise sites of the body within the optimum range and under the toxicity threshold (Wang 2009). Controlled drug delivery systems usually have been developed in polymer-based systems but in the last years by using also novel inorganic and hybrid materials-based systems (Yang et al. 2005).

16.2.1 Drugs Classification

For controlled-delivery systems, it is required to design carriers for the drugs instead drugs for carriers as Pillai et al. (2001) said. The increased the knowledge concerning drugs basic properties, and principles governing drug-biosystem interaction the better the chances to formulate suitable carriers and eventually more efficient DDS. The behavior of DDSs concepts to better understand, *in vivo* and *in vitro* dissolution testing emerged as the *in vitro–in vivo* correlation, that finally led to the birth of a biopharmaceutic classification system (BCS) (Amidon et al. 1995) which is based on two basic tenets of absorption: solubility and permeability (Pillai et al. 2001).

Drugs are classified as highly soluble if the maximum dose of drug is soluble in 250 ml (one glass) of water and highly permeable if drugs are more than 80% absorbed across the gastrointestinal membrane.

- Class I drugs are highly soluble and highly permeable;
- Class II drugs are poorly soluble and highly permeable;
- Class III drugs are highly soluble and poorly permeable;
- Class IV drugs are poorly soluble and poorly permeable (Pillai et al. 2001; Wilding 1999).

Determination of solubility in the pH range of 1–8 and correlating with maximum deliverable dose will help to selection of appropriate routes of administration and delivery technologies. Permeability can be determined experimentally by varying in levels of complexity from *in vitro* (shake flask method) to *ex vivo* (cell cultures) and further to *in vivo* (whole human body) (Braun et al. 2000). The recommended strategy during pharmaceutical development was to improve the solubility of the lead compound even if the permeability of the molecule is compromised as a result (Barker 2001).

16.2.2 DDSs Classification

As the inorganic mesoporous or magnetic and luminescent materials relatively easy can be associated with biocompatible polymers and/or other functionalities, in order to develop a CDDS, it is useful to be aweared about the main types and behaviour of polymeric DDS.

Starting from 1952 (Starr 2000) drug-delivery systems evolved, eventually reaching as called the novel drug-delivery systems NDDSs. Polymer NDDSs was classified on the basis of their *drug-release mechanism* and technology (Chien 1992; Pillai et al. 2001).

- Rate-preprogrammed CDDSs
  - Polymer membrane permeation
  - Polymer matrix diffusion
  - Microreservoir partition
- Physical-activated DDSs Osmotic-pressure-activated
  - Hydrodynamic-pressure-activated
  - Hydration-activated
  - Vapor-pressure-activated
  - Mechanically activated
  - Magnetically activated
  - Ultrasound-activated
  - Electrically activated
- Chemically activated DDSs
  - pH-activated
  - Ion-activated
  - Hydrolysis-activated
- Biochemically activated DDSs
  - Enzyme-activated
  - Biochemical-activated
- Feedback-activated DDSs
  - Bioerosion-regulated
  - Bioresponsive
  - Self-regulating
- Site-targeted DDSs
  - Passive targetting
  - Active targetting

Bajpai and coworkers (2008) reported a classification of polymeric DDS according to the *mechanism controlling the drug release* (Peppas and Khare 1993), as follows.

(1) Diffusion-controlled systems, as (a) Reservoir (membrane systems) or (b) Matrix (monolithic systems); (2) Chemically controlled systems that can either be (a) Bioerodible and biodegradable systems or (b) Pendent chain systems;

(3) Solvent-activated systems either by (a) Osmotically controlled systems or (b) Swelling-controlled systems, and (4) Modulated-release systems. Nevertheless it can be added the fifth category (5) Combined mechanisms; because it could be observed drug delivery devices that act by a combination of above mentioned mechanisms. Bajpai schematically further explain cited basics mechanisms of control (Bajpai et al. 2008).

Diffusion-controlled systems comprise drugs which diffuse through polymer; the polymer may undergo subsequent biodegradation on exhaustion of the drug. When the drug is enclosed in a reservoir (core) as long as the drug is maintained in the core in a saturated state, release will follow zero-order kinetics until it is nearly exhausted. *Reservoir systems* are hollow devices in which an inner core of dissolved, suspended or neat drug is surrounded by a membrane. The drug transfer mechanism through the membrane is usually a solution-diffusion mechanism. Drug transport occurs first by dissolution of the drug in the membrane on one side followed by diffusion through the membrane and desorption from the other side of the membrane. *Matrix (monolithic) systems-* contain the drug as uniformly dissolved or dispersed in the host material. The matrix systems release behavior is of first-order with continuously decreasing release rate. This is due to the increasing diffusion path length and the decreasing area at the penetrating diffusion front as the matrix release proceeds (Bajpai et al. 2008).

In ***chemically controlled drug delivery systems***, the release of a pharmacologically active agent usually takes place in the aqueous environment by one or more of the following mechanisms: (i) Gradual biodegradation of a drug containing polymer system; (ii) Biodegradation of unstable bonds by which the drug is coupled to the polymer system; (iii) Diffusion of a drug from injectable and biodegradable microbeads. In *Bioerodible and biodegradable systems*, the polymer erodes because of the presence of hydrolytically or enzymatically labile bonds. As the polymer erodes, the drug is released to the surrounding medium. Erosion may be either surface or bulk erosion. The main advantages of such biodegradable systems are the elimination of the need for surgical removal, and they are non-toxic, non-carcinogenic and non-teratologic. In *pendant chain systems*, the drug molecule is chemically bonded to a polymer backbone and the drug is released by hydrolytic or enzymatic cleavage. The rate of drug release is controlled by the rate of hydrolysis. This approach provides an opportunity to target the drug to a particular cell type or tissue. The structure of these polymers can be modified by the incorporation of functional units to obtain a specific tissue affinity (Bajpai et al. 2008).

Swelling-controlled systems are hydrogel based. Hydrogels consist of macromolecular chains crosslinked to create a tangled mesh structure, providing a matrix for the entrapment of drugs. When hydrogels come in contact with a thermodynamically compatible solvent, the polymer chains do relax (Shukla et al. 2003). This happens when the characteristic glass–rubber transition temperature of the polymer is below the temperature of experiments. Swelling is the macroscopic evidence of this transition. The dissolved drug diffuses into the external receiving medium, crossing the swollen polymeric layer formed around the hydrogel. When the hydrogel contacts the release medium, the penetrant water molecules invade the hydrogel surface and thus a moving front is observed that clearly separates the unsolvated glassy polymer region ahead of the front from the swollen and rubbery hydrogel phase behind it. Just ahead of the front, the presence of solvent plasticizes the polymer and causes it to undergo a glass-to-rubber transition (Davidson and Peppas 1986). Regarding the drug transport, the following possibilities arise: (i) If the glass transition temperature Tg of polymer is well below the experimental temperature, the polymer will be in the rubbery state and polymer chains will have a high mobility that allows easier penetration of the solvent into the loaded hydrogel and subsequent release of the drug molecules into the release medium (Grinsted et al. 1992). This results in Fickian diffusion, (Case I), which is characterized by a solvent (or drug) diffusion rate, R_{diff}, slower than the polymer chain relaxation rate R_{relax} ($R_{\mathrm{diff}} < R_{\mathrm{relax}}$). (ii) If the experimental temperature is below T_g, the polymer chains of hydrogels are not sufficiently mobile to permit immediate penetration of the solvent into the polymer core. The latter situation gives rise to a non-Fickian diffusion process which includes Case II and anomalous diffusion, respectively, depending on the relative rates of diffusion and chain relaxation (for Case II, $R_{\mathrm{diff}} > R_{\mathrm{relax}}$, and for anomalous diffusion, $R_{\mathrm{diff}} \sim R_{\mathrm{relax}}$) (Bajpai et al. 2008).

In modulated-release systems, the drug release is controlled by external *stimuli*, named also triggering signals or simple *triggers,* such as temperature, pH, ionic strength, magnetic, electric or ultrasonic field, electromagnetic radiation or UV light etc.

Temperature-sensitive release systems. Temperature, more precise temperature gradient, is the most widely utilized stimulus for a variety of triggered or pulsatile drug delivery systems. Primary, the use of temperature as a signal has been justified by the fact that the actual body temperature often deviates from the physiological value (37°C) in the presence of pathogens or pyrogens. This deviation can be a useful stimulus to activate release of therapeutic agents from various temperature-responsive drug delivery systems for diseases accompanied by fever. Drug-delivery systems responsive to temperature utilize various polymer properties, including the thermally reversible transition of polymer molecules, swelling change of networks, glass transition and crystalline melting (Anal 2007).

pH sensitive release systems. A polyelectrolyte is a macromolecule that dissociates to give polymeric ions on dissolving in water or other ionizing solvent. Because of the repulsion between charges on the polymer chain, the chain expands when it is ionized in a suitable solvent. If the solvent prevents ionization of the polyelectrolyte, the dissolved chain remains in a compact, folded state. The interplay between hydrophobic surface energy and electrostatic repulsion between charges dictates the behavior of the polyelectrolyte (Bromberg 2003). All pH sensitive polymers contain pendant acidic or basic groups that either accept or donate protons in response to the environmental pH (Bartil et al. 2007). Swelling of a hydrogel increases as the external pH increases in the case of weakly acidic (anionic) groups, but decreases if the polymer contains weakly basic (cationic) groups. When the ionic strength of the solution is increased, the hydrogel can exchange ions with the solution. By doing so, the hydrogel maintains charge neutrality and the concentration of free counter ions inside the hydrogel increases. An osmotic pressure difference between the hydrogel and the solution arises and causes the gel to swell. When the ionic strength is increased to high levels (1–10 M), the hydrogel will shrink. This is due to the decreasing osmotic pressure difference between the gel and the solution until the solution osmotic pressure equilibrates the osmotic pressure inside the gel.

Magnetic-sensitive release systems. It is well known that all living organisms, including animals and humans, contain magnetic particles that act as magnetic receptors (Bahadur and Jyotsnendu 2003) and the magnetism and magnetic materials have a strong role to play in health care and biological applications (Mykhaylyk et al. 2001; Kim et al. 2001; Jordan et al. 1999; Poeckler-Schoeniger et al. 1999). The combination of fine particles and magnetism in the field of biology and biomaterial has been found useful in sophisticated biomedical applications including gene and drug delivery, and magnetic intracellular hyperthermia treatment of cancer (Shinkai et al. 1999). A treatment method that involves the administration of a magnetic material composition containing single-domain magnetic particles consists in the application of an external static or alternating magnetic field which has to be active at targeted place. The magnetic material composition (drug carrier) causes the triggered release of therapeutic agents at target place. The triggering effects of an external applied magnetic field consists either in structural alterations that cause the leaking of the drug or a thermal effect which triggers the drug release. Pankhurst and co-workers studied the physical principles underlying some current biomedical applications of magnetic nanoparticles, such as the relevant physics of magnetic materials and their responses to applied magnetic fields (Pankhurst et al. 2003; Bajpai et al. 2008; Robin 2002). They studied cancer treatment by *electromagnetically activated* "nanoheaters". Heating cancer cells in a clinical setup can be achieved by various technologies such as focused ultrasound, radio frequency, thermal radiation, lasers and magnetic nanoparticles. The size of the magnetic nanoparticles lies in the range 5–90 nm. To turn these particles into heaters, they are subjected to an oscillating electromagnetic field, where the field direction changes periodically. Application of the magnetic field generates a directional force on each magnetic particle. As the magnetic field oscillates at high frequency, the average force on the particle is zero. The energy of the oscillation is converted into heat, raising the temperature of the nanoparticles and their biological material.

Electric-sensitive release systems consist in electrically responsive delivery systems and they are prepared from polyelectrolytes (polymers that contain a relatively high concentration of ionizable groups along the backbone chain) and are thus pH-responsive as well. Under the influence of an electric field, electroresponsive hydrogels generally deswell or bend, depending on the shape and orientation of the gel. The gel bends when it is parallel to the electrodes, whereas deswelling occurs when the hydrogel lies perpendicular to the electrodes.

The used way or ***method for delivering drugs*** into systemic circulation plays also a significant role in the design and formulation of NDDSs. In addition to the commonly used oral and injection routes, drugs can also be administered through other means, including transdermal, transmucosal, ocular, pulmonary or implantation (Hughes 2005).

In the ***peroral route***, for example, the gastrointestinal physiology influences the design of peroral DDSs and in particular small intestine and colon, where the drug can be effectively absorbed (Fix 1999). The controlled-release approaches, in this case, are the osmotic DDS (Verma et al. 2000) due to the continuous presence of water in the gastrointestinal tract (GIT). For activating and controlling the release in a zero order fashion (a constant release of drug independent of concentration and drug loading in the DDS) has justified the development of matrix-controlled release systems based on pseudo-zero-order or first-order drug-release kinetics.

The ***transdermal route*** of drug administration (Naik et al. 2000) showed efficacy by nasal mucosa which is considered to be the most permeable and also the lung which offers opportunity for systemic delivery because of their enormous surface area and rich blood supply (Gonda 2000). For these it has been required to design aerosol particles of desired characteristics. In transdermal route the use of microchips as controlled drug-delivery devices, a concept that originated from Langer's lab at the Massachusetts Institute of Technology, USA, is reported (Santini et al. 2000). Pillai and coworkers (2001) reported a technology based on tiny silicon or polymeric chips containing hundreds or thousands of micro-reservoirs, each of which can be filled with any combination of bioactive agents, and the complex release patterns can be achieved with programmed microprocessors. In tissue engineering, controlled release concepts are being applied for the delivery of growth factors to nurture the cells encapsulated in biocompatible polymers (Tabata 2000).

Parenteral administration (injection), which is the option for orally undeliverable drugs, has advanced as systemic and local drug-delivery (Cleland et al. 2001). This way make possible to deliver the active curing agents, by conjugating to polymers, or by entrapping them in colloidal carriers.

16.2.3 Pharmacokinetics, PK, and Pharmacodynamics, PD

During 100 years of evolution, the better appreciation and integration of pharmacokinetic (in other words "how does the body deal with drugs?" (Pillai et al. 2001), and pharmacodynamic ("what does the drug do in the body?" (Pillai et al. 2001), principles in design of drug delivery systems (DDSs) contributed to outstanding treatment metamorphose toward molecular medicine. The whole concept of therapeutics has been redefined from mainly symptomatic treatment to preventing and curing the cause of disease. Drug delivery has also graduated from conventional pills to sustained/modulated and controlled release by sophisticated programmable (intelligent) delivery systems. Consequently drug delivery has become more specific from systemic to organ, tissue and cellular targeting.

PK and PD concepts are merely attempts to understand mathematically the movement of the drug in the body [PK] and the drug's issues with receptors [PD] respectively. Each drug is characterized by its own PK–PD profile on the basis of the physicochemical properties, conformation and other structural attributes that govern the transport within the body and across various barriers. Extremely difficult is to delivery drug to the brain, because challenging the significant barriers, as called, blood–brain barrier (BBB) and the blood cerebrospinal fluid barrier. This task has been attempted by targeting drug- formulations to specific transporters in the BBB (Bruke et al. 1999). The discovery of special transporters has opened the scope for ***cellular delivery*** of drugs and macromolecules (Lindgren et al. 2000). New chemical approaches in gene therapy allowed for the transfer of genetic information to target cells enabling them to synthesize the protein encoded by the gene (Mountain 2000). As well as the route of administration, the number and type of biological barriers, a drug has to overcome, also governs the design of delivery systems (Pillai et al. 2001).

Compared with the general 'sigmoid' PK–PD profile for conventional molecules, the PK–PD relationship of biopharmaceuticals is complicated by short biological half-life, instability, multiple biological actions and operation of compensatory regulatory events in the body. Once the PK–PD relations have been established, plasma levels can be substituted for therapeutic effects that can aid in setting PK bioequivalence standards (Pillai et al. 2001).

Chronotherapeutics, the study of pharmacological response as a function of physiological mood, physical activity, biological rhythms etc. has redefined drug-delivery strategies in the treatment of diabet, ulcers, cardiac attacks, asthma, smoking cessation and other addictive illnesses (Cullander and Guy 1992). Since 1999 Breimer affirmed that the translation of a PD (pharmakodynamic)-dependent delivery pattern into improved therapeutic outcome is a major challenge for drug-delivery research (Breimer 1999).

Very recently seems that the more sophisticated farmacologic principle of chronotherapy has come back in the reserher's attention. The current advances in chronobiology and the knowledge gained from chronotherapy of selected diseases strongly suggest that "the one size fits all at all times" approach to drug delivery is no longer substantiated, at least for selected bioactive agents and disease therapy or prevention (Youan 2010). Thus, Youan (2010) showed that there is a critical and urgent need for chronopharmaceutical research (e.g., design and evaluation of robust, spatially and temporally controlled drug delivery systems that would be clinically intended for chronotherapy by different routes of administration). A brief overview of current drug delivery system intended for chronotherapy was reported. Youan found that there are three major hurdles for the successful transition of such system from laboratory to patient bedside that include the challenges to identify adequate (i) rhythmic biomaterials and systems, (ii) rhythm engineering and modeling, perhaps using system biology and (iii) regulatory guidance. Ohdo (2010) has brought into attention the fact that 24 h rhythm is demonstrated for the function of physiology and the pathophysiology of diseases. The effectiveness and toxicity of many drugs vary depending on dosing time. Such chronopharmacological phenomena are influenced not only by the pharmacodynamics but also pharmacokinetics of medications. The underlying mechanisms are associated with 24 h rhythms of biochemical, physiological and behavioral processes under the control of circadian clock. Thus, the knowledge of 24 h rhythm in the risk of disease plus evidence of 24 h rhythm dependencies of drug pharmacokinetics, effects, and safety constitutes the rationale for pharmacotherapy (Ohdo 2010). Chronotherapy is especially relevant, when the risk and/or intensity of the symptoms of disease vary predictably over time as exemplified by allergic rhinitis, arthritis, asthma, myocardial infarction, congestive heart failure, stroke, and peptic ulcer disease. However, the drugs for several diseases are still given without regard to the time of day. Identification of a rhythmic marker for selecting dosing time will lead to improved progress and diffusion of chronopharmacotherapy. To monitor the rhythmic marker such as clock genes it may be useful to choose the most appropriate time of day for administration of drugs that may increase their therapeutic effects and/or reduce their side effects. In Ohdo opinion, to produce new rhythmicity by manipulating the conditions of living organs by using rhythmic administration of altered feeding schedules or several drugs appears to lead to the new concept of chronopharmacotherapy. From viewpoints of pharmaceutics, the application of biological rhythm to pharmacotherapy may be accomplished not only by the appropriate timing of conventionally formulated tablets but the special drug delivery system to synchronize drug concentrations to rhythms in disease activity Ohdo said. New technology for delivering medications precisely in a time-modulated fashion, underlying mechanisms and usefulness, is overviewed from viewpoint of chronopharmacology and chronotherapy (Ohdo 2010).

16.3 Silica Based Nanoparticles as Controlled Drug Release Systems

Owing to high chemical and thermal stabilities, large surface areas and good compatibilities with other materials including biosystems, porous silica is gainging more and more applicability including drug delivery area. The new developed sol–gel technology offers practically countless possibilities for incorporating biologically active agents within silica gel and for controlling their release kinetics from the gel matrix (Li et al. 2004c). Amorphous colloidal and random porous silica has been firstly proposed as a drug delivery system on the basis of silica its own embedding abilities and biocompatibility (Barbe et al. 2004) as a direct mixture of both silica sol and the drug. Often that results in heterogeneous dispersion of drug in the gel volume affecting the drug release rate, reproducibility and control (Tourne-Peteilh et al. 2003). The discovery of ordered mesoporous silica opened the possibility of new developments in various fields of chemistry, such as host-guest synthesis of nanostructured materials.

16.3.1 Silica a Sol-Gel Derived Material, Ordered Mesoporous Silica, Hybrids

The syntheses of mesoporous materials, with regularly or in purpose designed sized pores by means of the supramolecular templating of a sol–gel process represent a starting point in the design of functional nanostructured materials.

Ariga et al. (2007) in a excellent review present, in a well documented and elegant way, *the flexibility of the synthesis and formulation methods* for designing and developing *the mesoporous nanospaces*, if the fundamental principles of supramolecular and coordination chemistry are taken as the leading concept. It was revealed that the structural dimensions of mesoporous materials permit access by functional supermolecules, including coordination complexes, and control of their functionality can be achieved by variation of pore geometry.

The hybridization of inorganic precursor allows the development of innovative multifunctional materials having complex architectures and combining properties of organic, biological and inorganic components (Kickelbick 2003; Schubert and Husing 2005; Kickelbick and Schubert 2001). Hybrid inorganic–organic materials are produced, by using organic siloxane precursors, when chemically active groups, containing ***non-hydrolysable Si–C bonds***, are covalently linked to the inorganic framework of mesoporous materials (Di Pasqua et al. 2007; Macquarrie 1996; Burkett et al. 1996; Lim and Stein 1999; Brunel 1999; Park and Komarneni 1998; Macquarrie et al. 2001; Brunel et al. 2001) but also can be obtained as hybrid composites, as physical mixtures (Ionescu et al. 2004).

One unique approach to the preparation of mesoporous organic–inorganic hybrids is synthesis of periodic mesoporous organosilicates (PMO), which was independently initiated by three groups: Inagaki group (Inagaki et al. 1999, 2002; Kappor and Inagaki 2006; Asefa et al. 1999 and Stein group (Melde et al. 1999) in 1999 using organic molecules bearing multiple alkoxysilane groups, such as bis(triethoxysilyl) ethene and bis(triethoxysilyl)benzene. Various organic components can be introduced as a PMO pore wall. Modifications of the mesopore framework and interior have been widely studied (Vinu et al. 2005) leading to facile accommodation of functional supramolecular materials including coordination complexes. In the co-condensation method (direct synthesis), an organosilane is hydrolytically condensed with conventional silica sources such as tetramethyl orthosilicate and tetraethyl orthosilicate (Maschmeyer et al. 1995; Lim and Stein 1999; Walcarius et al. 2004). The one-pot pathway of the co-condensation method provides several advantages such as homogeneous distribution of the functional groups and short preparation time. Grafting (post synthesis) is one modification method for pre-synthesized mesoporous silica through direct reaction of organosilanes with the silica surface (Burkett et al. 1996; Mercier and Pinnavaia 2000). Distribution of functional groups is sometimes non-uniform but this method does not compromise the framework structure of the parent mesoporous materials (Ariga et al. 2007).

The ordered mesoporous silica materials, have gained "celebrity", it can be said, due to their spectacular appearance and popularity, being continuously in attention.

They are also continuously designed, developed and used in unnumbered fundamental and experimental issues. We believe it is worth to give a brief historical remembering of scientists, who for the first time brought them in attention, synthesizing and giving them a name, associated with some essential characteristics (Ariga et al. 2007). This will be farther helpful because discussing various issues they are involved in, everything will be better understood.

In 1990, Kuroda and coworkers first reported the preparation of mesoporous silica with uniform pore size distribution through intercalation of cetyltrimethylammonium cations into the layered polysilicate kanemite followed by calcination to remove the organic moiety. These materials were named FSM-16 (**F**olded **S**heet **M**aterials) (Yanagisawa et al. 1990; Inagaki et al. 1993). Later, Mobil scientists supplied materials having large uniform pore structures, high specific surface area, with specific pore volume and hexagonal geometry (MCM-41, MCM coming from **M**obil **C**omposition of **M**atter) (Kresge et al. 1992; Beck et al. 1992), cubic geometry (MCM-48) (Vartuli et al. 1994) or lamellar geometry (MCM-50) (Dubois et al. 1993). Tanev et al. prepared HMS (**H**exagonal **M**esoporous **S**ilica) using a neutral amine as template (Tanev and Pinnavaia 1995) and Bagshaw et al. similarly synthesized a disordered mesoporous material designated as MSU-1 (**M**ichigan **S**tate **U**niversity) using polyethylene oxide (PEO) (Bagshaw et al. 1995). Stucky and coworkers developed highly ordered large pore mesoporous silica SBA-15 (**S**anta **B**arbara **A**morphous) with thicker pore walls and a two dimensional hexagonal structure using a amphiphilic triblock copolymer of poly(ethylene oxide) and poly(propylene oxide) in highly acidic media (Zhao et al. 1998a, b). The same group also prepared MCF (**M**eso **C**ellular **F**orm) type materials where triblock copolymers stabilizing oil in water microemulsions were used as templates (Schmidt-Winkel et al. 1999). Apart from these materials, mesoporous silica materials with various abbreviated names have been developed continuously (Yue et al. 1997; Kelleher et al. 2001; Che et al. 2003; Liang et al. 2005). Beside mentioned MCM, SBA-15, MCF etc. materials, Davidson has reviewed the abbreviations used in the literature, for mesoporous silicas, with regular mesopore diameters between 20 and 500 Å, as defined by IUPAC. SBA type silicas, from University of California at **S**anta **B**arbara, have also other members then SBA-15 that present different symmetries, such as, triblock SBA-12: 3-D hexagonal copolymer; SBA-16: cubic; SBA-11: cubic; SBA-11: cage-type, cubic. KIT-1 silica is coming from **K**orea Advanced **I**nstitute of Science and **T**echnology and MSU-X is coming from **M**ichigan **S**tate **U**niversity. Last two are mesoporous silicas disordered materials (Davidson 2002). We obtained silica matrices by sonocatalysed sol gel process with narrow pore distribution (Savii et al. 2009) and also by using Butymethyl pyridinium Telrafluoroborate Ionic Liquid as additive, by progressively substituting it for usually alcoholic solvent (Putz et al. 2010).

Here, it has to be mentioned that more or less porous and ordered, ***the amorphous state of silica*** mesoporous matrix is the only one that is suitable to be used in biological purposes, because crystalline silica was found to be toxic. Santarelly and coworkers (2004) found that the incubation of human endothelial cells with crystalline – silica leads to alteration typical of apoptosis. Anyway further studies are required to elucidate the precise mechanism responsible for crystalline – silica induced apoptosis in endothelial cells (Santarelli et al. 2004).

16.3.2 Matrix – Drug Interactions

Understanding the sensing-responsive behaviour, in the case of silica based CDDSs, with deep implications in the loading and release processes of a drug it is helpful in order to provide the intimate mechanism of modulated release kinetics. Both loading and release are closed to the matrix (carrier) – drug interactions.

As the wall of the pore contains free silanol groups, in the case of bare porous silica resulted from sol-gel synthesis, these latter can react only with the appropriate organic functional group present in molecules of drugs, those having –OH affinities for, otherwise the possible bonds to be developed are relatively weak. It was found that functionalisation of mesoporous materials can change surface character, including hydrophilic/hydrophobic properties and/or chemical affinities toward molecules to be further attached. As example could be implied the reported cases of functionalized MCM41 (Qu et al. 2006; Munoz et al. 2003; Zeng et al. 2005, 2006; Ramila et al. 2003; Wang 2009) and SBA-15 (Doadrio et al. 2006; Song et al. 2005) where the these changes from bare to functionalized silica are influencing drug loading and delivery rate. *In vitro* conditions behavior of a series of mesoscopically ordered silicas with different pore sizes, pore connectivity, and pore geometry were monitored. The degree of drug loading was found to be dependent on the specific surface area and the pore diameter of the host matrix. Organic modified mesoporous silica drug carriers properties, including texture ones (specific area and porosity) are significantly affected by functionalisation. The introduction of functional groups on the surface, to have specific host–guest chemical interactions with drugs, will result in a selective character of the carrier matrix and that will improve the control over drug load and over the releasing pattern of the guest drug, a critical parameter for clinical applications release (Ramila et al. 2003).

16.3.2.1 Sustained Release, Kinetics

The drug release process can be divided into four consecutive steps: the imbibition of release medium into the microspherical system driven by osmotic pressure arising from concentration gradients, drug dissolution, drug diffusion through the continuous matrices of microgels due to concentration gradients, and drug diffusional and convective transport within the release medium (Wang 2009). One or more of these steps can control the drug release process (Li et al. 2004a, b, c).

Generally, in the case of porous silica, ***a sustained release mechanism*** could be met. Many controlled-release products are designed on the principle of embedding the drug in a porous matrix. Liquid penetrates the matrix and dissolves the drug, which then diffuses into the exterior liquid (Marty et al. 1982). Wiegand and Taylor (1959) and Wagner (1959) showed that, for many controlled-release preparations reported in the literature, the percentage of drug released versus time shows a linear apparent first-order rate.

Higuchi tried to relate the drug release rate to the physical constants based on simple laws of diffusion. Release rate from both a planar surface and a sphere was considered. The analysis suggested that in the case of spherical pellets, the time required to release 50% of the drug was normally expected to be 10% of the time required to dissolve the last trace of solid drug in the center of the pellet Higuchi (Higuchi 1961, 1963) was the first to derive an equation to describe the release of a drug from an insoluble matrix as the square root of a time-dependent process based on Fickian diffusion, Eq. 16.1:

$$Q_t = [2DS\varepsilon(A - 0.5S\varepsilon)]^{1/2} \times t^{1/2} = k_H \times t^{1/2} \quad (16.1)$$

where, Qt is the amount of drug released in time t, D is the diffusion coefficient, S is the solubility of drug in the dissolution medium, ε is the porosity, A is the drug content per cubic centimeter of matrix tablet, and k_H is the release rate constant for the Higuchi model.

Higuchi drug release equation can be written also as, Eq. 16.2:

$$Q = [(2A - \mathrm{C_s})\mathrm{DC_s}]^{1/2} \times t^{1/2} \quad (16.2)$$

where A,D,$\mathrm{C_s}$ are constants, so, it can be simplified as, Eq. 16.3:

$$\mathrm{Q = K_H t^{1/2}} \quad (16.3)$$

Higuchi equation, amount released per surface area, Q, versus square root of time, t, should yield a straight line.

For granular matrices Higuchi's equation becomes, Eq. 16.4:

$$Q = \left[D_{\varepsilon/\tau}(2A - \varepsilon\mathrm{C_s})\mathrm{DC_S}\right]^{1/2} \times t^{1/2} \quad (16.4)$$

where ε is porosity and τ is tortuosity. The studying real systems, the Higuchi release experimental pattern could present a deviation from ideal Higuchi release pattern. Mathematical method proposed to quantitatively express the deviation from Higuchi kinetics (Gohel et al. 2000) may be extended to other systems. As shown, the most of drug delivery system based on mesoporous silica can be described using the Higuchi diffusion model. In general, in the case of a matrix in which the release is governed by the diffusion of the drug through the matrix, the release kinetics follows the conventional Higuchi relationship (n = 0.5). Most of the work on the kinetics of the release of drugs from mesoporous carrier materials is frequently described using Higuchi relationship and shows a two-step release profile composed of an initial burst followed by slow release (Doadrio et al.

2004, 2006; Andersson et al. 2004, 2005; Kim et al. 2001). This could be attributed to physical and/or chemical character of drug entrapping mechanism. Drug release from mesoporous materials is diffusion control, which is similar to other drug delivery systems such as hydrogels or biopolymers. The rate constant of drug release kinetics follows the order of MCM-41 > SBA-15 > MCM-48 > HMS (Wang 2009).

16.3.2.2 Modulated Release, Targeting-Triggering

Compared with the sustained release system, ***the stimuli-responsive controlled-release system*** in other words ***modulated release mechanism*** stimulated by so called *triggers* can achieve a site-selective, controlled-release pattern, which can improve the therapeutic efficacy (Slowing et al. 2008). Slowing showed that such carrier material should be, first of all *biocompatible*, to be able to high loading/encapsulation of desired drug molecules, to present zero premature release of drug molecules, to have ability to site directing by recognizing cell type or tissue specificity, to be able *to release* of drug molecules in a controlled manner (a proper rate of release to achieve an effective local concentration). It is also important that the empty carrier, after releasing their cargo, to be eliminated from biologic system in order to avoid accumulation and unexpected toxic effects. Several natural and synthetic biodegradable organic materials, such as polymeric nanoparticles, dendrimers, and liposomes, have been used as "smart" drug delivery systems that can controllably release pharmaceutical drugs in aqueous solution upon the structural degradation of the carrier, triggered by various chemical factors, under physiological conditions. Such DDS systems have been prepared, but it was difficult to achieve "zero" premature release of drugs in these structurally unstable "soft" materials (Kwon et al. 2007).

Silica is often the material of choice to enable the biological use of inorganic nanoparticles (Bottini et al. 2007; Gerion et al. 2007; Graf et al. 2006) as CDDS Slowing et al. (2008) said. For controlled release applications, it has been shown that *silica* posses some suitable properties. Silica is *biocompatible, biodegradable* at some extent, *able to* store and *gradually release* therapeutically relevant drugs like antibiotics (Meseguer-Olmo et al. 2006; Radin et al. 2004; Kortesuo et al. 1999a). In hybrid architectures, silica is also able to enhance the biocompatibility of several drug delivery systems, such as magnetic nanoparticles (Dormer et al. 2005; Arruebo et al. 2006; Zhao et al. 2005) biopolymers (Allouche et al. 2006) and micelles (Huo et al. 2006; Slowing et al. 2008). Porous silica, which are comprised of a porous structure with huge surface areas and countless of random or ordered empty channels that are able to absorb/encapsulate relatively large amounts of bioactive molecules, have been intensively studied. Many synthesis methods, including those based on surfactant-templated synthesis, originating in 1992 (Kresge et al. 1992) were applied, many have explored the functionalization and utilization of these materials for various applications, such as catalysis (Gruen et al. 1997) separation and sensors (Unger et al. 2000). But there was no report on the utilization of these materials for controlled release and delivery application until the early 2000s. This was explained by Slowing and coworkers (2008) to be due to the lack of morphology control, when using conventional synthesis methodologies.

16.3.3 Biocompatibility, Biodegradability, Excretion

In order to avoid the toxicity risks associated with accumulation, following delivery of the drug payload, biodegradability is a key parameter for the acceptance of drug delivery systems containing silica particles, both nano and micron-sized. The biocompatibility of mesoporous silica nanomaterials, MSN, with and without surface functionalization has been tested by different methods. Growth rates for cells exposed to MSN were found to be similar to the ones of cells grown in the absence of MSN (Gerion et al. 2007; Slowing et al. 2006).

Concerning the long-term biocompatibility and the biodegradation of MSN area of research there is a lot of work to be done. It was found in one report (Kortesuo et al. 1999a) on the long-term biocompatibility of silica implants in animals, indications that these materials did not present any toxic side effects for a period of 42 days.

Finnie et al. (2009) presented studies (both *in vitro* and *in vivo*) demonstrating that silica implants are degradable materials (Falaize et al. 1999; Radin et al. 2002; Viitala et al. 2005a) even so there remains a widespread misconception that silica is insoluble in aqueous conditions.

Silica undergoes hydrolysis to form silicic acid, $Si(OH)_4$. *In vivo*, following the degradation of silica gel granules, silicic acid was found to diffuse through the blood stream or lymph and was excreted in the urine at a rate of 1.8 mg silicon per day (Lai et al. 1998).

The reaction to form silicic acid is catalysed by OH^-, hence the rapid increase in hydrolysis rate with increasing pH. At pH = 9 the hydrolysis rates shows an increase in excess of three orders of magnitude comparing pH 2. Between pH 2 and 9, the silica equilibrium solubility, which depends on the nature of the silica, remains relatively constant, but increases sharply with increasing pH above 9 (Iler 1979). Silica could be of different nature, regarding structure, particle size and preparative history, in function of required properties the silica matrix as drug carrier has to meet. A recent study (Finnie et al. 2009) has shown few new aspects regarding biodegradability of porous sol–gel silica microparticles in physiological buffers. For nonporous, amorphous silica, the equilibrium concentration of silicic acid is ~70 ppm silica at 25°C, crystalline silica is ten times less soluble, but sol–gel derived highly porous silica nano- and microparticle systems present a solubility around 120 ppm (Iler 1979). By using dynamic (flow) conditions, Finnie and coworkers (2009) have studied the relative rates of dissolution of several types of sol gel synthesized porous microparticles in comparison with commercial sample of silica microparticles with similar surface area. They used two types of experiments, an 'open system' configuration to simulate the silica in vivo degradation and a 'closed system's configuration in which the addition of serum proteins creates better conditions in order to correlate the *in vitro* data with the in vivo situation. By manipulating the process variables, especially the silica precursor's nature, the sol gel derived samples were obtained as 'polymeric silica', PS, 'colloidal silica', CS, and 'waterglass silica', WS. The obtained samples differed each from other by both the particle internal structure (i.e. porosity and surface area) and chemical structure (degree of condensation) which could influence the dissolution rate. In 'open system' regime the dissolution rate appeared relatively constant with time, with the exception of the WS particles. The CS sample was slowest to dissolve (8%), followed by commercial sample (11%) and PS sample (26%) in 6 h. After 40% had been dissolved, WS sample dissolution dropped at 24% of the initial rate, the observation being correlated with particle size dispersion, smaller particles are expected to dissolve more rapidly than the larger ones. Regarding the influence upon silica dissolution, silicon connectivity was calculated from MAS NMR data as: $C = \Sigma n * Qn$, where n is the number of links to a neighbouring silicon and Qn represents the proportion of silicon centres with n links to other silicon atoms ((e.g. Q1 denotes a silicon with one Si–O–Si bond only), whereas Q4 denotes a silicon with four Si–O–Si linkages (fully condensed silica)). The degree of condensation, to facilitate the comparison, is defined by Viitala et al. (2005b) (Q2 + Q3), where Qn are defined as above. It was found that the dissolution rate trends down with increasing extent of condensation if monitored using Q4, this is not entirely consistent. While the polymeric sample has Q4, Q3 and Q2 proportions very similar to the waterglass sample, the dissolution rate is considerably slower. This is further exemplified by the only slight difference in connectivity and condensation observed between the water glass and polymeric sample. In contrast to what Viitala et al. (2005b) have postulated, the decrease in dissolution rate, seems not to be caused by an increase in the condensation (or connectivity) of the silica network. Rather, there is a clear trend linking the BET surface area with dissolution rate Finnie said. This appears reasonable, given that dissolution is essentially a hydrolysis reaction, and increased surface area implies enhanced exposure of the surface to the reactant (i.e. water). Dissolution experiments conducted for polymeric, waterglass and commercial silica in simulated gastric fluid at pH = 2 sowed negligible dissolution (0.5% after 6 h exposure), as expected given the dramatic difference in hydrolysis rates between pH 2 and 7.4 (Iler 1979). This corresponds with the observations of Begu et al. (2007) who showed that impermeable silica shells deposited on liposomes were rapidly (40–60 min) dissolved at pH = 7.4 but remained intact at pH = 1.2. The effect of increasing the colloidal silica sample size in closed system results in a critical concentration above which saturation effects start to become apparent, and lower increases in dissolution rate would be expected for larger samples at the same flow rate. It was also observed that the mass of silica dissolved increased with increasing sample mass, the proportion of the sample dissolved decreased with increasing sample mass. The authors concluded, the both observed effects of connectivity and sample size together with the dependence of the dissolution rate on the flow rate, suggest that the dissolution of the silica particles will vary depending on their location in the body and biological parameters such as the degree of blood irrigation and the quantity injected (or ingested). It was apparent that over the micron size range, neither particle size nor pore diameter has a significant effect on dissolution rates (Finnie et al. 2009). As in the open system, experiments conducted in closed system showed the dissolution rate increased in the order colloidal\comercial\polymeric silica, with 67%, 88% and 96% respectively of the samples dissolved in 36 h (Finnie et al. 2009). The corresponding dissolution of silica samples in 10% bovine serum showed a similar pattern, but slowed of approximately 20–30% comparing dissolution medium without bovine serum (Finnie et al. 2009). Similar results were obtained by Falaize et al. (Falaize et al. 1999; Kortesuo et al. 1999a) who determined rates of dissolution of sol–gel silica xerogel particles with an average size of 600 nm, in simulated body fluid (SBF) with various supplements. Serum proteins such as albumin do absorb strongly on oxide particles (Kresge et al. 1992; Gruen et al. 1997) and it was suggested that the protein attached on the particle surface have an inhibitory effect on dissolution (Lee et al. 2010).

The *in vivo* biodistribution of nanoparticle and its following mechanisms of biodegradation and/or *excretion* determine the feasibility and applicability of nano-delivery platform in the practical clinical translation. Synthesis of the highly positive charge, near-infrared fluorescent mesoporous silica nanoparticles (MSNs) that demonstrate rapid hepatobiliary excretion, for use as traceable drug delivery platforms of high capacity. Souris et al. (2010) reported incorporation of MSNs with near-infrared fluorescent dye indocyanine green (ICG) via covalent or ionic bonding, to derive comparable constructs of significantly different net surface charge. *In vivo* fluorescence imaging and subsequent inductively coupled plasma-mass spectroscopy of harvested tissues, urine, and feces revealed markedly different uptake and elimination behaviors between the two conjugations; with more highly charged moieties (+34.4 mV at pH 7.4) being quickly excreted from the liver into the gastrointestinal tract, while less charged moieties (−17.6 mV at pH 7.4) remained sequestered within the liver. Taken together, these findings suggest that charge-dependent adsorption of serum proteins greatly facilitates the hepatobiliary excretion of silica nanoparticles, and that nanoparticle residence time *in vivo* can be regulated by manipulation of surface charge (Souris et al. 2010).

Degradation of superparamagnetic core–shell nanoparticles following treatment implying this kind of materials was investigated by histochemical methods. Liver sections were stained for iron detection with Prussian blue (Beduneau et al. 2009). Positive staining with Prussian blue was visualized primarily in association with Kupffer cells at 2 h after injection. With time increasing, the Prussian blue staining became weak, indicating the degradation of SPIO intracellularly. It is known that hydrolytic enzymes in lysosomes may degrade the intracellular SPIO through low pH exposure and iron ions are incorporated into the hemoglobin pool (Thorek et al. 2006).

16.3.4 Silica Based Materials Toxicity

Pethushkov and coworkers present in their recent work (Petushkov et al. 2010) a survey of the cytotoxicity of silica nanomaterials, highlighting examples in which the properties such as size or surface functionalization could be correlated with cytotoxicity. From the beginning was stated what Peters and coworkers showed, that the human endothelial cells possess a large capacity for the internalization of particulate matter in the nanometer scale (Peters et al. 2004). Even if it is known that, generally the nanoparticles may increase their toxicity or change the mechanism by which they induce toxicity, it cannot necessarily be extrapolated to engineered nanoparticles (Colvin 2003; Petushkov et al. 2010) This is why the studies of the toxicity of nanomaterials has to be done for each particular case, they being complex because of both nanomaterials contribution (added function, size- and shape-dependent properties) and biologic part (e.g. cells) contribution that may affect toxicity. Taking cells-silica possible interaction, as example, it is necessary to show that the toxic effects result eventually in cell death. Cells can die by either of two major mechanisms: necrosis or apoptosis (Kanduc et al. 2002). *Necrosis*, which can occur in a matter of seconds, is the death of the cells through external damage usually mediated via destruction of the plasma membrane or the biochemical supports of its integrity. The other major form of cell death, *apoptosis*, is much slower and is based on the concept of programmed cell death. Apoptosis requires from a few hours to several days, depending on the initiator. The manifestations of apoptosis both biochemical and morphological, are unique and are completely different from those of necrosis (Proskuryakov et al. 2005). In the case of soluble drugs, the level of toxicity usually is expressed by LD50 indicator, which value is the dose lethal to 50% of the cells (Petushkov et al. 2010). Asefa and coworkers (Di Pasqua et al. 2008) and Martens and coworkers have proposed that for materials in the solid phase, the conventional metric of cytotoxicity, LD50, to be changed with Q50 parameter that measure the number of particles needed to inhibit normal cell growth by 50%. Chen and coworkers found that determining whether nanomaterials are toxic depended greatly on diverse parameters, such as the structural properties of particles, their dosage forms, and their intended uses. The location of particles relative to the cells was found to influence their cytotoxicity. Particles that were internalized into cells caused different cytotoxicity than particles mantling the cell membrane (Petushkov et al. 2010).

The interaction of mesoporous silica materials with cells was the subject of several recent studies. The effect of surface functionalization of MCM-41 on cellular uptake (Slowing et al. 2006) was investigated and showed that the cellular uptake varied with surface functionality in that the LD50 increased with decreasing ξ-potential. The unfunctionalized MCM-41 was more toxic on a per particle basis toward human neuroblastoma (SK-N-SH) than aminopropyl- and mercaptopropyl-functionalized MCM-41 and that the most toxic mesoporous silica materials studied were those with the largest BET surface areas (Di Pasqua et al. 2008). Positively charged MCM-41 was shown to enhance cellular uptake and increased surface charge did not increase cytotoxicity (Chung et al. 2007). Recent study suggested that mesoporous silica inhibited cellular respiration (Tao et al. 2008). As we already have shown, it was reported that in contrast to amorphous silica (generally thought to be safe) crystalline silica materials are potential human carcinogens. Extensive studies of crystalline silica materials such as quartz dusts have shown that physicochemical properties such as particle size, shape, surface area, and surface chemistry all impact toxicity (Fubini 1997, 1998; Fubini et al. 1999; Giovine et al. 2002).

Mesoporous silica cytotoxicity experiments have resulted in various findings concerning the extent of its toxicity, related to experimental conditions. Amorphous silica materials are known to cause hemolysis of mammalian red blood cells, raising serious biosafety concerns over the use of amorphous silica for intravenous drug delivery. It is widely accepted that the hemolytic activity of silica is related to surface silanol groups (Slowing et al. 2009). Ordered MSNs on the other hand have a low surface density of silanol groups for MSNs caused by the unique honeycomb-like structure with arrays of 3 nm mesopores resulting in low hemolytic behavior (Tao et al. 2009). Hoet and coworkers investigated the cytotoxicity of amorphous silica nanoparticles (sizes varied from 14 to 335 nm in diameter) in endothelial cells (EAHY926). They determined that smaller silica nanoparticles are more toxic when the particle morphologies are similar and the doses are expressed in mass concentration. The 14 and 16 nm silica nanoparticles have ED50 = 33 and 47 mg cm^{-2}, respectively, compared to the 104 and 335 nm silica particles that have ED50 = 1,095 and 1,087 mg cm^{-2}, respectively. The surface area was also an important factor in toxicity and when the toxicity was compared in terms of surface area (mg cm^{-2}), the toxicity was similar for different sized silica particles. The mechanism for cell death was also investigated and it was found that, for smaller particles, necrosis is observed within a few hours. Exposure to crystalline silicates is known to be dangerous because of their cytotoxic potential that leads to pulmonary inflammation and pulmonary fibrosis when the particles are inhaled (Rushton 2007). It was observed a big difference between toxic behaviour of freshly quartz surfaces and aged ones. Hochstrasser and Antonini and others have shown that reactive surface radicals that are present on the fresh surfaces of crystalline silica are responsible for induced toxicity. These surface radicals can react with other molecules, such as CO_2, and produce new radicals (Hochstrasser and Antonini 1972; Vallyathan et al. 1988).

Few general facts can be understood from reported findings regarding silica toxicity. First, regarding bare silica, it was shown that silica materials states, either amorphous or crystalline, and random or ordered mesostructures, they toxic behavior differs. Secondly, by functionalizing silica-based nanoparticles, the mechanisms and level of toxicity induced in cells can be changed. Mesoporous silica shows high compatibility with biological media at concentrations adequate for pharmacological applications (Petushkov et al. 2010). In order to reach at the correct toxic potential of silica based materials it is necessary to completely and thoroughly characterize the materials being studied.

16.3.5 Sinthesis and Formulation of Silica Based CDDSs

Mishra et al. (2010) reported basic documented principles regarding synthesis methodologies of nanomaterials. Nanotechnology-based synthetic methods are most commonly developed on the basis of two rational designs: top-down or bottom-up engineering of individual components (Alexis et al. 2008).

The top-down process involves starting with a larger object and breaking it up into nanostructures through different methods, as etching, grinding, ball milling etc. As examples of top-down principle adopted are silicon microfabrication and photolithography. The main disadvantages of the top-down method are time consuming character, also they generates considerably broader particle size distribution, La-Van et al. (2003) said.

The bottom-up technique refers to synthesis based on atom-by-atom or molecule-by-molecule arrangement in a controlled manner, which is regulated by thermodynamic means (Keck et al. 2008). The process takes place through controlled chemical reactions, either gas or liquid phase, resulting in nucleation and growth of nanoparticles. Bottom-up techniques (like supercritical fluid antisolvent techniques, precipitation methods etc.) create heavily clustered masses of particles that do not break up on reconstitution (Shrivastava 2008; Mishra et al. 2010).

16.3.5.1 Synthesis Variables – Release Behavior

It was shown that the properties, affecting (silica) host – guest (drug) interactions with outstanding implication in loading – release of the cargo, could be influenced by synthesis techniques implicite synthesis variabile. The liberation rates into water of nifedipine from the nifedipine-silica composites was used by Bottcher (1998) to demonstrate that the releasing behavior of nifedipine from the silica matrix can be controlled to a high degree by the sol-gel technique process parameters. The sol concentrations, the rate of gel formation, gelling temperatures during sol-gel process, were observed to have a considerable influence on composite structure and releasing behavior. This observed relationship preparation procedure-product behavior is an indirect one. The direct influences, and their effects, eventually measurable, can be found step by step relating the processing variable to resulted material properties (structure, morphology, texture etc.) and then relating synthesized material specific properties to material behavior in the specific application it was meant to. Higher sol concentrations and gelling temperatures (i.e., high gel formation rates) means fast gelation, higher porosity of the composite gel, amorphous state (solubility) of the drug preservation and consequently to a faster rate of release. Slow gelation leads to partially drug crystallisation (insolubility) corresponding to slower release.

Carrier material particle size and shape, porosity and surface area, each of it and also in correlation, could play an important role all along drug loading – delivery chain steps, including elimination of empty inorganic carrier. This is why it is useful to know how paticle size and shape and porosity at their turn influence the DDS behavior and how they can be tailored in order to obtain a desired result. Bottcher observed an inverse proportionality relationship between release rate and particle size (in micrometric range), in case of nifedipine-silica composites by soaking samples in water under standard conditions. The observed behavior was confirmed by the decrease in the activation energies, calculated from the Arrhenius plot, referring to particle size range of (160–250 μm), due to reduced surface-to-bulk ratio as the grain size increases. Before liberation of the drug can occur, it must diffuse from within the increasingly large grain volume to the composite surface (Bottcher 1998). Chemical or physical changes to the silica matrix, induced by reaction chemical parameters manipulation, during sol-gel synthesis also can influence the releasing rate. For example, it was studied a chemical change by alkoxide precursor partial substitution (10; 30 and 50 vol%) MeSi(OEt)3 for Si(OEt)4. This resulted in successive slowing of the nifedipine liberation from the corresponding composites. Bottcher found few explanations for this behavior. First, the effect could be due to the increasing hydrophobic character of the matrix and a diminished ratio of mesopores and nanopores within the matrix, as methilethoxi- precursor amount increased. Minimum liberation was observed at 30 vol% MeSi(OEt)3). Second, slow release could be due to different mechanisms in the gelation and drying processes (Bottcher 1998).

The composite structure can also be physically altered by the addition of external additives such as sol-soluble low or high molecular weight compounds during silica-ibuprofen (I) composite preparation. For example (low molecular weight) sorbitol substitution at 20 wt.% for silica causes an increase rate of release. Alternatively (high molecular weight) PEG (50 wt.%) addition to silica progressively retarded the drug liberation. The explanations were connected to the composite particle destruction and to the crystallization of ibuprofen in non-destructed particle's areas (Bottcher 1998). Generally, the (I)-composites can be considered as matrix containing uniformly distributed drug. Liberation of the drug occurs through penetration of solvent into the pores, cracks and interparticular spaces of the matrix. The drug slowly dissolves into the permeating fluid phase and diffuses from the system along the solvent-filled capillary channels. For such matrix systems there are four general categories, differing in the method of solution of the drug within the matrix and in the mechanism of diffusion Bottcher said. For example, the category when the drug is dispersed in the matrix and diffusion occurs through solvent-filled pores, the constant is given by Higuchi eq. described for granular materials. It can be seen that the liberation rate depends on the diffusion constant, D_s, of the drug in the solvent, and increases with: (a) the ratio of porosity ε to tortuosity τ for the matrix; (b) the solubility, C_s, of the drug in the solvent used; and (c) the initial drug content C_d within the composite.

It was found (Bottcher 1998) that there is good correlation between drug release amount *and* $t^{1/2}$ for the (I)-silica-PEG composites. The small deviations from linearity after long leaching times are caused by the observed destruction of the composite grains. Due to the higher viscosity of the PEG-water mixture, diffusion occurs more slowly than for pure water and the releasing rates are lower than in

PEG-free composites. In polymer-free (I)-silica composites, diffusion model is only valid for the initial period. The course of further liberation can be described by zero-order kinetics. Because the diffusion rate of (I) is lower than its dissolution rate, when a saturated solution of (I) is formed the drug is able to form crystalline precipitates in the pores. Therefore, liberation is only controlled by the diffusion constant and is directly proportional to the leaching time. From a pharmaceutical point of view, it is desirable that a drug should have such a pattern of releasing behavior, Bottcher (1998) said.

Mesoporous monodispersed spheres (490–770 nm) of MCM-41 silica functionalized by 3-aminopropyltriethoxysilane grafting and nonfuntionalised randomly polymerized silica matrix were also studied as a potential carrier for controlled drug release, using ibuprofen as a test drug (Manzanoa et al. 2008). In the case of amine-functionalized MCM-41 microspheres, Manzano et al. observed a significantly slower drug release rate than irregularly shaped powders, which should have facilitate drug delivery control over a longer time period. A reasonable and interesting conclusion was inferred by Vallet-Regi and coworkers, consisting in the affirmation that from an industrial point of view, it is worthless to expend time, money and effort designing a determined particle morphology because, they demonstrated that for non-functionalized MCM-41 materials, drug release rate shows little dependence on carrier morphology (Manzanoa et al. 2008).

In summary, it has been proved (Bottcher 1998) that the liberation of nifedipine from sol-gel composites can be controlled to a high degree, and it is possible to produce composites giving uniform drug release over a long period of time and to adapt the releasing behavior to the pharmacokinetic demands. The liberation rate is influenced by the grain size of the composites, by modification of the silica matrix with hydrophobic monoalchylated alkoxide and by low or high molecular weight sol-soluble additives. In order to obtain comparable measurements it is necessary to achieve reproducible release rates, the grain size as well as the gelling and drying conditions of the composites must be defined and to rigorously use identical preparation conditions (Bottcher 1998).

As it was shown, mesoporous silicas are currently widely and continuously studied including as carrier matrices in drug delivery applications. Even so, the studies concerning the release behavior of bare silica are relatively few. Ma and coworkers (2010) studied mesoporous silica MCM-41 and MCM-41 like materials, by using modified synthesis experimental conditions to obtain larger pores and particles of ordered material, as 200–400 nm in diameter, which also are suitable for endocytosis by human cells. They started from the finding that when the mesoporous nanoparticles loaded with drugs get into the blood circulation, it will be faced with the pressure pulses about every 20 s due to the pressure gap between the systolic blood pressure (SBP) and diastolic blood pressure (DBP). The influence of pressure pulse on drug delivery has been studied and found that the delivery rate of ibuprofen in a simulating body fluid solution increases dramatically under the pressure pulse. Other studies, also deals with controlled release behavior of mesoscopic ordered silica (Izquierdo-Barba et al. 2009) and sol-gel derived materials (Lukowiak and Strek 2009).

Tamanoi and coworkers presents in detail the recent progress of utilizing silica nanomaterials as a delivery vehicle for nucleic acid-based reagents, citing as well few research works on the subject of utilizing silica as CDDSs based on the fact that it is able to store and gradually release therapeutics such as antibiotics (Abeylath and Turos 2008; Meseguer-Olmo et al. 2002, 2006) and other clinically-relevant compounds (Lu et al. 2007; Kortesuo et al. 1999a, b; Hsiao et al. 2008).

Surface functionalization of the silica is employed, with the precise scope, in order to enhance the interaction between the drug and the support. Seems that not always is so. Rosenholm and Linden discovered that in many cases, the effectiveness of the introduced surface functions is much lower than what could be expected, and the release rate from surface functionalized silica is often not very different from that of the bare silica support, suggesting that the drug–support interactions are weaker than assumed under physiologically relevant conditions (Rosenholm and Linden 2008a, b). They studied the adsorption of a model acidic drug, salicylic acid, to amino-functionalized mesoporous silica both from organic solvents, and from water as a function of pH, in order to rationalize these findings. It is shown that the nature of the organic solvent has a great influence on the loading degree, which however is more pronounced for the pristine silica materials due to absence of strong drug–support interactions. More importantly, the net effective surface charge of the adsorbent was found to control the adsorption process in water, and remaining silanols on the silica surface after functionalization have a marked influence on the drug–support interactions. Their results tried to explain the relatively minor influence of amino groups on the release of acidic drugs reported in the literature, and gives a rational basis for optimization of support–drug interactions.

16.3.5.2 Tayloring Porosity, Fractal Character

It is an interesting thing to see, that almost all efforts being done in order to obtain a corresponding carrier, the work eventually is resumed to create and preservate as big as possible "nothing", meaning as big as possible specific free volume, inside of porous silica matrix, this being necessary in order to maximize of drug cargo loading, to be further delivered at targeted location, such way to minimize the volume/mass ratio of CDDSs. As the polymeric silica and/or hybrid matrices are obtained as porous materials, to be a host for a plethora of guests, the strategies to control porosity is a crucial task. Brinker and coworkers (1990, 1993) opened another perspective, a peculiar one, to control sol-gel process parameters in order to tailoring the gel porous properties. Sol-gel derived materials often exhibit a random structure, with fractal character within a certain space domain, in other words they remain similar to itself at any length scale of observation within this space range (Mandelbrot 1982). Brinker and coworkers (1993) proposed a strategy, which, we believe, deserves a special attention, in order to control porosity (Kickelbick 2003). This is based on the physics and chemistry principle of sol-gel processing (Brinker et al. 1990). Depending on the synthesis conditions, during sol-gel processing, by following different mechanism of condensation, a wide spectrum of structures ranging from weakly branched polymers to fully condensed particles may be prepared from silicon alkoxides. It is unanimous accepted that acid catalysed sol-gel process ideally could develop almost linear silica chains by cluster-cluster aggregation mechanism, and basic catalysed sol-gel leads to almost fully condensed particles following the monomer-cluster aggregation mechanism, highly favored by the silica solubility in high pH solutions. The distinction between polymers and particles can be made from **S**mall **A**ngle **S**cattering, SAS, studies, by using **X**-ray (SAXS), **n**eutons (SANS) etc. that probe structure on the ~0.1–50 nm length scale or **D**iffusion **L**ight **S**cattering (DLS) for other scales. The Porod region, in a certain wave vector (length scale) limits, reflects the primary particles structural characteristics, and P, is the Porod slope of the plot of log scattered intensity (Iq) versus log scattering wave vector, Q; $\mathrm{LogIq} = F(Q)$, where, Eq. 16.5

$$Q = (4\pi/\lambda)\sin(\theta/2), \tag{16.5}$$

and λ is neutron beam wavelength and θ is scattering angle. The Porod region is limited by Guinier region of scattered intensity plot, at in higher scale (smaller wave vector) reflecting more or less the bigger aggregates structural characteristics. At higher values of wave vector and smaller scales, Porod region is neighboring by Bragg region of scattered intensity plot. Porod slope, P is related to the mass fractal dimension, D_f, by equation $P = D_f - 2D$ (Schaefer 1988). Fractal models and the structure of materials, where D_f is fractal dimension, and $D = 3$ is the dimension of space. For uniform (non-fractal) objects such as dense particles, the mass fractal dimension $D_f = 3$ and surface fractal dimension $D_s = 2$, P reduces to -4. For mass fractal objects, $D = D_m$ and $P = -D$. For surface fractal objects, $D = 3$ and $P = D_s - 6$. The mass fractal dimension D_f relates an object's mass M to its (size) radius r_c $M \sim r_c^{Df}$ (Mandelbrot 1982), where for mass fractal objects, D_f is less than the embedding dimension of space $D = 3$. The surface fractal dimension D_s relates an objects area, A, to its size $A \propto r_c^{Ds}$ (Mandelbrot 1982): where for $D = 3$ the surface fractal dimension is $2 < D_s < 3$. In three dimensions space, the non fractal object density is a constant within all objects volume; the density of a mass fractal cluster decreases with increasing distance from its center of mass, $\rho \propto 1/r_c^{(3-D)}$. Because density is inversely related to porosity, this relationship requires that, unlike Euclidean objects, fractal objects become more porous as their size increases. It was show that this property may be exploited to tailor the pore structure of films deposited from fractal precursors (Scherer 1992; Brinker and Scherer 1990). Virtually, all particle packing concepts utilize dense particles that are packed together to create pores of a size related to the primary particle size (Leenars et al. 1984; Larbot et al. 1986; Anderson et al. 1988). Ideally, when (spherical) ***dense particles*** are assembled into a colloidal crystalline lattice, only ***monosized*** spherical particles ***packing always results in a maximum porosity of about 33%*** (theoretically calculated which do not depend of sphere diameter). The question is how it is possible that porous silica aerogels, for example, to attend such a huge free volume, more than 99% as and low density as 0.003% $\mathrm{g{\cdot}cm^{-3}}$. A *strategy* that is generally applicable to the wide range of polymeric sols including silica ones, characterized by a mass fractal dimension, is ***aggregation***. This strategy depends on the scaling relationship of size and mass of a fractal object, causing the density of an individual fractal cluster to decrease (porosity to increase) with cluster size r_c (Brinker and Scherer 1990). When the individual fractal clusters comprising the sol are concentrated during slipcasting or evaporation, this porosity is incorporated in the film or membrane providing that: (1) the clusters do not completely interpenetrate and (2) that there exists no monomer or low molecular weight species that are able to fill-in the gaps of the fractal cluster (Logan et al. 1992), sol-gel films with tailored microstructures. The ease of interpenetration of clusters depends on their mean number of intersections: the fewer the number of intersections, the greater the ease of interpenetration. Mandelbrot (1977) has shown that for two clusters with fractal dimensions D_{f1} and D_{f2} and size r_c, placed in the same region of space, the mean number of intersections $M_{1,2}$ is expressed as:

$$M_{1,2} \propto r_c^{Df1+Df2-D} \tag{16.6}$$

where D is the dimension of space (D = 3). Thus, if each structure has a fractal dimension less than 1.5 (e.g., from acid catalysed precursors sols) the probability of intersection decreases indefinitely as r_c, increases. One refers to these structures as *mutually transparent:* during film or membrane formation they should freely interpenetrate one another as they are forced into close proximity, first by the increasing concentration, and then by the capillary pressure accompanying drying. Alternatively, if the mass fractal dimensions exceed 1.5, (e.g., from base catalysed precursors sols) the probability of intersection increases with r_c. These structures are *mutually opaque*. During film or membrane formation, they do not interpenetrate as they are concentrated. Consequently, for opaque fractals, we expect both the percent porosity and mean pore size of the membrane to increase with r_c. The efficacy of the aggregation approach shows that for films formed from polymeric sols characterized by $D_m = 2.4$, the volume fraction porosity increases monotonically with the average size r_c of the polymers comprising the sol. In addition to the volume fraction porosity, the surface area and pore size also increase with the aging time employed to grow the fractal clusters to different sizes prior to film deposition. The porosity of a film depends on the porosity of the individual clusters and their extents of interpenetration during film formation. Although a reduction in D_f increases the porosity of a cluster, it also facilitates interpenetration. Thus highly porous films require an intermediate value of D_f: generally $2.0 < D_f < 3.0$ as Brinker and Scherer (1990) said. Low values of D_f promote interpenetration of polymers during film formation. Interpenetration can completely mask the porosity of the individual polymers. For example, for $D < 2$, quite dense films (5 vol% porosity) with small pores (~1.0 nm) are prepared over a wide range of aging times (Brinker and Scherer 1990) i.e., there is no dependence of the film porosity on polymer size. This situation is beneficial for the preparation of ultrathin membranes on porous supports, because aging can be employed to grow polymers large enough to be trapped on the support surface without suffering an increase in pore size (Brinker et al. 1993).

Brinker and Scherer suggest *another strategy* that is also generally applicable to the wide range of polymeric sols including silica ones, characterized by a mass fractal dimension, that is ***condensation***. The criterion for opaque or transparent fractals established by Eq. 16.6 is based on the assumption that instantaneous and irreversible reaction or *sticking* occurs at each point of polymer chain intersection. Thus, as the points of intersection are reduced, the polymers more easily interpenetrate one another – they become transparent to one another. Polymer transparency is also promoted by reducing the probability of sticking at each point of intersection. Chemically, this is achieved by reducing the condensation rate. Instantaneous and irreversible sticking is chemically equivalent to an infinite reaction (condensation) rate. Of course in reality the probability of sticking is much less than unity and is influenced by catalyst concentration, temperature, and reactivity of the terminal ligands. For example, for silicate polymers prepared from tetraethoxysilane by a sol-gel process, the condensation rate is strongly dependent on the sol pH. If the condensation rate is reduced, the probability of sticking at each point of intersection is reduced. This promotes polymer interpenetration, leading to denser films with smaller pore sizes. An additional (and perhaps more influential) consequence of the condensation rate is that it largely dictates the stiffness of the network and thus its extent of collapse during drying. A reduction in the condensation rate causes the network to be more compliant fostering its collapse in response to the capillary pressure created at the final stage of drying. It seems that the primary effect of the condensation rate is not to alter the polymer structure but rather to influence the extent of polymer interpenetration and collapse during membrane deposition. Clearly results, that a reduction in the condensation rate promotes the collapse of the structure during drying. For films it is possible to achieve pore sizes in the range appropriate for separations based on size exclusion. Such extensive collapse of the gel network has not been observed for bulk specimens. Although the condensation rate can be minimized, the considerably longer processing times associated with bulk xerogel formation allow considerably more condensation to take place, leading to stiffer structures that resist collapse. It can be concluded that the sol-gel process affords many strategies for the preparation of inorganic or a hybrid material, with pore sizes in the range appropriate for membranes and sensors, and allows thin films to be fabricated using simple procedures. Pores of controlled size can be created in a nanocompozite by removal of the dispersed organic phase. An alternate composite structure that permits the introduction of pores of a well-controlled size is one in which the dispersed phase is itself microporous.

Villalobos and coworkers studied "How fast can drug molecules escape from a finite fractal matrix-type release system?" via Monte Carlo computer simulations (Villalobos et al. 2006a, b) and the consequences of the spatial distribution of components in pharmaceutical matrices type Menger sponge on the drug release kinetic (Villalobos et al. 2009) Menger sponges (all the samples having the same fractal dimension, $D_f = 2.727$, with different values of random walk dimension, D_w [2.028, 2.998]) were employed as models of drug delivery devices with the aim of studying the consequences of matrix structural properties, unusually characterized, specifically by fractal dimension D_f, on drug release performance. Menger sponges is a three-dimensional extension of the Cantor (1882) set and Sierpinski (1916) carpet. It and it was first described by Karl Menger (1926) while exploring the concept of topological dimension. The Menger sponge simultaneously exhibits an infinite surface area and encloses zero volume (Mandelbrot 1982). Villalobos and coworkers also studied the effects of drug concentration and spatial distribution of the medicament, in porous solid dosage forms, on the kinetics and total yield of drug release (Villalobos et al. 2006a, b). From the simulation results it was found that the drug-excipient ratio is a factor that determines the release mechanism from a matrix system. It was also observed that the N_{leak}/N_{total} ratio is directly related to the surface/volume ratio of a matrix device. Since the N_{total} value is a function of the matrix size, the size of the matrix system affected the drug release profile too. Finally, the critical modeling of drug-release from matrix-type delivery systems was considered important in order to understand the implicated transport mechanisms, and to predict the effect of the device design parameters on the release rate (Villalobos et al. 2006a, b). The drug release process from Menger sponges was characterized by a non-Fickian behavior (Villalobos et al. 2006a, b).

Computer simulations are useful tools in the study of drug release from matrix platforms, especially because they can describe the profile through the whole process. The used equation by Villalobos and al. is a useful model to describe the release profiles generated from both fractal and Euclidian structures, and can be properly applied to the one-dimensional case but future efforts are necessary to relate theoretical values with experimental results, such way determination to give a straightforward idea of the internal structure of the platform and vice versa (Villalobos et al. 2009).

16.3.5.3 Intracellular Uptake of MSN

Morphology and texture properties of silica matrix play an important role in developing good properties in order to accomplish the suitability as CDDS. Conventionaly prepared mesoporous silica results in amorphous chunks and particles with different shapes and sizes. The polydispersity and the amorphous nature of the mesoporous silicas present a major challenge in understanding and controlling the mass-transport properties at the nano-level, which is of fundamental importance for drug delivery and controlled release in biological systems (Slowing et al. 2008).

Cell internalization of silica particles active pharmacologic agent bearing is in fact the biggest challenge. What are the implications of nanoparticles size and shape in the behavior of silica based CDDS for intracellular delivery will be envisaged by few examples.

While the *mesoporous silica microspheres* are potentially useful for many non-biological functions, they are not suitable for many important biotechnological and biomedical applications, such as: agents for gene transfection or carriers for intracellular drug delivery because, for example, mammalian cells cannot efficiently engulf large particles via endocytosis. Mesoporous silica microspheres are within the size window of bacteria and could potentially trigger acute immune response Slowing et al. (2008) showed.

In exchange, *mesoporous silica nanospheres* exhibit suitable properties as CDDS (Slowing et al. 2008). The relevant example could be the *intracellular uptake of MSN*. It was showed in order for a material to be efficiently uptaken by nonphagocytic cells, the particle size of the material needs to be in the sub-micron scale (Mayor and Pagano 2007; Rejman et al. 2004) and to have proper surface properties that can have favorable interactions with the drug molecules to achieve high loading. This would allow the release of drug with high local concentrations at the site of interest intracellularly. The particle size and surface property of MSN have to be adjusted with precision for intracellular controlled release applications. Cellular uptakes of molecules are often facilitated by the *specific binding* between these species and membrane-bound receptors and by *constitutive "adsorptive" endocytosis* or by fluid phase pinocytosis (Xing et al. 2005).

The fate of injected nanoparticles can be controlled by adjusting their size and surface characteristics to effectively deliver drug to the targeted tumor tissue; nanoparticles must have the ability to remain in the bloodstream for a considerable time without being eliminated (Cho et al. 2008).

Conventional surface unmodified nanoparticles are usually caught in the circulation by the reticuloendothelial system, such as the liver and the spleen, depending on their size and surface characteristics (Larsen et al. 2000). One of the advantages of nanoparticles is that their size is tunable. The size of nanoparticles used in a drug delivery system should be large enough to prevent their rapid leakage into blood capillaries but small enough to escape capture by fixed macrophages that are lodged in the reticuloendothelial system, such as the liver and spleen. Cho et al. reported that the size of the sinusoid in the spleen and fenestra of the Kuffer cells in the liver varies from 150 to 200 nm and the size of gap junction between endothelial cells of the leaky tumor vasculature may vary from 100 to 600 nm (Yuan et al. 1995; Cho et al. 2008).

Silica particles are known to have a great affinity for the head-groups of a variety of phospholipids (Mornet et al. 2005) such as functionalized silica will be easily adsorbed on the cell surfaces, further being internalized inside cell by endocytosis.

Slowing and others have demonstrated that MSN can be efficiently endocytosed in vitro by a variety of mammalian cells including cancer (HeLa, CHO, lung, PANC-1), non-cancer (neural glia, liver, endothelial), macrophages, stem cells (3TL3, mesenchymal) and others (Slowing et al. 2006; Huang et al. 2005; Mayor and Pagano 2007; Rejman et al. 2004; Xing et al. 2005; Mornet et al. 2005; Radu et al. 2004; Chung et al. 2007; Giri et al. 2005; Huang et al. 2005) The uptake of MSN is fast. Particles are often observed inside the cells within 30 min of introduction.

The cellular uptake efficiency and kinetics together with the correlation between the particle morphology and aggregation of two kind, spherical and tube-shaped particles, of mesoporous silica nanomaterials (MSNs) were investigated

(Trewyn et al. 2008) by Trewyn et al. in a cancer cell line (CHO) and in a noncancerous cell line (fibroblasts). The findings they reached were that both materials showed less efficient endocytosis for fibroblast, specifically, the rate of endocytosis for spherical particles was significantly faster (100% uptake in 180 min) than that of tube-like (360 min to reach 100% endocytosis particles). This difference in endocytosis kinetics was attributed to two variables, different size and aggregation ability between the MSN particles with different shapes. In other experiment three different morphologies, as hollow silica nanotubes, hollow silica nanospheres and solid silica nanoparticles were employed as supports for immobilization of lysozyme (Qing-Gui et al. 2008). The comparative study indicated that the amount of immobilized lysozyme on solid silica nanoparticles was 186 mg/g silica, while that of immobilized lysozyme within hollow silica nanotubes reach up to 351 mg/g silica and hollow silica nanospheres could reach up 385 mg/g silica. The hollow silica nanospheres represented the highest immobilization ability, whereas the specific activity of immobilized lysozyme on silica nanospheres (1.38×106 unit/g) was slightly lower than that of immobilized lysozyme on silica nanotubes (1.46×10^6 unit/g). This phenomenon was explained by the relatively airtight structure of hollow silica nanospheres. Compared to solid silica nanoparticles, silica materials with hollow structure and large pore size would facilitate the enzymatic immobilization (Qing-Gui et al. 2008).

The effectiveness of a drug therapy is dictated by several factors, including the rate or extent of the drug molecule's penetration into, and permeation through, the body tissue and cells to reach a site of action. Many barriers prevent the drug from reaching the site of action. They can be grouped according to size, including the organism itself on a 1–2 m scale, an organ representing 1–100 mm scale, tissue in the 1–100 μm size, and extracellular and intracellular space of only 1–100 nm. Koo and al defines nanomedicines as delivery systems, ideally in the nanometer size range of 1 –100 nm (National Science and Technology Council Committee on Technology 2005) containing encapsulated, dispersed, adsorbed, or conjugated drugs and imaging agents. Nanomedicines have a size range that allows them to be injected without occluding needles and capillaries and are ideal for targeted drug delivery and medical imaging due to the pathophysiology of certain disorders such as cancer and inflammation. The nanocarrier systems possess multiple desirable attributes, including the special ability to target tumors and inflammation sites that have permeable vasculature (Koop et al. 2005).

Targeting is often defined in terms of ***passive and active*** targeting where ***passive*** has come to refer to the ***accumulation of drug*** due to physicochemical or pharmacological factors naturally occurring in the body. A***ctive*** targeting is due to a (more or less) ***specific interaction*** between delivery system and cell or tissue component (Garnett 2001).

Passive targeting occurs due to extravasation of the nanoparticles at the diseased site where the microvasculature is leaky. Passive delivery refers to nanoparticle's transport through leaky tumor capillary fenestrations into the tumor interstitium and cells by passive diffusion or convection (Haley and Frenkel 2008). What is important to understand, passive targeting proceeds based on natural developed environmental gradients as a consequence of difference between the healthy and ill regions of the organism. The drug transport driving force consists in the fact that in vicinity of target place the drug accumulates and transport vector follows the rules imposed by the law governing local mechanism of transport in the given case. In order that nanoparticle to be internalized in cells by passive targeting, there are several transport mechanisms. The driving force of passive transport by diffusion, for example, is the concentration gradient. Nanoparticles that satisfy the size and surface characteristics requirements for escaping reticuloendothelial system capture have the ability to circulate for longer times in the bloodstream and also they have a greater chance of reaching the targeted tumor tissues resistance (Cho et al. 2008). The unique pathophysiologic characteristics of tumor vessels enable macromolecules, including nanoparticles, to selectively accumulate in tumor tissues. Examples of such diseases, where passive targeting of nanocarriers can be achieved, are tumor and inflamed tissues. Tumor vascular leakiness is the result of increased angiogenesis and the presence of cytokines and other vasoactive factors that enhance permeability. Tumor angiogenesis is characterized by vessels with irregular diameters and branching, and tumors lacking defining structures of vasculature such as arterioles, capillaries, or venules (Oeffinger and Wheatley 2004). Vascular endothelial growth factor (VEGF) and the angiopoietins are critical in regulating the balance between the leakiness associated with the defective endothelial linings of tumor vessels and vascular growth, maturation, and regression (Holash et al. 1999; Brown et al. 1997). This imbalance makes tumor vessels highly disorganized and dilated with numerous pores, and that leads in *enhanced permeability and retention effect*, EPR, which constitutes an important mechanism by which macromolecules, including nanoparticles, with a molecular weight above 50 kDa, can selectively accumulate in the tumor interstitium (Cho et al. 2008). Hobbs showed that the majority of solid tumors exhibit a vascular pore cutoff size between 380 and 780 nm (Hobbs et al. 1998) and tumor vasculature organization may differ depending on the tumor type, its growth rate and microenvironment (Jain 1998).

Particles need to be of a size much smaller than the cutoff pore diameter to reach to the target tumor sites. By contrast, normal vasculature is impermeable to drug associated carriers larger than 2–4 nm compared to free, unassociated drug molecules (Drummond et al. 1999; Fu et al. 1998; Firth 2002).

Cho et al. brought in attention another contributor to passive targeting, the unique microenvironment surrounding tumor cells, which is different from that of normal cells. Fast-growing, hyperproliferative cancer cells show a high metabolic rate, and the supply of oxygen and nutrients is usually not sufficient for them to maintain this. Therefore, tumor cells use glycolysis to obtain extra energy, resulting in an acidic environment. The pH-sensitive liposomes are designed to be stable at a physiologic pH of 7.4, but altered to release active drug in target tissues in which the pH is less than physiologic values, such as in the acidic environment of tumor cells. Additionally, cancer cells express and release unique enzymes such as matrix metalloproteinases, which are implicated in their movement and survival mechanisms (Cho et al. 2008).

By improving the passive targeting strategies, nanoparticle systems are able to improve the biodistribution of cancer drugs, to increase their circulation time in the bloodstream, to improve and selectively carry their loaded active drugs to cancer cells by using the unique pathophysiology of tumors, such as their enhanced permeability and retention effect and the tumor microenvironment and to accumulate in cells without being recognized by P-glycoprotein, one of the main mediators of multidrug resistance, resulting in the increased intracellular concentration of drugs (Cho et al. 2008).

Active targeting involves drug delivery to a specific site based on molecular recognition. One such approach is to couple a ligand to a nanoparticle NP so that the ligand can interact with its receptor at the target cell site (Haley and Frenkel 2008). Nanoparticles, to be internalized in cell, by active targeting, they have to bind a specific receptor, that helps the drug transport to proceed even against the rules imposed by natural local mechanism of transport. Localized diseases such as cancer or inflammation not only have leaky vasculature but also overexpress some epitopes or receptors that can be used as targets. Nanomedicines can also be actively targeted to these sites. During DDSs formulation process, ligands meant to specifically bind to surface epitopes or receptors, preferentially overexpressed at target sites, have been coupled to the surface of long circulating nanocarriers (Missailidis et al. 2005; Medina et al. 2005; Mitra et al. 2005; Farokhzad et al. 2004; Schiffelers et al. 2004; Li et al. 2004a, b; Gosk et al. 2004; Maruyama et al. 2004; Omori et al. 2003; Pastorino et al. 2003; Dharap et al. 2003; Fonseca et al. 2003; Ni et al. 2002; Ishida et al. 2001; Moody et al. 1998)

Ligand-mediated active binding to sites and cellular uptake are particularly valuable to therapeutics that are not taken up easily by cells and require facilitation by fusion, endocytosis, or other processes to access their cellular active sites (Willis and Forssen 1998).

Active targeting can also enhance the distribution of nanomedicine within the tumor interstitium (Drummond et al. 1999). Active targeting has been explored to deliver drugs into resistant cancer cells (Sapra and Allen 2003).

An important consideration when selecting the type of targeting ligand is its immunogenicity. For example, whole antibodies that expose their constant regions on the liposomal surface are more susceptible to Fc-receptor-mediated phagocytosis by the MPS (Harding et al. 1997; Koo et al. 2005; Metselaar et al. 2002). Koo et al. give some examples of targeting ligands and their triggers are systematically listed (Koop et al. 2005).

Various methods have been employed to covalent and/or noncovalent couple ligands to the surface of the nanocarriers with reactive groups. Common covalent coupling methods involve formation of a disulfide bond, cross-linking between two primary amines, reaction between a carboxylic acid and primary amine, reaction between maleimide and thiol, reaction between hydrazide and aldehyde, and reaction between a primary amine and free aldehyde (Nobs et al. 2004). Noncovalent binding by physical association of targeting ligands to the nanocarrier surface has the advantage of eliminating the use of rigorous, destructive reaction agents. However, there are potential problems, such as low and weak binding and poor control of the reactions and the ligands may not be in the desired orientation after binding (Koo et al. 2005).

Here, we will give some examples, of controlling targeting- triggering couples in drug delivery processing especially implying silica based sensing (smart) CDDSs.

16.3.6 Sensitive – Responsive Release in Controlled Drug Delivery Systems

Physical, chemical, and biochemical processes have been used to stimulate site-specific and/or time-dependent delivery from drug carrier systems. First published account of an actively-triggered release mechanism appeared in 1978 with the report of Yatvin, in which was described a CDDS that could be thermally induced to release their cargo (Yatvin et al. 1978). Until the early 1990s triggered release aspect of controlled drug delivery had been largely ignored, being a weak link in the drug delivery chain. Primary scope of target DDSs was to enhance both permeation and retention effect. However, the heart of triggered drug release research and development efforts consists in the issue of improving drug bioavailability, i.e. improving the extent to which a drug escapes its carrier and enables its interaction with the biological target (Thompson 2001). By applying active targeting strategies it is also possible to amplify the specificity by using ligands or antibodies directed against selected tumor targets and to overcome or reduced the resistance against drugs (Cho et al. 2008).

Important triggering approaches rely on triggering mechanisms induced by local expressed states, as thermal, pH-ionic-redox, light, magnetic, sonic. Triggered release has tremendous potential for improving the efficacy of drug carriers by providing control mechanisms that can regulate the place, time, and schedule over which a drug can be deployed. Among sensitive-responsive (intelligent) CDDS will be exemplified some, especially referring to porous and magnetic silica based ones and hybrids including multiresponsive CDDSs.

16.3.6.1 Thermo- Responsive Drug Release Profile

The hexagonally ordered mesoporous silicas with different pore sizes (10, 17, 30 nm) materials were tested as carrier, in smart controlled drug (Indomethacin (I)) release using the thermoresponsive poly(*N*-isopropylacrylamide) (PNIPAm) hybrid nanoporous structures during stepwise temperature changes between 25°C and 40°C (Chang et al. 2004).

The results were consistent with the positive squeezing mechanism: when the temperature was increased and maintained at 40°C, nearly constant release pattern was observed. No significantly sustained, but rapidly decreased release was observed on the temperature change to 25°C. The thermosensitive release profile could be also explained by the nanodiffusion mechanism. At low temperature, the drug is trapped in the polymer and in the porous structures, and the polymers are swelled to prevent the drug from being significantly released into the media. When the temperature is increased above the lower critical solution temperature (LCST) of PNIPAm, the polymers shrink to squeeze the drug into the porous channels, and open the pore structure. The authors observed that the overall delivery of the drug into the media is controlled by diffusion through the porous channels. In the nanodiffusion mechanism, diffusion in the nanoporous channels of different sizes, rather than in the surface area or in the gel phase, is controlling mechanism. The diffusion rate in the small nanochannels is severely hindered while that in large pore channel approaches the rate of bulk diffusion. A slight variation in pore dimension and the gel density along the pores will cause the fine nanoporous channels to be clogged before the drug has a chance to be extruded. The total loaded amount and released amount after 100 h of indomethacin showed that the ~90 mg of indeomethacin was loaded per g of material and 58–82% of the drug released for 100 h depending on pore size. Thermosensitive polymer hybrid nanoporous materials showed a sustained positive thermoresponsive drug release profile in which the overall release amount was controlled by change of the pore channel size (Chang et al. 2004).

The combination based on Fe_3O_4 magnetic nanoparticle as the core, mesoporous silica as the sandwiched layer, and thermo-sensitive P(NIPAM-*co*-NHMA) copolymer as the outer shell resulted in a temperature controlled drug release system (Liu et al. 2009).The hydrophilic co-monomer content affected the volume phase-transition temperature (VPTT) of this composite microsphere and the behavior of the temperature-triggered drug release. Zn(II) phthalocyanine tetrasulfonic acid ($ZnPcS_4$), a well-known photodynamic therapy (PDT) drug, was used as a model drug to assess the release system. The results demonstrated that the drug release behavior was dependent on the temperature and had a close correlation with the VPTT. Above the VPTT, the drug release rate was much faster than that below the VPTT, which showed potential application in tumor therapy (Liu et al. 2009).

16.3.6.2 pH and Ionically Redox Responsive Release Profile

A redox-controlled drug delivery system that is based on MSN capped with cadmium sulfide nanoparticles (CdS) as gatekeepers (Lai et al. 2003) was developed. In this system, CdS was chemically attached to MSN through a disulfide linker, which is chemically labile and could be cleaved with various disulfide reducing agents. The release properties and biocompatibility of the CdS–MSN system were evaluated *in vitro* by using different concentrations of reducing agents. After addition of a disulfide reducing agent the CdS–MSN systems has released in some extent the active agent they were carrying (Slowing et al. 2008) sensor was developed by Martinez-Manez and co-workers (Casasus et al. 2004). Their system is comprised of a diaminoethylenepropylsilane derivative grafted at the openings of the pores and a mercaptopropyl group anchored inside of the porous channels of the mesoporous silica. The signal transduction mechanism is based on a colorimetric reaction between blue squaraine dye and the mercaptopropyl groups inside of the channels. The authors showed that the agglomeration/packing of the oligoamine-based "gatekeepers" could be manipulated by different pH values and anions. In the case of the pH-controlled gating, an acidic solution of squaraine was introduced to the sensor, the colour was not changed, and the gate was observed to close because of the oligo-amines were protonated in the acidic solution causing a swelling between the gatekeepers. In a basic solution, a color bleaching of squaraine was observed; indicating the gate was opened and the thiol groups inside were accessible to react with the dye molecules (Casasus et al. 2004; Slowing et al. 2008).

By combining of mesoporous silica materials as solid supports and supramolecular assemblies as gatekeepers has resulted in organic/inorganic hybrid materials with improved functionalities. Lin and coworkers (Radu et al. 2004) developed a dendrimer-capped MSN gene transfection system and demonstrated that this system could be used as both drug delivery and gene transfection agent. Another example of supramolecular gates for mesoporous silica was developed by Stoddart, Zink and co-workers. Their system is comprised of a series of electrochemical redox-controlled pseudorotaxanes attached to mesoporous silica, which can be switched on and off (Hernandez et al. 2004). Kim and co-workers (Park et al. 2007) also reported a pseudorotaxane-based MSN controlled release system triggered by pH. Several mesoporous silica nanoparticles, MSN, based nanodevices have been achieved through the development of photochemical, pH responsive, and redox active gatekeepers (Slowing et al. 2008).

The systems based on gatekeeping concept have the advantage of using a variety of chemical entities (like nanoparticles, organic molecules, or supramolecular assemblies) as "gatekeepers" to regulate the encapsulation and release of drug molecules.

A pH-controlled delivery system has been developed based on hollow mesoporous silica spheres using pH-sensitive polyelectrolyte multilayer coated on the spheres as a switch to store and release gentamicin molecules (a model drug) (Zhu et al. 2007). The polyelectrolyte layers with an average thickness of 12 nm were coated on hollow mesoporous silica spheres through a layer-by-layer technique. Gentamicin molecules were successfully stored in these spheres by means of adjusting the gentamicin solution from pH 2 to pH 8. The storage capacity can reach 614.8 mg/g (34.11%) at an initial gentamicin concentration of 60 mg/ml. The controlled release of gentamicin molecules from this system has been achieved by simply changing the pH value in the release media. Therefore, this type of material is of potentials for the controlled drug release applications Zhu said.

DOX-loaded silica nanostructures exhibiting superb drug release behavior, which is controlled by varying the pH was reported by Tan et al. (2011). Porous hollow silica nanoparticles (HSNS) have been successfully fabricated by a novel combination of stabilizing condensation and dynamic self-assembly. Further, encapsulation of organic molecules into HSNS and their controlled release were investigated by using Doxorubicin (DOX) as a model. HSNS show excellent drug delivery properties. The authors believe, that their current approaches will doubtlessly open many possibilities toward biological and technological applications of silica nanomaterials.

16.3.6.3 Photo Responsive Release Profile

Light-sensitivity is an attractive phenomenon for developing advanced DDS capable of precise external modulation of the site. Light-sensitive inorganic substrate systems that can help achieve improved control of the loading and release of guest substances have been highlighted recently (Haraguchi 2007; Sortino 2008; Johansson et al. 2008; Aznar et al. 2007, 2009; Vivero-Escoto et al. 2009; Wang 2009).

Fujiwara and co-workers accomplished, for the first time, the photocontrolled reversible release of drug molecules from coumarin-modified MCM-41 (Mal et al. 2003a and b). In their work, a photoresponsive coumarin derivative was grafted on the pore outlet of silica MCM-41. Irradiation of UV light longer than 310 nm wavelength to this coumarin-modified MCM-41 induced the photodimerisation of coumarin to close the pore outlet with cyclobutane dimer. The irradiation to the dimerized-coumarin-modified MCM-41 with shorter wavelength UV light around 250 nm regenerates the coumarin monomer derivative by the photocleavage of cyclobutane dimer, and guest molecules, phenanthrene, included inside are released from the pore void.

Johansson reported light-activated CDDS, and mesoporous silica (Johansson et al. 2008) is highly versatile support or frameworks for functional materials where a desired function (such as energy transfer, electron transfer, or molecular machines) is induced by molecules deliberately placed in specific regions of the structure. The relatively gentle templated sol–gel synthesis methods allow a wide variety of molecules to be used, and the optical transparency of the framework is very suitable for studies of light-induced functionality. In order to develop active materials that can trap and release molecules from the pores upon command Zink and coworkers (Johansson et al. 2008) used three types of functionality. First, photo-induced energy transfer was used to verify that molecules can be placed in specific spatially separated regions of the framework; fluorescence resonance energy transfer is used as a molecular ruler to measure quantitatively the distance between pairs of molecules. Secondly, photoinduced electron transfer was used to obtain fundamental information about the electrical insulating properties of the framework and finally, two types of molecular machines, a light-driven impeller and a light activated nanovalve were developed. Both machines contain moving parts attached to solid supports and do useful work. The valves trap and release molecules from the mesopores. The impellers expel molecules from the pores. Applications of the materials to drug delivery and the release of drug molecules inside living cells is described.

Controlled release of guests from mesostructured silica nanoparticles has been accomplished using supramolecular nanovalves constructed from [2]rotaxanes and [2]pseudorotaxanes (Johansson et al. 2008; Hernandez et al. 2004; Leung et al. 2006; Saha et al. 2007; Nguyen et al. 2006, 2007). When these machines, as redox active nanovalve, are positioned on the surface of nanoparticles, the $CBPQT^{4+}$ rings are able to gate the entrances to mesopores and the valves are closed. The movement of the rings away from the pore opening can be induced by reducing agents to unblock the pore entrances and open the valves. Their versatility was demonstrated by reconfiguring the system for light activation (Nguyen et al. 2007). Photosensitizers that are powerful reductants in their photo-excited states are capable of reducing the $CBPQT^{4+}$ rings to cause the movement. By attaching them to the surface of the nanoparticles, a photo-induced electron transfer reduced the ring and activated the valve. Specifically, 9-anthracenecarboxylic acid (ACA) and RuII(bpy) 2(bpy(−CH2OH)2 were used as photosensitizers. Upon irradiation with the appropriate wavelength of light (351 nm for ACA, 457 nm for Ru(bpy)3), the excited photosensitizer transferred an electron to a nearby CBPQT4+ ring to induce dissociation and open the nanovalve (Johansson et al. 2008). In order to prevent electron transfer back from the $CBPQT^{3+}$ ring to the oxidized photosensitizer a sacrificial reducing agent, triethanolamine, is used, as following eqs.:

$$\text{Photosensitizer}\bullet + \text{CBPQT}^{4+} \rightarrow \text{Photosensitizer} + \text{CBPQT}^{3+}$$

$$\text{Photosensitizer}^{+} + \text{Triethanolamine} \rightarrow \text{Photosensitizer} + \text{Products}$$

The source of energy used to operate the machine is light and thus enables the machine to be operated by remote control. Photo-induced electron transfer, in this case between the donor and the moving acceptor, through the intervening solvent causes the chemical changes that open the valves (Johansson et al. 2008).

A great deal of knowledge can be found in the reported research dealing with synthesis of azobenzene-functionalized mesoporous inorganic materials (Alvaro et al. 2005; Besson et al. 2005; Liu et al. 2004) and with detailed investigations into azobenzene cis/trans isomerization in mesostructured silica (Sierocki et al. 2006).

To study a ***light activated impellers*** for releasing, molecules azobenzenes were positioned in the pore interior, with one end bonded to the pore wall, and the other end free to undergo large amplitude photo-driven motion (Johansson et al. 2008). The azobenzene-derivatized mesoporous silica nanoparticles were filled with either coumarin 540 (C540) or Rhodamine 6 G (R6G). The photo-driven expulsion of the dye molecules from the pores was monitored by fluorescence spectroscopy. Under 457 nm illumination, the continuous isomerization of two conformers occurs reversibly, stimulating the untethered azobenzene terminus to move back and forth and expelling the loaded molecules from the pores. In the absence of light activation, dye molecules are trapped in the pores because the static azobenzene machines congest the inner pore channels and prevent the escape of molecules (Johansson et al. 2008). Based on obtained results, the findings were used to create an effective ***intracellular*** drug delivery system. It was obtained a CDDSystem of anticancer drugs into living cells on demand using azobenzene derivatized nanoparticles (350 nm in diameter) (Lu et al. 2008). It was demonstrated that the intracellular release of guest molecules was sensitized by light excitation of the azobenzenes. When the treated cells were irradiated with 413 nm light, a wavelength that stimulates continuous cis/trans isomerization, the cell nuclei were stained red. No staining of the cells was observed when they were irradiated with 676 nm light, a wavelength that the azobenzene molecules do not absorb. When cells were loaded with particles containing the anticancer drug camptothecin (CPT), apoptosis was induced using light stimuli. In the absence of light stimuli, the CPT remained inside the particles and the cells were not damaged. Illumination, however, promptly expelled the drug molecules from the particles, which induced nuclear fragmentation and cell death (Lu et al. 2008). As functions attached nanoparticles, organic molecules have been shown to be able to serve as gatekeepers for mesoporous silicas as well as photoactivable.

Ferris et al. (2009) and Mal et al. (2003) demonstrated the photo-responsive pore size control of mesoporous silica by azobenzene and coumarin modification. Azobenzene groups are toxic according to the Federal Drug Administration, and this limits the application of such DDS to topical formulations. Thus, coumarin groups may be better than azobenzene for photo-responsive pore size control of biomaterials. Irradiation with UV light (>310 nm) induces the photo-dimerization of coumarin to close the pore with a cyclobutane dimer. Guest molecules such as phenanthrene can neither enter nor escape from the individual pores of the MBG. However, irradiation with shorter wavelength UV light (~250 nm) cleaves the coumarin dimer to regenerate the coumarin monomer, the pores are opened, and the guest molecules can be released. This material with an "open–close double doors" system can be used in the photoswitched controlled release of included compounds.

Tanaka and co-workers first reported reversible photo-triggered controlled release mesoporous silica (MCM-41) system that could function in organic solvents (Mal et al. 2003). Their system took advantage of a photodimerization reaction of coumarin. The pores of MCM-41were functionalized with 7-[(3-triethoxysilyl) propoxy]coumarin. The reversible dimerization of this coumarinic functionality was large enough to close and open the mesoporous opening of MCM-41. Silica cholestane and phenanthrene could be encapsulated and released in n-hexane solution (Mal et al. 2003).

Functionalized mesoporous bioactive glasses (MBG) with photoactive coumarin demonstrate photoresponsive dimerization resulting in reversible gate operation. Irradiation with UV light (>310 nm) induced photo-dimerization of the coumarin-modified MBG, which led to the pores' closing with cyclobutane dimers and trapping of the guest phenanthrene in the mesopores. However irradiating the dimerized-coumarin-modified MBG, with shorter wavelength UV light (250 nm) regenerates the coumarin monomer derivative by the photo-cleavage of cyclobutane dimers, such that trapped guest molecules are released from the mesopores (Lin et al. 2010).

Modification of mesoporous silica nanoparticles using the bifunctional strategy, post-synthesis grafting, and backfilling strategy in order to make them suitable for drug delivery applications was reported (Lu et al. 2007). The modified nanoparticles were able to deliver the water insoluble drug camptothecin into different types of human cancer cells (Johansson et al. 2008).

Controlled release of guests from mesostructured silica nanoparticles, using also supramolecular nanovalves constructed from [2]rotaxanes and [2]pseudorotaxanes has been accomplished by others (Nguyen et al. 2005, 2006, 2007; Hernandez et al. 2004; Leung et al. 2006; Saha et al. 2007). The source of energy used to operate the machine is light and thus enables the machine to be operated by remote control. Photo-induced electron transfer, in this case between the donor and the moving acceptor, through the intervening solvent causes the chemical changes that open the valves.

Silica–gold (SiO_2–Au) nanoshells are a new class of nanoparticles that have a silica dielectric core, which is surrounded by a gold shell. The peak extinction of SiO_2–Au nanoshells are very easily tunable to absorb or scatter light strongly within the wavelengths of 650–900 nm that is commonly known as the NIR region (Averitt et al. 1999). This region is of significant biological importance and hence nanoshells are currently being investigated for use in the NIR region for a variety of biomedical applications

A photothermal modulated drug delivery system in which near IR light of a specific wavelength can be used to induce the collapse of a polymeric matrix loaded with model drug molecules and proteins was reported by Bikram et al. (2007).

They have embedded SiO_2–Au nanoshells of varying concentrations within NIPAAm-co-AAm hydrogels. Moreover, these nanoparticles were designed to strongly absorb near IR light at ~808 nm, the emission wavelength of the laser used in these studies. The drug delivery profiles of methylene blue, insulin, and lysozyme were then obtained by photothermal irradiation of the resulting nanoshell-composite hydrogels (Bikram et al. 2007).

16.3.6.4 Magnetic Responsive Release Profile

Polymers that respond to the presence or absence of magnetic fields can exist as free chains in solution, immobilized to surfaces, or crosslinked within networks (represent the majority consisting in gels swollen with complex fluids (Roy et al. 2010; Zrinyi, 1997, 2000; Filipcsei et al. 2007; Barsi et al. 1996; Pyun 2007).

Inorganic magnetic (nano)particles (physically entrapped within or covalently immobilized) into a three-dimensional crosslinked network, result in such materials whose shape and size are distorted reversibly and instantaneously, in the presence of a non-uniform magnetic field (Szabo et al. 1998; Zrinyi et al. 1997; Wang et al. 2006; Sewell et al. 2008). In uniform magnetic fields there is a lack of magnetic field–particle interactions, but particle–particle interactions arise from the creation of induced magnetic dipoles. Particle assembly within the surrounding polymer matrix can lead to dramatic transformations in material properties (Roy et al. 2010).

When between polymer chains and magnetic particles there are non-covalent interactions magnetic nanoparticles (Fe3O4) – polimer nanocomposites, could suffer controllable changes in shape when exposed to a magnetic field. In a non-uniform field, such a magnetic hybrid composite particles aggregate and reversibly return to initial state when the magnetic field is removed.

In the case that the polymer chains are covalently immobilized to the surface of magnetic particles, Pyun and co-workers (2007) have shown that diverse range of mesoscale morphologies could be observed, ranging from randomly entangled chains, field aligned 1D mesostructures, and nematic-like liquid crystal colloidal assemblies whose behavior could be adaptive to environmental conditions (Roy et al. 2010; Zhang et al. 2004; Suzuki and Kawaguchi 2006; Brugger and Richtering 2007; Xulu et al. 2000; Benkoski et al. 2008; Keng et al. 2007; Bowles et al. 2007; Benkoski et al. 2007; Korth et al. 2006). Hu et al. reported an attempt to accelerate drug release from magnetic-sensitive silica nanospheres by controlled bursting to a therapeutically effective concentration by a high-frequency magnetic field (HFMF) (Hu et al. 2008). The magnetic-sensitive silica nanospheres, with particle sizes about 50 nm are able to release specific amounts of drug in a burst manner via short exposure to a HFMF. The HFMF accelerates the rotation of magnetic nanoparticles deposited in the silica matrix with generated heat energy and subsequently enlarges the nanostructure of the silica matrix to produce porous channels that cause the drug to be released easily. By taking these magnetic-responsive controllable drug release behaviors, the magnetic silica nanospheres can be designed for controlled burst release of therapeutic agents for especially urgent physiological needs (Hu et al. 2008).

16.3.6.5 Ultrasonic Responsive Release Profile

Ultrasound has been shown to trigger drug release by raising the local temperature or causing cavitation (Husseini and Pitt 2008). Both processes can increase the permeability of cell membranes and accelerate polymer degradation (Husseini and Pitt 2008).

Ultrasound-sensitive vehicles have the potential to treat tumorigenic cancers due to their invasive character, ability to penetrate deeply into the human body, and ease of control (You 2010).

There are many advantages to ultrasound, US, that make it ideal for the delivery of therapeutics and the stimulation of tissues—perhaps the most important of which is that US is a mechanical and yet non-invasive means of delivery that can be applied to a very wide range of therapeutics and target sites.

Cavitation bubbles, the primary and observable effect of insonation (Susslick, Mason Lorimer) lead to fluid convection, high fluid shear stresses, shock waves, free radicals, and high temperatures, all of which can produce significant biological effects, both beneficial and harmful to cells (Pitt 2008). The beneficial effects, with respect to drug and gene delivery, include the loosening of cell-to-cell junctions, the permeabilization (and even poration) of cell membranes, the stimulation of stress response (or other) pathways in cells, the release of drugs and genes from various carriers, the deposition of heat, and the activation of some chemicals by free radicals. The detrimental effects include unwanted cell injury and death, and unwanted degradation of the drugs and polynucleotides. Thus, one of the challenges to ultrasonic drug and gene delivery is to find the correct balance of ultrasonic parameters that maximizes helpful and minimizes harmful effects in order to create a functional therapeutic window (Pitt 2008).

A complete and precise summary of the potential applications for ultrasonic enhancement of drug and gene delivery remains challenging as the mechanisms are multiple and not fully characterized (Ferrara 2008). From an engineering viewpoint, the methods by which ultrasound facilitates drug and gene delivery can be summarized as direct changes in the biological or physiological properties of tissues facilitating transport, direct changes in the drug or vehicle increasing bioavailability or enhancing efficacy, and indirect effects by which ultrasound acts on the vehicle to produce changes in the surrounding tissue. While there are common mechanisms between these methods, the engineering challenges associated with the optimization of ultrasound systems and drug and vehicle design are unique for each method. Promising examples of each of these methods can be identified at this time (Ferrara 2008). The thermal effects of low ultrasound intensities have been associated with increases in perfusion and changes in drug-carrying vehicles. At higher ultrasound intensities, these thermal effects can locally ablate tissue.

Therapeutic effects associated with ultrasound could be enhanced by various mechanisms, as enhanced membrane transport. Changes in cell membrane permeability, by *in vitro* and *in vivo* insonation (Prentice et al. 2005; Tachibana et al. 1999; Ward et al. 2000) were observed when small holes are produced within cell membranes by the collapse of a microbubble and production of a jet. Pores produced within the cell membrane may be transient (facilitating successful therapeutic delivery) or permanent (resulting in cell death) (Ward et al. 1999, 2000). Momentum transfer from the propagating wave to delivery vehicles can result in the translation of vehicles across a blood vessel and facilitate vascular targeting. These radiation force and streaming effects can also be combined with other therapeutic mechanisms (e.g. heating, cavitation) to further enhance drug delivery (Ferrara 2008).

Lipids and polymers can be engineered to self-assemble into vehicles that protect or concentrate drugs, particularly those with systemic toxicity. The mechanical or thermal properties of ultrasound can release the drug from the vehicle in the selected region of interest. Engineering such vehicles to stably incorporate the drug or gene during circulation and yet effectively release the contents at the desired site remains challenging (Ferrara 2008; Husseini and Pitt 2008).

Enhanced delivery to tumors is a similarly important application of ultrasound-enhanced drug and gene delivery, since transport of drugs within solid tumors can be very limited. Rapoport and coworkers reported an increase in localised tumor delivery of up to 5–10 fold by using sonic activation (Rapoport et al. 2007; Husseini and Pitt 2008).

The ultrasound thermal and mechanical bioeffects as relevant safety issues for *in vitro* and in vivo applications, including drug delivery, are also discussed by Wu and coworkers (2008) focusing on ultrasound effects interaction with cells. It was shown that in proposed mechanism of drug internalization the cells are "massaged" by the shear stress generated by the microstreaming pattern and the repeated "massage" may make the cells more permeable to "foreign" particles such as DNA and drugs to make their uptake relatively easier. When the change of cell membrane is reversible it is called reparable sonoporation. Also it is possible that cavitation effects may render the cells permanently deformed, by a non-reversible process. Wu's opinion was that even if currently the efficiency of targeting drug delivery in vitro and in vivo using sonoporation is still relatively low, acoustic radiation force has a great potential to direct vectors toward the target cells in a controllable fashion (Lum 2006) using nanoparticle carriers. More in vitro experimental work and computational simulations are needed to better understand the physics of sonoporation (Wu et al. 2008).

16.3.6.6 Multifunctional Silica CDDS – Release Profile

"Haglund et al. (2009) showed that multifunctional nanoparticles hold great promise in CDDS for drug/gene delivery, being able simultaneous" to diagnostic and to cure, the complex function named theragnostics. Such system include, for example, the use of core materials that provide in vivo imaging and opportunities for externally modulated therapeutic interventions. In principle, multicomponent particles, composed by highly specialized layers to accomplish complex and potentially decision-making tasks, based on emitted signals by chemistry and physics of processes occurring as an interaction between nanoparticle system and biologic environment, can act as nanomedical intelligent systems. The steps to be accomplish in order to attain such a task are consisting in initial cell recognition, cell entry, re-targeting to the appropriate sub-region of the cell where the therapeutic package can be localized, potentially control of the therapeutic process through feedback systems. The premises suggest that the design and developing process of a sophisticated nanomedical platform system can be conduct also step by step, firstly designing each subsystem able to accomplish one task at the time, and then the most difficult step is to full integrate these interacting subcomponents. Specific examples of sub-components developed for specific applications are described in literature and also some multicomponent CDDS, actually dual, encompassing two functions, as luminescent labeled magnetic targeting system which can be monitored concerning, circulation in the body system, accumulation near or in the target organ and elimination route after releasing the active cargo. Nanomedicine is a fundamental nanotechnology approach because it approaches medicine in a bottoms-up rather than top-down approach and performs parallel-processing medicine at the single cell level. Current research indicates that multifunctional medical devices can now be constructed at the nanoscale. Much of the design of these nanomedical devices can be guided by biomimetic studies of nanostructures. This has been performed through the construction of multilayered nanostructures, where layer has a unique function. Initially, core materials must be carefully selected based on their potential for diagnostic and therapeutic applications. Specifically, core material properties that enhance disease detection capabilities are clinically advantageous. Cytotoxicity of the core material and subsequent biocoatings is critical to evaluate due to the uncertainty of their stability in biological environments. In addition, functional layers and delivery methods are also critical for the many functions of nanomedical systems. In order to achieve cellular delivery, various biomolecules have been used to guide nanostructures to diseased cells (Haglund et al. 2009).

Stucky group (Suh et al. 2009) reported in excellent review important issues related to multifunctional nanoparticle systems (MFNPSs) This review covers the physical and biological aspects involved in nanoparticle systems having multiple functions such as optical and magnetic resonance imaging capabilities with incorporated bioactive molecules. Recent examples are covered based on a simple but logical categorization scheme.

Ghosh et al. (2010) reports the synthesis and alternating magnetic field dependent actuation of a remotely controllable, multifunctional nano-scale system and its marked biocompatibility with mammalian cells. Monodisperse, ***magnetic nanospheres based on thermo-sensitive polymer network*** poly(ethylene glycol) ethyl ether methacrylate-copoly(ethylene glycol) methyl ether methacrylate were synthesized using free radical polymerization. Synthesized nanospheres have oscillating magnetic field induced thermo-reversible behavior. The developed CDDS exhibited an excellent intracellular uptake and enhanced biocompatibility. Nanosphere exposure did not inhibit the extension of cellular processes (neurite outgrowth) even at high iron concentrations (6 mM). Bikram et al. synthesized hydrogels with SiO_2–Au and studied the release mechanism (Wong et al. 2008).

Silica–gold (SiO_2–Au) nanoshells are a new class of nanoparticles that consist of a silica dielectric core that is surrounded by a gold shell. Au shell peak extinctions are very easily tunable over a wide range of wavelengths (near infrared (IR) region). Irradiation of SiO_2–Au nanoshells at their peak extinction coefficient results in the conversion of light to heat energy.

A ***photothermal*** modulated drug delivery system was designed and synthesized as a composite by combination of the inorganic part: (photosensitive)SiO_2–Au nanoshell- with (temperature-sensitive) hydrogel. By varying Au nanoshell concentrations in SiO_2–Au system it was initiate a temperature change with light. As temperature-sensitive component N-isopropylacrylamide-co-acrylamide (NIPAAm-co-AAm) hydrogels were obtained, which exhibit a lower critical solution temperature (LCST) slightly above body temperature (~50°C). The resulting composite hydrogels had the extinction spectrum of the SiO_2–Au nanoshells in which the hydrogels collapsed reversibly in response to temperature (50°C) and laser irradiation. The degree of collapse of the hydrogels was controlled by the laser fluence as well as the concentration of SiO_2–Au nanoshells. Modulated drug delivery profiles for methylene blue, insulin, and lysozyme were achieved by irradiation of the drug-loaded nanoshell-composite hydrogels, which showed that drug release was dependent upon the molecular weight of the therapeutic molecule (Bikram et al. 2007).

Mishra and coworkers obtained a *in vitro* drug delivery response of polyethylene glycol (PEG)-functionalized magnetite activated with a stable ligand, folic acid, and conjugated with an anticancer drug, doxorubicin. The drug-release behavior of PEG-functionalized and folic acid–doxorubicin-conjugated magnetic nanoparticles was characterized by two stages involving an initial rapid release, followed by a controlled release (Zhang et al. 2008). Wong et al. describe the synthesis, characterisation and surface-modification of magnetic nanoparticles and a poly (*N*-isopropylacrylamide) microgel, followed by the assembly and characterisation of magnetic nanoparticles on the microgel (Wong et al. 2008). They had in attention the ***thermoresponsive and magnetic*** properties in prepared hybrid core–shell structures, specifically the amount of heat transfer from the magnetic core onto the thermosensitive (loaded) microgel (for the subsequent heat-triggered release of drugs). Inductive heat study reveals that the heat generated by the magnetic nanoparticles is sufficient to cause the collapse of the microgel above its volume phase transition temperature. Successful confinement of positively and negatively charged magnetic nanoparticles between polyelectrolyte layers is achieved using the layer-by-layer deposition onto the microgel. Dynamic light scattering measurements show (i) the presence of each layer successfully deposited, (ii) the preservation of thermoresponsivity in the coated microgel, and (iii) that the magnetic nanoparticles do not get detached during the phase transition of the microgel. Electrophoresis measurements confirm charge reversal at every stage of layering of polycations, polyanions and magnetic nanoparticles (Wong et al. 2008).

Yang et al. obtained a multifunctional CDDS starting from nonporous silica encapsulating Fe3O4 microspheres. A further layer of ordered mesoporous silica through a simple sol–gel process was added. The surface of the outer silica shell was further functionalized by the deposition of YVO4:Eu^{3+} phosphors, realizing a sandwich structured material with ***mesoporous, magnetic and luminescent properties***. The multifunctional system was used as drug carrier to investigate the storage and release properties using ibuprofen (IBU) as model drug by the surface modification (Yang et al. 2005).

Gan et al. developed a stimuli-sensitive mesoporous silica nanoparticle (MSNs)-based hybrid "gate-like" ensembles capable of performing specific programmed release mode represent a new generation delivery system in recent years. A magnetic and reversible pH-responsive, MSNs-based nanogated ensemble was fabricated by anchoring superparamagnetic Fe3O4 nanoparticles on the pore outlet of MSNs via a reversible boronate esters linker (Gan et al. 2011).

The pH-driven "gate-like" effect was studied by in vitro release of an entrapped model dexamethasone from the pore voids into the bulk solution at different pH values. The results indicated that at pH 5–8, the pores of the MSNs were effectively capped with Fe3O4 nanoparticles and the drug release was strongly inhibited. While at pH 2–4, the hydrolysis of the boroester bond took place and thus resulted in a rapid release of the entrapped drug. And by alternately changing the pH from 3 to 7, these Fe3O4 cap gate could be switched "on" and "off" and thereby released the entrapped drug in a pulsinate manner (in small portions). Additionally, this nanogated release system exhibited good magnetic property, high cell biocompatibility and cellular uptake for MC3T3-E1 cells. The present data suggest that it is possible to obtain simple and very effective pH-driven pulsinate release using these Fe3O4-capped-MSNs, and this new platform represents a promising candidate in the formulation of in vivo targeted delivery of therapeutic agents to low pH tissues, such as tumors and inflammatory sites, Gan said. Chen et al. obtained PNIPAm-Fe3O4 hybrid microgels were thermosensitive, exhibiting a reversible swelling and deswelling behavior as a function of temperature. The PNIPAm-Fe3O4 hybrid microgels also show superparamagnetic behavior at room temperature (300 K) (Chen et al. 2011).

16.4 Conclusions

In the chapter it was presented some considerations concerning advance in (bio) sensing materials, mainly destined to be used in controlled drug delivery systems, named also intelligent materials, developments. The CDDS's is relatively new and hot research field and inorganic materials, specifically mesoporous and magnetic or luminescent nanostructures ones are even newer and increasingly developing topics in the area. During CDDS history (bio) polymers were used as drug carriers, because of their biocompatibility and other suitable characteristics. Nevertheless, among so many appropriate qualities as drug sensing – responsive carrier polymers showed few weaknesses such as fair resistance in the body environment, low mechanical resistance etc. Due to the huge expertise of scientific community concerning hybrid materials, that encompasses a combination of their moieties desired properties, it came out the idea to use inorganic components in order to strengthen and improve the DDS structure and behavior. This is why we considered useful to start with describing essentials of polymer based CDDS, to be aware about the place and role to be played by mesoporous and magnetic silica as part of CDDS. More than that we have tried to reveal from literature examples if and how the required properties as a brick in the complex architecture of a CDDS can be meet in inorganic silica and composites. Further it was given examples with the reported accomplishments in the silica based CDDS area of research.

In the chapter we referred also to peculiar basics in order to design and synthesize materials and formulate CDDS architectures such as coordination chemistry principles and fractal character of materials structure.

It can be clearly seen that the use of nanoparticulate pharmaceutical carriers essentially contributed to enhancing the *in vivo* efficiency of many drugs. The current engineering pharmaceutical nanocarriers, in some cases, revealed some aspects that leaded to some skeptical opinions. According Vallet-Regi and coworkers, from an industrial point of view, it is worthless to expend time, money and effort designing a determined particle morphology because, they demonstrated that for non-functionalized MCM-41 materials, drug release rate shows little dependence on carrier morphology (Manzanoa et al. 2008). Rosenholm and Linden discovered that in many cases, the effectiveness of the introduced surface functions is much lower than what could be expected (Rosenholm and Linden 2008a, b).

As Torkilin have said, looking into the future of the field of drug delivery, we have to think about the development of the next generation of pharmaceutical nanocarriers, combining different properties and allowing for multiple functions, as new "smart" generation of DDS (Torchilin 2007).

What remains and always will be important, Torkilin said, is that the new generations of DDS to come have to simultaneously or sequentially demonstrate the following basic properties: (1) Circulate and stay long in the blood; (2) Specifically target the site of the disease; (3) Respond local stimuli characteristic of the pathological site or respond externally applied stimuli; (4) Provide an enhanced intracellular delivery of an entrapped drug in case the drug is expected to exert its action inside the cell; (5) Provide a real-time information about the carrier (and drug) biodistribution and target accumulation as well as about the outcome of the therapy due to the presence within the structure of the carrier of a certain reporter moiety (Torchilin 2007).

The obvious final conclusion is that there is a lot of work to be done in the field of CDD systems, in order to systematically gain knowledge and experience to understand and being able solve the current and coming issues.

Acknowledgement Authors thanks Romanian Academy, and Romanian Ministry of Education and Research for supporting the present work through the CNCS-UEFISCDI (former CNCSIS-UEFISCSU) project < Quantification of The Chemical Bond Within Orthogonal Spaces of Reactivity. Applications on Molecules of Bio-, Eco- and Pharmaco- Logical Interest>, Code PN II-RU-TE-2009-1 grant no. TE-16/2010-2011.

References

Abeylath SC, Turos E (2008) Expert Opin Drug Deliv 5:931–949
Alexis F, Rhee JW, Richie JP, Radovic-Moreno AF, Langer RR, Farokhzad OC (2008) Urol Oncol 26:74–85
Allouche J, Boissiere M, Helary C, Livage J, Coradin T (2006) J Mater Chem 16:3120–3125
Alvaro M, Benitez M, Das D, Garcia H, Peris E (2005) Chem Mater 17:4958–4964
Amidon GL, Lennernas H, Shah VP, Crison JR (1995) Pharm Res 12:413–420
Anal KA (2007) Recent Pat Endocr Metab Immune Drug Discov 1:83–90
Anderson MA, Gieselman MJ, Xu QJ (1988) Membr Sci 39:243–258
Andersson J, Rosenholm J, Areva S, Linden M (2004) Chem Mater 16:4160–4167
Andersson J, Areva S, Spliethoff B, Linden M (2005) Biomaterials 26:6827–6835
Ariga K, Vinu A, Hill JP, Mori T (2007) Coord Chem Rev 251:2562–2591
Arruebo M, Galan M, Navascues N, Tellez C, Marquina C, Ibarra MR, Santamaria J (2006) Chem Mater 18:1911–1919
Asefa T, MacLachlan MJ, Coombs N, Ozin GA (1999) Nature 402:867–871
Averitt RD, Westcott SL, Halas NJ (1999) J Opt Soc Am B 16:1824–1832
Aznar E, Casasus R, Garcia-Acosta B, Marcos MD, Martinez-Manez R (2007) Adv Mater 19:2228–2231

Aznar E, Marcos MD, Martinez-Manez R, Sancenon F, Soto J, Amoros P, Guillem C (2009) J Am Chem Soc 131:6833–6843
Bagshaw SA, Prouset E, Pinnavaia TJ (1995) Science 269:1242–1244
Bahadur D, Jyotsnendu G (2003) Sadhana 28:639–56
Bajpai AK, Shukla SK, Bhanu S, Kankane S (2008) Prog Polym Sci 33:1088–1118
Barbe C, Bartlett J, Kong LG, Finnie K, Lin HQ, Larkin M, Calleja S, Bush A, Calleja G (2004) Adv Mater 16:1959–1966
Barker S (2001) Drug Discov Today 6:75–77
Barsi L, Buki A, Szabo D, Zrinyi M (1996) Prog Colloid Polym Sci 102:57–63
Bartil T, Bounekhel M, Cedric C, Jerome R (2007) Acta Pharm 57:301–314
Beck JS, Vartuli JC, Roth WJ, Leonowicz ME, Kresge CT, Schmitt KD, Chu CTW, Olson DH, Sheppard EW, McCullen SB, Higgins JB, Schlenker JL (1992) J Am Chem Soc 114:10834–10843
Beduneau A, Ma Z, Grotepas CB, Kabanov A, Rabinow BE, Gong N et al (2009) PLoS ONE 4(2): e4343
Begu S, Pouessel AA, Lerner DA, Tourne-Peteilh C, Devoisselle JM (2007) J Control Release 118:1–6
Benkoski JJ, Bowles SE, Korth BD, Jones RL, Douglas JF, Karim A, Pyun J (2007) J Am Chem Soc 129:6291–6297
Benkoski JJ, Bowles SE, Jones RL, Douglas JF, Pyun J, Karim A (2008) J Polym Sci Part B: Polym Phys 46:2267–2277
Besson E, Mehdi A, Lerner DA, Reye C, Corriu RJP (2005) J Mater Chem 15:803–809
Bikram M, Gobin AM, Whitmire RE, Jennifer L (2007) J Controlled Release 123:219–227
Bottcher H (1998) J Sol-Gel Sci Technol 13:277–281
Bottini M, D'Annibale F, Magrini A, Cerignoli F, Arimura Y, Dawson MI, Bergamaschi E, Rosato N, Bergamaschi A, Mustelin T (2007) Int J Nanomedicine 2:227–233
Bowles SE, Wu W, Kowalewski T, Schalnat MC, Davis RJ, Pemberton JE, Shim I, Korth BD, Pyun J (2007) J Am Chem Soc 129:8694–8695
Braun A, Hammerle S, Suda K, Rutishauser BR, Gunthert M, Kramer SD, Allenspach HW (2000) Eur J Pharm Sci 11(2):S51–S60
Breimer DD (1999) J Control Release 62:3–6
Brinker CJ, Scherer GW (1990) Sol-gel science. Academic, San Diego
Brinker CJ, Ward TL, Sehgal R, Raman NK, Hietala SL, Smith DM, Hua DW, Headley TJ (1993) J Membr Sci 77:165–179
Bromberg I (2003) Curr Pharm Biotechnol 4:39–49
Brown LF, Detmar M, Claffey K, Nagy JA, Feng D, Dvorak AM et al (1997) In: Goldberd ID, Rosen EM (eds) Regulation of angiogenesis. Birkhauser Verlag, Basel, pp 233–269
Brugger B, Richtering W (2007) Adv Mater 19:2973–2978
Bruke M, Langer R, Brem H (1999) In: Mathowitz E (ed) Encyclopaedia of controlled drug delivery, vol 1. Wiley, New York, pp 184–212
Brunel D (1999) Microporous Mesoporous Mater 27:329–344
Brunel D, Fajula F, Nagy JB, Deroide B, Verhoef MJ, Veum L, Peters JA, Bekkum AV (2001) Appl Catal A 213:73–82
Burkett SL, Sims SD, Mann S (1996) Chem Commun 11:1367–1368
Cantor G (1882) Math Ann 21:545–591
Casasus R, Marcos MD, Martinez-Manez R, Ros-Lis JV, Soto J, Villaescusa LA, Amoros P, Beltran D, Guillem C, Latorre J (2004) J Am Chem Soc 126:8612–8616
Chang JH, Kim KJ, Shin YK (2004) Bull Korean Chem Soc 25(8):1257
Che S, Garcia-Benett AE, Tokoi T, Sakamoto K, Kunieda H, Terasaki O, Tatsumi T (2003) Nat Mater 2:801
Chen T, Cao Z, Guo X, Nie J, Xu J, Fan Z (2011) Polymer 52:172–179
Chien YW (1992) In: Chien YW (ed) Novel drug delivery systems. Marcel Dekker, New York, pp 1–42

Chinnayelka S, McShane MJ (2005) Anal Chem 77:5501–5511
Cho K, Wang X, Nie S, Chen (Georgia) Z, Shin DM (2008) Clin Cancer Res March 1(14):1310–1316
Chung TH, Wu SH, Yao M, Lu CW, Lin YS, Hung Y, Mou CY, Chen YC, Huang DM (2007) Biomaterials 28:2959–2966
Cleland JL, Daughtery A, Mrsny R (2001) Curr Opin Biotechnol 12:212–219
Colvin VL (2003) Nat Biotechnol 21:1166–1170
Cullander C, Guy RH (1992) Adv Drug Deliv Rev 8:291–329
Davidson A (2002) Curr Opin Colloid Interface Sci 7:92–106
Davidson GWR III, Peppas NA (1986) J Control Release 3:243–258
Dharap SS, Qiu B, Williams GC, Sinko P, Stein S, Minko T (2003) J Control Release 91:61–73
Di Pasqua L, Testa F, Aiello R, Cundari S, Nagy JB (2007) Microporous Mesoporous Mater 103:166–173
Di Pasqua AJ, Sharma KK, Shi YL, Toms BB, Ouellette W, Dabrowiak JC, Asefa T (2008) J Inorg Biochem 102:1416–1423
Doadrio AL, Sousa EMB, Doadrio JC, Pariente JP, Izquierdo-Barba I, Vallet-Regi M (2004) J Control Release 97:125–132
Doadrio JC, Sousa EMB, Izquierdo-Barba I, Doadrio AL, Perez-Pariente J, Vallet-Regi M (2006) J Mater Chem 16:462–466
Dormer K, Seeney C, Lewelling K, Lian G, Gibson D, Johnson M (2005) Biomaterials 26:2061–2072
Drummond DC, Meyer O, Hong K, Kirpotin DB, Papahadjopoulos D (1999) Pharmacol Rev 51:691–743
Dubois M, Gulik-Krzywicki T, Cabane B (1993) Langmuir 9:673–680
Ehrlich P (1954) Int Arch Allergy Appl Immunol 5:67–86
Eisenberg R (2011). Celebration of the 50th anniversary "Voices of Inorganic Chemistry" video series. *ACS Matters*, February 1, http://acswebcontent.acs.org/newsletter/110201.html. Accessed June 18, 2011
Falaize S, Radin S, Ducheyne P (1999) J Am Ceram Soc 82:969–976
Farokhzad OC, Jon S, Khademhosseini A, Tran TN, Lavan DA, Langer R (2004) Cancer Res 64:7668–7672
Ferrara KW (2008) Adv Drug Deliv Rev 60:1097–1102
Ferris DP, Zhao YL, Khashab NM, Khatib HA, Stoddart JF, Zink JI (2009) J Am Chem Soc 131:1686–1688
Filipcsei G, Csetneki I, Szilagyi A, Zrinyi M (2007) Adv Polym Sci 206:137–189
Finnie KS, Waller DJ, Perret FL, Krause-Heuer AM, Lin HQ, Hanna JV, Barbe CJ (2009) J Sol-Gel Sci Technol 49:12–18
Firth JA (2002) J Anat 200:541–548
Fix J (1999) In: Mathowitz E (ed) Encyclopaedia of controlled drug delivery, vol 2. Wiley, New York, pp 698–728
Fonseca MJ, Jagtenberg JC, Haisma HJ, Storm G (2003) Pharm Res 20:423–428
Fu BM, Adamson RH, Curry FE (1998) Am J Physiol 274 H2062–2073
Fubini B (1997) Environ Health Perspect 105:1013–1020
Fubini B (1998) Ann Occup Hyg 42:521–530
Fubini B, Zanetti G, Altilia S, Tiozzo R, Lison D, Saffiotti U (1999) Chem Res Toxicol 12:737–745
Gan Q, Lu X, Yuan Y, Qian J, Zhou H, Lu X, Shi J, Liu C (2011) Biomaterials 32:1932–1942
Garnett MC (2001) Advanced Drug Delivery Reviews 53:171–216
Gerion D, Herberg J, Bok R, Gjersing E, Ramon E, Maxwell R, Kurhanewicz J, Budinger TF, Gray JW, Shuman MA, Chen FF (2007) J Phys Chem C 111:12542–12551
Ghosh S, GhoshMitra S, Cai T, Diercks DR, Mills NC, Hynds DL (2010) Nanoscale Res Lett 5:195–204

Giovine M, Pozzolini M, Fenoglio I, Scarfi S, Ghiazza M, Benatti U, Fubini B (2002) Toxicol Ind Health 18:249–255
Giri S, Trewyn BG, Stellmaker MP, Lin VSY (2005) Angew Chem Int Ed 44:5038–5044
Gohel MC, Panchal MK, Jogani VV (2000) PharmSciTech 1(4):31
Gonda I (2000) J Pharm Sci 89:940–945
Gosk S, Vermehren C, Storm G, Moos T (2004) J Cereb Blood Flow Metab 24:1193–204
Graf C, Dembski S, Hofmann A, Ruehl E (2006) Langmuir 22:5604–5610
Grinsted RA, Clark K, Koenig JL (1992) Macromolecules 25:1235–1241
Gruen M, Lauer I, Unger KK (1997) Adv Mater 9:254–257
Guo J, Yang WL, Deng YH, Wang CC, Fu SK (2005) Small 1:737–743
Gupta RB, Kompella UB (2006) Nanoparticle technology for drug delivery. Taylor & Francis Group, New York/London
Haglund E, Seale-Goldsmith MM, Leary JF (2009) Ann Biomed Eng 37:2048–2063
Haley B, Frenkel E (2008) Urol Oncol 26(1):57–64
Haraguchi K (2007) Curr Opin Solid St M 11:47–54
Harding JA, Engbers CM, Newman MS, Goldstein NI, Zalipsky S (1997) Biochim Biophys Acta 1327:181–92
Hernandez R, Tseng HR, Wong JW, Stoddart JF, Zink JI (2004) J Am Chem Soc 126:3370–3371
Higuchi T (1961) J Pharm Sci 50:874–875
Higuchi T (1963) J Pharm Sci 52:1145–1149
Hobbs SK, Monsky WL, Yuan F, Roberts WG, Griffith L, Torchilin VP et al (1998) Proc Natl Acad Sci USA 95:4607–4612
Hochstrasser G, Antonini JF (1972) Surf Sci 32:644–664
Hoffman AS (2008) J Control Release 132:153–163
Holash J, Maisonpierre PC, Compton D, Boland P, Alexander CR, Zagzag D et al (1999) Science 284:1994–1998
Hsiao JK, Tsai CP, Chung TH, Hung Y, Yao M, Liu HM, Mou CY, Yang CS, Chen YC, Huang DM (2008) Small 4:1445–1452
Hu SH, Liu TY, Huang HY, Liu DM, Chen SY (2008) Langmuir 24(1): 239–244
Huang DM, Hung Y, Ko BS, Hsu SC, Chen WH, Chien CL, Tsai CP, Kuo CT, Kang JC, Yang CS, Mou CY, Chen YC (2005) FASEB J 19:2014–2016
Hughes GA (2005) Nanomed Nanotechnol Biol Med 1:22–30
Huo Q, Liu J, Wang LQ, Jiang Y, Lambert TN, Fang E (2006) J Am Chem Soc 128:6447–6453
Husseini GA, Pitt WG (2008) Adv Drug Deliv Rev 60:1137–1152
Iler RK (1979) Chemistry of silica: solubility, polymerisation, colloid and surface properties and biochemistry. Wiley, New York
Inagaki S, Fukushima Y, Kuroda K (1993) J Chem Soc Chem Commun 8(8):680–682
Inagaki S, Guan S, Fukushima Y, Ohsuna T, Terasaki O (1999) J Am Chem Soc 121(41):9611–9614
Inagaki S, Guan S, Ohsuna T, Terasaki O (2002) Nature 416:304
Ionescu C, Savii C, Balasoiu M, Popovici M, Enache C, Kuklin A, Islamov A, Kovalev Y, Almasy L (2004) Acta Periodica Technologica 35:95–101
Ishida O, Maruyama K, Tanahashi H, Iwatsuru M, Sasaki K, Eriguchi M et al (2001) Pharm Res 18:1042–1048
Iwai K, Maeda H, Konno T (1984) Cancer Res 44(5):2115–2121
Iyer SS, Haddad YM (1994) Int J Press Vessels Pip 58(3):335–344
Izquierdo-Barba I, Sousa E, Doadrio JC, Doadrio AL, Pariente JP, Martinez A, Babonneau F, Vallet-Reg M (2009) J Sol-Gel Sci Technol 50:421–429
Jain RK (1998) J Control Release 53:49–67
Johansson E, Choi E, Angelos S, Liong M, Zink JI (2008) J Sol–Gel Sci Technol 46:313–322
Jordan A, Scholz R, Wust P, Fohling H, Felix R (1999) J Magn Magn Mater 201:413–419

Kanduc D, Mittelman A, Serpico R, Sinigaglia E, Sinha AA, Natale C, Santacroce R, Di Corcia MG, Lucchese A, Dini L, Pani P, Santacroce S, Simone S, Bucci R, Farber E (2002) Int J Oncol 21:165–170
Kappor MP, Inagaki S (2006) Bull Chem Soc Jpn 79:1463–1475
Keck C, Kobierski S, Mauludin R, Muller RH (2008) DSOIS 24:124–128
Kelleher BP, Doyle AM, O'Dwyer TF, Hodnett BK (2001) J Chem Technol Biotechnol 76:1216–1222
Keng PY, Shim I, Korth BD, Douglas JF, Pyun J (2007) ACS Nano 1:279–292
Kent J, Lewis D, Sanders L, Tice T (1987) Microencapsulation of water soluble active polypeptides, US Patent 4:675:189
Kickelbick G (2003) Prog Polym Sci 86:83–114
Kickelbick G, Schubert U (2001) Monatsh Chem 132:13–30
Kim DK, Zhang Y, Kehr J, Klason T, Bjelke B, Mohammed M (2001) J Magn Magn Mater 225:256–261
Kono K, Takagishi T (1996) In: Salamone JC (ed) Polymeric materials encyclopedia. CRC Press, Boca Raton
Koop OM, Rubinstein I, Onyuksel H (2005) Nanomed Nanotechnol Biol Med 1:193–212
Kortesuo P, Ahola M, Karlsson S, Kangasniemi I, Yli-Urpo A, Kiesvaara J (1999a) Biomaterials 21:193–198
Kortesuo P, Ahola M, Karlsson S, Kangasniemi I, Kiesvaara J, Yli-Urpo A (1999b) J Biomed Mater Res 44:162–167
Korth BD, Keng P, Shim I, Bowles SE, Tang C, Kowalewski T, Nebesny KW, Pyun J (2006) J Am Chem Soc 128:6562–6563
Kresge CT, Leonowicz ME, Roth WJ, Vartuli JC, Beck JS (1992) Nature 359:710–712
Kwon IK, Jeong SH, Kang E, Park K (2007) Nanoparticulate drug delivery for cancer therapy, Chapter 19 in *Cancer Nanotechnol*. In: Nalwa HS, Webster T (eds) American Scientific Publishers, Stevenson Ranch, CA 333–344
Lai W, Ducheyne P, Garino J (1998) In: Legeros RZ, Legeros JP (eds) Proceedings of the 11th international symposium on ceramics in medicine, vol 11. World Scientific Publishing Co, New York, p 383
Lai CY, Trewyn BG, Jeftinija DM, Jeftinija K, Xu S, Jeftinija S, Lin VSY (2003) J Am Chem Soc 125:4451–4459
Lancok A, Bezdicka P, Klementova M, Zaveta K, Savii C (2008) Acta Phys Pol A 113(1):577–581
Langer R, Folkman J (1976) Nature 263:797–800
Langer R, Peppas NA (2003) J AIChE 49:2990–3006
Larbot A, Alary JA, Babre JP, Guixard C, Cot L (1986) Mater Res Sot Symp Proc 73:659
Larsen AK, Escargueil AE, Skladanowski A (2000) Pharmacol Ther 85:217–229
La-Van D, McGuire T, Langer R (2003) Nat Biotechnol 21:1184–1191
Lee PW, Hsu SH, Wang JJ, Tsai JS, Lin KJ, Wey SP, Chen FR, Lai CH, Yen TC, Sung HW (2010) Biomaterials 31:1316–1324
Leenars AFM, Keizer K, Burggraaf AJ (1984) J Mater Sci 10:1077–1088
Leung KCF, Nguyen TD, Stoddart JF, Zink JI (2006) Chem Mater 18:5919–5928
Li H, Yan GP, Wu SN, Wang ZJ, Lam KY (2004a) J Appl Polym Sci 93:1928–1937
Li L, Wartchow CA, Danthi SN, Shen Z, Dechene N, Pease J et al (2004b) Int J Radiat Oncol Biol Phys 58:1215–1227
Li ZZ, Wen LX, Shao L, Chen JF (2004c) J Control Release 98:245–254
Liang YC, Hanzlik M, Anwander R (2005) Chem Commun 4:525–527
Lim MH, Stein A (1999) Chem Mater 11:3285–3295
Lin HM, Wang WK, Hsiung PA, Shyu SG (2010) Acta Biomater 6(8):3256–3263
Lindgren M, Hallbrink M, Prochiantz A, Langel U (2000) Trends Pharmacol Sci 21:99–103
Liu N, Dunphy DR, Atanassov P, Bunge SD, Chen Z, Lopez GP, Boyle TJ, Brinker CJ (2004) Nano Lett 4:551–554
Liu XY, Gao CY, Shen JC, Mohwald H (2005) Macromol Biosci 5:1209–1219

Liu C, Guo J, Yang W, Hu J, Wang C, Fu S (2009) J Mater Chem 19:4764–4770
Logan DL, Ashley CS, Brinker CJ (1992) Mater Res Soc Symp Proc 271:541–546
Lu R (2004) Cellular drug delivery, principle and practice. Humana Press Inc, Totowa
Lu J, Liong M, Zink JI, Tamanoi F (2007) Small 3:1341–1346
Lu J, Choi E, Tamanoi F, Zink J I (2008) Small 4(4):421–426
Lukowiak A, Strek W (2009) J Sol-Gel Sci Technol 50:201–215
Lum AFH, Borden MA, Dayton PA, Kruse DE, Simon SI, Ferra KW (2006) J Control Release 111:128–134
Macquarrie DJ (1996) Chem Commun 16:1961–1962
Macquarrie DJ, Brunel D, Renard G, Blanc AC (2001) Stud Surf Sci Catal 135:312
Mal NK, Fujiwara M, Tanaka Y (2003) Nature 421:350–353
Mal NK, Fujiwara M, Tanaka Y, Taguchi T, Matsukata M (2003a) Chem Mater 15:3385–3394
Mal NK, Fujiwara M, Tanaka Y (2003b) Nature 421:350–353
Mandelbrot BB (1977) Fractals, form, and chance. Freeman, San Francisco
Mandelbrot B (1982) The fractal geometry of nature. W.H.Freeman & Co, New York
Manzanoa M, Aina V, Arean CO, Balas F, Cauda V, Colilla M, Delgado MR, Vallet-Regi M (2008) Chem Eng J 137:30–37
Marty JP, Puisieux F, Carstensen JT (1982) J Pharm Sci 71:749–752
Maruyama K, Ishida O, Kasaoka S, Takizawa T, Utoguchi N, Shinohara A et al (2004) J Control Release 98:195–207
Maschmeyer T, Rey F, Sankar G, Thomas JM (1995) Nature 378:159
Mayor S, Pagano RE (2007) Nat Rev Mol Cell Biol 8:603–612
Medina OP, Kairemo K, Valtanen H, Kangasniemi A, Kaukinen S, Ahonen I et al (2005) Anticancer Res 25:33–42
Melde BJ, Holland BT, Blanford CF, Stein A (1999) Chem Mater 11:3302–3308
Menger K (1926) Allgemeine Raume und Cartesische Raume. I., Communications to the Amsterdam Academy of Sciences (English translation reprinted in Edgar GA (ed) (2004) Classics on fractals, studies in nonlinearity. Westview Press. Advanced Book Program, Boulder)
Mercier L, Pinnavaia TJ (2000) Chem Mater 12:188–196
Meseguer-Olmo L, Ros-Nicolas MJ, Clavel-Sainz M, Vicente-Ortega V, Alcaraz-Banos M, Lax-Perez A, Arcos D, Ragel CV, Vallet-Regi M (2002) J Biomed Mater Res 61:458–465
Meseguer-Olmo L, Ros-Nicolas M, Vicente-Ortega V, Alcaraz-Banos M, Clavel-Sainz M, Arcos D, Ragel CV, Vallet-Regi M, Meseguer-Ortiz C (2006) J Orthop Res 24:454–460
Metselaar JM, Mastrobattista E, Storm G (2002) Mini-Rev Med Chem 2:319–329
Mishra BM, Patel BB, Tiwari S (2010) Nanomed Nanotechnol Biol Med 6:9–24
Missailidis S, Thomaidou D, Borbas KE, Price MR (2005) J Immunol Methods 296:45–62
Mitra A, Mulholland J, Nan A, McNeill E, Ghandehari H, Line BR (2005) J Control Release 102:191–201
Moody TW, Leyton J, Gozes I, Lang L, Eckelman WC (1998) Ann NY Acad Sci 865:290–296
Mornet S, Lambert O, Duguet E, Brisson A (2005) Nano Lett 5:281–285
Motornov M, Roiter Y, Tokarev I, Minko S (2010) Prog Polym Sci 35:174–211
Mountain A (2000) Trends Biotechnol 18:119–128
Munoz B, Ramila A, Perez-Pariente J, Diaz I, Vallet-Regi M (2003) Chem Mater 15:500–503
Mykhaylyk O, Cherchenka A, Ilkin A, Dudchenka N, Ruditsa V, Novoseletz M et al (2001) J Magn Magn Mater 225:241–247
Naik A, Kalia YN, Guy RH (2000) Pharm Sci Technol Today 3:318–326
National Science and Technology Council (2005) Committee on Technology, The National Nanotechnology Initiative: research and development leading to a revolution in technology and industry, Washington (DC)7 Office of Science and Technology Policy
Nguyen TD, Tseng HR, Celestre PC, Flood AH, Liu Y, Stoddart JF, Zink JI (2005) Proc Natl Acad Sci USA 102:10029–10034
Nguyen TD, Leung KCF, Liong M, Pentecost CD, Stoddart JF, Zink JI (2006) Org Lett 8:3363–3366

Nguyen TD, Liu Y, Saha S, Leung KCF, Stoddart JF, Zink JI (2007) J Am Chem Soc 129:626
Ni S, Stephenson SM, Lee RJ (2002) Anticancer Res 22:2131–2135
Nobs L, Buchegger F, Gurny R, Allemann E (2004) J Pharm Sci 93:1980–1992
Oeffinger BE, Wheatley MA (2004) Ultrasonics 42:343–347
Ohdo S (2010) Adv Drug Deliv Rev 62:859–875
Omori N, Maruyama K, Jin G, Li F, Wang SJ, Hamakawa Y et al (2003) Neurol Res 25:275–279
Pankhurst QA, Connolly J, Jones SK, Dobson J (2003) J Phys D Appl Phys 36:167–81
Park M, Komarneni S (1998) Microporous Mesoporous Mater 25:75–80
Park C, Oh K, Lee SC, Kim C (2007) Angew Chem Int Ed 46:1455–1457
Pastorino F, Brignole C, Marimpietri D, Cilli M, Gambini C, Ribatti D et al (2003) Cancer Res 63:7400–7409
Pathak Y, Thassu D (2009) Drug delivery nanoparticles formulation and characterization. Swarbrick J, Informa Healthcare USA, Inc, New York
Peppas NA (2004) Adv Drug Deliv Rev 56:1529–1531
Peppas NA, Colombo P (1997) J Control Release 45:35–40
Peppas NA, Khare AR (1993) Adv Drug Deliv Rev 11:1–35
Peters K, Unger RE, Kirkpatrick CJ, Gatti AM, Monari E (2004) J Mater Sci Mater Med 15:321–325
Petushkov A, Ndiege N, Salem AK, Larsen SC (2010) Toxicity of silica nanomaterials: zeolites, mesoporous silica, and amorphous silica nanoparticles, Chapter 7. Elsevier, Amsterdam
Pillai O, Dhanikula AB, Panchagnula R (2001) Curr Opin Chem Biol 5(4):439–446
Pitt WG (2008) Adv Drug Deliv Rev 60:1095–1096
Poeckler-Schoeniger C, Koepke J, Gueckel F, Sturm J, Georgi M (1999) Magn Reson Imaging 17:383–392
Popovici M, Gich M, Roig A, Casas L, Molins E, Savii C, Becherescu D, Sort J, Surinach S, Munoz JS, Baro MD, Nogues J (2004a) Langmuir 20:1425–1429
Popovici M, Savii C, Niznansky D, Subrt J, Vecernikova E, Enache C, Ionescu C (2004b) Acta Per Technol 1450–7188(35):121–129
Popovici M, Ghich M, Niznansky D, Roig A, Savii C, Casas L, Molins E, Zaveta K, Enache C, de Sort S, Brion J, Chouteau G, Nogues J (2004c) Chem Mater 16:5542–5548
Popovici M, Savii C, Enache C, Niznansky D, Subrt J, Vecernikova E (2005) J Optoelectron Adv Mater 7(5):2655–2664
Popovici M, Savii C, Niznansky D, Enache C, Ionescu C, Radu R (2006) Mater Struct Chem Biol Phys Technol 13(2):79–81
Prentice P, Cuschierp A, Dholakia K, Prausnitz M, Campbell P (2005) Nat Phys 1:107–110
Proskuryakov SY, Gabai VL, Konoplyannikov AG, Zamulaeva IA, Kolesnikova AI (2005) Biochemistry (Mosc) 70:1310–1320
Putz AM, Ianăşi C, Dascălu D, Savii C (2010) Int J Eng Sci 1:79–88
Pyun J (2007) Polym Rev 47:231–263
Qu FY, Zhu GS, Huang SY, Li SG, Qiu SL (2006) Chemphyschem 7:400–406
Qing-Gui X, Tao X, Zou HK, Chen JF (2008) Chem Eng J 137:38–44
Radin S, Falaize S, Lee MH, Ducheyne P (2002) Biomaterials 23:3113–3122
Radin S, El-Bassyouni G, Vresilovic EJ, Schepers E, Ducheyne P (2004) Biomaterials 26:1043–1052
Radu DR, Lai CY, Jeftinija K, Rowe EW, Jeftinija S, Lin VSY (2004) J Am Chem Soc 126:13216–13217
Ramila A, Munoz B, Perez-Pariente J, Vallet-Regi MJ (2003) Sol-Gel Sci Technol 26:1199–1202
Rapoport N, Gao ZG, Kennedy A (2007) J Natl Cancer Inst 99:1095–1106
Rejman J, Oberle V, Zuhorn IS, Hoekstra D (2004) Biochem J 377:159–169
Ren N, Dong AG, Cai WB, Zhang YH, Yang WL, Huo SJ, Chen Y, Xie SH, Gao Z, Tang Y (2004) J Mater Chem 14:3548–3552
Robin Y (2002) Cancer treatment by electromagnetic activated nanoheaters. Eurek Alert-Nanotechnol in Context 1–3

Rosenholm JM, Linden M (2008a) J Control Release 128:157–164
Rosenholm JM, Linden M (2008b) J Control Release 128:157–164
Roy D, Cambre JN, Sumerlin BS (2010) Polymer Science 35:278–301
Rushton L (2007) Rev Environ Health 22(4):255–272
Saha S, Leung KCF, Nguyen TD, Stoddart JF, Zink JI (2007) Adv Funct Mater 17:685–693
Santarelli L, Rechioni R, Moroni F, Marcheselli F, Governa M (2004) Cell Biol Toxicol 20:97–108
Santini JT Jr, Richards AC, Scheidt R, Cima MJ, Langer R (2000) Angew Chem Int Ed Engl 39:2396–2407
Sapra P, Allen TM (2003) Prog Lipid Res 42:439–462
Savii C, Popovici M, Enache C, Subrt J, Niznansky D, Bakardjieva S, Caizer C, Hrianca I (2002) Solid State Ionics 151(1–4):219–227
Savii C, Almasy L, Ionescu C, Szekely KN, Enache C, Popovici M, Sora I, Nicoara D, Savii GG, Resiga DS, Subrt J, Stengl V (2009) Proc Appl Ceram, PAC 3(1–2):59–64
Schaefer DW (1988) MRS Bull 13:22–27
Scherer GW (1992) J Non Cryst Solids 147–148:363–374
Schiffelers RM, Ansari A, Xu J, Zhou Q, Tang Q, Storm G et al (2004) Nucleic Acids Res 32:e149
Schmidt-Winkel P, Lukens WW, Zhao D, Yang P, Chmelka BF, Stucky GD (1999) J Am Chem Soc 121:254
Schubert U, Husing N (2005) In synthesis of inorganic materials. VCH-Wiley, Weinheim, pp 20–116
Sewell MK, Fugit KD, Ankareddi I, Zhang C, Hampel ML, Kim DH, et al (2008) PMSE Prepr 98:694–695
Shinkai M, Yanase M, Suzuki M, Honda H, Wakabayashi T, Yoshida J et al (1999) J Magn Magn Mater 194:176–184
Shrivastava S (2008) Dig J Nanomater Biostruct 3:257–263
Shukla S, Bajpai AK, Bajpai J (2003) Macromol Res 11:273–82
Sierocki P, Maas H, Dragut P, Richardt G, Vogtle F, Cola L, Brouwer FAM, Zink JI (2006) J Phys Chem B 110:24390–24398
Sierpinski W (1916) Crhebd Seanc Acad Sci Paris 162:629–632
Slowing I, Trewyn BG, Lin VSY (2006) J Am Chem Soc 128:14792–14793
Slowing II, Vivero-Escoto JL, Wu CW, Lin VSY (2008) Adv Drug Deliv Rev 60:1278–1288
Slowing I, Wu CW, Vivero-Escoto JL, Lin VSY (2009) Small 5:57–62
Song SW, Hidajat K, Kawi S (2005) Langmuir 21:9568–9575
Sortino S (2008) Photochem Photobiol Sci 7:911–924
Souris JS, Lee CH, Chend SH, Chen CT, Yang C, Ho JA, Mou CY, Lo LW (2010) Biomaterials 31:5564–5574
Starr C (2000) Patient Care 15:107–137
Suh WH, Suh YH, Stucky GD (2009) Nano Today 4:27–36
Suzuki D, Kawaguchi H (2006) Colloid Polym Sci 284:1443–1451
Swarbrick J (2009) Drugs and the pharmaceutical sciences a series of textbooks and monographs, A Series of Textbooks and Monographs Executive Editor James Swarbrick Vol 191:Drug Delivery Nanoparticles Formulation and Characterization, edited by Yashwant Pathak and Deepak Thassu, Informa Healthcare USA, Inc., New York, 2009, International Standard Book number-10: 1-4200-7804-6 (Hardcover) International Standard Book Number-13: 978-1-4200-7804-6 (Hardcover)
Szabo D, Szeghy G, Zrinyi M (1998) Macromolecules 31:6541–6548
Tabata Y (2000) Pharm Sci Technol Today 3:80–89
Tachibana K, Uchida T, Ogawa K, Yamashita N, Tamura K (1999) Lancet 353:1409
Tan S, Wu Q, Wang J, Wang Y, Liu X, Sui K, Deng X, Wang H, Wu M (2011) Microporous and Mesoporous Materials 142(2–3):601–608
Tanev PT, Pinnavaia TJ (1995) Science 267:865–867

Tao ZM, Morrow MP, Aseta T, Sharma KK, Duncan C, Anan A, Penefsky HS, Goodisman J, Souid AK (2008) Nano Lett 8:1517–1526
Tao ZM, Toms BB, Goodisman J, Asefa T (2009) Chem Res Toxicol 22:1869–1880
Taqieddin E, Amiji M (2004) Biomaterials 25:1937–1945
Thompson DH (2001) Adv Drug Deliv Rev 53(3):245
Thorek DLJ, Chen AK, Czupryna J, Tsourkas A (2006) Ann Biomed Eng 34:23–38
Torchilin VP (2007) Pharm Res 24(12):2333–2334
Tourne-Peteilh C, Brunel D, Begu S, Chiche B, Fajula F, Lerner DA, Devoisselle JM (2003) New J Chem 27:1415–1418
Trewyn BG, Nieweg JA, Zhao Y, Lin VSY (2008) Chem Eng J 137:23–29
Unger KK, Kumar D, Grun M, Buchel G, Ludtke S, Adam T, Schumacher K, Renker S (2000) J Chromatogr A 892:47–55
Vallyathan V, Shi X, Dalal NS, Irr W, Castranova V (1988) Am J Respir Dis 138:1213–1219
Vartuli JC, Schmitt KD, Kresge CT, Roth WJ, Leonowicz ME, McCullen SB, Hellring SD, Beck JS, Schlenker JL, Olson DH, Sheppard EW (1994) Chem Mater 6:2317–2326
Verma RK, Mishra B, Garg S (2000) Drug Dev Ind Pharm 26:695–708
Viitala R, Jokinen M, Tuusa S, Rosenholm JB, Jalonen H (2005a) J Sol-Gel Sci Technol 36:147–156
Viitala RM, Jokinen M, Maunu SL, Jalonen H, Rosenholm JB (2005b) J Non Cryst Solids 351:3225–3234
Villalobos R, Cordero S, Vidales AM, Domnguez A (2006a) Phys A 367:305–318
Villalobos R, Vidales AM, Cordero S, Quintanar D, Domnguez A (2006b) J Sol-Gel Sci Technol 37:195–199
Villalobos R, Dominguez A, Ganem A, Vidales AM, Cordero S (2009) Chaos Solitons Fractals 42:2875–2884
Vinu A, Hossain KZ, Ariga K (2005) J Nanosci Nanotechnol 5:347–375
Vivero-Escoto JL, Slowing II, Wu CW, Lin VSY (2009) J Am Chem Soc 131:3462–3463
Wagner JG (1959) Drug Stand 27:178–186
Walcarius A, Etienne M, Lebeau B (2004) Chem Mater 15:2161–2173
Wang S (2009) Microporous Mesoporous Mater 117:1–9
Wang B, Siahaan T, Soltero R (2005) Drug delivery: principles and applications. Wiley, Hoboken\Canada
Wang G, Tian WJ, Huang JP (2006) J Phys Chem B 110:10738–10745
Ward M, Wu JR, Chiu JF (1999) J Acoust Soc Am 105:2951–2957
Ward M, Wu JR, Chiu JF (2000) Ultrasound Med Biol 26:1169–1175
Wiegand RG, Taylor JD (1959) Drug Stand 27:165–171
Wilding IR (1999) Eur J Pharm Sci 8:157–159
Willis M, Forssen E (1998) Adv Drug Deliv Rev 29:249–271
Wong JE, Gaharwar AK, Muller-Schulte D, Bahadur D, Richtering W (2008) J Colloid Interface Sci 324:47–54
Wu J, Nyborg WL (2008) Adv Drug Deliv Rev 60:1103–1116
Xing X, He X, Peng J, Wang K, Tan W (2005) J Nanosci Nanotechnol 5:1688–1693
Xulu PM, Filipcsei G, Zrinyi M (2000) Macromolecules 33:1716–1719
Yanagisawa T, Shimizu T, Kuroda K, Kato C (1990) Bull Chem Soc Jpn 63:988–992
Yang Q, Wang SH, Fan PW, Wang LF, Di Y, Lin KF, Xiao FS (2005) Chem Mater 17:5999–6003
Yatvin MB, Weinstein JN, Dennis WH, Blumenthal R (1978) Science 202:1290–1292
You JO, Almeda D, Ye GJC, Auguste DT (2010) J Biol Eng 4:15
Youan BBC (2010) Adv Drug Deliv Rev 62:898–903
Yuan F, Dellian M, Fukumura D et al (1995) Cancer Res 55:3752–3756
Yue Y, Sun Y, Gao Z (1997) Catal Lett 47:167–171
Zeng W, Qian XF, Zhang YB, Yin J, Zhu ZK (2005) Mater Res Bull 40:766–772
Zeng W, Qian XF, Yin J, Zhu ZK (2006) Mater Chem Phys 97:437–441
Zhang J, Rana S, Srivastava RS, Misra RDK (2008) Acta Biomaterialia 4:40–48

Zhang J, Xu S, Kumacheva E (2004) J Am Chem Soc 126:7908–7914
Zhao D, Feng J, Huo Q, Melosh N, Fredickson GH, Chmelka BF, Stucky GD (1998a) Science 279:548
Zhao D, Huo Q, Feng J, Chmelka BF, Stucky GD (1998b) J Am Chem Soc 120:6024–6036
Zhao W, Gu J, Zhang L, Chen H, Shi J (2005) J Am Chem Soc 127:8916–8917
Zhu Y, Shi J (2007) Microporous and Mesoporous Materials 103:243–249
Zrinyi M (1997) Trends Polym Sci 5:280–285
Zrinyi M (2000) Colloid Polym Sci 278:98–103
Zrinyi M, Barsi L, Szabo D, Kilian HG (1997) J Chem Phys 106:5685–5692

Index

M.V. Putz (ed.), *Carbon Bonding and Structures: Advances in Physics and Chemistry*, Carbon Materials: Chemistry and Physics 5, DOI 10.1007/978-94-007-1733-6,

D